Insect Populations
In theory and in practice.

Insect Populations
In theory and in practice.

19th Symposium of the Royal Entomological Society
10–11 September 1997
at the University of Newcastle

Edited by

J.P. Dempster and I.F.G. McLean

KLUWER ACADEMIC PUBLISHERS

DORDRECHT / BOSTON / LONDON

Library of Congress Cataloging in Publication Card Number: 98-071844

ISBN 0 412 83260 7

Published by Kluwer Academic Publishers,
P.O. Box 17, 3300 AA Dordrecht, The Netherlands

Sold and distributed in North, Central and South America
by Kluwer Academic Publishers,
101 Philip Drive, Norwell, MA 02061, U.S.A.

In all other countries, sold and distributed
by Kluwer Academic Publishers Group,
P.O. Box 322, 3300 AH Dordrecht, The Netherlands.

Printed in Great Britain

Contents

Preface

This book comprises the papers presented at the 19th Symposium of the Royal Entomological Society held at the University of Newcastle in September 1997. The subject of the Symposium, Insect Populations, was a particularly interesting one, because few controversies in science can have lasted longer than that concerning the processes that determine the size and stability of insect populations. Many of the arguments about the role of density-dependent processes in population regulation that were voiced in 1967, at the Society's last Symposium on this topic, still rumble on, in spite of a further 30 years of research. The aim of this Symposium was to bring together an international group of scientists working on insect populations, to review our current knowledge and to see how far we can go in reconciling different population theories.

In selecting speakers, the convenors have tried to cover the main opposing views, and to air new findings from recent theoretical and field population studies. The resulting contributions published in this book provide a series of reviews which, taken together, identify areas of agreement and those where further research is required. We hope that the book will appeal to a wide range of entomologists and ecologists who wish to learn more about population dynamics.

The success of the Symposium owed much to many people. The convenors wish to record their thanks to the President of the Society, Professor W. M. Blaney, for opening the Symposium; to the 28 authors of the presented papers; to the Chairmen of the four sessions of the meeting, Dr A. D. Watt, Professor P. W. Price, Professor M. P. Hassell and Mr I. P. Woiwod, for their contributions; and to the anonymous referees who reviewed each paper. The local organiser, Dr G. R. Port and his colleagues at Newcastle University, and the Registrar of the Society, Mr G. G. Bentley and his staff, all did sterling work to ensure that the Symposium ran smoothly. We thank them for making our stay in Newcastle so enjoyable.

Jack Dempster and Ian McLean

Contributors

C. S. Awmack
Department of Entomology and Nematology, IACR–Rothamsted, Harpenden, Hertfordshire, AL5 2JQ, UK

C. J. Briggs
Department of Integrative Biology, University of California, Berkeley, California 94720, USA

R. T. Clarke
Institute of Terrestrial Ecology, Furzebrook Research Station, Wareham, Dorset, BH20 5AS, UK

T. R. Collier
Department of Entomology, University of Arizona, Arizona 85721, USA

T. P. Craig
Life Sciences, Arizona State University–West, PO Box 37100, Phoenix, Arizona 85069-7100, USA

M. J. Crawley
Department of Biology, Imperial College at Silwood Park, Ascot, Berkshire SL5 7PY, UK

P. J. den Boer
Kampsweg 52, 9418 PG Wijster, The Netherlands

J. P. Dempster
The Limes, The Green, Hilton, Huntingdon, Cambridgeshire, PE18 9NA, UK

A. F. G. Dixon
School of Biological Sciences, University of East Anglia, Norwich, Norfolk, NR4 7TJ, UK

G. W. Elmes
Institute of Terrestrial Ecology, Furzebrook Research Station, Wareham, Dorset, BH20 5AS, UK

O. Fincke
Department of Zoology, University of Oklahoma, Norman, Oklahoma 73019, USA

H. C. J. Godfray
Department of Biology and NERC Centre for Population Biology, Imperial College at Silwood Park, Ascot, Berkshire SL5 7PY, UK

I. A. Hanski
Department of Ecology and Systematics, Division of Population Biology, PO Box 17 (Arkadiankatu 7), FIN-00014 University of Helsinki, Finland

M. P. Hassell
Department of Biology, Imperial College at Silwood Park, Ascot, Berkshire SL5 7PY, UK

M. E. Hochberg
École Normal Supérieure, Université Piere et Marie Curie, Laboratoire d'Écologie, (CNSR-URA 258), 46 Rue d'Ulm, 75230 Paris 05, France

M. D. Hunter
Institute of Ecology, University of Georgia, Athens, Georgia 30603-2202, USA

P. Kindlmann
Faculty of Biological Sciences, University of South Bohemia and Institute of Entomology, Czech Academy of Sciences, Branisovská 31, CS 37005, Ceské Budejovice, Czech Republic

S. R. Leather
Department of Biology, Imperial College at Silwood Park, Ascot, Berkshire SL5 7PY, UK

I. F. G. McLean
Joint Nature Conservation Committee, Monkstone House, City Road, Peterborough PE1 1JY, UK

C. B. Müller
Department of Biology, Imperial College at Silwood Park, Ascot, Berkshire SL5 7PY, UK

W. W. Murdoch
Department of Ecology, Evolution and Marine Biology, University of California, Santa Barbara, California 93106, USA

R. M. Nisbet
Department of Ecology, Evolution and Marine Biology, University of California, Santa Barbara, California 93106, USA

T. Ohgushi
Institute of Low Temperature Science, Hokkaido University, Sapporo 060–0819, Japan

P. W. Price
Department of Biological Sciences, Northern Arizona University, Flagstaff, Arizona 86011-5640, USA

J. Roland
Department of Biological Sciences, University of Alberta, Edmonton, Alberta T6G 2E9, Canada

P. Rothery
Institute of Terrestrial Ecology, Monks Wood, Abbots Ripton, Huntingdon, Cambridgeshire PE17 2LS, UK

N. A. Straw
Forestry Commission Research Agency, Alice Holt Lodge, Wrecclesham, Farnham, Surrey GU10 4LH, UK

J. A. Thomas
Institute of Terrestrial Ecology, Furzebrook Research Station, Wareham, Dorset BH20 5AS, UK

E. van der Meijden
Institute of Evolutionary and Ecological Sciences, University of Leiden, Van der Klaauw Laboratory, Kaiserstraat 63, 2311 GP Leiden, The Netherlands

Introduction

The study of population dynamics is central to the science of ecology and is fundamental to many applied problems in agriculture and conservation, but it is a subject that has been plagued by controversy for over 50 years. The arguments have mainly concerned the ways in which natural populations are stabilized and the role of density-dependent processes in this stabilization. At times, these arguments have been so acrimonious that many people have been deterred from entering the debate, thus reducing the progress that has been made.

One of the problems in the past has been a lack of precision in the terminology used, so that arguments have been caused by misunderstandings in what people mean by terms such as regulation, equilibrium, and stability. At the same time, real differences in opinion have sometimes been ignored on the grounds that they are simply semantic, whilst other workers have met criticisms of their views by then redefining terms, thus shifting the goal posts. Over the years, many researchers have come to think that continuation of the debate is futile, and have withdrawn into their own field-based or theoretical studies without concerning themselves with generalization of their findings. As a result, a gap has developed between theory and practice, and too few studies have been made which test the basic concepts underlying population theory. Future progress depends on bridging this gap, and this Symposium was intended as a step along that road.

With these problems in mind, we took the unusual action of trying to agree definitions of the more important terms with the contributors to the Symposium, before the meeting started. All contributors had the opportunity to discuss these definitions and agreed that if they needed to deviate from them, they would say why this was necessary and then give their own definitions. We may not finish up agreeing with one another, but we can try to ensure that differences are real, and not simply the result of loose terminology.

Undoubtedly, if population ecology was starting today as a new scientific discipline, we could find more acceptable definitions for many of these terms. Thus it would be preferable sometimes to replace old deterministic definitions with stochastic ones. However, although usage of some terms has tended to change over the years, we have tried wherever possible to keep their original meanings. This both safeguards the priority of the originator of an idea, and of its critics, and reduces misunderstandings when reviewing old and new literature. Inevitably, agreement has required some compromises, but the outcome of these deliberations is seen in the list of definitions on page xix.

Three terms require special comment: namely regulation, limitation and stability. These terms make use of words that are in regular English usage and so care is required to ensure that they are used only within the more limited definitions required in ecology. All three terms are needed in order to distinguish between different concepts of population control. Population stability is used to describe a relative constancy in population size; the population showing no trend, and population fluctuations being in a narrower range than expected from random variation. No mechanism is implied by the use of the term, as increased stability can result from many kinds of process, which may or may not be related to population density. In contrast, regulation and limitation both refer to increased stability brought about by density-dependent processes. Regulation describes the return of a population to a theoretical equilibrium density, whereas limitation describes the reduction in population growth as density approaches the carrying capacity of the environment. These two views of how density-dependent processes are likely to determine population size are very different. Regulation requires a feedback of density on population growth, so that any deviation (up or down) of population size away from the equilibrium density is countered by an increase or decrease in mortality, reproduction or dispersal. Population limitation results from an increased resistance to population growth as density approaches the ceiling set by available resources. Unlike regulation, limitation operates only at relatively high densities and its effect can only be downwards; it can not affect the probability of extinction at low densities. It is important to recognize that both the equilibrium and the ceiling are dynamic, in that they may change from generation to generation as a result of environmental variation.

In planning the programme for this Symposium, we tried to include speakers whose views reflect the full range of opinions on how natural populations are stabilized. Whilst these speakers will stress their differences in opinion, many of these views may not be mutually exclusive. We are still not in a position where we can be certain that all populations of all insect species are stabilized in the same way. All ecologists rightly aspire to producing a generalized theory, akin to, say, the theory of evolution, but we are still some way away from this; perhaps not surprisingly in view of the huge array of lifestyles adopted by insects. More progress may be made if we are a bit less dogmatic in our approach than our predecessors were.

Since the last Symposium of the Society, on Insect Abundance in 1967, (Southwood, 1968), computers have become readily available to researchers, and we now have a larger number of long-term data sets for analysis. As a result, theoretical studies have expanded enormously. Perhaps the most important recent theoretical development has been the recognition that different species may have very different population structures. In particular, the concept of metapopulations has brought some of the ideas of island biogeography into population ecology. Metapopulations are assemblages of local populations inhabiting discrete habitat patches and connected by migration. Local populations have a substantial risk of extinction, so that metapopulations persist regionally in a balance between local extinctions and recolonizations. Obviously the scale over which a metapopulation can exist depends on the mobility of the species, and ideally we need to know this when planning a field population study. Unfortunately, the majority of past studies do not provide information to enable one to judge the integrity of the population being studied. This would not matter if immigration and emigration to and from the population were quantified, but few researchers have attempted this. Ilkka Hanski will discuss the importance of scale in insect population dynamics in Chapter 1.

Although population theories have been elaborated and extended, the basic arguments about the processes causing population stability and persistence that were voiced in 1967 are still with us today. In 1967, there were three main hypotheses to explain the observed relative stability of insect populations, namely those generally associated with the names of Andrewartha and Birch (1954), Nicholson (1954), and Milne (1957a), and recent developments of their ideas are covered in Chapters 2, 3 and 4 by Mike Hassell, Piet den Boer and Jack Dempster, respectively.

The arguments that have occupied our science for so long really rest on a hierarchy of basic questions, and we suggest that you bear these in mind when reading this book. Firstly, how important are density-dependent processes in governing the extent of fluctuations in population size, and the persistence of populations? Secondly, if they are important, how are these density-dependent processes likely to act? Do they provide a mechanism for regulating populations around a theoretical equilibrium density, and if so, can both intraspecific competition and predation act in this way? Alternatively, is competition the only process capable of consistently acting in a density-dependent manner, and is it capable of regulating a population about an equilibrium density, or simply of limiting the upward growth of populations?

Related to all these questions is how different trophic levels interact with one another. In 1960, Hairston *et al.* published a paper in which they proposed that plants, decomposers and carnivores are all limited by their food resources (i.e. by the trophic level below), whilst herbivore populations are kept below the limits of their food by predators (i.e. by the trophic level above). Others criticized these views (e.g. Murdoch, 1966), pointing out that they were the result of a superficial but false impression of an abundance of edible plant food given by a predominantly green world; but nevertheless the views of Hairston *et al.* met widespread acceptance amongst ecologists. Obviously, if their views are correct, they suggest that population limitation might be more common in all plants, decomposers and carnivores, but not in herbivores. Several of our contributors will refer to this question of top-down versus bottom-up control of populations, and Charles Godfray, and Bill Murdoch and their colleagues deal specifically with top-down effects of natural enemies on their prey populations in Chapters 6 and 7. Are such effects density dependent, and can they regulate prey populations? Also, are natural enemies of any greater importance in the population dynamics of herbivores than at other trophic levels?

Changes in the 'quality' of individuals with changing population density have long been recognized in some types of insect. Some are brought about by physiological and behavioural changes resulting from competition, but others are the result of genetic changes from differential survival of genotypes with changing density. This topic of density-dependent changes in the quality of individuals is covered in Chapter 8 by Simon Leather and Caroline Awmack, and we are especially grateful to them for taking on this task at very short notice.

The potential importance of density dependence has greatly influenced the thinking of ecologists over the years, so much so that many have concentrated on simply attempting to detect density dependence in population data, on the assumption that density dependence can be equated with regulation. Detection of density dependence in population data is actually not as straightforward as was at one time thought. In 1967, correlation techniques were being used, but a battery of alternative methods are now available, and these are discussed in Chapter 5 by Peter Rothery.

At the time of our last Symposium on this topic, population studies of insects with discrete generations tended to be aimed at developing life tables covering all developmental stages of the study insect, and then using *k*-factor analysis (Morris, 1959; Varley and Gradwell, 1960) to investigate the roles of separate factors in accounting for population size. The life-table approach proved valuable in identifying the cause of density-dependent factors and the stage of the life cycle at which they were operating, but as we shall see, the detection of density dependence is simply a first step in identifying processes that might regulate or limit population size. We then need to determine whether they are actually capable of affecting population change from one generation to the next.

The life-table approach was particularly useful in identifying the impacts of mortality factors on population size, but it was far less useful in studying the impacts of those processes affecting recruitment. More recent studies have largely corrected this imbalance, and contributors to this book give several examples in which variations in fecundity are largely responsible for changes in numbers from one year to the next. Adult behaviour, particularly at the time of egg laying, can play a dominant role in determining the use of resources and interactions with other organisms by their offspring. Modern population theory has to take cognizance of such behavioural traits.

So much for theoretical considerations. The second half of this book concentrates on case studies. These have been grouped into three main themes. The first considers the extent to which we can generalize about the population dynamics of closely related taxa. Repeatedly during this Symposium, we have heard how populations of individual species are constrained by their phylogenies – their lifestyles, specific food or climatic requirements, evolved interactions with other organisms, etc. One might expect, then, to find characteristic patterns in the population dynamics of closely related taxa, as indeed has been found in previous reviews of the population dynamics of particular insect groups (e.g. Dempster, 1963, 1983; Price *et al.*, 1995). In Chapter 9, Tony Dixon and Pavel Kindlmann consider the population dynamics of Aphididae, by re-analysing data for tree-living species. Then, Nigel Straw, and Jeremy Thomas and his colleagues review population studies of the Tephritidae and of the genus *Maculinea* (Lycaenidae), respectively, in Chapters 10 and 11. If we cannot generalize about the population dynamics of these closely related taxa, there is little hope of doing so more broadly across the Insecta, or indeed, across the animal kingdom.

Next, we have two chapters comparing the dynamics of the populations of two well-worked species in different geographical locations, namely, the cinnabar moth, *Tyria jacobaeae* (Eddy van der Meijden, Roger Nisbet and Mick Crawley, Chapter 12) and the winter moth, *Operophtera brumata* (Jens Roland, Chapter 13). Are the same processes operating on their populations in different locations, or do different environments impose fundamental differences on their population dynamics?

Then, we have four contributions covering the population dynamics of less well-known groups of insects. Peter Price and his colleagues describe their long-term studies of gall-forming sawflies in Chapter 14; Ian McLean reports on research on a gall-forming psyllid (Chapter 15); Takayuki Ohgushi describes his long-term study of a herbivorous ladybird (Coccinellidae) (Chapter 16); and finally, we have our only case study of a generalist predator, in Ola Fincke's description of her work on a dragonfly community in Chapter 17.

It is disappointing how few studies have been made of the population dynamics of predators and parasitoids. The latter, in particular, are such a diverse and successful group of insects that they deserve far better coverage. There are very few field studies of populations of phytophagous insects that have not considered insect parasitoids, but research has mainly concentrated on their impacts on their prey populations, rather than on their own population dynamics. Theoretical studies have commonly assumed that parasitoid populations are coupled with those of their prey, but this assumption is certainly not true for all parasitoids (see Godfray *et al.*, Chapter 6).

We think that we have put together an exciting set of contributions, covering a wide range of insect species. To be honest, we did not expect to get a meeting of minds between holders of different views, but we do hope that readers of this book will receive a clear picture of where the subject stands today, and will be stimulated into thinking about what research is required to bridge the gap between theory and practice.

Jack Dempster and Ian McLean

Definitions of terms

CARRYING CAPACITY: the maximum number of individuals which can be supported through their life cycle by the resources within the area of habitat occupied by a population.

CEILING: the upper limit of population density set by the carrying capacity of the habitat. This is a dynamic 'ceiling' which varies with changes in carrying capacity.

COMPETITION: the endeavour of two or more individuals to gain the resources that they require, when the supply of those resources is insufficient for both or all. For competition to occur, one individual gaining its needs must reduce the chances of others doing likewise, so that competition must be density-dependent.

DENSITY DEPENDENCE: a decrease in the rate of population growth as population density increases, brought about by either a proportional increase in mortality and/or emigration, or a decrease in fecundity and/or immigration. [Strictly, this is direct density dependence, to distinguish it from inverse and delayed responses.]

EQUILIBRIUM: the long-term stationary probability distribution of population density, to which populations tend to be driven, as a result of density-dependent regulation.

INTRINSIC RATE OF INCREASE: the maximum rate of increase of a population in a given physical and biotic environment, when free from density-dependent constraints.

METAPOPULATION: an assemblage of local breeding populations inhabiting discrete habitat patches and connected by migration. It is usual for local populations in a metapopulation to have a substantial risk of extinction, and 'classical metapopulations' persist regionally in a balance between local extinctions and recolonizations of empty habitat patches.

POPULATION: a group of interacting individuals of a species (breeding and competing with one another), usually associated with a particular area, and at least partially isolated from other populations of the same species.

POPULATION LIMITATION: the density-dependent reduction in population growth brought about by intraspecific competition as density approaches the carrying capacity of the habitat.

POPULATION REGULATION: the return of a population to an equilibrium density, following departure from that density, as a result of density-dependent processes.

POPULATION STABILITY: a relative constancy in population size, i.e. the population shows no trend, and population fluctuations are in a narrower range than expected from random variation. [Some authors define stability in terms of inertia, resilience and persistence, but these features all follow directly from lack of fluctuation.]

Part One

1

Spatial structure and dynamics of insect populations

Ilkka Hanski

1.1 INTRODUCTION

A strong tradition in insect population ecology has its roots in the early conceptual and theoretical analyses of population regulation (Nicholson, 1933, 1954; Nicholson and Bailey, 1935; Smith, 1935; Elton, 1949; Lack, 1954) and in the classical experiments of A. J. Nicholson on laboratory populations of blowflies (Nicholson, 1950, 1954, 1958). The key factor (Varley and Gradwell, 1960; Morris, 1963a) and life table analyses (Watt, 1961; Morris, 1963a) were developed for detailed studies of insect populations, to elucidate the forces that cause populations to oscillate and to unveil the identity and the strength of the density-dependent processes that would nonetheless ensure population regulation. Insect populations causing economic damage in forests (e.g. *Diprion hercyniae*, Neilson and Morris, 1964), cultivated fields (*Brevicoryne brassicae*, Hughes, 1963), and orchards (*Cydia pomonella*, Geier, 1964) were obvious (and well funded) targets for such studies (Clark *et al.*, 1967). These studies were typically concerned with large populations, and the task was to describe, analyse and model the mechanisms that contribute to local population regulation. Many species are especially abundant only periodically, in which case the task is also to understand the processes that keep the density at a low level for long periods of time (Clark, 1964). A few populations are cyclic, and entomologists have asked which mechanisms might maintain the regular cyclic component in their oscillations (Baltensweiler, 1964).

Along with this dominant tradition of emphasizing local density-dependent processes in insect population ecology, another research tradi-

Insect Populations, Edited by J. P. Dempster and I. F. G. McLean. Published in 1998 by Kluwer Academic Publishers, Dordrecht. ISBN 0 412 83260 7.

tion has survived, though in a much less well developed form. Andrewartha and Birch (1954) were early champions of the view that density-dependent population processes are not the key to understanding insect population dynamics, and that the essence of long-term persistence of populations is not captured by the classical population models. Instead, they maintained, insect populations are greatly affected by the prevailing environmental conditions and relative (density-independent), rather than absolute, shortage of resources – though exactly how this would lead to long-term persistence was not made sufficiently clear. More recently, Den Boer (1968, 1981, Chapter 3 in this volume), Strong (1983, 1986b) and Stiling (1987, 1988) have expressed similar or related views about density dependence and insect population ecology. Dempster (1983, Chapter 4 in this volume), following Milne (1957a, 1961), has concluded that insect populations are rarely regulated by natural enemies, but are limited by the availability of resources. The key process in both 'regulation' and 'limitation' is density dependence, and the difference between the two does not appear to me to be as great as suggested by Dempster (1983).

There is no need to return here to the controversy, much of which is now outdated, about the significance of density-dependent processes in population ecology in general and in insect population ecology in particular (Sinclair, 1989; Hanski, 1990; Turchin, 1995; Hassell, Chapter 2 in this volume). Simple logic dictates that some density dependence must occur, at least occasionally, in the dynamics of every species, though not necessarily in the dynamics of every population (sink populations are one exception; see section 1.2 below). How frequently populations are regulated, in the sense that they have settled down to a stationary probability distribution of densities, is another matter. Real populations are affected by environmental changes at many temporal scales as well as by the endogenous density-dependent processes, and populations may never, strictly speaking, reach a stationary distribution of population densities. Small populations may not have time to reach a quasi-stationary distribution (Nisbet and Gurney, 1982; Stelter *et al.*, 1997) because of their high probability of extinction. Nonetheless, at least some populations of all species are affected at least occasionally by density dependence, which is required for long-term persistence at the regional scale (Hanski *et al.*, 1996a).

My purpose in this chapter is to focus on the spatial structure and the scale of persistence of insect populations. We should recognize that not all populations are structured in space in the same manner, and that different research approaches may be needed for the effective study of species with different types of spatial structure. As my own research in insect ecology has been largely restricted to the study of species with highly fragmented spatial structure, this chapter is biased in that direction. Hassell (Chapter 2) presents a clear account of population regulation in local populations, and Dempster (Chapter 4) elaborates on the concept of population limitation. In this chapter, the term 'population regulation' is used in the sense of a population having a tendency, due to density-dependent processes, to approach a stationary distribution of population densities, though it may not have time to reach the stationary distribution before the environment has changed or the population has gone extinct. In practice, population regulation is tested indirectly by testing for the type and strength of density dependence in temporal population dynamics (Turchin, 1995).

1.2 THREE TYPES OF POPULATIONS AND POPULATION STRUCTURES

Populations are spatially structured on many scales. Over the past 20 years, much work has been done on the role of small-scale spatial structure in the dynamics of local insect populations (Hassell, Chapter 2). Two key ideas in this research have been linking the behaviour that causes the small-scale spatial structure to the ensuing dynamics (Hassell and May, 1985); and the generally stabilizing effect of aggregated spatial distribution of individuals in local populations (May, 1978; Hassell, 1978, Chapter 2 in this volume; de Jong, 1979; Hanski, 1981, 1990; Atkinson and Shorrocks, 1981). This chapter considers spatial structure at the next higher level, at the level of metapopulations consisting of weakly connected local populations with more-or-less independent dynamics (Hanski and Gilpin, 1997). To start with a simple classification of populations, consider the variation that occurs among populations in three basic population dynamic parameters: the intrinsic rate of population increase, r; the variance of the intrinsic rate of population increase, V; and the carrying capacity or the population ceiling, K. Assigning a small or large value to these parameters leads to three types of population (Table 1.1).

1.2.1 Classical populations

Populations with large r and large K represent what we might call classical populations. These populations grow rapidly when small, which means that their dynamics show strong density dependence, and their expected size is large, which means that they tend to have a small risk of extinction. The actual extinction risk, however, also depends critically on the third parameter, the variance of the intrinsic rate of population increase, V. Using a recent extinction model as a guide (Box 1), we find that what really matters is the ratio r/V, with large values increasing the expected lifetime of a population. Classical populations thus have a large value of K and a large value of r/V. Classical population models are deterministic ($V=0$) and hence represent an extreme case.

Insect population structure

Table 1.1 Three types of populations in terms of the values of the intrinsic rate of population increase, r, its variance, V, and the population ceiling, K

	Small r	Large r	
		Small V	Large V
Large K	Sink populations	Classical populations	Local populations in metapopulations
Small K			

---

1. EXPECTED LIFETIME OF A POPULATION IN A MATHEMATICAL MODEL

The following model is due to Lande (1993), Foley (1994, 1997) and Middleton *et al.* (1995). In this stochastic discrete model, density dependence operates only when the population reaches the ceiling, as envisioned by Milne (1957a, 1961), otherwise the population performs a random walk until it happens to go extinct. The dynamics are governed by the following equations:

$$n_{t+1}=n_t+r_t \quad \text{if} \quad 0 \leq n_{t+1} \leq k$$
$$n_{t+1}=k \quad \text{if} \quad n_{t+1}>k$$
$$n_{t+1}=0 \quad \text{if} \quad n_{t+1}<0$$

where n_t is the natural logarithm of population size, k is the logarithm of the population ceiling, K (a reflecting upper boundary), and r_t is a normally distributed random variable with mean r and variance V. Using a diffusion approximation, it can be shown that the expected time to extinction of a population initially at the ceiling is given by:

$$T(K)=K^c/cr[1-(1+ck)/e^{ck}]$$

where $c=2r/V$. For positive r and large ck the expected time to extinction scales asymptotically as:

$$T(K) \approx K^c/cr$$

The logarithm of the extinction rate as a function of the ceiling is then given approximately by:

$$\log e(K) \approx -\log T(K) \approx \log cr - ck$$

which summarizes the effects of r, c ($=2r/V$) and k ($=\log K$) on the probability of population extinction.

Having a small risk of extinction, classical populations naturally invite studies on the population dynamic mechanisms that contribute to population regulation (Hassell, Chapter 2). These studies belong to the traditional approach to insect population ecology outlined in the introduction to this chapter.

1.2.2 Sink populations

When the value of the intrinsic growth rate is small, even small populations with no adverse density effects grow on average only slowly, if at all. Furthermore, the apparent positive growth rate at low density may be due to immigration, rather than to local birth rate exceeding local death rate plus emigration. Such populations are commonly referred to as sink populations (Pulliam, 1988; Stacey *et al.*, 1997). Watkinson and Sutherland (1995) coined the term pseudo-sink for populations with $r>0$ but whose size is augmented by immigration. True sink populations, in contrast, go extinct deterministically in the absence of immigration ($r<0$). Regardless of whether r is negative or slightly positive, the important difference from classical populations is that we cannot ignore the population dynamic consequences of migration. It would be misleading to assume that immigration equals emigration, because in sink and pseudo-sink populations immigration in fact exceeds emigration and may be critical for long-term persistence. Naturally, migration may also affect the dynamics of classical populations (Ims and Yoccoz, 1997; Stacey *et al.*, 1997), which in this context are called source populations, but the role of immigration is likely to be more important the smaller the value of r.

Sink populations represent an interesting exception to the general role of density dependence in population regulation. Paradoxically, a sink population may be regulated in the sense of having a stationary distribution of population densities, even though there is no local biological density dependence: changes in population size are not necessarily affected by the present nor past numbers in the population. The solution to this anomaly is that the sink population size is regulated by density-dependent processes taking place in the source population(s) (Hanski *et al.*, 1993a).

The value of r may be low because of low habitat quality. For some ecologists, it might not make much sense to study sink populations in suboptimal habitat. However, if a large fraction of local populations are sink or pseudo-sink populations of one type or another, we should not simply ignore them. Sink populations may add greatly to local species diversity, and they pose interesting evolutionary issues (Holt and Gaines, 1992; Kawecki and Stearn, 1993; Holt, 1996). Furthermore, although long-term persistence of the species is generally reduced by an increasing fraction of sink populations, in some cases (large V) even metapopulations consisting entirely of sink populations may persist for a long time (Kuno, 1981; Hanski *et al.*, 1996a). Even more importantly, environmental conditions change over time, and it is possible that long-term persistence of a metapopulation depends critically on pseudo-sink populations with small *r* but also smaller *V* than in source populations. Thomas *et al.* (1996) describe a convincing example in which a population of Edith's check-erspot butterfly (*Euphydryas editha*) functioned as a source in most years but performed catastrophically badly in an exceptional year; the metapop-ulation survived because of the presence of a pseudo-sink population.

1.2.3 Local populations in classical metapopulations

The third main population type in Table 1.1 is associated with a small population ceiling and/or small value of *r/V*; in both cases, populations have a high risk of extinction (Box 1). The population ceiling is small when populations are restricted to small patches of suitable habitat. In this case, though *r* may be large and the populations may be much affected by density dependence, they are nonetheless prone to go extinct for many reasons because they are chronically small. Given a high extinction rate of local populations, it is clear that a comprehensive understanding of the long-term dynamics of the species cannot be based only on studies of local density-dependent processes; we have also to consider the factors and processes that affect the occurrence of the species at a larger spatial scale. These factors and processes include the size distribution of habitat patches, scaling of expected population lifetime with patch size, strength of between-patch interaction via migration, and the minimum viable metapopulation size in finite patch networks (Hanski and Gilpin, 1997).

1.3 LOCAL POPULATION SIZES IN INSECTS

1.3.1 Many, or even most, insect species are locally rare

What kinds of populations and spatial structures do we find in insects? The easy answer is that examples can be found of practically any kind of population structure that we can think of. It is more difficult to provide a quantitative answer, partly because insect ecologists have traditionally focused on species with classical population structures (Clark *et al.*, 1967), partly because they have often completely neglected the spatial structure, as in studies based on time-series analyses of population densities (Hassell *et al.*, 1989a; Woiwod and Hanski, 1992; Wolda and Dennis, 1993 and refer-ences therein). The latter studies are not invalid, but they test for density dependence at an unspecified spatial scale and without attempting to dis-criminate between the three population types shown in Table 1.1.

If, instead of considering particular species and populations, we examine the abundances of species in insect communities, one is impressed by the large numbers of species that are locally rare (Fig. 1.1; Preston, 1948; Williams, 1964; May, 1975). Admittedly, most community samples indicate only the relative abundances of species, and it is often difficult or impossible to know how sparse the rare species actually are, and indeed to know what sort of populations they represent. Most insects are fairly mobile, and a large sample of individuals is likely to include many species with no breeding population at all at the sampling site. 'Populations' of non-breeding vagrant individuals have been aptly called 'tourists' (Moran and Southwood, 1982). Other locally rare species are represented by sink populations, which may breed locally but whose presence is more or less dependent on immigration from elsewhere. In yet other cases, even the rarest local species are represented by populations that are just very small: local populations in metapopulations. The most species-rich insect communities occur in tropical forests, and these communities are renowned for large numbers of rare species (Erwin, 1982; Hammond, 1992).

Figure 1.2 gives four examples of species-abundance distributions in local dung beetle communities. In all cases, there is an indication of a bimodal distribution, which can be explained by the communities consisting both of classical populations of the more common species as well as of species represented by tourists, by sink populations, or by local populations in classical metapopulations (Hanski and Cambefort, 1991). Bimodality is presumed to arise because of 'minimum viable size' for classical populations; populations falling below this size are expected to go extinct rapidly, or their persistence is due to migration and metapopula-

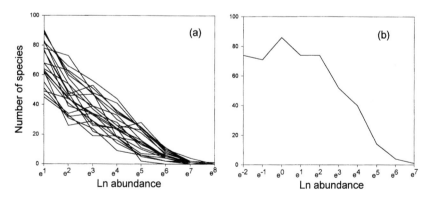

Fig. 1.1 (a) Species abundance distributions for moths sampled at 22 similar forest localities in southern and central Finland. (b) The corresponding distribution for the average abundances of the species in the pooled sample (data courtesy of the Finnish Environment Research Institute).

tion processes. In a thorough community study of *Aphodius* dung beetles near Oxford, I found multiple lines of evidence indicating that a quarter of the 22 species in the local community were represented by tourists, sink populations or extinction-prone local populations in metapopulations (Hanski, 1979).

1.3.2 Many locally rare species are also rare elsewhere

To summarize the previous section, many local populations of insects are small for one reason or another. The obvious next question is whether the locally rare species might be more abundant in some other communities, and indeed in how many local communities the species occur.

It is implicit in the concept of a sink population that the species has source populations (classical populations) elsewhere, which may be critical for the long-term persistence of the species within a larger region. A good field entomologist knows where to look for particular species, and in this

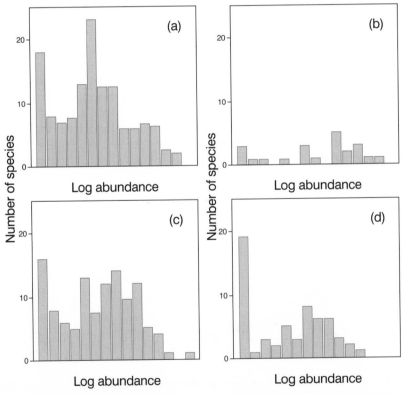

Fig. 1.2 Species abundance distributions for four dung beetle communities: (a) tropical savanna in Ivory Coast; (b) north temperate pasture in the UK; (c) montane pasture in the Alps; and (d) tropical forest in Panama (from Hanski and Cambefort, 1991).

sense even the rare species typically have their strongholds. But it does not follow that all species have large populations even at those sites where their occurrence is most predictable. Many species are uncommon every-where, and they are likely to persist only regionally as metapopulations.

Figure 1.1 shows the species-abundance distributions for moths sam-pled at 22 similar forest sites across south and central Finland, as well as the respective distribution for their average abundance across all the sites. If locally rare species tend to be common elsewhere, and *vice versa*, the variance of the distribution for the average abundances would be smaller than the variance for individual communities. But this is not the case (Fig. 1.1); on the contrary, in the pooled sample the standard devia-tion of the logarithm of average abundance was 0.899, greater than the average value in individual communities, 0.662. The pooled sample included many more species (439) than the average number in individual communities (201), indicating that a large number of locally rare species are also rare in the sense that they occur only in a small fraction of the total area, as has been found in many previous studies (Hanski *et al.*, 1993b; Lawton, 1993; Gaston, 1994). Figure 1.3 gives an example of the occurrence of ground beetles within an area of 30 km², which was sam-pled exceptionally thoroughly. Roughly half of the species occurred in less than 10% of the total area. The regionally rare species may be rare because their habitat is scarce, or they may be rare for other reasons. In both cases, most ground beetle species in Fig. 1.3 have strikingly patchy distributions at the scale of 30 km².

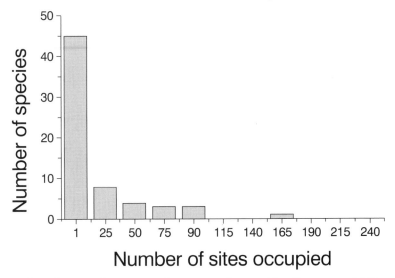

Fig. 1.3 Distribution of occupancy frequency of ground beetles amongst 240 sam-pling sites of 250×250 m² each, placed systematically within a continuous area of 30 km² of heterogeneous landscape in southern Finland (H. Kinnunen, personal communication).

1.4 AN EXAMPLE OF SPECIES WITH CHRONICALLY SMALL POPU-
LATIONS

As an example of insect species consisting entirely of small local populations and hence having a classical metapopulation structure, I use the
Glanville fritillary butterfly (*Melitaea cinxia*), which I have studied in
Finland since 1991, with several co-workers. This study provides a good
example of the spatial population structure in an insect species, for the
simple reason that we have been able to locate and describe practically all
the local populations of the species within its entire range in Finland, in
the Åland islands, within an area of 50×70 km (Fig. 1.4). We have
mapped all the meadows which are suitable for the Glanville fritillary
and which are >25 m² in area (Hanski *et al.*, 1995a). In Åland, there are
roughly 1600 such meadows. The exact number varies a little from one
year to another, because there is some turnover of meadows due to disturbance, grazing and overgrowth.

Let us examine the four criteria by which the Glanville fritillary persists
as a classical metapopulation in the Åland islands (Hanski *et al.*, 1995a).

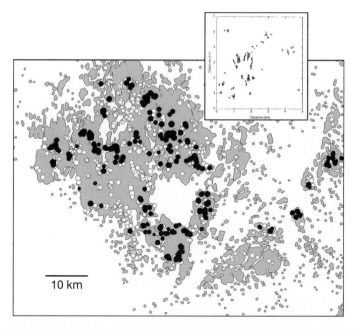

Fig. 1.4 Distribution and abundance of the Glanville fritillary butterfly (*Melitaea
cinxia*) in the Åland islands in south-west Finland. Each dot represents a suitable
habitat patch (dry meadow) for the butterfly, with the size of the dot being proportional to patch area. Filled dots were occupied by the butterfly in autumn
1996. The small map shows a close-up of an area of 5×6 km, giving the true
shapes and areas of the patches in this region.

1.4.1 Discrete small local populations

The suitable habitat for the Glanville fritillary consists of dry meadows with one or both of the larval host plants, *Plantago lanceolata* and *Veronica spicata*. These meadows occur as discrete small patches, with the mean, median and maximum areas of 0.13, 0.03 and 6.80 ha, respectively ($n=1502$; Hanski *et al.*, 1995a). The meadows typically verge on forests or cultivated fields, which are completely unsuitable for the butterfly. We have estimated that 60–80% of butterflies spend their entire lifetime in the natal meadow (Hanski *et al.*, 1994; Kuussaari *et al.*, 1996), with emigration rate decreasing with increasing patch size, increasing abundance of nectar flowers, percentage of patch boundary bordering with forest and increasing butterfly density (Kuussaari *et al.*, 1996). In other words, although there is substantial emigration especially from small meadows, the local populations are far from ephemeral aggregates of individuals in a panmictic population. The local populations represent an important entity in the spatial structure of the species (Fig. 1.4).

1.4.2 Even the largest local populations have a substantial risk of extinction

During the period 1993–97, the very largest local populations have had 100–150 larval groups, roughly corresponding to 1000 adult butterflies (the larvae live in large groups of full sibs; Hanski *et al.*, 1995a). The vast majority of local populations have fewer than five larval groups, or less than 50 adult butterflies (Hanski *et al.*, 1995a). It is not surprising that these very small populations have a high risk of extinction, but even the largest populations are far from secure. In as short a period as 6 years, we have witnessed the extinction or near-extinction of the two largest local populations ever recorded in this metapopulation. In both cases, high rates of parasitism by the specialist braconid parasitoid *Cotesia melitaearum* greatly contributed to drastic population declines (Lei and Hanski, 1997). Overall, of the 515 local populations in existence in 1993, only 121 populations persisted continuously until 1996. During the same period, we have recorded a total of 341 recolonizations of empty meadows, with many of the newly established populations rapidly going extinct again.

Given the small size of most local populations, one might assume that the primary mechanism of population extinction is simply demographic stochasticity. This is not so. We have found strong evidence for a variety of mechanisms of population extinction operating in the Glanville fritillary metapopulation. In some cases the entire habitat patch has been destroyed, leading to a catastrophic extinction. More commonly, environmental stochasticity in the form of minor droughts and grazing by cattle has greatly contributed to local extinctions (Hanski *et al.*, 1996b). The metapopulation-level analogues of demographic and environmental sto-chasticity play a key role at the regional level (see below, and section 1.6). Increased immigration has been observed to reduce the risk of local extinction (the rescue effect; Hanski *et al.*, 1995b), and we have found evidence for the Allee effect, inverse density dependence at low densities, due to increased emigration from low-density populations and lowered mating success of females at low density (Kuussaari *et al.*, 1998). Two specialist parasitoids attack the Glanville fritillary, and parasitism by the braconid *C. melitaearum* has been found to increase the local extinction rate (Lei and Hanski, 1997, 1998). We have also found that high extinction risk of local populations is associated with low levels of enzyme and DNA heterozygosity (Saccheri *et al.*, 1998), most probably reflecting inbreeding depression in small populations.

In summary, many different causes and mechanisms contribute to local extinctions in the Glanville fritillary metapopulation, the overall extinction rate is high, and even the largest local populations have a high risk of local extinction.

1.4.3 Movements and recolonization

We have conducted two large mark–recapture studies on the Glanville fritillary to study movement behaviour (Kuussaari *et al.*, 1996) and movement distances (Hanski *et al.*, 1994). The results of these studies demonstrate that most butterflies leaving the natal meadow move only a short distance, less than 1 km, and only a small fraction of butterflies move a few kilometres (Fig. 1.5). In agreement with these results, recolonization of empty habitat patches is greatly affected by isolation, and all observed recolonizations have occurred in patches within 5 km from the nearest local population (Fig. 1.5). In the Åland islands, the mean nearest-neighbour distance between habitat patches is only 240 m (median 128 m, maximum 3870 m; Fig. 1.4), and hence the patches are not generally too isolated to prevent recolonization. At any one time, however, many patches that are relatively isolated from the existing populations (Fig. 1.4) have a low probability of recolonization within a short period of time.

1.4.4 Asynchronous local dynamics

An intensive study of the Glanville fritillary conducted in a 50-patch network in 1991–93 revealed largely asynchronous local dynamics (Hanski *et al.*, 1995a). Subsequent studies have shown that in this network, where the habitat patches and hence host populations are located relatively close to each other, the interaction between the butterfly and its specialist parasitoid *C. melitaearum* is an important source of asynchrony in butter-

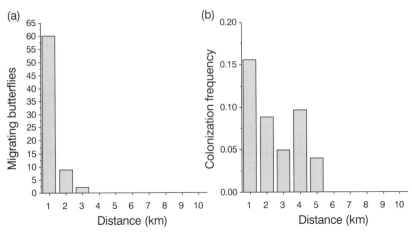

Fig. 1.5 (a) Frequency distribution of movement distances in a mark-recapture study of the Glanville fritillary within an area of 5×5 km (Hanski *et al.*, 1994). 1737 butterflies were marked, of which 741 were recaptured and 72 had moved to another meadow. (b) Frequency distribution of the observed colonization distances from the nearest local population to an empty patch in the pooled data for 1993 and 1994 (I. Hanski, unpublished data).

fly dynamics (Lei and Hanski, 1997). The host–parasitoid interaction appears to be unstable locally, and parasitism significantly increases the extinction rate of the host (Lei and Hanski, 1997, 1998).

This is not the whole story, however. More comprehensive studies covering the entire Åland islands have revealed spatially correlated changes in population sizes at a larger spatial scale (Fig. 1.6). The most likely explanation of these large-scale correlated changes is some effect of weather; these patterns cannot be attributed to movements of the butterfly nor to host–parasitoid dynamics. One possibility is localized heavy showers during early larval development, which might wash young larvae off the host plant and cause high mortality. The opposite is also possible: during drought years local showers in late summer may save larvae from starvation (during the driest years host plants dry out completely at the driest sites). Another possible cause of spatially correlated changes in population sizes is related to the spatial distribution of the two host plants, *P. lanceolata* and *V. spicata*. The latter species occurs primarily in the north-western part of Åland, roughly where the population decline was most marked in 1993–94 and where the recovery was most obvious in 1995–96 (Fig. 1.6). It is possible that in some years *Plantago*-feeding caterpillars do better than *Veronica*-feeding caterpillars, and *vice versa*, which might contribute to the observed spatial patterns. This would be another example of interaction between weather perturbations and habitat quality, previously demonstrated for the American checkerspot butterflies (Ehrlich and Murphy, 1987; Weiss *et al.*, 1993).

One should not overemphasize (nor underemphasize) the large-scale patterns in population change. Within regions of general population increase, or decrease, there have been deviating populations showing the opposite trend. The spatial patterns have been different in different years. In 1994–95 spatial correlation in population change was relatively weak, whereas in 1995–96 there was a distinct geographical trend in the change, western populations increasing and eastern populations declining (Fig. 1.6). It is notable that within the entire Åland islands, not all populations have so far moved in one direction, though undoubtedly this rule will find an exception in the future.

1.4.5 The message

In summary, the Glanville fritillary butterfly in Finland satisfies the four necessary conditions for a species to persist as a classical metapopulation (Hanski *et al.*, 1995a), in a balance between stochastic local extinctions and recolonizations. Can we therefore explain the distribution and abundance of this species with minimal reference to density-dependent processes, in the spirit of Den Boer's (1968) "spreading of risk" concept, and the ideas of Andrewartha and Birch (1954)?

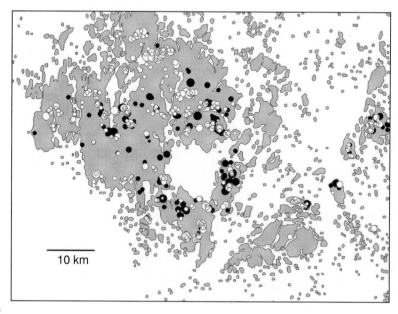

10 km

a)

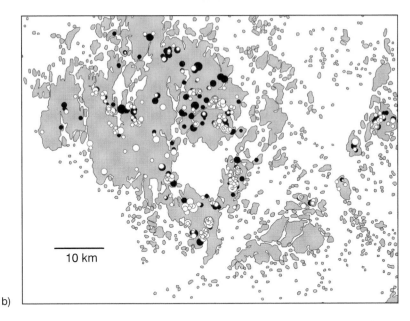

b)

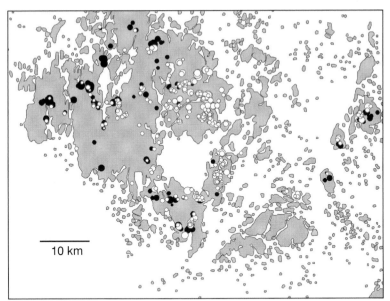

c)

Fig. 1.6 Spatially correlated changes in population sizes of the Glanville fritillary. The three panels show the changes in population sizes in (a) 1993–94, (b) 1994–95, and (c) 1995–96. Open symbols indicate population declines, filled symbols population increases. The size of the symbol is proportional to relative change in population size (I. Hanski, unpublished data).

The answer to this question is no, as there is no lack of evidence for local density dependence in the Glanville fritillary. The most obvious density-dependent mechanism in local populations is starvation due to food shortage, which occurs commonly when a group of larvae has consumed all the host plants within their reach. Less commonly, larvae have been observed to run out of food at the level of local populations (M. Kuussaari, personal communication). Density dependence is a complex issue, however, because the larvae have to stay in large groups to be able to diapause successfully (M. Kuussaari, personal communication). Another density-dependent process is parasitism by the specialist parasitoid *C. melitaearum*. The incidence of parasitism in local butterfly populations increases with increasing host population size, because the probability of parasitoid colonization increases and the probability of parasitoid extinction decreases with increasing host population size (Lei and Hanski, 1998). Parasitism by *C. melitaearum* characteristically brings the largest host populations to a small size in only a few years (Lei, 1997). Parasitism may also cause local host extinctions (Lei and Hanski, 1997), hence parasitism is not a purely stabilizing process in local dynamics.

The important point is that despite these density-dependent processes, local populations have a high risk of extinction for the many reasons described above. At the metapopulation level there is also obvious density dependence due to the fixed number of suitable habitat patches in the Åland islands. The Glanville fritillary metapopulation exemplifies Den Boer's (1968) "spreading of risk" concept in the sense that there is substantial asynchrony in the spatial dynamics (Fig. 1.6). In terms of the extinction model in Box 1, such spreading of risk means smaller variance of the population growth rate when the growth rate is calculated for the metapopulation as a whole. But spreading of risk cannot be the process whereby there is no large overall change in the numbers, and no trend in the size of the metapopulation as a whole; this is due to both local and metapopulation-level density dependence.

1.5 OTHER INSECT METAPOPULATIONS

The Glanville fritillary butterfly in Finland serves as a well investigated insect example of species with a classical metapopulation structure (for comparable mammalian, avian and amphibian examples see Moilanen *et al.*, 1998; Verboom *et al.*, 1991; Sjögren Gulve and Ray, 1996, respectively). The Glanville fritillary is by no means unique, however, and with increasing interest in metapopulation studies in general (Hanski and Simberloff, 1997) the number of other insect examples is increasing rapidly (Table 1.2). There are many butterfly studies in particular that have demonstrated a metapopulation structure of one type or another (reviewed by Thomas, 1994; Thomas, C. D., 1995; Hanski and Kuussaari, 1995; Thomas and Hanski, 1997). Other studies that have reported many of the results also found for the Glanville fritillary include beetles living in large decaying tree trunks (Siitonen and Martikainen, 1994; see also Hanski and Hammond, 1995), grasshoppers and bush crickets on dry meadows (Kindvall and Ahlén, 1992; Köhler, 1996), ground beetles in fragmented heathlands (de Vries, 1996), mycophagous beetles (Whitlock, 1992; Ingvarsson *et al.*, 1997) and gall-forming flies (Eber and Brandl, 1994, 1996). The results of Dempster *et al.* (1995a) on the entire insect community inhabiting the flower-heads of *Centaurea nigra* and *Arctium minus* also support the notion of metapopulation persistence of extinction-prone local populations.

1.6 METAPOPULATION THEORY AND INSECT POPULATION DYNAMICS

1.6.1 Classical metapopulations: the Levins model

Species which are structured into small local populations, such as the Glanville fritillary in Finland and the other insect examples in Table 1.2, comprise classical metapopulations. Local populations have a high risk of extinction, but local dynamics are more or less independent and hence extinctions do not occur simultaneously. Local extinctions are compensated for by recolonization of currently unoccupied habitat patches, and the metapopulation may persist at the regional scale, at a stochastic balance between extinctions and colonizations. The essence of classical metapop-

Table 1.2 A sample of studies reporting metapopulation structures and dynamics in insects

Species	Order	Country	Reference
Melitaea cinxia	Lepidoptera	Finland	Hanski *et al.* (1995a)
Melitaea diamina	Lepidoptera	Finland	Wahlberg *et al.* (1996)
Mellicta athalia	Lepidoptera	England	Warren (1991)
Euphydryas aurinia	Lepidoptera	England	Warren (1994)
Euphydryas editha	Lepidoptera	USA	Harrison *et al.* (1988)
Proclossiana eunomia	Lepidoptera	Belgium	Baguette and Nève (1994)
Hesperia comma	Lepidoptera	UK	Thomas and Jones (1993)
Plebejus argus	Lepidoptera	UK	Thomas and Harrison (1992)
moths on islands	Lepidoptera	Finland	Nieminen (1996)
saproxylic beetles	Coleoptera	Finland	Siitonen and Martikainen (1994)
Phalacrus substriatus	Coleoptera	Sweden	Ingvarsson (1997)
Bolitotherus cornutus	Coleoptera	USA	Whitlock (1992)
ground beetles	Coleoptera	Netherlands	de Vries (1996)
Metrioptera bicolor	Orthoptera	Sweden	Kindvall (1995)
Urophora cardui	Diptera	Germany	Eber and Brandl (1994)

ulation dynamics is captured in Levins's (1969, 1970) metapopulation model. Considering the metapopulation as a population of local populations inhabiting an infinitely large network of habitat patches, Levins constructed a model for changes in the fraction of currently occupied patches, $P(t)$. In the Levins model, changes in P are given by

$$dP/dt = cP(1-P) - eP \qquad (1.1)$$

where c and e are the colonization and extinction rate parameters, respectively. For a discussion of the model assumptions and interpretation, see Hanski (1997). The main value of the Levins model is in drawing our attention to the hallmark of classical metapopulation dynamics, large-scale persistence of species in spite of frequent population turnover.

The equilibrium value of P is obtained by setting the right-hand side of Equation 1.1 equal to zero,

$$\hat{P} = 1 - (e/c) \qquad (1.2)$$

It is apparent that the model may have, at most, one positive equilibrium point, the value of which increases with decreasing value of the parameter combination e/c. When the value of e/c equals one, the metapopulation goes extinct. To appreciate the biological significance of this threshold, consider a previously empty patch network that has just been colonized, with only one local population in existence. P is therefore very small, and Equation 1.1 simplifies to $dP/dt = (c-e)P$ (the nonlinear term cP^2 is negligibly small). Condition $e/c < 1$ implies that, during the lifetime of the original population, given by $1/e$, it has to produce at least one new population, via colonization, for the metapopulation to persist. This condition is familiar from epidemiological models (Anderson and May, 1991), which are in many ways similar to metapopulation models for free-living organisms. We will return below to the role of patch isolation in affecting metapopulation persistence.

It is occasionally assumed that persistence in classical metapopulations does not require density dependence. This is not correct. There is implicit density dependence in the Levins model at the local level, and explicit density dependence at the metapopulation level, and in general it is easy to see that some local density dependence is required for long-term persistence of classical metapopulations (Hanski *et al.*, 1996a; Hassell, Chapter 2 in this volume). However, what the model does not imply is local populations that would necessarily last for a long time even if the dynamics are affected by local density dependence. Small populations in particular are vulnerable to extinction for a large number of reasons, despite density dependence. This point is amply illustrated by the results for the Glanville fritillary (section 1.4).

The Levins model nicely captures the fundamental idea of long-term persistence of a metapopulation in a stochastic balance between local extinctions and recolonizations, but the model is unrealistic in many details (Harrison, 1991, 1994). Most obviously, habitat patches are not identical in real fragmented landscapes; they are not equally connected to each other (because species have typically limited migration ranges; Fig. 1.5); and the real patch networks studied by ecologists consist of a finite, and often quite a small, number of habitat patches. All these features have been incorporated into the incidence function model (Hanski, 1994a; Box 2), which otherwise retains the spirit and assumptions of the Levins model. These features of real metapopulations are discussed below in some detail. Gyllenberg *et al.* (1997) describe an alternative modelling approach, which relaxes the simple presence/absence description of local populations, employed both in the Levins model and in the incidence function model.

1.6.2 Patch area and isolation effects

Habitat patches in real fragmented landscapes typically differ in size, and in how well they are connected to other patches and populations.

---

2. THE INCIDENCE FUNCTION MODEL

The incidence function model (Hanski, 1994a) is a stochastic metapopulation model with the following differences from the Levins model (Equation 1.1): the habitat patches may differ in size, the patches do not need to be equally connected, and the model assumes a finite patch network. The key idea is to assume simple relationships between the probability of extinction (E) and patch size, and the probability of colonization (C) and patch isolation, and to assume that, at equilibrium, the long-term probability of patch occupancy, called the incidence, is given by:

$$J = C/(C+E)$$

This will hold if the patches have constant though patch-specific extinction and colonization probabilities, and independent dynamics (Hanski, 1994a). Plugging the assumptions about patch size-dependent extinction rate and isolation-dependent colonization rate into this formula leads to a nonlinear regression model of the incidence as a function of patch sizes and isolations. This model can be parameterized with information on patch occupancy and population turnover as functions of patch sizes and isolations. The parameterized model can be used to iterate metapopulation dynamics in real and hypothetical fragmented landscapes. For details see Hanski (1994a, b, 1997) and ter Braak *et al.* (1998).

The patches can also differ in terms of habitat quality. From the point of view of colonization–extinction dynamics, the effects of patch area and isolation are often the dominant effects. To start with patch area, the probability of population extinction almost always increases with decreasing patch size (Williamson, 1981; Diamond, 1984; Schoener and Spiller, 1987; Hanski, 1994b), evidently because small patches tend to have small and extinction-prone local populations (Schoener and Spiller, 1987; Kindvall and Ahlén, 1992; Hanski *et al.*, 1995a). All extinction models tell us that small populations have a higher risk of extinction than large ones (Box 1; Lande, 1993; Foley, 1994, 1997, and references therein). Considering colonizations, numerous field studies have demonstrated that the probability of colonization of empty patches decreases with increasing isolation from existing local populations (Fig. 1.5; Thomas *et al.*, 1992; Hanski *et al.*, 1995a). The obvious explanation is the limited migration capacity of most species.

Given these effects, of patch area on extinction and patch isolation on colonization, we can predict that the long-term probability of occupancy of a particular habitat patch increases with decreasing isolation and with increasing area. Thus, other things being equal, large and little isolated patches have a greater chance of being occupied, at any given time, than small and isolated patches. These patterns have been widely observed in empirical studies on insects (Hanski, 1994b).

Metapopulation theory (Hanski, 1997) predicts that the effects of patch area and isolation also occur at the level of patch networks, not only at the level of individual habitat patches. Thus the fraction of occupied patches, P, is predicted to be smaller in networks with small and isolated patches than in networks with large and well-connected patches. Table 1.3 reports empirical results for the Glanville fritillary supporting this prediction. This result has an important practical implication: with increasing habitat loss and fragmentation, there comes a point when the average patch area is too small and the patches are too isolated for the metapopulation to persist in the network.

1.6.3 Colonization–extinction stochasticity

The Levins model assumes an infinite number of habitat patches, though in practice it also provides a reasonable approximation for networks with many patches. In the case of small networks the stochastic nature of extinctions and colonizations becomes increasingly important. A small metapopulation with only a small number of local populations may go extinct 'by chance', that is, all local populations may happen to go extinct simultaneously, though by the Levins model or by some other deterministic model the equilibrium patch occupancy is positive. This form of sto-

Table 1.3 Effects of average habitat patch size and patch density within areas of 4 km² on the proportion of patches occupied (P) by the Glanville fritillary. For the purpose of this analysis, the entire study area shown in Fig. 1.4 was divided into 2 × 2 km squares. For details see Hanski *et al.* (1995a)

Patch size			Patch density		
Average area (ha)	Occupancy		Number of patches (per 4 km²)	Occupancy	
	n	P		n	P
<0.01	23	0.24	1	61	0.21
0.01–0.1	138	0.24	2–3	70	0.32
0.1–1.0	88	0.40	4–7	58	0.25
>1.0	6	0.56	>7	66	0.41

Effects of average patch size and patch density on occupancy (P) were tested with ANOVAs on ranks, using the four patch size and density classes shown in the table. Both effects were highly significant (area: $F_{3,251}=5.69$, $p=0.001$; density: $F_{3,251}=4.21$, $p=0.006$; no significant interaction).

chasticity, which is analogous to demographic stochasticity in single populations, has been called extinction–colonization stochasticity (Hanski, 1991).

Data on the Glanville fritillary are used again here to illustrate the consequences of colonization–extinction stochasticity for metapopulation persistence. For the purpose of the following analysis, the network of *ca* 1600 habitat patches (Fig. 1.4) has been divided into semi-independent patch networks, which are relatively isolated from each other and which therefore have relatively independent metapopulations (Hanski *et al.*, 1996b). Figure 1.7 shows the fraction of such networks that were occupied at one point in time as a function of the number of patches in the network. When there were <10 patches per network, only one third of the networks were occupied, whereas occupancy reached practically 100% when the network had 15 or more patches (Fig. 1.7). This result agrees well with theoretical predictions (Hanski *et al.*, 1996c). Some other butterfly examples discussed by Thomas and Hanski (1997) also support the conclusion that a viable network of small, well connected patches with extinction-prone populations should have at least 15–20 patches. Results for the specialist parasitoids of checkerspot butterflies suggest that on average more than 20–30 host populations may be needed for the persistence of the parasitoid metapopulation (Hanski, unpublished data). This figure is higher than the corresponding figure for the host butterflies, possibly because of temporal variation in the number of host populations.

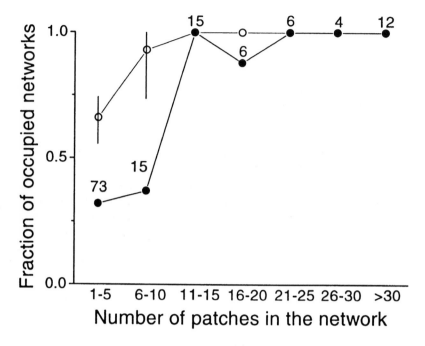

Fig. 1.7 Fraction of patch networks occupied by a metapopulation of the Glanville fritillary in 1993, as a function of the number of patches in the network. The networks have been arranged in groups with 1–5, 6–10, etc. patches. The number of networks in each class is given by the small number in the figure. Open dots give the null hypothesis obtained by randomizing the positions of the existing local populations amongst the habitat patches (average, minimum and maximum values in 10 randomizations; modified from Thomas and Hanski, 1997).

1.6.4 Rescue effect and alternative equilibria

Metapopulation models that implicitly or explicitly incorporate the effects of immigration on local dynamics demonstrate the possibility of complex spatial dynamics in fragmented landscapes, with alternative stable equilibria for the same set of parameter values (Hanski, 1985; Hastings, 1991; Gyllenberg and Hanski, 1992; Gyllenberg *et al.*, 1997). Hanski *et al.* (1995b) and Kuussaari *et al.* (1998) have shown that both the Allee effect and the rescue effect occur in the Glanville fritillary, hence it is not entirely surprising that alternative stable equilibria in the spatial dynamics of this species are also indicated (Fig. 1.8). The presence of such complex dynamics limits our ability to predict the long-term behaviour of the metapopulation. Stochastic perturbations may push the metapopulation from the domain of one equilibrium to the domain of the alternative equilibrium, which would be seen as a sudden and possibly long-lasting change in metapopulation size without a permanent change in the environment.

1.7 FINAL REMARKS

Against the background of this limited review of spatial structures, and especially the metapopulation structure in insects, I suggest that one reason why the notorious debate about density-dependent processes in insect population dynamics has been slow to die away is persistent confusion about the relationships between density dependence, population regulation and the spatial scale of population persistence. The research tradition de-emphasizing the role of density-dependent regulation in insect population dynamics appears to have drawn justification from observations of lack of long-term population persistence (Andrewartha and Birch, 1954; Dempster, 1983) as well as of apparent lack of density dependence in empirical time-series data (Gaston and Lawton, 1987; Stiling, 1987; Den Boer and Reddingius, 1989). The latter issue is now settled by the studies of Hassell *et al.* (1989a), Woiwod and Hanski (1992)

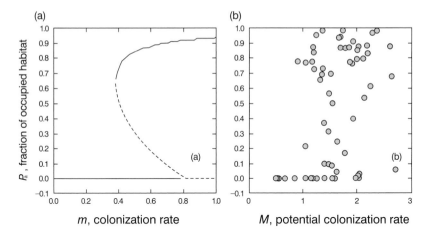

Fig. 1.8 Bifurcation diagrams for the fraction of occupied habitat (P) with increasing colonization rate (m). (a) A theoretical result based on the model described by Hanski and Gyllenberg (1993), using parameter values in their Fig. 3d.
Continuous lines represent stable equilibria, broken lines are unstable equilibria.
(b) Empirical result for 65 semi-independent patch networks of the Glanville fritillary butterfly. The occupied fraction of the pooled patch area, P_A, was used as the dependent variable. The potential colonization rate, M, gives the average value of the expected immigration rate to the patches, calculated on the assumption that all patches are occupied (because M measures the potential, not actual, rate of colonization). For more details see Hanski *et al.* (1995b)

and Wolda and Dennis (1993), all of whom find a high incidence of significant density dependence when sufficiently long time series are analysed. As to the lack of long-term persistence, small populations and populations with large variance of the intrinsic growth rate have a high risk of extinction, even though their dynamics were affected by regulating density-dependent processes. Strong density dependence may even increase the risk of extinction by generating complex dynamics with a large amplitude of oscillations. Therefore, though populations do not persist for a long time within finite positive limits without density dependence, even a high risk of local extinction is no evidence against density dependence nor against population regulation in the sense of a population having a tendency to approach a stationary distribution of densities.

If one seeks for contrasts in population dynamics, the interesting one is perhaps not so much between populations with strong versus weak density dependence, but between species which persist as classical populations versus species which persist as metapopulations. The *Viburnum* whitefly discussed by Hassell (Chapter 2 in this volume) is an excellent example of the former, whereas the Glanville fritillary (section 1.4) is a good example of the latter. Persistence is always relative, of course, as no population persists for ever, but there is nonetheless an important difference between persistence times measured in generations versus those measured in tens or hundreds of generations. The spatial population structure may be different even in the same species in different environments (Thomas and Harrison, 1992; Hochberg, 1996a). Our study of the Glanville fritillary has been conducted in a naturally fragmented landscape at the northern edge of the species' geographical range – we might have found something quite different had we been studying the species in a landscape with more continuous habitat in the centre of its range (Brown, J.H., 1995).

For some ecologists, a metapopulation consisting of local populations with a high rate of population turnover should be analysed by pooling data for the entire metapopulation. This approach would yield a 'population' resembling the classical population: it would not be difficult to demonstrate density dependence, the 'population' is regulated in the sense that measures of population size approach a stationary distribution, and the 'population' would persist for a long time. All this is fine if one is simply interested in demonstrating population regulation leading to long-term persistence at some spatial scale. But simply ignoring the spatial population structure to obtain better behaved time series would throw away much of the mechanistic understanding of the dynamics that the metapopulation approach may yield. The spatial population structure makes a difference, because it dramatically restricts interactions amongst individuals. For instance, demographic stochasticity and inbreeding depression are significant causes of population extinction in the Glanville fritillary. They would not play the same role if the population structure was not highly fragmented. The empirically observed patterns in habitat occupancy discussed in section 1.6 could not be explained by pretending that we have a single panmictic population whereas, in fact, the population structure is highly fragmented.

I have here emphasized the value of explicitly accounting for the spatial structure in analyses of population dynamics, because of the increased mechanistic understanding that this approach is likely to yield. That said, it is also instructive to reflect on the fundamental similarities of population dynamic concepts and theories, whether applied to local populations or to metapopulations. Metapopulation models are typically equilibrium models at the metapopulation level, as are classical models at the population level, with strong density dependence due to limitation by the available habitat patches. Metapopulation dynamics may include inverse density dependence, which may lead to alternative stable equilibria (section 1.6). Metapopulation dynamics may also include delayed density dependence due to predation, inducing oscillatory behaviour at the metapopulation level (Taylor, 1988). Strength of interaction amongst populations in a metapopulation typically declines with distance, which may lead to complex spatial dynamics (Nee *et al.*, 1997), but so does the strength of interaction amongst individuals in populations of many sessile organisms. The very same models can be used for metapopulations consisting of local populations and for populations consisting of territorial individuals, e.g. Lande's (1988) use of Levins's model. Metapopulation dynamics are affected by forms of stochasticity analogous to those seen in populations of individuals, namely colonization–extinction stochasticity (corresponding to demographic stochasticity) and regional stochasticity (corresponding to environmental stochasticity; Hanski, 1991). In this sense, moving to a higher spatial scale, from local populations to metapopulations, has involved renaming the study objects, rather than changing the fundamental theoretical framework.

Classical insect population ecology developed largely in response to economic concerns about pest species (Clark *et al.*, 1967). The current interest in insect metapopulation ecology also has a practical incentive, though now from the opposite direction. It is increasingly realized that habitat loss and fragmentation pose the greatest threat to the survival of species (Morris, 1995), and metapopulation ecology has been awarded the challenge of predicting the population biological consequences of habitat fragmentation. One well studied insect group in particular, butterflies, has played a key role in the development of current metapopulation biology (Thomas and Hanski, 1997).

ACKNOWLEDGEMENTS

I thank Jack Dempster, Chris Thomas and Peter Turchin for helpful comments on the manuscript.

2

The regulation of populations by density-dependent processes

Michael P. Hassell

2.1 INTRODUCTION

The concept of population regulation by density-dependent processes underpins most ecological theory, and yet has been the focus of more heated argument and discussion than any other idea in ecology. People have argued over the definitions of population regulation, the necessity for density dependence for population persistence, the frequency with which density dependence occurs in natural systems, and the best ways to detect density dependence and to identify regulated populations. Although there is now a broad consensus that a regulated population has a long-term, stationary probability distribution of population densities, implying some mean level around which the population fluctuates (Turchin, 1995), there is still disagreement about the processes involved in achieving this. Despite the simplicity of the ideas, ecologists somehow continue "to muddle the basic issues of the very existence of their study objects" (Hanski *et al.*, 1993a).

Although implicit in the writings of Malthus (1798) and Darwin (1859), and in Verhulst's (1838) logistic model of population growth, it was Howard and Fiske (1911) who gave the first clear exposition of the ideas of density dependence: "A natural balance can only be maintained through the operation of facultative agencies which effect the destruction of a greater proportionate number of individuals as the insect in question increases in abundance". These ideas were later reformulated and brought to a much wider audience by Nicholson (1933) and the term 'density-dependent factors', introduced soon after by Smith (1935) in

Insect Populations, Edited by J. P. Dempster and I. F. G. McLean. Published in 1998 by Kluwer Academic Publishers, Dordrecht. ISBN 0 412 83260 7.

place of 'facultative agencies'. (Smith also coined inverse 'density-dependent factors' for Howard and Fiske's agencies, which 'average to destroy a certain gross number of individuals each year" independently of population density.) This simple idea that density dependence was crucial for population persistence was readily accepted by some – Elton (1949) wrote that "It is becoming increasingly understood by population ecologists that the control of populations ...... is brought about by density-dependent factors" – but was strongly opposed by others.

This discord reached its peak in the 1950s (the Cold Spring Harbor Symposium volume of 1957 excellently summarizes the polarized views of the time). Like Howard and Fiske, and Smith, the principle protagonists were insect ecologists: A. J. Nicholson maintained that populations could only persist if regulated by density-dependent processes, while H. G. Andrewartha and L. C. Birch denied this central role for density dependence and assumed that most populations persist by a balancing of various density-independent processes. Joining the fray, G. C. Varley and T. Reynoldson championed Nicholson's views, W. R. Thompson supported Andrewartha and Birch, while A. Milne proposed an intermediate position in which competition for resources was the only reliable density-dependent process. This debate was largely carried out in the absence of good comparative data on the incidence of density dependence in different taxa. One empirical study that held centre stage for a long time was the classic long-term study of Davidson and Andrewartha (1948a, b) on the population dynamics of *Thrips imaginis* feeding on roses (Fig. 2.1).

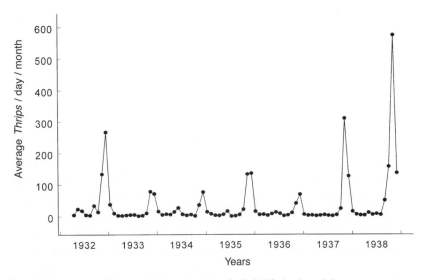

Fig. 2.1 Mean monthly population counts of adult *Thrips imaginis* on roses at Adelaide, Australia. (Data from Davidson and Andrewartha, 1948a).

Davidson and Andrewartha found that they could explain 84% of the variance in the log of the peak numbers each year from four weather terms: the number of day degrees up to 31 August, day degrees in September and October, day degrees in August of the previous season, and the rainfall in September and October. It was Andrewartha and Birch (1954), however, in reviewing this study, who drew the really contentious conclusion: "not only did we fail to find a 'density-dependent factor' *but we also showed that there was no room for one*" [my italics]. Several people have pointed to the unsoundness of this conclusion (e.g. Kuenen, 1958; Smith, 1961; Varley *et al.*, 1973). Indeed, within a closed population, the high correlation between the favourability for population growth and subsequent peak populations each year requires density dependence of some kind to bring the populations down to a relatively constant level (the winter minima) each year, irrespective of the peak abundance (Fig. 2.2).

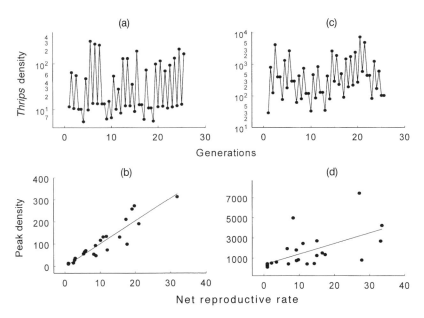

Fig. 2.2 Numerical examples showing the importance of density dependence for the correlation of peak population densities with a random net reproductive rate. Peak summer densities (P_t) are obtained from $P_t = \lambda W_{t-1}$ where W_{t-1} are the previous winter numbers and λ is a normally distributed reproductive rate. The following winter minima (W_{t+1}) are obtained from $W_{t+1} = P_t(1 + a\,P_t)^{-b}$ where a and b define the density dependence. (a) Numerical example with $a=0.1$ and $b=0.9$ (strong density dependence). (b) Corresponding correlation between peak densities and the winter to peak reproductive rate ($r^2=0.87$). (c) Numerical example with $a=0.6$ and $b=0.3$ (weak density dependence). (d) Correlation corresponding to (c) between peak densities and the winter to peak reproductive rate ($r^2 = 0.36$).

Nowadays, far more information is available, either where the actual density-dependent process (e.g. predation, competition) has been identified and quantified, or where one can infer density dependence from time series of population abundance. But the debate continues, particularly amongst insect ecologists who often find the identification of density-dependent processes elusive. For example, Dempster (1983), reviewing 24 life-table studies on temperate Lepidoptera, failed to find any density dependence in eight of them. More recently, Stiling (1987) surveyed 63 published life-table studies on insects, in almost half of which no density dependence was reported by the original authors. He concluded that many insect populations are not regulated by density-dependent factors acting across the usual range of population densities. Failures of detection, however, should not necessarily be taken as evidence of absence. For example, Hassell *et al.* (1989a) showed that the data set used by Stiling gives encouraging evidence of widespread density dependence provided that the studies are long enough. More generally, there are a variety of statistical problems associated with detecting density-dependent relationships (e.g. Maelzer, 1970; Vickery and Nudds, 1984; Royama, 1992; Holyoak and Lawton, 1993; Wolda and Dennis, 1993; Dennis and Taper, 1994; Holyoak, 1994b; Rothery, Chapter 5 in this volume). The poor detection rate could therefore stem from some amalgam of inadequate data sets and weak statistical tests.

These failures to detect density dependence, combined with the often striking levels of fluctuation in population time series, has prompted some to the view that many populations are not regulated in the 'classical' sense. Instead, the relationships between survival and density are generally thought to be poorly defined or, as coined by Strong (1984, 1986a, b), 'density vague': "Explicit density dependence has, at most, trivial variance and spans all densities ...... Vague density relationships ...... have the logically necessary feature of a decrease in *per-capita* performance at high densities, but emphasize a lack of change in performance over medial densities, where many populations spend most time. At very low densities, several alternatives are possible (density-dependence, density-independence, or inverse density-dependence)" (Strong, 1986a). Explicit and precisely defined density dependence is dismissed as a convenience in developing population models "which have proved disappointingly sterile for ecology". This challenge to conventional theory has been supported by some insect ecologists (Stiling, 1987, 1988; Wolda, 1989), and rejected by others (Hassell and Sabelis, 1987; May and Hassell, 1989; Berryman, 1991; Hanski *et al.*, 1993a; Turchin, 1995). Figure 2.3 summarizes some of these issues and shows the consequences of various kinds of density relationships. Starting with a theoretical density-dependent relationship (line in Fig. 2.3a), and the resulting perfectly stable population dynamics (line in Fig. 2.3b), variance is now added (points in Fig.

2.3a), which drives the fluctuations around the equilibrium population density (points in Fig. 2.3b). This is exactly the situation discussed by Southwood (1967), in which he emphasized how a density-dependent relationship could both regulate and, with sufficient variance, also be the key factor (Varley and Gradwell, 1960) driving population change. Fig. 2.3c–e portrays Strong's notion of density vagueness: density dependence occurs at very low (an Allee effect) and at high population densities, while density independence occurs in the intermediate range of densities. The dynamics now depend on the average population densities. If average densities are high enough to fall within the region of clear density dependence, the population is quite tightly regulated by the obvious density dependence (Fig. 2.3d). If at intermediate levels, however, the population is much more affected by the density independence, but still persists due to the occasional 'forays' into the regions of density dependence (Fig. 2.3e). The question of how frequently density dependence must act for a population to persist is discussed later.

In these examples, the mechanism of population regulation is readily interpretable in terms of the underlying density-dependent relationship and any associated variance. Adopting density vagueness as "an ecologically realistic alternative to the mathematical convenience of explicit density-dependence" (Strong, 1986a) is not necessary in order to interpret the dynamics, and hinders attempts to gain a better mechanistic understanding of population dynamics. Stochasticity can have crucial effects on population dynamics (Chesson, 1994), but laying variance aside and concentrating on the effects of the underlying deterministic relationship need not be a sign of naiveté (Griffiths *et al.*, 1984); it is simply that, in the absence of variation, the way that many processes contribute to dynamics becomes much more transparent.

While the accumulated collection of insect life-table data has not resolved the arguments about density dependence, a clearer picture is now emerging from the analysis of large numbers of insect time series in which a given developmental stage has been counted over many generations. One of the most impressive of these data sets comes from a survey over the past 30 years organized by Rothamsted Experimental Station, in which the abundances of moths and aphids have been sampled using light traps and suction traps, respectively, at sites throughout Great Britain. Woiwod and Hanski (1992) have analysed 5715 of these time series of 94 species of aphids and 263 species of moths. All the time series were over 10 years in length, and many were over 20 years. Using the tests of Bulmer (1975) and Pollard *et al.* (1987) for detecting density dependence, they found density dependence to be both pervasive and common, strengthening the argument that previous failures to detect density dependence were due to statistical problems and shortcomings in the data. With Bulmer's test, they found significant density dependence

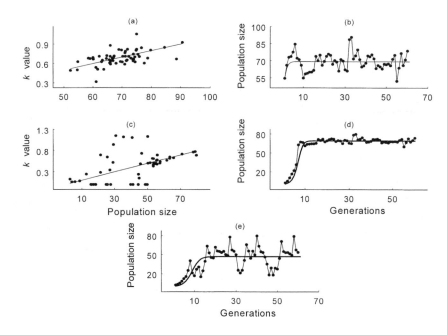

Fig. 2.3 The consequences of different kinds of density dependence for population regulation. (a) Density dependence with mortality expressed as a k value. The line represents the density-dependent relationship, $s=\exp(-\alpha N)$ where $\alpha=0.01$ and s is the fraction of n surviving density dependence, and the points come from the same relationship with normally distributed noise ($\mu=0$; $\sigma=0.01$). (b) The consequences of each of these for population dynamics where $N_{t+1}=\lambda N_t \exp(-\alpha N_t)$ and the net rate of increase, $\lambda=2$. Notice that the density-dependent relationship represented by the points both regulates the population and is the key factor (Southwood, 1967). (c) The points represent a 'density vague' relationship, using the terminology of Strong (1986a), where the density dependence is well defined at low and high population densities, but is more or less absent at intermediate densities. (d) The consequences for population dynamics of the relationships in (c) with $\lambda=2$. (e) The same but now with $\lambda=1.6$. The difference arises from the different average population densities (see text for further details).

in 81% of aphid and 47% of moth time series, while using the more conservative Pollard test the equivalent figures were 69% and 29%, respectively. If analyses were restricted to time series of 20 or more years, the frequency of density dependence increased further: the Pollard test now found density dependence in 84% of the aphid time series and 57% of the moths. Neither test for density dependence performed well with populations showing long-term changes in characteristic abundance, for example in response to habitat change. When Woiwod and Hanski excluded time series with a significant linear trend in density over time, the frequency of density rose yet higher. The Pollard test detected density dependence in 87% of aphid time series and 67% of moth time series. Woiwod and Hanski's results thus suggest that the detection of density dependence in natural populations of insects may not be as intractable to study as many have feared.

The rich dynamics that can result from density dependence in simple, single-species population models are well understood (May, 1974; May and Oster, 1976), and need not be reviewed again here. Instead, this chapter explores the role of density dependence in single-species populations living in more complex, patchy habitats, where individuals are spatially distributed across units of resource (patches). A particular model framework, supported by empirical examples, illustrates how spatial processes can be important to the net density dependence acting on a population as a whole. At larger spatial scales, when a number of local populations are linked together by dispersal to form a metapopulation, there is the possibility of different population dynamics occurring at local and regional scales. Such models are useful in examining to what extent what P. J. den Boer (see Chapter 3 in this volume) has called the 'spreading of risk', that is inevitable in metapopulations, can be a mechanism promoting population persistence – even perhaps to the point of precluding the necessity for any density dependence at all (Den Boer, 1968; Den Boer and Reddingius, 1989; Reddingius and Den Boer, 1989).

2.2 POPULATIONS IN A PATCHY ENVIRONMENT

Let us consider a habitat containing discrete patches of food plants for the larvae of a herbivorous insect population. The insects have discrete generations, and in each generation the adult females disperse between patches, laying their eggs on the plants. The emerging larvae, and subsequently the pupae, complete their development within their natal patch prior to the next generation of adults emerging. If all the insects had the same probability of survival, irrespective of the number in their particular patch, the spatial structure of the environment would be largely irrelevant and a simple homogeneous model would suffice to describe the dynamics of the system. Density-dependent survival within patches, however, makes the spatial distribution of individuals very important to the survival of the population as a whole. Birch (1971) describes such a system involving the cactus-feeding moth *Cactoblastis cactorum*, feeding on prickly pears (*Opuntia* spp.). The distribution of *Cactoblastis* egg sticks per plant is highly aggregated, leading to high proportions of plants with more *Cactoblastis* than can be sustained.

Patchy, single-species systems of this kind have been explicitly modelled by De Jong (1979) in a way that shows clearly the relationship between spatial distribution, within-patch density dependence and the overall density dependence acting on the population. She assumed a habitat containing n patches, adults that lay a fixed complement of F eggs, and a within-patch density-dependent survival of the immature stages depending on the population density per patch. Two kinds of dispersal between plants were considered. (i) In her 'adult dispersal', the mated adult females that emerge from the different patches enter a dispersal phase during which they mix thoroughly, before going on to search for plants on which to oviposit. The resulting distribution of adults from plant to plant was assumed in turn to be either uniform, binomial, Poisson or negative binomial. Once on a plant, each female then lays her full complement of eggs; thus, only multiples of F eggs can occur on a plant. (ii) De Jong's so-called 'juvenile dispersal' could occur in two ways: either by the population of immature stages actively dispersing between plants or, as is more likely, by the adult females flitting from plant to plant and laying a single egg at each encounter with a plant. In this second case, it is the egg distribution, rather than the adult females, that is assumed to be either uniform, binomial, Poisson or negative binomial, and now the possible egg densities per plant are 0, 1, 2, 3, ..., instead of the 0, F, $2F$, $3F$, ... from the adult dispersal. These two scenarios lead to much the same model, but there are differences in the number of patches occupied by the insects (greater for juvenile dispersal) and in the variance in egg densities per patch (greater for adult dispersal, particularly at low total population densities), both of which can affect population dynamics.

The dynamics of these spatially distributed populations can be portrayed as follows (De Jong, 1979).

For adult dispersal:

$$A_{t+1} = nF \sum_{i=0}^{\infty} [ip(i)f(Fi)] \qquad (2.1a)$$

For juvenile dispersal:

$$J_{t+1} = nF \sum_{i=0}^{\infty} [jp(i)f(j)] \qquad (2.1b)$$

where A and J are the total population sizes of adults and juveniles, respectively, over all patches. The summation represents the average number of surviving offspring emerging from any one of the n patches, $p(i)$ and $p(j)$ are, respectively, the probabilities of having i adults or j juveniles in a patch (described by one of the four distributions above), and $f(\bullet)$ is a function giving the density-dependent probability of an egg surviving to an adult on a plant where either Fi or j eggs have been laid. De Jong chose the density-dependent function, $\exp(-\alpha x)$ for this survival (see Fig. 2.7 below), where α defines the strength of the within-patch density dependence, and $x=Fi$ in Equation 2.1a and $x=j$ in Equation 2.1b (alternative forms of density dependence could equally well have been used; Hassell and May, 1985).

The key components of this model affecting dynamics are (i) the spatial distribution of the individuals, (ii) the degree of within-patch density dependence, a, and (iii) the adult fecundity, F. Of the four different spatial distributions considered by De Jong, the uniform distribution is the most unlikely and the easiest dealt with: dividing the population equally amongst n patches has no effect on the dynamics. Equation 2.1b, for example, thus collapses exactly back to the equivalent non-spatial model:

$$J_{t+1} = FJ_t\exp(aJ_t) \tag{2.2}$$

whose rich dynamic properties, from stable points to chaos, are very well known (May, 1973; May and Oster, 1976). The remaining three distributions, the binomial, Poisson and negative binomial, all involve some variation in the number of individuals from patch to patch, and this, if strong enough, can have a profound influence on the dynamics of the population. Let us consider in particular the negative binomial distribution, defined by two parameters – the mean of the distribution and a parameter, k, that inversely defines the extent of clumping (most aggregated as $k \rightarrow 0$, becoming random or Poisson as $k \rightarrow \infty$), and which well describes the degree of spatial clumping for a wide range of natural populations (e.g. Harcourt, 1965; Southwood, 1976; Anderson and May, 1978; Atkinson and Shorrocks, 1984; Jones and Hassell, 1988; Hails and Crawley, 1992; Naeem and Fenchel, 1994), although there are problems with its indiscriminate use (Taylor et al., 1979). Several authors have stressed how many different biological mechanisms can give rise to the negative binomial distribution (e.g. Anscombe, 1959; Boswell and Patil, 1970; Southwood, 1976; Atkinson and Shorrocks, 1984). For example, it can arise from (i) true contagion in which the presence of one individual (eggs, for example) increases the likelihood of finding another egg in the same place; (ii) heterogeneous Poisson sampling in which eggs are laid randomly within patches but, due to variable patch attractiveness, the mean number of eggs laid per patch is gamma distributed or, alternatively, the patches may be equally attractive but the quality of females (i.e. the number of eggs they lay) varies in a gamma distributed way; and (iii) a compound distribution in which eggs are laid in clutches which themselves vary in size logarithmically (Green, 1986). Naeem and Fenchel (1994) have recently given an example of the latter: the distribution of feeding opportunities by a marine ciliate on wounded invertebrates is described by a negative binomial distribution obtained from a Poisson distribution of encounters with food and a logarithmically distributed duration of feeding once food is encountered.

The dynamic effect of having a negative binomial distribution of individuals in the De Jong model is most easily portrayed using a very convenient simplification (De Jong, 1979, 1982). De Jong found that if the detailed ('microscopic') parameters are redefined in terms of three 'macroscopic' parameters ($\lambda=F\exp(-\alpha F)$, $a=[1-\exp(-\alpha F)]/nK$ and $b=k+1$), the model then collapses, rather fortuitously and conveniently, to the much simpler model for density-dependent population growth:

$$N_{t+1} = \lambda N_t[1+aN_t]^{-b} \qquad (2.3)$$

Here n is the population size, b defines the strength of density dependence (e.g. contest when $b=1$ and scramble when $b\to\infty$), and a is a scaling constant indicating the population size when density-dependent effects become marked (Hassell, 1975; Hassell *et al.*, 1976). The properties of Equation 2.3 are well known. Stability hinges only on the values of λ and b and, depending on these, the population may be monotonically or oscillatorily stable, or show limit cycles or chaos (Fig. 2.4).

In other words, the detailed model (Equation 2.1) and the phenomenological model (Equation 2.3) are identical as long as the macroscopic parameters of total population growth and density dependence have the

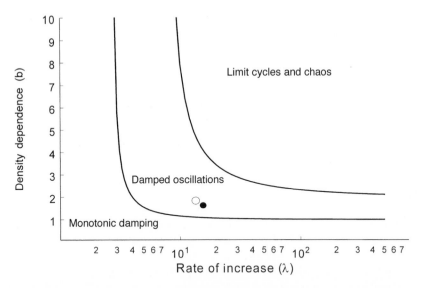

Fig. 2.4 The stability boundaries from the model in Equation 2.3 in terms of the density dependence parameter b, and the population rate of increase λ. The boundaries also apply to the models in Equation 2.1 if the following definitions are made: $\lambda=F\exp(-\alpha F)$, $a=[1-\exp(-\alpha F)]/nk$, and $b=k+1$. The density dependence is now solely in terms of the degree of spatial aggregation measured by k. The superimposed points represent the predicted stability of *C. chinensis* (solid circle) and *C. maculatus* (hollow circle) (see text).

precise definitions above in terms of the microscopic parameters of fecundity, spatial aggregation and within-patch density dependence. The particular advantage of identifying this simplification is that the stability boundaries in Fig. 2.4 now also define the stability properties of De Jong's model. The vertical axis becomes a measure of spatial aggregation $(b=k+1)$, and the overall net rate of increase of the population (λ) on the horizontal axis depends on both the fecundity per female (F) and the within-patch density dependence (a). With these definitions, it is clear that the degree of spatial aggregation of the insects markedly affects the overall amount of density dependence acting on the population. This is further illustrated by the numerical examples in Fig. 2.5. Broadly speaking, increasing the level of aggregation limits the number of patches in which density dependence operates so that the remainder of the patches become a refuge from this mortality, which therefore stabilizes the population. As the aggregation becomes less marked (k increasing), more and more patches suffer density dependence and the overall feedback acting on the population increases. This causes cycles and higher order behaviour to appear due to overcompensation provided the insect's rate of increase is high enough.

De Jong's model has been outlined in some detail as an example of how density dependence and spatial patchiness can interact together to regulate populations. The problems of detecting the net density dependence per generation arising in this way have been discussed by Hassell

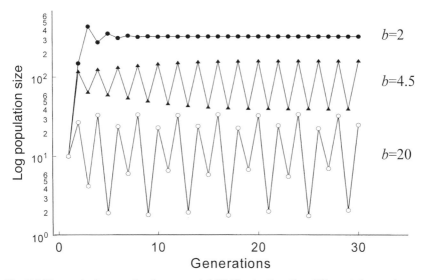

Fig. 2.5 Numerical examples from model (3) illustrating the different dynamics that occur as the density dependence or spatial aggregation (measured as $b=k+1$) changes as indicated ($\lambda=20$, $a=0.01$).

(1987), who showed that random variations in k, d or a can all make the density dependence acting on the population as a whole difficult to detect using the usual procedures (see also Dempster and Pollard, 1986). Although De Jong's model is primarily a heuristic tool for exploring how spatial aggregation can interact with other demographic parameters affecting population dynamics, the model can also be extended to represent the dynamics of some real populations. The following two sections illustrate this from a laboratory study on bruchid beetles and a field study on whiteflies.

2.3 TWO EXAMPLES

2.3.1 Bruchids in the laboratory

Some bruchid beetles, particularly *Callosobruchus chinensis* and *Callosobruchus maculatus*, have been popular subjects for exploring the population dynamics of insects in simple laboratory microcosms (e.g. Utida, 1950, 1953, 1967; Fujii, 1968, 1975; Bellows, 1982a, b; Hassell *et al.*, 1989b; Shimada and Tuda, 1996). The adults lay their eggs on the surface of stored product pulses, such as azuki and black-eyed beans. On hatching, the first instar larvae burrow into the bean and pass through four instars, before pupating in a cell near the surface of the bean. The mean generation times (between egg hatch and mean survivorship of the ensuing adults) is about 24 days for *C. chinensis* and 27 days for *C. maculatus* (Bellows, 1982b). Because they are so easy to culture in the laboratory, more time series have been published for these species than for any other insects, with the exception of species of *Drosophila* and *Tribolium*. Most of these have come from populations kept in very small cages (often petri dishes) with a fixed amount of beans uniformly distributed on the floor and replaced regularly. The two time series for *C. chinensis* and *C. maculatus* shown in Fig. 2.6 were obtained from larger arenas (460×460×85 mm) containing 50 equidistantly placed small, plastic pots as 'patches', each with a single bean replaced weekly. Adult beetles could enter pots only via a small hole in the lid, from which escape was impossible. Further details of the design and method of sampling are given by Hassell *et al.* (1989b).

Age-structured populations in an aseasonal environment and with more or less continuous breeding are best modelled using systems of time-lagged differential equations describing the rate of change of numbers of individuals in specified stages of the life cycle. These techniques were first applied to insect population dynamics by Nisbet and Gurney (1982, 1983) and Gurney and Nisbet (1985), and since then have been widely used to model insect population dynamics in continuous time (e.g. Gurney *et al.*, 1983; Murdoch *et al.*, 1987; Godfray and Hassell, 1989; Godfray and Chan, 1990; Gordon *et al.*, 1991; Briggs and Godfray, 1996).

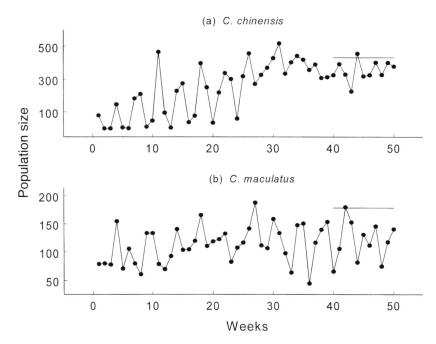

Fig. 2.6 Time series of (a) C. *chinensis* and (b) C. *maculatus* measured as total adults emerging per week, from 50 patches as described in the text. The horizontal lines between weeks 40 and 50 indicate the predicted equilibrium levels from model (2.4) (N^*_{CHI}=432.3 and N^*_{MAC}=178.7), as discussed in the text.

Here we simplify greatly by assuming discrete generations which (at least to some extent) is justified by the two populations in Fig. 2.6 showing clear signs of oscillations of approximately one host generation period. Gurney and Nisbet (1985) show how such apparent generation cycles can arise if a stage-structured population is regulated by density-dependent larval survival, and if the adults have a high fecundity and short longevity (both of which apply to the bruchid systems).

Once discrete generations are assumed, the bruchid populations can readily be represented by a version of the De Jong 'adult dispersal' model:

$$A_{t+1} = nFH\left[\sum_{i=0}^{\infty} ip(i)\exp(-diFH)\right] \tag{2.4}$$

Here, A_{t+1} is the total adults in generation $t+1$, n (=50) is the number of patches in the arenas, and the term in square brackets is the average number of progeny surviving in any one of these patches. In generation *t*, the dispersing adult females, having located a patch, stay there and lay their full complement of *F* eggs per adult (assuming equal sex ratios), of which a proportion *H* hatch (average hatched eggs per adult: $FH_{mac}=11.3$; $FH_{chi}=8.0$). The distribution of these adults in the 50 patches is strongly aggregated, and is well described by the negative binomial distribution ($k_{mac}=0.84$; $k_{chi}=0.63$), which is therefore used to define $p(i)$, the probability of having *i* adult females in a patch. Finally, a number of density-dependent processes are operating in this system (*F* and *H*), but the most marked of these, and the only one that affects the dynamics appreciably, is strong density-dependent survival of larvae within beans, described by the same exponential expression used by De Jong (Fig. 2.7).

The predicted dynamics of the *C. chinensis* and *C. maculatus* populations are shown by the equilibrium levels superimposed on Fig. 2.6, and by the two points overlain on the stability diagram in Fig. 2.4 (see Hassell *et al.*, 1989b for further details) which shows both populations as locally stable and falling more or less on the boundary between the regions of monotonic damping and damped oscillations. The regulation of this population thus depends, as in the De Jong model, on the way that the spatial distribution interacts with the within-patch density-dependent survival (egg-to-adult within beans) to produce a net density-dependent effect on the whole population. Less spatial aggregation would therefore be predicted to enhance the cyclic behaviour of the two populations.

In the next section, the same model framework is further extended to describe the dynamics of a long-term field study.

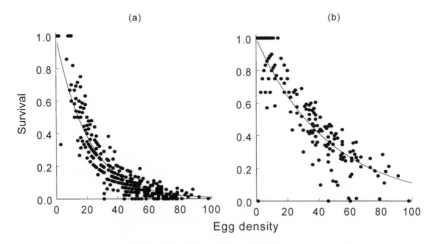

Fig. 2.7 Density-dependent survival of (a) *C. maculatus* and (b) *C. chinensis* from egg to adult within a bean. The fitted curves come from the model $s=\exp(-aN)$ where *n* is the hatched egg density per bean and *a* defines the severity of density dependence (after Hassell *et al.*, 1989b, in which full details are given).

2.3.2 Whiteflies in the field

During the 1950s and 1960s, a number of long-term life-table projects were initiated on a range of populations that either were univoltine or reproduced in discrete annual cohorts. The principle aim in this body of work was to identify (i) the main determinants of population fluctuation – the 'key factor(s)' (Varley and Gradwell, 1960; Morris, 1963b; Podoler and Rogers, 1975) – and (ii) the density-dependent processes contributing to population regulation. The techniques were designed to analyse insect population data (e.g. Varley and Gradwell, 1960; Richards and Waloff, 1961; Morris, 1963b; Klomp, 1966; Harcourt, 1971; Southwood and Reader, 1976; Dempster, 1982), but have subsequently been used on fish (Elliott, 1984), birds (Blank *et al.*, 1967; Krebs, 1970; Southern, 1970; Watson, 1971) and mammals (Sinclair, 1973). The longest lasting of all these is Southwood's study of the viburnum whitefly (*Aleurotrachelus jelinekii*) which he commenced in 1962 and which still continues today, albeit with a break of a few years in the middle. It is a study that can be treated as a single-species system and, like the bruchids, is also well described using De Jong's model. The first phase of the study was on the populations on three relatively isolated viburnum bushes (*Viburnum tinus*) at Silwood Park, Berkshire between 1962 and 1978 (Southwood and Reader, 1976, 1988; Reader and Southwood, 1984, 1989). The work was then discontinued, but resumed in 1984 on two of the bushes, A and B (Hassell *et al.*, 1987; Southwood *et al.*, 1989).

The life cycle of the whitefly is relatively straightforward. There is a single generation per year, with adults emerging in May or early June. Reader and Southwood (1984) found very little dispersal of adults following emergence, and concluded that the great majority oviposit on their natal bush; each bush was thus regarded as an isolated population. The eggs hatch after about 4 weeks and the larvae then remain on their respective leaves throughout development. The fourth and final instar is reached by November; the larvae then become quiescent over winter before resuming feeding in the spring. Southwood and Reader (1976) developed a comprehensive sampling schedule to estimate the densities of all stages between egg and adult, as well as the number of leaves on the bush, so that population densities could also be expressed per leaf. In addition, in each year since 1968, detailed information was collected on the spatial distribution of the immature stages from leaf to leaf. This involved randomly choosing 100 leaves on which eggs had been laid and following each of these cohorts regularly through the season to determine the spatial distribution of survival of the immature stages.

For simplicity, only the population dynamics on bush A will be considered here. At the start of the study in 1962, the population was at a very low level (Fig. 2.8) following the severe winter of 1962–63 (the whitefly generations span two calendar years, but for notational ease the different generations will be referred to just by the year in which the eggs are laid). The population then increased and between 1965 and 1970 fluctuated between approximately five and nine eggs per leaf; this was followed by a whitefly outbreak, and the population then fluctuated between about 30 and 50 eggs per leaf between 1972 and 1978. When sampling resumed again in 1984 the population was once more at a much lower level (12 eggs per leaf), and continued to fall until 1988. Since then it has fluctuated between two and eight eggs per leaf. This time series can be viewed in two quite different ways: (i) as a population that increased more or less continually between 1962 and 1978, followed by a dramatic decrease during the interval years and relative stability at a low level since then; or (ii) as a population that has been relatively stable at a low level except for the outbreak years between 1971 and sometime in the period 1979 to 1983. The analysis below throws some light on these alternatives and how density dependence operates in this system.

The whitefly populations fit well into the general mould of De Jong's model: the generations are entirely discrete, and in each generation the adults move about the bush, laying eggs on leaves (= patches) on which the immature stages then complete their development without further movement between patches. Unlike the bruchid experiment, however, where females were forced to lay their eggs within a single patch (as in De Jong's adult dispersal model), the ovipositing whitefly females flit from leaf to leaf laying only one or a few eggs at a time. The juvenile dispersal model (2.2b) is thus the more appropriate:

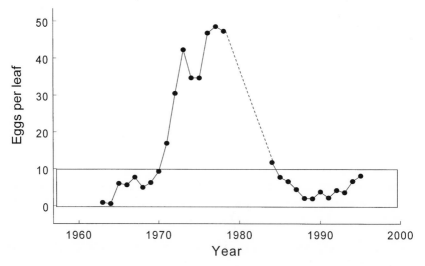

Fig. 2.8 The population changes in whitefly eggs per leaf on bush A. (After Southwood *et al.*, 1989, to which additional data have been added.)

$$\frac{E_{t+1}}{n} = Fs \left[\sum_{j=0}^{\infty} jp(j)f(j) \right] \qquad (2.5)$$

Here E_{t+1} is the egg population in generation $t+1$, n is the number of leaves on the bush, F (=55) is the potential fecundity per adult whitefly, and s is the fraction of these eggs that are actually laid; Fs is thus the net reproductive rate per adult. The term s has been thought either to be density-independent (Southwood and Reader, 1976; Hassell *et al.*, 1987) or density-dependent (Southwood *et al.*, 1989) depending on the amount of data available for analysis. Taking the 27 years of data now available from 1963 to 1994 reveals the patterns in Fig. 2.9. During the seven out-break years between 1971 and 1978, there is a very clear density-depen-dent relationship between s (expressed as the k value; Varley and Gradwell, 1960) and the potential numbers of eggs per leaf, but in the remaining, lower density years there is no apparent relationship. A threshold density is therefore introduced into the model (assumed at $E_t/n=10$) separating the years in which s is density-dependent or densi-ty-independent.

The term within the square brackets in Equation 2.5 gives the average number of adults emerging per leaf. It thus depends on the spatial distri-bution of eggs laid per leaf, $p(j)$, and the function, $f(j)$, giving the within-

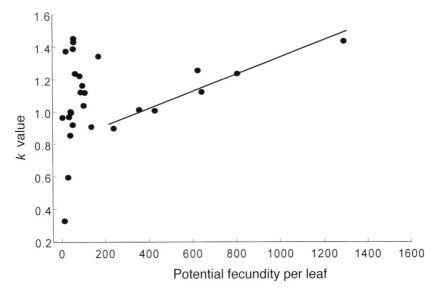

Fig. 2.9 The relationship between the potential fecundity of whiteflies per year and the proportion of these that are actually laid as eggs, expressed as a k value ($k=\log_{10}$[adults emerging×55]−$\log_{10}$[eggs laid]. The regression line ($y=0.82+0.0005x$; $r^2=0.91$) applies to the seven outbreak years discussed in the text.

leaf density-dependent survival from egg to adult. As in the bruchid example above, the probability distribution of eggs is well described by the negative binomial, mean $k=0.36+0.04$ (SE). The form of the density-dependent function for leaf-to-leaf survival from egg to adult, $f(j)$, was determined from the spatial data of 100 leaves sampled each year since 1968. Interestingly, in some years the relationship is inversely density-dependent (Fig. 2.10a), while in others it is directly density-dependent (Fig. 2.10b). Once again, there appears to be a difference between the outbreak years and the remainder: the significant relationships in the outbreak years are all inversely density-dependent, while in the remaining years the significant relationships are all density-dependent (Southwood *et al.*, 1989). Just why this should be so is far from clear. A pathogen affecting the fourth instar nymphs seems to be the main cause of the density dependence; perhaps during the outbreak years the numbers on the highest density leaves are high enough to create a sink effect and thereby enhance survival enough to mask the density-dependent effects of the pathogen (Southwood *et al.*, 1989). A second threshold density, again at $E_t/n=10$, is therefore introduced into Equation 2.5, above which the spatial density dependence is negative (given by the average slope of the outbreak years), and below which there is direct density dependence (given by the average slope of the remaining years). The different relationships for the outbreak and non-outbreak years produce interesting dynamic predictions, particularly the alternative stable states shown in Fig. 2.11. The lower stable equilibrium ($E^*_{LOW}=6$), corresponds well with the average population levels observed during the non-outbreak years, and is maintained entirely by the spatial density dependence operating when $E_t/n<10$. Above this threshold, the population 'escapes' and moves to a higher stable equilibrium ($E^*_{HIGH}=46$), which now corresponds well to the population levels reached during the outbreak years. The population escape from the lower densities is the result of the inverse spatial density dependence at $E_t/n<10$, and the subsequent regulation at the higher level is due to the strong density dependence in the eggs-laid term, s, during the outbreak years.

The rather complex regulation of the whitefly population therefore appears to depend on the combination of straightforward density dependence acting at the whole population level in years of abundance, and a much more subtle density dependence acting spatially within the population in low-density years. As in the example of the bruchid populations given above, the De Jong model is a good framework for disentangling these temporal and spatial effects.

2.4 PERSISTENCE IN METAPOPULATIONS

In the De Jong model (1), the different habitat patches contain different initial population densities depending on the egg-laying behaviour of the

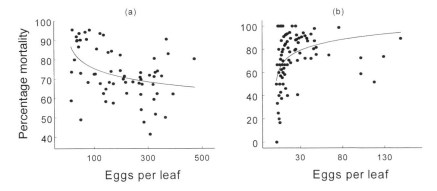

Fig. 2.10 Two examples of spatially density-dependent mortality operating between the time of whitefly eggs being laid and the subsequent adults emerging on 100 leaves on bush A. (a) Inverse density dependence in 1976 ($r=0.41$; $p<0.001$); (b) direct density dependence in 1992 ($r=0.44$; $p<0.001$).

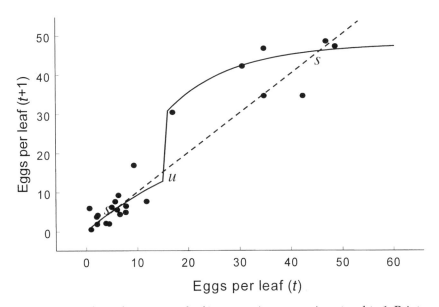

Fig. 2.11 Map of egg densities per leaf in successive generations t and $t+1$. Points show the observed data from Fig. 2.8. Solid line shows the predictions from the model in Equation 2.5 with the parameters as discussed in the text. Locally stable (S) and locally unstable (U) equilibria occur where the solid line intersects the broken 45° line.

adults. But in each generation, the individuals from all the patches mix thoroughly and the spatial distribution is then 'recreated' anew. No patch, therefore, can have any degree of independent temporal dynamics over a period of generations. A very different picture emerges, however, if the total habitat is on a scale that encompasses a number of more or less discrete groupings of individuals interconnected by some limited degree of dispersal. The whole ensemble of these linked, local populations now becomes a metapopulation (Hanski and Gilpin, 1997; Nee *et al.*, 1997; Hanski, Chapter 1 in this volume). Iwao (1971) described just such a system. Populations of the phytophagous lady beetle *Epilachna vigintioctomaculata* have discrete annual generations and, because of the patchy distribution of suitable habitats, are divided into distinct subpopulations. These are interconnected by adult movement which, rather than involving complete mixing amongst all subpopulations, is limited to a small range. Iwao concludes that "The numbers in the ...... subpopulations tended to be stabilized through population interchange among them, density-dependent adult mortality ...... and larval competition for food".

In the original concept of a metapopulation (Levins, 1968), space was implicit; each local population was effectively linked to all others, so that the spatial positions of each local population were effectively ignored. This kind of metapopulation model also deals first and foremost with occupancy of the different habitats rather than local population densities as a continuous variable (Taylor, 1988, 1990; Hanski, 1994c; Hanski and Gilpin, 1997). The metapopulation model explored in this section differs in being spatially explicit and made up of local populations with discrete generations, set in an environment of interconnected habitats. It therefore inevitably involves at least two quite different spatial scales: a local, within-habitat scale and a regional, metapopulation scale. Characteristically, the regional scale will be much larger than the typical scale of movement of individuals during their lifetime. The limited amounts of mixing between local populations means that, over a period of generations, asynchronous dynamics between local populations becomes a possibility, and it is this that underlies many of the rich spatial dynamics shown by many metapopulation models (e.g. Hassell *et al.*, 1991a, 1994; Sole and Valls, 1992; Comins *et al.*, 1992; Swinton and Anderson, 1995).

Asynchronous dynamics also emerge in these models as a mechanism promoting persistence of the population as a whole. This then raises the important question of whether or not metapopulations can persist even in the absence of any form of density dependence. The principal champion of this idea of 'spreading of risk', P. J. den Boer, argues that local extinctions and colonizations can allow a metapopulation of density-independently fluctuating local populations to persist, if not forever, then at least for so long that the distinction between density-dependent and density-independent persistence becomes academic (Den Boer, 1968, 1971; Reddingius and Den Boer, 1970, 1989; Den Boer and Reddingius, 1989; Den Boer, Chapter 3 in this volume). Their simulations to illustrate this rely heavily on the built-in constraint that the net reproductive rate of the population [$\ln R = \ln(N_{t+1}/N_t)$] over a period of time be close to zero (Latto and Hassell, 1987; Den Boer, 1991). For Den Boer (1971), "the longer a population is observed to persist the closer the average $\ln R$ will be to zero; this is a statistical truism which itself has nothing to do with 'regulation'". For many others the fact that $\ln R$ is close to zero may have everything to do with regulation by density-dependent processes. In this section, we cut through this argument by removing this constraint and just ask the following straightforward and specific questions:

is some density dependence necessary for the long-term persistence of a metapopulation of local populations each modelled as a random walk, or is the metapopulation structure alone sufficient to enable persistence for long periods of ecological time?

if density dependence is needed, how infrequently can it act and populations still persist?

To tackle these questions, Hanski *et al.* (1996a) considered a metapopulation in which the local populations are represented by a random walk between zero and a 'ceiling', K, and are interconnected by dispersal which declines with distance in a negative exponential way. Were there only a single local population, we would have:

$$n_{t+1} = n_t + r_t \tag{2.6}$$

where n_t and n_{t+1} are the $\log_e$-transformed population sizes in successive generations t and $t+1$, and r_t is a normally distributed random variable with given mean and variance (Foley, 1994). The population is assumed to go extinct if $n < 0$ and population sizes cannot exceed a ceiling ($k = \ln K$); if n_{t+1} is greater than k, then it is assumed immediately to be 'reflected' by an equal amount below k (i.e. $n_{t+1} \rightarrow 2k - n_{t+1}$). Density dependence is thus completely absent whenever the n_{t+1} is below k, and only appears in those generations when the population has reached this ceiling. Such an all-or-nothing form of density dependence is unrealistic, but it does provide a convenient means of generating a pure random walk between limits, and does enable one to quantify the level of density dependence operating on the population as a whole in a direct and simple way – as the probability of a local population 'hitting' its ceiling per generation, $P(K_{hit})$.

Hanski *et al.* (1996a) consider a variety of metapopulation models (both analytical and simulation) based on these local dynamics. In the most detailed of these, a number of patches with explicit spatial co-ordinates are randomly distributed in an environment, and in each time interval a fraction emigrates from a patch and is distributed among the other patches in such a way that the distribution of migration distances is negative exponential. Thus the absolute number of migrants to any patch j from patch i depends both on the distance between the two patches and on how many patches there are at different distances from patch i. In each simulation the level of density dependence, $P(K_{hit})$, is also scored, and the metapopulation deemed to persist if it survives for 1000 intervals.

The following conclusions emerged.

Long-term persistence is always associated with density dependence and there is a clear positive relationship between the time to extinction and the level of density dependence. For example, in one series of simulations with the most detailed model (Table 1 in Hanski *et al.*, 1996a), the average value of $P(K_{hit})$ in the cases of long-term persistence was 0.14. These values are lower than required in a comparable, single population on its own, but still represent density dependence occurring in 14% of the habitats per generation. Metapopulation structure therefore contributes to persistence, but does not preclude the necessity for density dependence.

In some persisting metapopulations, however, the incidence of this density dependence can be much lower (the minimum value of $P(K_{hit})$ in the series of simulation mentioned above was 0.014). In general, persistence with only rare episodes of density dependence occurs in one rather special case – where the population turnover is very high due to high rates of local extinctions and colonization. In these cases, detection of the density dependence could be difficult. For example, Hanski *et al.* (1996a) found from one series of simulations that $P(K_{hit})$ of about 0.05 was detected in only 40% of time series of longer than 20 intervals using Bulmer's (1975) method to detect density dependence. Detection of density dependence in metapopulation studies is clearly a problem fraught with difficulties.

Finally, and in contrast to the usual view that real metapopulations consist of both sources and sinks, the simulations show that it is possible for a metapopulation to persist even if all the local populations are sinks. In other words, all the local populations would show a deterministic trend to extinction if isolated, and only persist due to the positive effects of movement between habitats.

2.5 CONCLUSIONS

While insect ecologists have held centre stage in arguments about the role and detection of density-dependent processes, in some other taxa density dependence is so obvious that it excites little comment (Godfray and Hassell, 1992). Many plants and sessile animals, for example, clearly compete for space at high population densities (e.g. Ross and Harper, 1972; Paine, 1974; Pacala and Silander, 1987; Rees *et al.*, 1996), as do territorial birds (e.g. Wiens, 1989). Likewise, herbivorous mammals often exhaust their food supply when numbers are high, and then suffer unambiguous density-dependent mortality (e.g. Clutton Brock, 1982). Nowadays, whatever the taxon, most ecologists would argue that the persistence of any plant or animal population for many generations in an environment results from some form of density-dependent feedback on net population growth as population densities rise, either by a reduction in fecundity or immigration, or by an increase in mortality or emigration. Without density dependence, simple population models undergoing a random walk rapidly diverge outside realistic limits. But ecological systems are complex, and random processes in the real world will operate in much more subtle ways than a straightforward random walk of the population as a whole. Many environments will exist as a network of patches, in which local populations are interconnected by some degree of dispersal. Asynchrony within such metapopulations clearly promotes persistence, and it therefore remains a valid question to ask for what periods of time real populations could persist in such cases under the influence of random processes without any density dependence (e.g. Den Boer, 1971). Model results show that long-term persistence in a metapopulation is not possible without some density dependence, although the incidence of this may be very low in some rather extreme conditions, particularly if there is a very high rate of local population turnover due to frequent local extinctions and colonizations. In such cases, however, it will be important not to neglect the effects of demographic stochasticity in enhancing extinction rates and increasing the importance of density dependence for persistence.

3

The role of density-independent processes in the stabilization of insect populations

P. J. den Boer

3.1 REGULATION AND NATURAL CONTROL

When I became involved in population dynamics in the 1950s, two concepts concerning the fluctuations of insect numbers dominated the literature: the hypothesis of regulation or governing of numbers introduced by Nicholson (1933), and the idea of natural control advanced by Thompson (1929, 1939). Nicholson supposed insect populations to be in a state of balance with the environment, i.e. special processes – later recognized as feedback processes (e.g. Wilbert, 1962, 1971; Bakker, 1964, 1971) – would return the density to an equilibrium which was determined by the average environmental conditions vital to the species. Thompson, who was an applied entomologist acquainted with biological control of insect pests, supposed insect numbers to be kept down by predators and parasitoids. This idea of natural control was clearly inspired by successes in the biological control of pest insects which had been inadvertently introduced into the USA. By introducing specific predator or parasitoid species from the pests' areas of origin, it appeared possible in some cases to keep such pests below the level where they cause serious damage to the relevant crops.

Hence, natural control, as a generalization derived from biological control, gave a possible explanation for the phenomenon that, under natural conditions, outbreaks of insect species are the exception rather than the rule. The balance of animal numbers, however, was entirely theoretical,

Insect Populations, Edited by J. P. Dempster and I. F. G. McLean. Published in 1998 by Kluwer Academic Publishers, Dordrecht. ISBN 0 412 83260 7.

although illustrated by Nicholson (1957) with the help of laboratory experiments with blowflies. How can one show, for instance, that there is an equilibrium density in a natural insect population? And is such a supposed equilibrium density a constant, or, as supposed by Wilbert (1962), a variable density level that changes with changing environmental conditions? How can one estimate the values of such a variable equilibrium level in a natural population (see e.g. Wolda, 1989, 1991)? It is often assumed that the existence of an equilibrium level of density is a logical necessity when numbers are kept within positive limits over a long period of time by density-dependent processes. Nicholson (1933) probably had this in mind when stating (page 133) that "Populations must exist in a state of balance for they are otherwise inexplicable".

Although I consider the few successful cases of biological control to be legitimate field experiments supporting the hypothesis of the occurrence of natural control, I am also aware of the counter-arguments. Outbreaks of insect pests occur in spite of the presence of specific predators and/or parasitoids (and not only in crops, orchards and forest plantations). Added to this, difficulties in biological control arise if introduced enemies are so successful as to almost eradicate the prey species, thus causing the extinction of the controlling agent, so that it must be repeatedly reintroduced (e.g. Van Lenteren and Woets, 1988). In my opinion, these counter-arguments illustrate not only that biological control does not always succeed, but also that natural control is less self-evident than was initially believed. The lesson to be learnt from biological control, therefore, is not that insect populations are naturally controlled, but that predators and parasitoids are often, but not always, able to reduce high densities of prey species for a period of time. However, this does not prevent outbreaks or local extinctions. Although natural control is not a general principle, it gives a realistic explanation for the often effective reduction of high densities of many natural insect populations. However, at the same time, such a reduction in density increases the chance of a population reaching dangerously low densities in the future (underpopulation: Andrewartha and Birch, 1954; Den Boer, 1968), and may ultimately cause extinction.

3.2 RETURN FROM LOW DENSITIES

The hypothesis that insect numbers are in a state of balance with environmental conditions (Nicholson, 1933) requires not only that high densities return to some mean (equilibrium) level, but also that low densities return to the same level. Nicholson (1937) thought that a return of population density to the equilibrium level from either above or below was brought about by intraspecific competition: at high densities, competition would be so severe that mortality increased and reproduction decreased, thus the density would be lowered until a level without significant competition was reached. At low densities, the animals would profit from the absence of competition by more easily obtaining enough of their necessary resources (in the first instance, food) to reduce mortality and increase reproduction, so that density would increase until a level of significant competition was reached again. Milne (1957b, 1962) assumed that intraspecific competition would operate only at high densities, when one individual obtaining its needs reduces the chances of others doing likewise. At lower densities, numbers would be determined by density-independent and/or imperfectly density-dependent processes. Milne claimed that density dependence is restricted to intraspecific competition at high densities, a view that stemmed from the observation that specific predators and parasitoids can keep down an increase in the density of their prey only after they have produced a new generation of adults. This delay in the response of specific enemies to a change of prey density causes the fluctuations in their number to lag behind those of their prey. In the deterministic regulation models of Nicholson and Bailey (1935), this resulted in the fluctuations in prey numbers showing an ever-increasing amplitude, i.e. the opposite of regulation. Nicholson (1937) supposed that intraspecific competition would not show this lag in a numerical response, at least not so far as mortality from competition was concerned. Later it was found, for example by East (1974), that generalist predators can also react without a time lag.

In view of the undeniably great effects of changes in physical conditions, especially weather, on insect numbers (e.g. Uvarov, 1931; Andrewartha and Birch, 1954), I considered that these theoretical discussions about a possible regulating mechanism were not very fruitful. I would expect physical conditions to be able to change insect numbers at least as vigorously as density-dependent processes, i.e. density-independent processes would determine the greater part of the variation in numbers. Moreover, I doubted whether, at low densities, environmental conditions will always be favourable, i.e. allow numbers to return to some mean (equilibrium?) density. To test the latter hypothesis, I checked for three species whether the mean egg production per female was always high enough to return low densities to the geometric mean (which is often considered to give a reasonable indication of the expected value of the equilibrium density). These species were selected because their population density was claimed to be regulated by a clearly indicated mechanism, and densities had been estimated for more than 10 years. It was evident that, in all three populations at the majority of densities below the geometric mean, egg production per female was too low to return density to this level in one generation (Figs 3.1–3.3). Indeed, in many years egg production per female would have needed to be many times higher than the maximum possible fecundity of the females; see section 2.3.7 of Den Boer and Reddingius (1996).

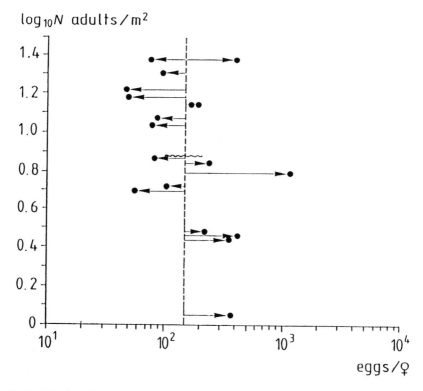

Fig. 3.1 Egg production (averaged per female) of winter moth (*Operophtera bruma-ta*) in Wytham Wood, Oxford, in the period 1950–68, that would have been need-ed in each generation to compensate for total generation mortality (black dots), plotted against density of reproducing adults. The required egg production would return density to its supposed equilibrium value (the geometric mean); the geometric mean of density over these 18 years is 7.6 adults/m² canopy ($\log_{10}=$ 0.88, horizontal wavy line). Sex ratio is assumed to be 1:1. Vertical broken line represents actual egg production per female, assumed to be about constant at 150 eggs. Data from Varley *et al.* (1973).

It may be objected that density might return to its mean level after some delay, i.e. not always in the next year, but after 2 or 3 years. I also checked this possibility (Table 2.2 of Den Boer and Reddingius, 1996) and, indeed, after an average of 3 years, decreasing densities turned into increasing ones, or *vice versa*. However, this delayed density dependence is exactly similar to what would be expected if densities fluctuated ran-domly (e.g. Cole, 1951, 1954). In addition, the supposed delayed density-dependent reactions tended to destabilize the fluctuations by increasing their amplitude, as is demonstrated in the time series of Fig. 3.4. In these

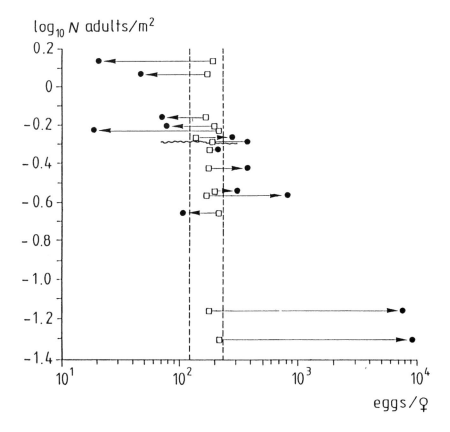

Fig. 3.2 Egg production (averaged per female) of pine looper (*Bupalus piniarius*) at Hoge Veluwe, the Netherlands, in the period 1950–63; relationship plotted as in Fig. 3.1. The geometric mean of density over these 14 years was 0.52 adults/m² canopy ($\log_{10} = -0.28$, horizontal wavy line). Sex ratio is assumed to be 1:1. Open squares between two vertical broken lines represent actual mean egg production per female as estimated by Klomp (1966) for each generation. A weak correlation ($r_s = -0.20$, NS) between density and mean egg production did not compensate for generation mortality. See caption to Fig. 3.1. Data from Klomp (1966).

field data the difference between highest and lowest density (log-range) was greater (by the permutation test of Reddingius and Den Boer, 1989) than could be expected in more than 5–10% of the time series in which the same net reproduction values succeed each other randomly (except for larvae in August of the pine looper of Klomp, 1966) (Den Boer and Reddingius, 1989; but see also Den Boer, 1990a). Apparently, random walks of densities resulting from net reproduction values of the same magnitude, but in a different order (i.e. independent of the preceding

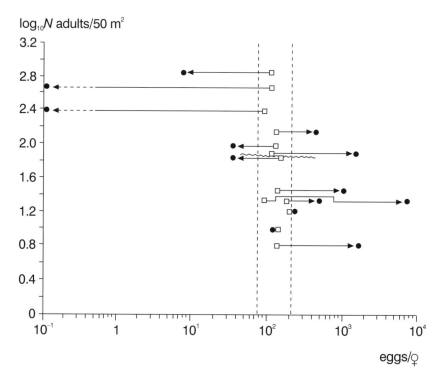

Fig. 3.3 Egg production (averaged per female) of the carabid beetle *Calathus melanocephalus* at Dwingelder Veld, Drenthe, the Netherlands between 1965 and 1978; relationship plotted as in Figs 3.1 and 3.2. Black dots show egg production per female needed to maintain density at the geometric mean over these 13 years: 75.9 adults per 50 m^2 heath (log$_{10}$=1.88, horizontal wavy line). Open squares between two vertical broken lines represent actual egg production in the field as derived from relationship between eggs in ovaries and eggs laid according to Table 3 of Van Dijk (1986). Significant correlation (r_s=–0.54, P=0.05) between density and egg production did not compensate for generation mortality. Note that at two of the higher densities of adults, the number of old adults surviving from the preceding year is sufficiently high to maintain density at its mean level, so that no egg production would have been necessary. See further captions to Figs 3.1 and 3.2. Data from Baars and Van Dijk (1984).

densities) may often be more stable than time series observed in the field (see also Den Boer, 1991), because in random walks the delayed density-dependent effects are removed. As stated above, this destabilizing effect of delayed density dependence was evident in the deterministic models of Nicholson and Bailey (1935), and can also be expected to play some part in the more stochastic situation of field populations.

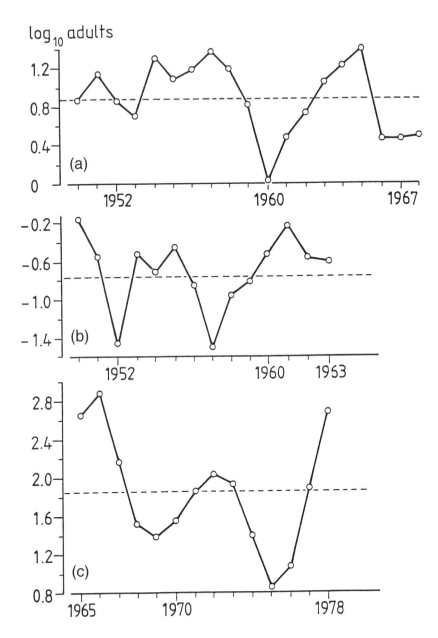

Fig. 3.4 Adult density ($\log_{10}$) in a number of succeeding years in (a) winter moth in Wytham Woods, Oxford (Varley *et al.*, 1973); (b) pine looper at Hoge Veluwe, the Netherlands (Klomp, 1966), $\log_{10}$ reproducing adults; (c) carabid beetle *Calathus melanocephalus* at Kralo Heath, Dwingelder Veld, the Netherlands (Baars and Van Dijk, 1984).

If insect populations were regulated by some feedback mechanism, density would fluctuate (after some initial period after founding) between constant limits, i.e. the log-range (log highest density minus log lowest density) would not increase with time (Fig. 4.13 of Den Boer and Reddingius, 1996). Den Boer (1991) showed that in all time series longer than 10 years, including those claimed to be regulated, the log-range increased over time to the same degree as in random walks of density. Hence, in none of these was density kept within limits, as could have been expected with densities in a state of balance with environmental conditions.

Nevertheless, regulation of density is not impossible in field populations. If, in each generation, both the production of reproducing adults is sufficiently high to compensate for the losses in the preceding generation, and high densities are effectively controlled, such a population might be considered regulated. Kluyver (1971) convincingly showed that a population of great tits on the isle of Vlieland was regulated according to these criteria. Even taking away 40% of the eggs or young did not disturb the stability of this population, where density was effectively restricted by territorial behaviour in each generation. It was a pity, however, that the situation studied by Kluyver was artificial, insofar as the birds were breeding in nest boxes, in which survival of eggs and young is much better than in natural tree holes (see further Den Boer and Reddingius, 1996). Nonetheless, the possibility cannot be excluded that natural populations of some territorial species might be regulated if the production of reproducing adults is always high enough to compensate for losses in the preceding generation, and if the exchange of individuals with other groups of the same species, i.e. the turnover of individuals, is not greater than the changes of the number of autochthonous individuals. It would be worthwhile to check populations of territorial vertebrates for these conditions, but I do not expect insect populations will be found that can be considered regulated according to these criteria.

It must be noted that the tests developed in recent decades for overall density dependence in time-series data (Reddingius, 1971; Bulmer, 1975; Pollard *et al.*, 1987; Dennis and Taper, 1994) indicate only the occurrence of statistical density dependence; they do not tell us anything about the biological processes that may be involved. Although density dependence is a necessary condition for regulation to occur, it is not a sufficient condition. Den Boer (1990a) gives some examples of the application of such tests producing significant results without the operation of a density-dependent mechanism. For instance, after a sudden drop in density, Bulmer's (1975) first test often gives a significant result. Sometimes random walks of density give significant results in tests for density dependence, if by chance they do not show a clear trend for a large number of generations. Long series of unreliable estimates of density may give significant results with the test of Dennis and Taper (1994), e.g. data from light traps (Wolda and Dennis, 1993; section 6.5.3 in Den Boer and Reddingius, 1996). Averaged estimates from large and heterogeneous areas in long time series may give significant results with almost all of these tests; compare the *Panolis* data over 60 years in Den Boer (1990a). Hence, application of these tests gives an indication that some interesting processes might have been operating in the population concerned, but a careful analysis is required to decide whether or not these processes have anything to do with regulation of numbers, or even with some kind of overall density dependence. For instance, many correlations with density may be indirect or even chance events.

3.3 OVERALL QUANTITATIVE EFFECTS OF DENSITY-INDEPENDENT FACTORS

Although I do not deny the possibility of the occurrence of natural control over a period of some decades, and once a regulated population in the sense of Nicholson (1933) is actually found, I will not deny it, I am sure that density-independent forces have played a decisive role in the dynamics of the majority of insect populations observed by ecologists. In other words, I believe that neither natural control nor regulation of numbers gives a sufficiently general theoretical base for the explanation of the fluctuating patterns of insect numbers observed in field populations. In particular, physical factors (weather) will significantly influence insect numbers (Uvarov, 1931). Some examples are given below.

Because of an extreme drought in the summer and autumn of 1959, many populations of the caddis fly *Enoicyla pusilla* became extinct in drier parts of the Netherlands (Van der Drift, 1963). In a moist forest that I sampled with pitfalls, the catches of full-grown larvae of this caddis fly fell from about 11 000 in the spring of 1959 to seven in the spring of 1960, because of the almost total mortality of young larvae in the exceptionally dry September–October of 1959. In a much wetter forest, numbers fell from about 3400 to 50 (section 2.3.6 of Den Boer and Reddingius, 1996). In a moist winter, more than 95% of the over-wintering larvae of *Calathus melanocephalus* (and most probably of other autumn-breeding carabid beetles) become infected by fungi and die (Van Dijk and Den Boer, 1992), dramatically decreasing the number of reproducing adults in the next year. On the other hand, after a dry winter the number of reproducing adults usually increases. For a period of 17 years (1971–87), both the yearly increases and decreases in the number of reproducing adults fitted the alternation (with a period of about 3 years) of moist and dry winters, respectively (Van Dijk and Den Boer, 1992). Because insects are poikilothermic, all of their activities and processes are highly dependent on temperature. For instance, in *C. melanocephalus* (and most probably in other carabid species), four times more eggs are laid at 19 °C than at 8.5 °C (Van Dijk, 1979, 1982). See Uvarov (1931) for more examples.

After the severe winter of 1995–96, I observed a spectacular example of the phenomenon, well-known among entomologists, of the increase in numbers of many insect pests, most probably because over-wintering larvae and pupae did not die from fungal disease. In spring 1996, all around Wijster enormous outbreaks of winter moth, *Tortrix viridana*, a number of miners and some other moths occurred, which resulted in local defoliation of trees in many sites. Density-independent changes in the quality of the food, especially nitrogen content, can cause both outbreaks and crashes of insect numbers (Van Dijk, 1996; White, 1993).

In general, the variation between years of many density-independent factors affecting insect numbers may be much greater (sometimes even many times greater) than the variation of density-dependent processes, so that the effects of the latter are overwhelmed and cannot regulate numbers. An example of this is given by Den Boer (1986a). In the carabid *C. melanocephalus*, the effect on numbers of the density-dependent egg production is completely overwhelmed by the much greater variation in larval mortality, over years. In order to regulate numbers, the variation of density-dependent processes must be at least as great as the variation of all density-independent forces.

As we cannot make science by merely piling up examples, I will not give more examples of the huge impacts that density-independent factors may have on insect numbers. Instead, I will discuss the effects of heterogeneity in both populations and environment on insect numbers, which gives an alternative explanation for the often surprising stability over time of some insect populations.

3.4 THE INTERACTION GROUP AS A UNIT OF POPULATION

When I was first confronted with the hypothesis of the regulation of animal numbers, I realized that this idea stemmed from a mechanistic view of natural processes. I have no objections to mechanistic and deterministic hypotheses about natural processes, but I doubt whether such hypotheses can successfully explain the apparently restricted fluctuations of insect numbers over time. One can argue that a balance of animal numbers can be supposed to exist only if the individual animals either all have the same chance of survival and reproduction, or show a stable frequency distribution of the critical properties, e.g. a stable age distribution and/or a stable frequency distribution of reproductive rates. The population area should also be homogeneous, so that the individual animals can be distributed uniformly in space, and be affected to the same degree everywhere by relevant environmental factors. Although these necessary conditions can be relaxed somewhat without completely abolishing the possibility of population regulation, rapid and disorderly local changes of environmental conditions and the composition of populations, in my opinion, are not compatible with a balance of animal numbers over long periods of time.

My field experiences have convinced me that even in a seemingly homogeneous environment such as a uniform heathland, relevant environmental factors, e.g. temperature, can change over short distances (e.g. Den Boer and Sanders, 1970). Therefore, the environment of a natural population will consist of many microhabitats in which the fate of individual animals may be very different, and usually difficult to predict. Moreover, among iteroparous carabid populations studied in the field, we observed neither a stable age distribution (see e.g. Den Boer, 1979; Baars and Van Dijk, 1984), nor a stable frequency distribution of genotypes (e.g. Den Boer *et al.*, 1993); frequencies of both changed from year to year in a disorderly manner. Therefore, I did not see a rosy future for the application of the regulation hypothesis to natural insect populations, and this led me to think about possible alternative explanations for the relative stability of some insect populations. Such an alternative should incorporate the natural variability in the composition of field populations and in their environments.

Heterogeneity and variability should not be considered as just drawbacks of field data, to be averaged away in order to discover the underlying more deterministic rules of nature. They should be recognized as fundamental features of a natural situation. The chance of survival of a population may even be increased if variations between individuals within that population make it possible for them to cope with the variations in environmental conditions in space and time (Den Boer, 1968). For instance, in cold periods and at cold sites in their habitat, cold-resistant individuals will have a better chance of survival than more sensitive individuals. The same applies to heat-resistant individuals during warm periods and at hot sites in the habitat. In other words, the risk of dying from lethal temperatures is spread over individuals with different tolerances to extreme temperatures. The number of victims of such temperatures will be restricted to the unlucky sensitive ones, that were at the wrong site at the wrong time. This relationship between spreading of risk and chance to survive is the raison d'être of the insurance business.

In almost every natural population there will be sufficient genetic and/or phenotypic variability in sensitivity to relevant environmental variables to simulate a degree of risk spreading that resembles the situation met by an insurance company: the reduction of extreme losses, so that the fluctuation of losses over time is more or less stabilized around a certain value. Note that this value is not an equilibrium value, but a simple mean, because there is no mechanism operating that tends to return deviations from this mean, either in the payments of the insurance company, or in the losses from natural populations. In both cases, the time series result from a succession of chance realizations.

Reddingius and Den Boer (1970), Reddingius (1971) and section 4.3 of Den Boer and Reddingius (1996) demonstrated that stabilization of animal numbers by spreading of risk might occur when variability and/or heterogeneity of reproduction and/or mortality are introduced in otherwise unspecified simulation models. This suggests that this kind of stabilization may be quite general in field populations. To test this supposition, I sampled field populations in such a way that the degree of variability and/or heterogeneity in reproduction and mortality can be estimated and compared under different conditions. I decided first to concentrate upon the effects of heterogeneity in space, to see whether the fluctuations of numbers over time, in subpopulations at different sites of the same habitat, may differ as a result of differences in the reproduction and mortality of individuals.

The first problem to solve was: how to define a subpopulation as an adequate unit of population, i.e. as a group of individuals the composition (and local environmental conditions) of which differ from those of other subpopulations within the same population area. As we were working on the dynamics of carabid beetles, we already knew that both egg production and mortality are significantly determined by physical factors, such as temperature (e.g. Van Dijk, 1982). We also knew that the relevant physical factors are affected to a high degree by the structure of the vegetation (e.g. Den Boer and Sanders, 1970). Hence, we decided to sample differently structured sites within the same heathland, the Heath of Kralo (part of the National Park Dwingelder Veld) in the Netherlands. However, in this large heath area it was difficult to determine which individuals may be considered to belong together, i.e. to form a subpopulation that differs from other subpopulations.

This problem was attacked and solved by Martien Baars. Individual carabid beetles of two heathland species were radioactively marked and tracked with a sensitive scintillation detector. The directions of movement appeared to be arbitrary, and the distances covered fell into two classes, short distances, called random walks, and long distances in a more or less fixed, but arbitrary, direction, called directed walks. From the frequency distributions of directions and of the two groups of distances, the pattern of movement could be simulated in a computer programme. With this computer programme, the movements of a thousand individuals of two species, *Calathus melanocephalus* and *Pterostichus versicolor*, could be simulated, and the area covered by these movements estimated (Baars, 1979a). Using a thousand brand-marked beetles, released in small groups in the field and recaptured at different distances, the results of the simulations could be checked and were found to be sufficiently close to reality. At the same time, Baars (1979b) showed that year-catches in standard sets of pitfalls give reliable estimates of local densities. This was confirmed by Den Boer (1979) for four other species, with the help of extensive mark–recapture experiments in a small forest. The correlation between actual mean densities and year-catches in all cases tested is almost unity.

Ninety percent of the individuals of *C. melanocephalus* caught during a year (year-catch) in a standard set of pitfalls appeared to have moved around an area of about 2 ha. Such a group of beetles of the same species is called an interaction group, because these individuals are readily able to meet and interact. The other 10% of the year-catch are considered directed movers that had covered greater distances, i.e. had moved in from outside the interaction group and were thus connecting different interaction groups by exchange of individuals. To give some idea of the size of an interaction group: at the Heath of Kralo, an interaction group of *C. melanocephalus* consists of about 10 000 to 200 000 beetles, of which about 30 to 900 will be caught in a standard set of pitfalls in a year-catch (Fig. 2.2 of Den Boer and Reddingius, 1996). In the same way, it was established that the individuals in an interaction group of *P. versicolor*, the adults of which are twice the length of those of *C. melanocephalus*, move around an area of about 12 ha and at the Heath of Kralo will consist of about 30 000 to 130 000 individuals, of which about 250 to 1000 will be caught in a year-catch (Fig. 2.12 of Den Boer and Reddingius, 1996). An interaction group of the related *Pterostichus lepidus* lives in about the same area of 12 ha, but will consist of fewer individuals at most sites. Small beetles will tend to move over small areas and big beetles over large areas; for instance, an interaction group of *Carabus cancellatus*, another heathland species with individuals of about four times the length of *C. melanocephalus*, will cover an area of more than 700 ha.

3.5 SPREADING OF RISK IN SPACE

What are the results of comparing the patterns of fluctuation in numbers in different interaction groups over years, to test whether they are the best population units for studying the dynamics of carabid beetles? Our standard pitfall sampling enabled us, over many years, to estimate the annual relative densities (year-catches) of some carabid species occupying differently structured sites in the entire Heath of Kralo. Taken together, the interaction groups of each species can be considered a random sample of all interaction groups of that species covering the whole area and thus forming a multipartite population, i.e. a population that consists of a continuum of mutually merging interaction groups (section 9.2 of Andrewartha and Birch, 1984).

Let us take *Pterostichus versicolor* as an example (see Fig. 3.5 for details). Figure 3.6 shows the fluctuations in numbers over years of 10 interaction groups, and of the total (multipartite) population that can be considered to be composed of these 10 interaction groups. Section 4.4.3 of Den Boer and Reddingius (1996) showed that the limits between which the density was fluctuating (log-range=*LR*) in the multipartite population (consisting of six of these interaction groups sampled for at least 25 years) was reduced to 58% of that of an average interaction group (with mean *LR* values). Such an increased range stability for a multipartite population, as compared to an isolated interaction group, results in an increase in its expected survival time under stationary conditions, in the present case, an increase by a factor of >10 (see Den Boer, 1981).

For students who like to draw conclusions from the application of the current density dependence tests, it might be interesting to mention that

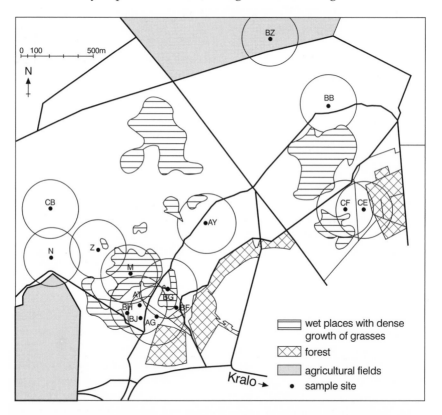

Fig. 3.5 Some interaction groups of the carabid beetle *Pterostichus versicolor* occupying areas of about 12 ha each (circles) around a standard set of pitfalls at Kralo Heath, Dwingelder Veld, the Netherlands. The heath area is uninterruptedly populated by this species. Radioactively marked beetles of this species cover about 180–190 m during a year (as the crow flies). For the carabid beetle *P. lepidus* these interaction groups will cover about the same areas.

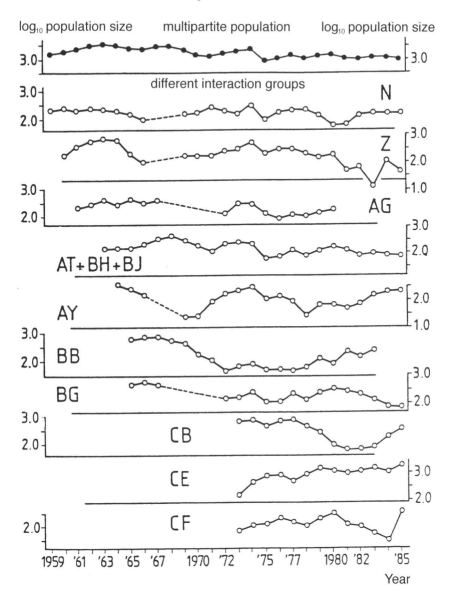

Fig. 3.6 Fluctuations of year-catches in standard sets of pitfalls in different inter-action groups (subpopulations) of the carabid beetle *Pterostichus versicolor* at Kralo Heath, Dwingelder Veld, the Netherlands. At top, fluctuations in the multi-partite population composed of these interaction groups. Capital letters indicate different sample sites pictured in Fig. 3.5. Data from Van Dijk and Den Boer (1992).

over many years the densities in one, but only one, of the interaction groups (group N in Fig. 3.6) scored significantly with Bulmer's (1975) first test, and almost significantly ($P=0.054$) with the test of Pollard *et al.* (1987). In my opinion, this was not because this interaction group, and none of the others, had been regulated by some density-dependent process(es), but because, by chance, the numbers of adults did not show a distinct trend during these 25 years. This suggests that density has in some way been kept between limits. From our work with this species, no indications of stabilizing density-dependent effects were found. On the contrary, both larvae and adults are polyphagous predators, individually hunting all kinds of available arthropods, so that intraspecific competition for food does not seem a very probable option, the more so because a number of carabid species of about the same size are living in the same sites and are hunting the same kinds of arthropods (Hengeveld, 1980).

The significant negative correlation between mean density and the mean number of eggs in the ovaries, as found in another interaction group (AG) by Baars and Van Dijk (1984), was an indirect one, and most probably resulted from changes in food quality which are known to have a strong effect on the egg production of carabid beetles (Van Dijk, 1996; personal communication), and which by chance might have been correlated with changes in density (see also Den Boer, 1986a). Intraspecific competition for food is not a very probable hypothesis in this case, because many kinds of potential prey species were found in this site (to judge from other arthropods both caught in these pitfalls and found as remnants in the intestines of *P. versicolor*; Hengeveld, 1980). Also, the density of *P. versicolor* in AG was relatively low as compared with that in other interaction groups sampled (Fig. 3.6).

It must be noted that the multipartite population of this species scored nearly significantly ($P=0.062$) with the permutation test of Reddingius and Den Boer (1989). As the multipartite population is heterogeneously composed of a number of different interaction groups, this will not result from the succession of net reproduction values being favoured by some density-dependent process(es). We should realize that the multipartite population shown is a rather arbitrary collection of interaction groups, but this is only used to illustrate the possible effects of spreading of risk in space. Hence, by chance, the combination of some interaction groups may have resulted in a relatively favourable succession of net reproduction values. However, this spreading of risk is not optimal, for the fluctuations of numbers in the different interaction groups are not independent: the coefficient of concordance (W of Kendall, 1955) being 0.22 ($P\approx0.01$). Strictly speaking, we cannot exclude the possibility that density-dependent processes, which affect the course of numbers over time, are also operating in some interaction groups, but we would not expect a significant stabilizing effect from such processes; see also Wolda and Dennis (1993); Wolda et al. (1994).

Figures 3.7 and 3.8 show that in the species *Pterostichus lepidus* (with interaction groups occupying about the same area as *P. versicolor*) and *Pterostichus diligens* (with interaction groups covering much smaller areas), the spreading of risk in space over interaction groups has similar results as in *P. versicolor* (33 and 40% reduction of *LR*, respectively, in the multipartite population). In these three species, the variation of net reproduction within interaction groups over years is about the same as the variation of net reproduction within years over interaction groups, and therefore the latter variation can approximately compensate for the former, provided there is sufficient exchange of individuals between interaction groups; see also section 4.4.3 and Table 4.6 of Den Boer and Reddingius (1996). The exchange of individuals between adjacent interaction groups is not known exactly, but is estimated to be between 5 and 20% of the individuals (see Den Boer, 1991), which is sufficient to level out differences in density fluctuations between adjacent interaction groups.

Although the above examples of spreading of risk in space were established by studying a number of interaction groups, similar results would have been found by studying habitat patches (e.g. Hassell, 1987) or subpopulations of a metapopulation (e.g. Hanski, 1991), or some other grouping in which individuals are expected to interact. The only necessary condition is that there is sufficient exchange of individuals between asynchronously fluctuating groups to contribute a certain degree of spreading of risk, by levelling out fluctuations of density and in preventing local extinctions (compare Den Boer, 1991). In fact, the subpopulations model of Reddingius and Den Boer (1970) is a general model, illustrating spreading of risk in space by exchange of individuals between groups, whether these groups are patches, interaction groups or subpopulations of a metapopulation. Note that, as shown by simulations, the degree of spreading of risk, i.e. the reduction of fluctuations in the total population by exchange of individuals, is not affected significantly by whether the exchange is density-dependent or density-independent.

In the species *Amara lunicollis* (Fig. 3.9) and *Calathus melanocephalus* (Figs 3.10 and 3.11) the resulting stability of the multipartite population is much smaller (15 and 23% reduction of *LR*, respectively), because the variation of net reproduction values within interaction groups over years is much greater than the variation within years over interaction groups, so that the latter variation cannot compensate for the former. In particular, in *C. melanocephalus* the different interaction groups fluctuate almost in parallel and have a coefficient of concordance (*W*) of 0.65 ($P<<<0.001$). For an explanation of these synchronous fluctuations in different interaction groups, rarely observed among carabid beetles, see Van Dijk and Den Boer (1992).

In spite of the significant negative correlation between mean density and numbers of eggs in ovaries found by Baars and Van Dijk (1984) in

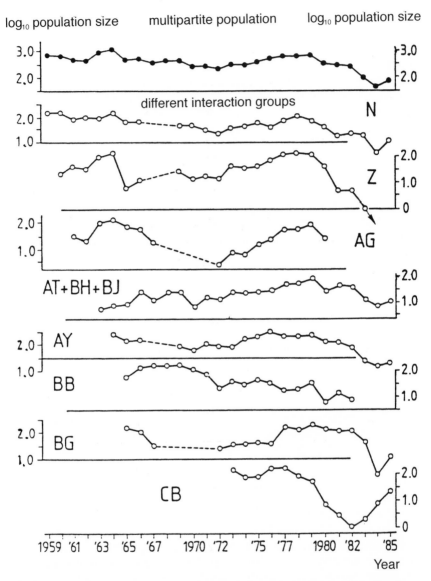

log₁₀ population size   multipartite population   log₁₀ population size

Fig. 3.7 As in Fig. 3.6, for the carabid beetle *Pterostichus lepidus* at Kralo Heath, Dwingelder Veld, the Netherlands. Areas of interaction groups are pictured in Fig. 3.5.

two interaction groups of *C. melanocephalus* (AG and [AT+BH+BJ], Fig. 3.10), neither the separate fluctuation patterns of any of the interaction groups sampled, nor that of the multipartite population, scored significantly with any of the tests for overall density dependence proposed in

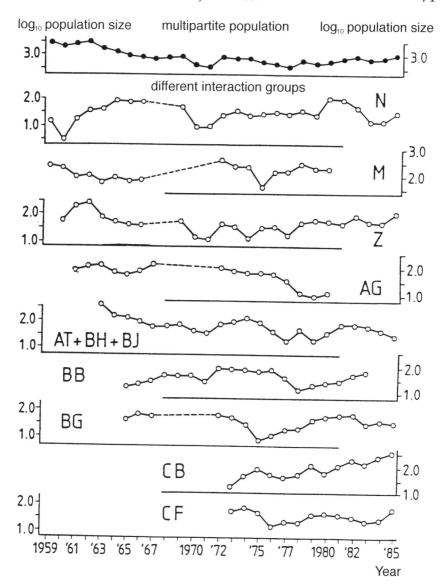

Fig. 3.8 As in Fig 3.6, for the carabid beetle *Pterostichus diligens* at Kralo Heath, Dwingelder Veld, the Netherlands. Areas of interaction groups will be about as pictured in Fig. 3.10.

the literature in recent decades. As expounded by Den Boer (1986a), this lack of density dependence in the fluctuation patterns of this species must be caused by the variation in survival of larvae over years being many times greater than the variation in size of egg production over

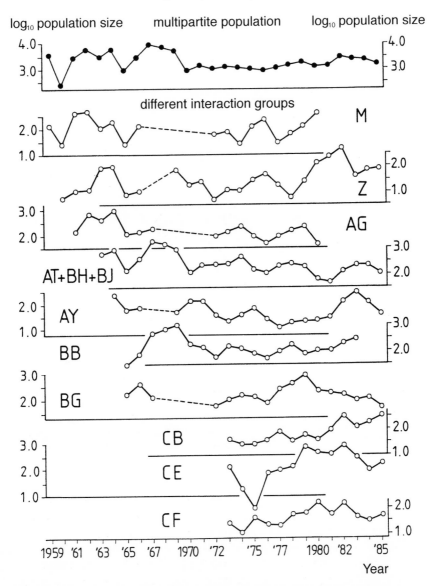

Fig. 3.9 As in Fig. 3.6 for the carabid beetle *Amara lunicollis* at Kralo Heath, Dwingelder Veld, the Netherlands. Areas of interaction groups are not exactly known for this species.

years, so that the latter variation is almost completely overwhelmed by the former. Just as expected in *P. versicolor* (see above), the density-dependent egg production in this species probably results from changes in the quality of food species.

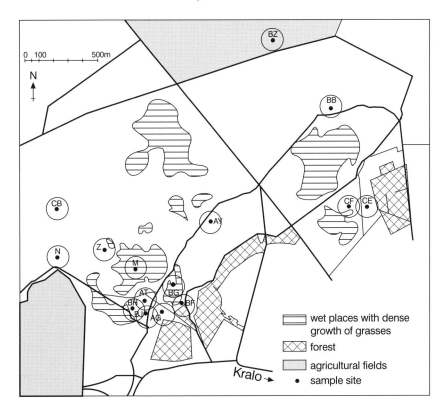

0 100   500m

N

BZ

BB

CB

CF CE

AY

N

Z

M

A
AT BG
BN BF
BJ AG

wet places with dense
growth of grasses

forest

agricultural fields

• sample site

Kralo →

Fig. 3.10 Some interaction groups of the carabid beetle *Calathus melanocephalus* at Kralo Heath, Dwingelder Veld, the Netherlands. Each interaction group occupies an area of about 2 ha, circles with a radius of 80 m around a standard set of pitfalls.

The importance of food quality is indicated by Fig. 3.12, which shows that low numbers of eggs in the ovaries not only coincide with high densities of adults, but also with outbreaks of the heather beetle (*Lochmaea suturalis*). These outbreaks severely damage the heather and result in low quality food for other phytophages on the heather (Den Boer, 1986b). As such phytophages (especially larvae) are important prey species for *C. melanocephalus*, the adults of this species necessarily consume low quality prey. In addition, the abundant larvae of the heather beetle are of little value, because these are more or less toxic to the adults of *C. melanocephalus*. Although they do not prefer the larvae of the heather beetle, adults can hardly avoid eating them, because they actively crawl in high numbers on the soil surface in search of undamaged heather plants. Consumption of these larvae results in almost complete cessation of egg production and sometimes even in premature death (section 5.5.4 of Den Boer and Reddingius, 1996).

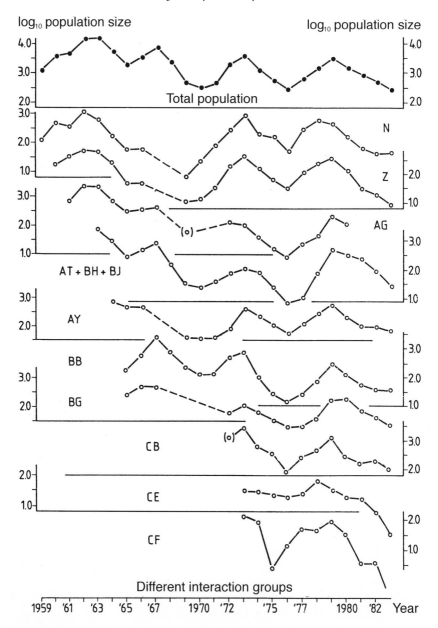

Fig. 3.11 As in Fig 3.6 for the carabid beetle *Calathus melanocephalus* at Kralo Heath, Dwingelder veld, the Netherlands. Areas of interaction groups are pictured in Fig. 3.10.

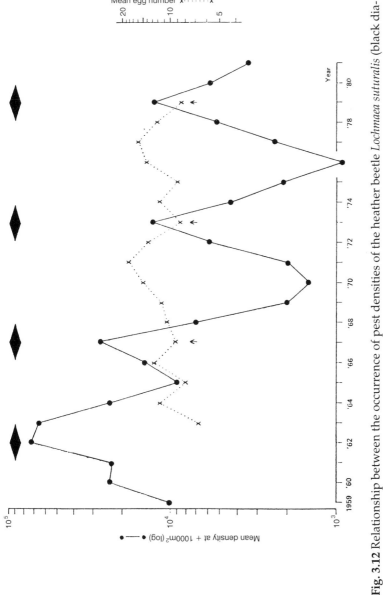

Fig. 3.12 Relationship between the occurrence of pest densities of the heather beetle *Lochmaea suturalis* (black diamonds above the graph), densities (year-catches; black dots), and mean numbers of eggs in ovaries in the carabid beetle *Calathus melanocephalus* (crosses) at Kralo Heath, Dwingelder veld, the Netherlands. Mean density (year-catch) is significantly correlated with mean number of eggs in ovaries ($r_s = -0.667$, $P=0.0076$). Model II regression (Bartlett) eggs/female=13.2117−0.16297 density/m².

The fact that a natural population is composed of subpopulations (inter-action groups) with differing frequencies of genotypes and age classes, liv-ing under different conditions, suggests that the fluctuations of a natural population cannot be satisfactorily represented by large-number models, neither deterministic nor stochastic (pseudodeterministic); see section 2.2.4 of Den Boer and Reddingius (1996). Thus, the argument sometimes advanced in favour of the use of large-number approximations, that the variation of the sum of random variables is the sum of the variances of those variables plus twice the sum of all covariances, does not apply to interaction groups of a multipartite population, because the changes in numbers in such groups are a mixture of additions and multiplications (section 2.2.5 of Den Boer and Reddingius, 1996). Partly due to this, we pre-fer to work with variances of net reproduction values, in the way demon-strated in Tables 4.4 and 4.6 of Den Boer and Reddingius (1996).

I realize very well that my work with interaction groups was more or less imposed on me by the sampling technique (pitfalls) used. In many other animal species it will not be easy to recognize the best units of pop-ulation to study, i.e. interaction groups of individuals largely averaging the effects of small local differences in conditions by their normal mobili-ty, such as occurs in carabid beetles. Nevertheless, with sufficient knowl-edge of the behaviour of the animals to be studied, it should be possible to find the right scale of population units for studies of population dynamics. In cases of doubt, such units of population are better too small (e.g. patches) than too large. Concerning the latter, it must be noted that many subpopulations of a metapopulation (Hanski, Chapter 1 in this vol-ume) may be multipartite populations covering a number of interaction groups. Considering the combined changes in numbers of differently fluctuating interaction groups as the fluctuation pattern of such a sub-population may lead one either to overlook the most important stabiliz-ing processes (spreading of risk), or to attribute this pattern wrongly to other (e.g. density-dependent) processes. Combining changes in num-bers of some almost identical, or largely overlapping, interaction groups (as occurred in *P. versicolor* in Fig. 3.5) may, in the worst case, result in too pessimistic an impression of the stability of the multipartite population. Since we will not usually be able to study all units of a natural population that can be distinguished, our impression of the stability of the multipar-tite population will in any case be too pessimistic. Hence, in these discus-sions we have tended to underestimate the effect of spreading of risk on the stability of natural populations.

3.6 SPREADING THE RISK OVER INDIVIDUALS

It will be evident that differences between the fates of individuals of the same group may have a stabilizing effect on the changes in numbers of such a group, because the risk is spread over individuals with different chances of death and/or reproduction. It is difficult, however, to illustrate this effect in a natural interaction group, because we do not have long series of reliable data on changes in the frequencies of genotypes or age classes in insect populations, comparable with the statistics on which insurance premiums are based. A first step in this direction was taken by Den Boer *et al.* (1993) by recording the changes in frequencies of two phenotypes (with a genetic basis) in interaction groups of the carabid beetle *Pterostichus oblongopunctatus*, but it appeared impossible to show directly the stabilizing effect of these changes, because the time series was too short to allow an analysis like that in Tables 4.4 and 4.6 of Den Boer and Reddingius (1996).

From Van Dijk (1979, 1982) we know that egg production of females of *P. versicolor*, and other carabid species, differs with age (see also section 2.3.9 of Den Boer and Reddingius, 1996), older beetles laying significantly more eggs on average than young beetles. Therefore, changes in age composition over years may affect the numbers of eggs laid, although it will not be easy to show such effects under field conditions. Even the fact that some of the first-year beetles survive the winter and reproduce again in the next year (iteroparity) must have a stabilizing effect on adult numbers. We can demonstrate this effect in two carabid species, in which old beetles were distinguished from young beetles (see Baars and Van Dijk, 1984), by simulating the situation in which no old beetles survived the first year's reproduction, i.e. are semelparous. Of course, the production of young beetles will be lower if the contribution of old beetles is neglected, so we had to compensate for these losses, as described in section 4.4.2 of Den Boer and Reddingius (1996), in order to keep simulated numbers comparable to those observed in the field. The results of these simulations are shown in Fig. 3.13: both LR and VarlnR in the one age-class simulation are about twice those of the values estimated in the field series (see Table 4.2 of Den Boer and Reddingius, 1996).

As natural populations usually show high genetic variability and natural environments are heterogeneous in many respects, the stabilizing effects of spreading the risk over individuals can hardly be overestimated, despite the fact that these effects are less easy to demonstrate than those shown by insurers for human populations. This kind of risk spreading must be very general, and together with the effects of spreading of risk in space, it could well have given many entomologists the impression that natural insect populations must be regulated. It should be noted that under certain conditions, iteroparity may result in statistical density dependence because of the transfer of quantitative effects from one generation to following ones. Therefore, Den Boer and Reddingius (1989) restricted tests to time series of species that are both univoltine and semelparous. I hope I have shown that, even without the assumption of a

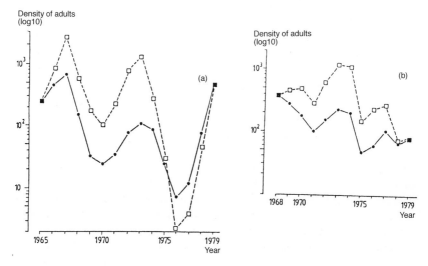

Fig. 3.13 Density fluctuations (year-catches) in field populations (interaction group 'grassy field', [AT, BH, BJ], see Figs 3.5 and 3.10) of two carabid beetles in which more than one year-class participates in reproduction (black dots), as compared with the fluctuations if only a single year-class had reproduced in such a way that mean net reproduction stayed the same (open squares). (a) *Calathus melanocephalus* (Figs 3.10 and 3.11); (b) *Pterostichus versicolor* (Figs 3.5 and 3.6). Data from Appendix of Baars and Van Dijk (1984).

general occurrence of regulation, stability of insect populations is not inexplicable (Nicholson, 1933).

Apart from spreading the risk over individuals, the risk of large changes in numbers may also be spread in other ways. For instance, a polyphagous predator spreads risk over different prey species, so reducing the chance of food shortage. The price for this is, of course, that a polyphagous predator cannot be expected to develop a highly specialized hunting behaviour. Something similar will apply to generalized feeders at all trophic levels.

3.7 THE STABILITY OF THE FLUCTUATIONS OF INSECT NUMBERS

Now we have seen that, apart from the remote possibility of strict regulation of numbers, other situations and processes may contribute to a certain degree of stability in numbers over time, we should ask ourselves "How stable are insect numbers in general?". As I have intensively studied mainly carabid populations, I can answer this question only for a number of carabid species in Drenthe. However, there are no reasons why carabid beetles should be so special as not to be representative of many other insect groups. Therefore, I believe my findings for carabid beetles may apply to insect species in general and possibly also to many other animal species.

In the course of 30 years, I have been able to follow the fluctuations in numbers for interaction groups of 64 of the more abundant carabid species in the province of Drenthe (Den Boer, 1985, 1990b). First of all, I established that the coefficients of net reproduction, estimated from different interaction groups of the same species, cover the same ranges of values (Mann–Whitney test), so that these values could be put together as a frequency distribution specific for the species. For none of the 64 species did the frequency distribution of net reproduction values differ significantly (χ^2 test) from the fitted log-normal (see Reddingius's Appendix to Den Boer, 1985). Therefore, the pattern of fluctuation of each of these species could be characterized by mean ln$R$ and SD ln$R$ of the fitted log-normal (examples given by Den Boer, 1990b). These data enabled me to simulate random walks of densities of the same lengths as those observed in the field for each of the 64 species, and compare the *LR* found in the field with the mean *LR* value of 500 simulated random walks of density of the same number of years, for each interaction group of that species.

In 38 of these species, the *LR* values of interaction groups, observed in the field during 8 years or longer, did not differ from the values expected when the numbers had fluctuated from year to year, as in random walks of densities (examples given by Den Boer, 1990b). In 22 other species, the numbers fluctuated significantly between even wider limits than expected with random walks of densities, and in only four species were the fluctuations in numbers in the field smaller than those in relevant random walks of densities, a result that could almost have been expected by chance ($P=0.06$).

From these data, I concluded that the extent of fluctuations of density in interaction groups of carabid species either are similar to, or greater than, those of random walks of densities of the species concerned. I conclude from this that interaction groups of carabid beetles, and most probably of other insect species, are not kept within limits (stabilized) by density-dependent (regulating) processes, but largely fluctuate under the influence of density-independent forces. As stated earlier, this does not mean that density-dependent processes, such as density-dependent egg production (Baars and Van Dijk, 1984), do not occur. I merely believe such processes usually do not contribute significantly to keeping density within limits (Varley *et al.*, 1973), i.e. they are taken up in the general fluctuation pattern that is dominated by density-independent processes.

If density-independent processes dominate, it would be expected that increases in the range of densities in time (IR) in random walks and in time series of densities from field populations would be similar. If one compares the *LR* value of the first 5 years of observation with that after 18 years (i.e. $IR = (LR_{18} - LR_5)/LR_5$), the increase of IR in time, for 13 random walks of densities, appears to vary between 0.34 and 2.446 (Den Boer, 1991). This variation is smaller, but not significantly so (Mann–Whitney, $P=0.082$), than for interaction groups of carabid beetles observed in the field (Den Boer, 1991). For a number of other time series from the literature, including other kinds of insects, and populations claimed to be regulated such as winter moth, pine looper, great tit, muskrat, bobwhite quail, etc. (Den Boer, 1991), IR increased in time at a rate similar to that in random walks (Mann–Whitney, $P=0.58$). The lowest IR value found so far was in the great tit population studied by Kluyver (1951): a value of 0.03, a population that might be considered regulated in a broad sense, i.e. each year sufficient young reproducing birds are produced (in nest boxes) to compensate for the losses in the current generation, and density is efficiently reduced by territorial behaviour.

In summary, I conclude that regulation of numbers in the sense of keeping density within limits (Varley *et al.*, 1973), if it exists, must be an exceptional phenomenon, particularly for insect populations which, being poikilothermic, are always strongly affected by density-independent forces (Uvarov, 1931). On the other hand, there are many generally occurring phenomena that can be considered aspects of heterogeneity in both population and environment, which might result in a smaller or greater stabilization of numbers over time, but these stabilizing forces are not connected in any way with an equilibrium density. Further arguments and data are given by Den Boer and Reddingius (1996).

ACKNOWLEDGEMENTS

I thank Hans Reddingius (Haren, the Netherlands) and Tom White (Adelaide) for comments and for correcting my English. This contribution is Communication No. 568 of the Biological Station, Wijster, the Netherlands.

4

Resource limitation of populations and ceiling models

J. P. Dempster

4.1 INTRODUCTION

The role of resources, particularly of food, in limiting population growth has been recognized by ecologists for more than 150 years. Indeed the first ceiling model, the logistic equation, describing population growth in a finite environment, is attributed to Verhulst (1838). In this early model,

$$dN/dt = Nr(1-N/K)$$

N is the population size, K the 'carrying capacity' of the habitat, and r the 'intrinsic rate of increase' of the population (see definitions on page xix). dN/dt is the actual growth rate of the population at any particular population size, and this declines to zero as N approaches K, and resources are fully used (Fig. 4.1). In the simple deterministic model that was first proposed, K is the upper asymptote of N, but it is of course possible for N to exceed K temporarily if, for example, there is a time lag in the response of dN/dt to changes in population size, or if stochastic variation of K is introduced into the model. The logistic equation demonstrates the density dependence of population growth rates as a result of intraspecific competition for resources, with growth declining to zero as the carrying capacity of the habitat is approached.

The term competition was used very loosely by early writers on population ecology, and indeed, Nicholson (1933, 1954), the originator of the equilibrium theory, included predation under this term. It is basically Milne's (1961) definition that we are using, and I will repeat it here because competition is so central to the main theme of this chapter.

Insect Populations, Edited by J. P. Dempster and I. F. G. McLean. Published in 1998 by Kluwer Academic Publishers, Dordrecht. ISBN 0 412 83260 7.

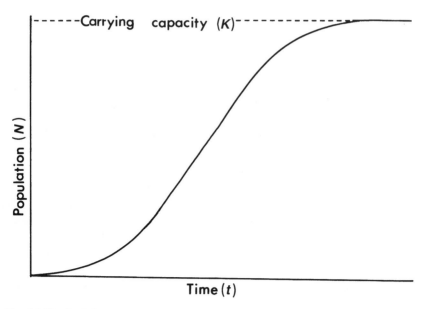

Fig. 4.1 The logistic curve.

Competition is the endeavour of two, or more, individuals to gain the resources that they require, when the supply of those resources is insufficient for both, or all. For competition to occur, one individual gaining its needs must reduce the chances of others doing likewise, so that competition must be density-dependent. Competition for resources is the only certain density-dependent process affecting all populations. It is an inevitable result of populations approaching the carrying capacity of their habitats. Milne (1957a) made another important contribution to our understanding of the way in which competition is likely to operate, by pointing out that, by definition, competition will come into effect only above the density at which one individual affects another's ability to obtain its needs. Thus, below that density, intraspecific competition can have no effect on mortality, reproduction or dispersal, and so cannot influence the likelihood of a population becoming extinct. With this in mind, Milne (1957a, 1962) proposed that populations fluctuate within certain limits which are imposed by the environment, the upper limit being determined by the carrying capacity of the environment, while the lower limit is determined by the density below which extinction is inevitable. He argued that any density dependence that might exist in the action of natural enemies was not sufficiently consistent or strong enough to prevent populations of their prey from approaching the carrying capacity.

This is a very different concept to that of regulation (Nicholson, 1933, 1954), in which any change away from the theoretical equilibrium density is countered by change in the overall growth rate of the population, thus reducing the likelihood of extinction.

Whilst Milne's papers did much to dispel the erroneous idea that intraspecific competition can cause density-dependent effects at all densities, they led to the view that competition can have an impact only at very high densities. Many writers, including Milne, consider that the majority of populations of most species exist at densities well below those where resources become limiting, so that for most of the time, intraspecific competition can play no part in their natural control. I believe that this is an over-simplification of the way in which resources actually limit population growth, and as I hope to demonstrate in this chapter, I consider that most insect populations are limited, for much of the time, by the resources available to them, even though their densities may often appear to be very low. Nevertheless, I think that Milne was correct in maintaining that competition for resources cannot regulate populations at low densities, and so cannot influence the probability of population extinction when numbers become low.

Resources may be in short supply and yet be quite unaffected by the number of insects searching for them. In such cases, the shortage is brought about as a result of poor searching ability relative to the distribution of the resource. This led Andrewartha and Browning (1961) to distinguish between a density-dependent, absolute shortage (when competition is occurring) and a density-independent, relative shortage (when competition is not occurring). This is an important distinction, although in practice it is often extremely difficult to determine when an insect species has reached a density at which it affects the availability of its resources. As with any process affecting populations, there may be considerable heterogeneity in the extent of crowding within a population, so that, initially, competition may have a small impact on the population, and so be difficult to detect. Nevertheless, there will always be a density below which the availability of resources will not be density-dependent. This does not mean that resource availability is unimportant below that density, but simply that it will be independent of density. As we shall see later, relative shortages of resources can have considerable impacts on population growth rates.

The likelihood of intraspecific competition occurring is frequently underestimated by entomologists. This is particularly true for phytophagous insects. Often, population limitation is a direct result of poor quality of available food (White, 1993), and this has led to the evolution of highly selective feeding preferences in many species. As a result, it is easy to overestimate the availability of food resources, and so underestimate the likelihood of competition. I shall come back to this point.

The effect that intraspecific competition has on a population depends on the behaviour of the species and the extent to which resources are partitioned between individuals. Nicholson (1954) recognized two types of competition, which he called 'contest' and 'scramble' competition. In contest competition, the winner obtains as much of the resource as it requires for survival and reproduction, and the loser relinquishes the resource to its successful competitors. Thus, some individuals high in the pecking order will obtain their needs, even when resources are in very short supply. In contrast, all individuals are approximately equal in scramble competition, and so a free-for-all results, which can lead to very high mortalities. Populations will fluctuate far more under the influence of scramble competition than with contest competition.

Sometimes, the availability of suitable food is sufficiently low to cause crashes in population size, through scramble-type competition, as seen, for example, in the cinnabar moth (*Tyria jacobaeae*), when it defoliates its food plant *Senecio jacobaea* (Dempster, 1971b, 1982: see also Van der Meijden *et al.*, Chapter 12 in this volume). Under such circumstances, high death rates are normally found only in the less mobile, larval stages of an insect's life cycle. In the more mobile adult stage of insects, emigration is more likely to occur than starvation (see Dempster, 1968).

Extensive defoliation of food plants, as seen in *Tyria*, is rather rare, but this does not mean that competition is rare. Larval food supply is often largely determined by the egg-laying behaviour of the parent insect, and as we shall see later, competition can occur at this earlier stage, without the dramatic effects of starvation. This, too, can lead to underestimation of the likelihood of competition having occurred.

Behavioural responses to density appear to have evolved in many species because of the harmful effects of scramble competition, and these can play so important a role in population dynamics that some ecologists treat them as a separate group of processes, distinct from resource limitation. I prefer to think of them as a substitution of competition for resources by competition for space. Thus, territorial behaviour is an extension of contest competition in which an individual competes for the space containing its needs for survival and reproduction rather than for specific resources. In many vertebrate animals, territorial behaviour can lead to very stable, low-density populations, but in insects it usually operates at a far smaller scale and simply results in a spacing out of the individuals in a population. Nevertheless, resource partitioning between individuals plays an important role in population limitation of many insect species, and an example of this is described below, in the orange-tip butterfly, *Anthocharis cardamines* (section 4.4).

4.2 POPULATION LIMITATION BY FOOD RESOURCES

There are many examples in the literature which show how population growth is tied to the availability of food resources (e.g. Dempster and

Pollard, 1981), and White (1993) presents convincing evidence that many animal populations are limited by an inadequacy in the quality of their food. He considers that shortage of nitrogen is commonly the factor limiting population growth. Of course, food quality and quantity are closely linked: if food quality is poor, larger quantities are required to provide an adequate diet.

Further evidence in support of the resource limitation of insect populations can be seen in a recent study (Dempster *et al.*, 1995a, b) of the flower-head faunas of two herbaceous plants, *Centaurea nigra* and *Arctium minus*. In this study, populations of nine phytophagous insects and seven insect parasitoids were monitored on more than 50 patches of each plant in an area of 5 km². *Arctium* is a monocarpic plant, individual plants taking several years to reach a flowering size, and then dying after flowering, but *Centaurea* is a long-lived perennial. However, in both species of plant there were large changes in the number of flower-heads per patch (i.e. in the food resource for the larvae of the phytophagous insects) between years.

Local extinctions were found to be a regular feature of all of these insect populations, especially on the smaller patches of plant, but these extinctions were rapidly made good by recolonization by adult insects.

Each year the sizes of all the phytophagous insect populations (log no./patch) were significantly correlated spatially, with their food resources (log no. flower-heads/patch) (see Chapter 10, Fig. 10.3), whilst all the parasitoid populations were similarly correlated with their food supply (log no. larval hosts). What is more, changes in population sizes between years (population growth) were significantly correlated with changes in their food resources, in six out of nine herbivores, and seven of seven parasitoids (column 5, Tables 4.1 and 4.2). In all cases, the regression coefficients were positive.

Multiple regression analysis was used to look for density dependence in population growth from one year to the next, with log change in resources as a second explanatory variable. Density was found to have a significant effect on the population growth rates of nine of the 16 species (Tables 4.1 and 4.2), and the regression coefficient was negative for all species.

The use of regression analysis to test for density dependence in population data can be criticized on the grounds that successive counts in a time series are not independent of each other, and that sampling errors in population estimates can cause a false identification of density dependence (see Rothery, Chapter 5 in this volume). The first objection does not apply to these data, as each data point is for a different plant patch, thus avoiding any serial correlation. The second criticism could be more important, and a simulation study using the actual data and their standard errors showed that sampling errors did, in fact, increase the levels of significance for the effects of density in the regression analyses. However, this effect was not large enough to bring the significance below the 5% level, except in one species, *Metzneria metzneriella* (Table 4.1).

Table 4.1 Data for *Centaurea nigra*: percentage variance accounted for by multiple regression of log change in numbers (y) against log initial density (number per flower-head or host, x_1) and log change in resources (flower-heads or hosts, x_2).

| Species | n | Density x_1 | | Change resource x_2 | | |
		x_1 alone	x_1 after x_2	x_2 alone	x_2 after x_1	x_1+x_2
Urophora jaceana	17	27.6*	30.6**	31.5*	34.5*	62.1**
Chaetorellia jaceae	16	16.0	11.3	54.0**	49.3***	65.3***
Isocolus jaceae	14	17.9	3.1	81.4***	66.6***	84.5***
Eucosma hohenwartiana	18	37.6**	39.8**	12.4	14.6*	52.2**
Metzneria metzneriella	18	5.9	22.3*	16.8	33.2*	39.1*
Eurytoma tibialis	17	62.0***	39.9***	31.5*	9.4*	71.4***
Habrocytus albipennis	4	20.8	2.9	95.2*	77.3	98.1
Habrocytus elevatus	9	59.4*	12.2	71.2**	24.0*	83.4***
Torymus cyanimus	6	58.4	12.5	67.0*	21.1	79.5
Tetrastichus sp.	15	52.1*	34.2**	36.3*	18.4*	70.5*

* $P=0.05$; ** $P<0.01$; *** $P<0.001$.

Table 4.2 Data from *Arctium minor*: percentage variance accounted for by multiple regression of log change in number (y) against log initial density (number per flower-head or host, x_1) and log change in resources (flower-heads or hosts, x_2)

| Species | n | Density x_1 | | Change resource x_2 | | |
		x_1 alone	x_1 after x_2	x_2 alone	x_2 after x_1	x_1+x_2
Cerajocera tussilaginis	14	39.5*	30.6***	56.9**	48.0***	87.5***
Tephritis bardanae	8	33.4	4.1	60.4*	31.1	64.5
Aethes rubigana	17	2.2	38.8***	36.7**	73.3***	75.5***
Metzneria lappella	16	49.6**	46.7**	17.5	14.6*	64.2***
Habrocytus albipennis	12	0.3	2.9	81.5***	84.1***	84.4***
Bracon minutator	12	23.3	22.0***	68.2***	66.9***	90.2***

* $P<0.05$; ** $P<0.01$; *** $P<0.001$.

One would not necessarily expect to find density dependence in the population growth rates of all of these insects. For example, *Tephritis bardanae* over winters as an adult, away from the *Arctium* patch on which it was reared, so it must find and colonize a new plant patch each year.

Thus, numbers on a patch in one year will be independent of those in the previous year, and so one would not expect to find density dependence in this species at the patch scale. A population of this species will cover many patches.

As has been shown by several authors, the likelihood of detecting density dependence increases with the number of generations studied. This is to be expected on purely statistical grounds, but if competition is responsible for the density dependence, failure to demonstrate it may simply reflect a high number of points below the threshold when competition comes into play.

The regression model of the combined effects of density and resources on population growth was statistically significant in 13 of the 16 species, with generally high amounts of the variation accounted for (last column, Tables 4.1 and 4.2). The three exceptions were those with the smallest number of data points ($n < 9$). In two of these, *Torymus cyanimus* (Table 4.1) and *Tephritis bardanae* (Table 4.2), the regression model was close to significance ($P = 0.09$), and in the third (*Habrocytus albipennis*), the regression was highly significant on *Arctium* ($n = 12$), but not on *Centaurea* ($n = 4$). There can be little doubt that populations of all these species are limited by resources, and this applies to herbivores and parasitoids alike.

4.3 'BOTTOM-UP' VERSUS 'TOP-DOWN' CONTROL

The parasitoids in the study described above attack the four species of Tephritidae (*Urophora*, *Chaetorellia*, *Cerajocera* and *Tephritis*). On individual patches of plants these parasitoids can cause high mortalities in their hosts, but in none of them could any density dependence be shown in their impact on their host populations, either spatially or temporally. Nor was there any significant improvement in the amount of the variation in population change accounted for by the multiple regressions, when log ($1-p$), the proportion surviving parasitism, was added to the regression analyses for the four tephritids, as a third explanatory variable. *Eurytoma tibialis* was considered by Varley (1947) to be capable of regulating the numbers of its host *Urophora jaceana* (see Straw, Chapter 10 in this volume), but clearly, these parasitoids actually play a very minor role in determining population change in their hosts from one year to the next, and certainly cannot regulate their numbers. Thus, there appears to be a one-way interaction between parasitoid and host; host numbers determine population size of the parasitoid, but not *vice versa*.

I think it is true to say that whenever food resources have been quantified, they have been shown to limit population size of the exploiting insects, whereas natural enemies rarely, if ever, appear to be capable of density-dependent regulation of their prey populations. This does not mean that natural enemies are incapable of greatly reducing the numbers of their prey (as the many examples of successful biological control show), but simply that their impact is not consistently density-dependent.

Resource limitation at one trophic level can have repercussions at two, or more, higher trophic levels. This can be seen in the population dynamics of *Cotesia* (*Apanteles*) *popularis*, a specific braconid parasitoid of cinnabar moth (Dempster and Pollard, 1981). *Cotesia* has a single generation each year. Adults lay their eggs into first- or second-stage caterpillars, and on hatching these feed within their host until the latter is fully grown, at which stage the host is killed. Normally, five or six larvae develop in a single host. The numbers of *Cotesia* followed those of its hosts (Fig. 4.2), but were particularly depressed in those years when cinnabar caterpillars defoliated their food plant, ragwort (*S. jacobaea*). In those years, many *Cotesia* larvae died when their hosts died of starvation, and there was a reduction in the mean number emerging per host, as host size was reduced. This was particularly marked in those years when the host's food ran out earliest, i.e. in years 3 and 8 (Fig. 4.2). Thus we have the situation where the numbers of *Cotesia* are being affected by changes, two trophic levels below, in the amount of ragwort present. Again, *Cotesia* appears to be having a negligible impact on the numbers of its host (Dempster, 1982).

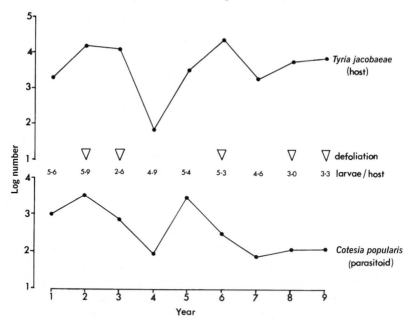

Fig. 4.2 Relationship between the number of fully grown larvae of *Cotesia popularis* each year and the number of hosts, cinnabar moth larvae, available to them. Arrows indicate the years when cinnabar larvae ran out of food; figures in parenthesis show the mean numbers of *Cotesia* larvae per parasitized host. Reproduced from Dempster and Pollard (1981) by courtesy of Springer-Verlag.

The superficial impression of a super-abundance of plant food, given by a 'green world', led Hairston *et al.* (1960) and their followers to claim that, whilst decomposers, plants, and carnivores are all limited by their food resources (i.e. by the trophic level below), herbivore populations are generally kept below the limits of their food by predators (i.e. by the trophic level above). In my opinion, there is no reason for thinking that insect herbivores are any different to other organisms in being limited by resources.

There are five features of insect biology that, in my opinion, have led to an underestimation of the importance of limitations in food availability to insect–herbivore populations, namely:

the very specialized needs of many insect herbivores;

the inadequate quality of much of the food that is available to them (see White, 1993);

the short time when suitable food is available, often as a result of lack of synchrony between the insect's life cycle and the food supply;

resource partitioning between individuals, which occurs in many species;

the extreme mobility of most insects, which frequently makes overcrowding transitory.

I will illustrate some of these features with the orange-tip butterfly, *A. cardamines*.

4.4 RESOURCE SELECTION AND PARTITIONING

Anthocharis cardamines is a widespread species in Britain. It has a single generation each year and its larvae feed on a range of cruciferous plants. The butterfly lays its eggs on the opening flowers of its food plant, and the newly hatched larva eats the egg shell, then feeds on the flowers, before feeding on the developing seed pods (siliquae) of the plant.

Numbers of the butterfly's young stages were recorded on a permanent transect in Monk's Wood National Nature Reserve each year, from 1982–93, together with the number of flower-heads of *Cardamine pratensis*, its only food plant in the wood (Dempster, 1997).

The relatively high numbers of *Cardamine* flower-heads in the wood each year, compared to the very low densities of adults present, would lead one to believe that there were always sufficient plants to support the butterfly's population, and indeed, only a small proportion of the flower-heads is ever used by the butterfly. Over the 12 years, it is estimated that only 12.1% of a total of 11 484 flower-heads that were present, were used. However, a detailed study of the butterfly's reproduction in the wood showed that not all flower-heads are acceptable or usable by the butterfly, and egg laying is, in fact, limited every year by a shortage of suitable flower-heads available to the ovipositing females.

The female butterfly is extremely selective in its choice of plant for egg laying, choosing mainly the largest flower-heads growing in open, sunny locations. Egg laying was heavily biased towards the largest plants, and in all years the smallest flower-heads, with less than five flowers, were completely ignored. Larvae often died from starvation on small flower-heads, and none completed their development on heads containing less than seven flowers. Added to this, larval survival was greatly reduced if the flower-head was more than 8 days old at the time of egg laying, because the seed pods become too tough for feeding.

Over the 12 years, 84.6% of eggs were laid on sections of the transect receiving more than an estimated 100 mW/cm²/day radiation. Only 46.4% of the available flower-heads were on those sections of the transect, and there were no significant differences in the mean sizes of flower-heads on sunny and shaded sections.

Females also tend to avoid flower-heads already carrying a conspecific egg, in response to an oviposition-deterring pheromone laid down at the time of egg laying (Dempster, 1992). This pheromone is water-soluble, and so its effect is short-lived in the field. However, only one larva ever survives on a single flower-head, because the larvae are cannibalistic, the first to hatch killing others that it encounters.

The numbers and sizes of *Cardamine* flower-heads fluctuate enormously between years. Numbers on the transect have varied between 153 and 3724, whilst the mean size of these flower-heads has varied between 6.0 and 13.2 flowers, over the 12 years. In 4 years, the mean number of flowers per head was less than seven, the estimated minimum number required to support a larva through its development.

The number of eggs laid each year closely follows these changes in the number of flower-heads. A significant correlation exists between log number of eggs and log number of flower-heads ($r=0.72$, $P<0.01$). The correlation becomes even closer if one uses the number of suitable flower-heads in the analysis, instead of total number ($r=0.92$, $P<0.001$). For this, a suitable flower-head was defined as one with more than seven flowers, growing in an open, sunny section of a ride (i.e. receiving an estimated radiation of >100 mW/cm²/day). This relationship is shown in Fig 4.3.

A test for a causal relationship underlying this correlation was made in 1987, by planting 413 suitable flowering *Cardamine* into the transect. The provision of extra plants resulted in a significant increase in the number of eggs laid, from 21 on the wild plants to a total of 192 ($t=2.58$, $P<0.05$, Fig 4.3).

Each year, adults were seen flying through the wood long after the last egg was laid there, and laying often continued outside the wood on other

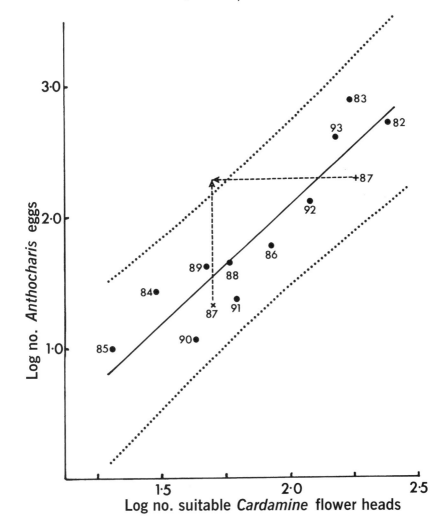

Fig. 4.3 Relationship between the numbers of eggs laid each year by *Anthocharis* females and the number of suitable *Cardamine* flower-heads available to them. The 95% confidence limits of the predicted numbers of eggs laid from the regression model ($y=-1.5278+1.8094x$) are shown. Two points are recorded for 1987: +, total flower-heads, including those planted; and ×, wild flower-heads only. The lines joining these points show how the provision of extra flower-heads has significantly increased the number of eggs above that estimated for wild plants alone. Reproduced from Dempster (1997) by courtesy of Springer-Verlag.

food plants, particularly, on *Alliaria petiolata*. In the year of the experiment, the last egg laid in the wood on a wild *Cardamine* was on 29 May 1987, but further planting prolonged egg laying until 19 June on the planted flower-heads. This suggests that females stayed in the wood, in that year, because of the provision of the extra plants.

The orange-tip butterfly is an active flyer, and evidence was obtained to suggest that the population in Monks Wood National Nature Reserve covers a much larger area than just the wood itself (Dempster, 1997). Adults move in and out of the wood, staying to breed when larval food plants are present in sufficient numbers.

An average-sized flower-head can support only one larva, and the butterfly reduces the likelihood of food shortage by appropriating flower-heads by means of an oviposition-deterring pheromone and by larval cannibalism.

Ecologists have long associated resource partitioning between individuals with higher animals, particularly vertebrates, but its widespread occurrence in insects is now becoming recognized. The use of pheromones at the time of egg laying, to mark and so to 'corner' food resources, was first identified in insect parasitoids, but it has now also been shown to occur in many phytophagous insects; in at least 25 genera, belonging to 15 insect families (Prokopy, 1981; McNeil and Quiring, 1983; Klijnstra and Schoonhoven, 1987; Kozlowski, 1989; Straw, 1989c; Imai *et al.*, 1990; Pittara and Katsoyannos, 1990). All are species that exploit patchily distributed resources, that can support only a limited number of individuals.

This study demonstrates several of the features listed on page 89. The butterfly is highly selective in its choice of larval food plant (size, age, insolation); there is a short period of time when a flower-head is usable (about 8 days); the flower-head resource is partitioned between individuals, so that only one larva has use of a flower-head; and all of this is achieved by a high mobility of the adult butterfly. However, the overriding impact of resources to this population would not have been appreciated, without a detailed understanding of the way in which the larval food resource is being used, and there have been few similar studies.

4.5 LIMITATION VERSUS REGULATION

Studies of the frequency of occurrence of density dependence in long-term population data have shown that it is present in many (but not all) data sets (Dempster, 1983; Stiling, 1987,1988; Woiwod and Hanski, 1992). However, the value of such studies is greatly reduced by the varied quality of long-term data. Frequently, population structure and movement have not been studied, and so the integrity of the population being investigated can not be assessed. Added to this, failure to detect density dependence may simply reflect inadequate or inappropriate data. Nevertheless, these studies do show that density dependence is a common feature of insect populations, and any population theory that does not take density-dependent processes into account will be incomplete. There is however, still disagreement amongst insect ecologists as to whether these processes are capable of regulating populations about an equilibrium density, or of simply limiting population growth.

Although the concept of population regulation about an equilibrium is still accepted by the majority of theoretical ecologists, it has never found universal acceptance by field ecologists. In my opinion, there are a number of reasons for this.

The idea of a 'balance in nature' is an attractive one, but in practice it is extremely difficult, perhaps impossible, to prove or disprove the existence of an equilibrium. Equilibrium values can be inferred only from population models, they cannot be measured in the real world. In practice, regulationists simply take the population mean as the equilibrium value, and identify regulation in model and field data by testing for increased stability. It is, of course, true that regulation will increase stability, but stability cannot be used as proof of regulation, because many types of process can lead to increased stability, including many that are independent of density (see Chapter 2). I think that this difficulty in testing the validity of the equilibrium concept undermines its usefulness in population ecology.

In my opinion, failure to distinguish between stability and regulation has been the main cause for the frequent redefining of the term regulation (e.g. Hassell, 1985; Murdoch, 1994); a problem that has been highlighted in recent years by a number of papers claiming that spatial patchiness (heterogeneity) in the attack of parasitoids can regulate populations of their hosts, even when no density dependence is detectable (e.g. Pacala and Hassell, 1991; Jones et al, 1993; but see also Dempster and Pollard, 1986). Heterogeneity has long been recognized as being capable of increasing population stability (Den Boer, 1968), but this is not regulation as defined here (page xx). Regulation requires that density has a temporal feedback upon population growth. Sometimes, spatial effects can result in temporal density dependence, but this is not necessarily so. In fact, it is generally rather unlikely that density-dependent processes in subpopulations will regularly survive the statistical summation into an overall population. Indeed, the stabilizing effect of heterogeneity acting through 'spreading of risk' actually depends on subpopulations fluctuating out of phase.

Finally, and perhaps most importantly, I think that more than 50 years of research on field populations has failed to demonstrate the universal presence of density-dependent processes that are capable of acting in the way required by the regulation theory. Although such processes frequently may contribute towards population stability, they are often so weak, or are density-dependent over such a small range of densities, that they will be incapable of consistently reversing population trends.

In contrast, a ceiling to population growth set by resources is a universal phenomenon in both plants and animals, and with knowledge of a species requirements, it can be measured in the field. Population limitation by resources has been demonstrated for very many species, and White (1993) clearly demonstrates the inadequacy of most food resources, especially plant food, to animals and shows the dominant role that this inadequacy has played in the evolution of feeding behaviour. Many species have evolved very specific food requirements, and competitive interactions can play a dominant role in their population dynamics, as we saw with the orange-tip butterfly. Resource partitioning between individuals can result in intraspecific competition at relatively low densities, but there will always be a threshold below which resource limitation is not density-dependent, and it can play no role in affecting the likelihood of population extinction.

In my opinion, ceiling models such as that proposed by Pollard and Rothery (1994) describe these processes far more realistically than those based on the equilibrium concept. Their simple stochastic model is based on a resource limit to population size, which cannot be exceeded, and a population growth rate that is influenced by resource availability. Simulations in which resource levels and intrinsic rates of increase were varied generally showed a strong synchrony between resource levels and population size. As resource levels and/or intrinsic rates of increase declined this synchrony was lost, and the model populations became highly unstable and liable to extinction, just as was found in the study of the flower-head faunas of *Centaurea* and *Arctium*. This is in sharp contrast with the findings from equilibrium models, in which high growth rates tend to result in instability, i.e. larger fluctuations about the equilibrium. In the resource limitation model, high growth rates keep the population close to the ceiling set by resources. The Pollard and Rothery model describes contest competition, which will lead to a closer synchrony between population and resource levels than will be obtained with scramble competition, but I think that it captures the important elements of population dynamics better than equilibrium models.

In conclusion, I believe that there is growing evidence that insect populations are consistently up against the ceiling set by resources, as seen by the strong correlations between population size and resource availability. Intraspecific competition is the ultimate cause of population decline as the population approaches the resource ceiling, but resource availability can also affect population growth rates at lower densities, through relative shortages. Growth rates will, of course, be affected by a range of density-independent and weakly or inconsistently density-dependent processes, but these will rarely have a dominant role in determining population size. If population growth rates become too low there is a high probability of population extinction.

ACKNOWLEDGEMENTS

The *Centaurea* and *Arctium* study was funded by the Natural Environment Research Council (Grant GST/02/472) as part of the Joint Agriculture and Environment Programme. Financial support for the orange-tip project was given by the Leverhulme Trust and the Natural Environment Research Council. The Nature Conservancy Council (since 1991, English Nature) kindly allowed me to do the orange-tip study in Monks Wood National Nature Reserve. I am also grateful to Ernie Pollard, Peter Rothery and Ian McLean for their constructive comments on this chapter.

5

The problems associated with the identification of density dependence in population data

Peter Rothery

5.1 INTRODUCTION

Density dependence occurs when the *per capita* growth rate of a population depends on its own density. This symposium defines direct density dependence as a decrease in the rate of population growth as population density increases, brought about by either a proportional increase in mortality and/or emigration, or a decrease in fecundity and/or immigration (as distinct from inverse and delayed responses).

Density dependence is central to the theory and application of population dynamics, and is used extensively in mathematical models to describe the effects of competition, predation and parasitism. In the absence of density dependence, populations either increase without limit, decline to extinction or display unbounded fluctuations. Some argue that density dependence is a logical consequence of long-term population persistence, so requiring no verification. Others have accepted a burden of proof, noting that populations go extinct locally and that theoretical populations can persist for long periods with no density dependence. Moreover, density dependence has often proved difficult to detect in insect populations (Dempster, 1983; Stiling, 1987, 1988). However, detecting density dependence is only a first step towards understanding population dynamics, and the problems of identification are wider than detection. There is also a need to identify the density-dependent biological mechanisms, the life-cycle stages and scales on

Insect Populations, Edited by J. P. Dempster and I. F. G. McLean. Published in 1998 by Kluwer Academic Publishers, Dordrecht. ISBN 0 412 83260 7.

which they operate, the strength of density dependence and interaction with density-independent factors, and to relate these effects to changes in overall population size.

This chapter discusses problems associated with the identification of density dependence from a statistical perspective, within a broader framework of analysing and modelling the dynamics of biological populations. The discussion applies to populations of both plants and animals, although the emphasis is on temperate-zone insects with one or more generations per year where reproduction is commonly interrupted during winter, and where populations often show large fluctuations in density. Section 5.2 discusses design and types of data, and section 5.3 describes some simple models which have been widely used to analyse density dependence. Section 5.4 reviews the problems associated with methods for detecting density dependence in a time series of population densities, and section 5.5 outlines some of the difficulties in modelling time-series data. Section 5.6 discusses some related problems arising in the analysis of density dependence by key-factor analysis of life-table data. Section 5.7 extends the discussion to spatio-temporal data and the analysis of spatial density dependence within a generation.

5.2 POPULATIONS AND DATA

5.2.1 Laboratory and field populations

Laboratory studies have several advantages for the study of density dependence. Firstly, closed populations can be kept in a controlled environment where it is possible to manipulate densities experimentally. Secondly, the effects of stochastic disturbances and spatial heterogeneity can be made relatively small. Thirdly, direct estimates of density can usually be obtained with relatively small measurement error. In many cases, density dependence is clear and the emphasis is on describing its form (Fig. 5.1). The main potential problem with laboratory studies is extrapolating from their results to field populations.

Field studies, on the other hand, often present problems for analysing density dependence. Firstly, the spatial extent of the population may be difficult to define. Secondly, there is often limited scope for manipulating density. Thirdly, both the biotic and abiotic environment can be very variable because of inter-specific interactions, environmental stochasticity and spatial heterogeneity. Fourthly, measurement error in population density may be relatively large or difficult to estimate. Observed patterns of density dependence in field data often contain a lot of random variation and there is a strong requirement for statistical methods to detect and describe underlying relationships (Fig. 5.2).

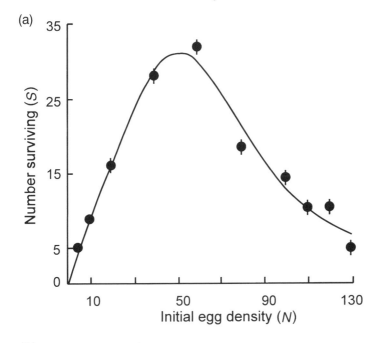

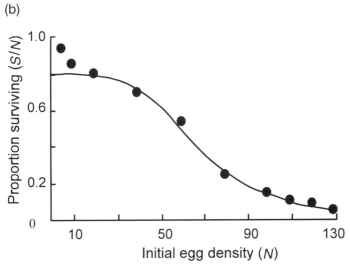

Fig. 5.1 Plots of density-dependent data from a laboratory experiment on survival of *Tribolium castaneum* showing: (a) the total number of adults ($S \pm$ S.E.) surviving from eggs kept at different initial densities (N); (b) the proportion (S/N) against initial density. Fitted model: $S = 0.8N/[1 + (0.0149N)^{4.21}]$. Reproduced with permission of Blackwell Scientific Publications from Bellows (1981).

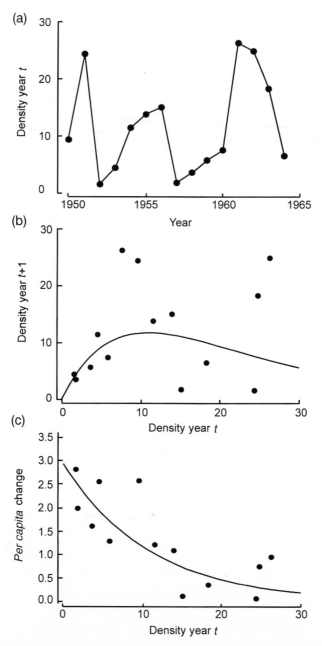

Fig. 5.2 Plots of density-dependent data from a field study on the pine looper over 15 years showing: (a) time plot of August larval density; (b) larval density in year $t+1$ against density in year t; (c) *per capita* change in larval density year t to $t+1$ against larval density year t (Klomp, 1966; Den Boer and Reddingius, 1989). Fitted model: $N_{t+1}=N_t\exp(1.08-0.092N_t)$.

5.2.2 Types of data

Densities may be recorded as (i) serial data with successive generations in a population at a given location, forming a time series; (ii) locational data collected at different sites or subpopulations to form a spatial series; (iii) spatio-temporal data with a time series for each of several different populations forming a space–time data matrix. This chapter deals mainly with single-species populations but the classification extends to multi-species systems.

Per capita change may be measured (i) between generations based on the number of individuals at a fixed developmental stage in two successive generations; (ii) within generations using the number of individuals at two successive developmental stages in the same generation; (iii) at regular time intervals, e.g. from one breeding season to the next.

In populations subject to intraspecific competition, it would be more meaningful to relate the *per capita* change to density measured as the number of individuals per unit of resource (e.g. Dempster *et al.*, 1995), but more often density is measured as numbers per unit area of habitat.

5.3 DYNAMIC STOCHASTIC MODELS

Models play a central role in detecting density dependence, describing patterns and exploring the consequences for population dynamics. This section gives a brief discussion of some simple models which have been widely used. These are presented for populations which consist of a single generation with no overlap between successive generations, typical of many temperate-zone insect species. The models apply to closed populations in a stochastic environment, but which are large enough so that the effect of demographic stochasticity can be ignored. The models described below are formulated in terms of the actual density in generation t, denoted by N_t, although in practice densities will usually be subject to measurement error.

5.3.1 Density independence

Density-independent population growth occurs when the *per capita* growth rate varies independently of density. The simplest stochastic model relates successive densities by

$$N_{t+1}=N_t\exp(r+Z_t)$$
$$\text{or}$$
$$\log_e N_{t+1}=\log_e N_t+r+Z_t \tag{5.1}$$

where Z_t reflects the environmental stochasticity in the *per capita* growth rate r. This dynamic noise is represented as a series of independent, random perturbations with mean zero and constant variance (σ^2). The model performs a random walk with expected drift r in which fluctuations in density are unbounded. After t time steps, log density has mean equal to $X_0 + rt$ and variance $\sigma^2 t$. Segments of a random walk with no expected drift ($r=0$) often show realized drift which causes high positive autocorrelation and apparent time trends.

5.3.2 Direct density dependence

Dynamic stochastic population models combine the effects of density-dependent factors with random density-independent factors. A general formulation with direct density dependence is

$$N_{t+1} = N_t f(N_t) \exp(Z_t)$$
or
$$\log_e N_{t+1} = \log_e N_t + \log_e(f) + Z_t \tag{5.2}$$

where the function f describes how the *per capita* growth rate depends on density, and Z_t is a random density-independent effect for environmental stochasticity. In this case the dynamic noise is added to the log *per capita* change, i.e. a vertical shift in the log net reproductive curve. Alternatively, the density-independent effect could be a lateral shift or an overall change in the shape of the curve (Royama, 1992). Typically, a deterministic model (i.e. no stochastic element) has equilibrium density, given by $f(N_E) = 1$, which may be stable, or unstable leading to limit cycles or chaos (May, 1976). In stochastic models, the notion of a stable equilibrium density is replaced by the mean of a quasi-stationary distribution of population density.

(a) Gompertz/first-order linear autoregressive model

In this empirical model successive densities are related by

$$N_{t+1} = A N_t^\beta \exp(Z_t)$$
or
$$\log_e N_{t+1} = \mu + \beta(\log_e N_t - \mu) + Z_t \tag{5.3}$$

where $\mu = \log_e A/(1-\beta)$. The density dependence is linear in log density with slope β, and is weakest for β close to 1 when the model reduces to a random walk with no expected drift, and strongest for β near -1. The case $\beta = 0$ is a random series. Log density varies about a mean μ with variance $\sigma^2/(1-\beta^2)$, for $-1 < \beta < 1$. The autocorrelation coefficient of lag h is equal to β^h and the partial autocorrelation coefficients of lag two or more are zero. The correlation between change in log density and previous log density is equal to $-[(1-\beta)/2]^{1/2}$. The model has the unrealistic property of unbounded growth at low density, but it often provides a reasonable empirical approximation over an observed range of densities.

(b) Ricker/logistic model

In this model successive densities are related by

$$N_{t+1}=N_t\exp(r+bN_t+Z_t)$$

or

$$\log_e N_{t+1}=\log_e N_t+r+bN_t+Z_t \tag{5.4}$$

where r is the intrinsic *per capita* growth rate (Ricker, 1954; May, 1976). The model embodies density-dependent feedback through a linear relationship between mean change in log density and density. The deterministic form has an equilibrium density $N_E=-r/b$, for $b<0$, and a wide range of nonlinear dynamic behaviour from a stable equilibrium ($0<r<2$) to limit cycles ($2<r<2.69$) and chaos ($r<2.69$). For $0<r<2$, the stochastic model has a similar autocorrelation structure to the Gompertz model with $\beta=1-r$. In particular, for $r=1$ the autocorrelation at all lags is close to zero, giving the appearance of a random series.

(c) Other models for density dependence

Bellows (1981) reviews a range of models for density dependence in terms of their theoretical properties and ability to describe observed patterns in 30 sets of insect data. Some well known and widely used models relating *per capita* change to density are as follows.

(1) $f(N)=R/(1+aN)^b$ (Hassell *et al.*, 1976)
(2) $f(N)=R\exp(-aN^b)$ (Royama, 1992)
(3) $f(N)=R/(1+aN^b)$ (Maynard Smith and Slatkin, 1973)

Varying the parameter b generates different patterns of density dependence from contest to scramble competition. Model (2) contains the Gompertz model ($b=0$) and the Ricker model ($b=1$) as special cases. The models can be made stochastic by adding random variation to the model parameters, but some care is needed to avoid creating unrealistic models (section 5.4.4d). For the Gompertz, Ricker and models (1) to (3), density dependence operates over the whole range of population density, but other formulations are possible. For example, Strong (1986a) argues for density-vague relationships characterized by lack of trend at medial densities, but with density-dependent changes away from the extremes of high and low density, and large variability in the *per capita* growth rate (Hassell, Chapter 2 in this volume). Another example, is a stochastic resource limitation model where the growth rate varies independently of density below a ceiling which is subject to stochastic variation (e.g. Pollard and Rothery, 1994).

5.3.3 Delayed density dependence

Models can allow for delayed density dependence by incorporating time lags. This extends the range of dynamic behaviour to include quasi-cycles with decline phases spanning two or more generations (Royama, 1992).

(a) Second-order linear autoregressive model

This is usually written in terms of log densities as

$$X_t = \mu + \beta_1(X_{t-1}-\mu) + \beta_2(X_{t-2}-\mu) + Z_t \qquad (5.5)$$

where $X_t = \log N_t$. Log density varies with mean μ and variance $\sigma^2(1-\beta_2)/\{(1+\beta_2)[(1-\beta_2)^2-\beta_1^2]\}$ provided that $-2<\beta_1<2$, $-1<\beta_2<1$ and $(1-\beta_2)^2>\beta_1^2$. Quasi-cycles occur when $\beta_2<-\beta_1^2/4$ and are sustained by the stochastic variation: the autocorrelation function is a damped sine wave with damping factor $(-\beta_2)^{1/2}$ and period $2\pi/\theta$, where $\cos\theta = |\beta_1|/2(-\beta_2)^{1/2}$. The lag 1 autocorrelation coefficient is equal to $\beta_1/(1-\beta_2)$, and the lag 2 partial autocorrelation coefficient is equal to β_2, with higher order partial autocorrelation coefficients equal to zero. This empirical model is sometimes used to test for delayed density dependence using time-series data (section 5.4.8), and to analyse quasi-population cycles (e.g. Moran, 1952, 1953a).

(b) Second-order Ricker/logistic model

The Ricker model extends to lag two as follows

$$N_t = N_{t-1}\exp(r+b_1N_{t-1}+b_2N_{t-2}+Z_t) \qquad (5.6)$$

(Turchin, 1990; section 5.4.8). The deterministic form of the model can generate sustained cycles.

(c) Predator–prey interactions

Models with delayed density dependence can be developed from simple models of predator–prey interactions. For example, consider the deterministic host–parasitoid model

$$H_{t+1} = RH_t f(P_t); \; P_{t+1} = H_t[1-f(P_t)] \qquad (5.7)$$

where H_t and P_t are densities of hosts and parasitoids in generation t, R is the intrinsic geometric growth rate of the host and f is the fraction of hosts not parasitized. The dynamics for the host are then

$$H_{t+1} = RH_t f(H_{t-1} - H_t / R) \tag{5.8}$$

The Nicholson–Bailey model is a special case with $f(P_t) = \exp(-aP_t)$, which reduces to a lag 2 Ricker model

$$H_{t+1} = RH_t \exp(aH_t / R - aH_{t-1}) \tag{5.9}$$

(Murdoch and Reeve, 1987). This model is unstable, producing cycles of increasing amplitude. Note that a model with random variation in R is different to the stochastic Ricker model with additive dynamic noise (section 5.3.2b). The example illustrates the idea of using lags in a single-species time series to represent the dynamics of a multi-species system (section 5.5.1; Takens, 1981).

5.4 DETECTING DENSITY DEPENDENCE USING SERIAL DATA

Various tests have been proposed for detecting density dependence in serial data. Typically, the observations are a time series of densities in some stage of the life cycle recorded at regular intervals. This section describes the main methods and discusses their properties and associated problems. For further discussion see Den Boer and Reddingius (1996).

5.4.1 Significance tests, dynamic processes and density dependence

A standard approach to detecting density dependence in time series data is to test the statistical null hypothesis (H_0) of a random walk model (section 5.3.1). A test statistic is chosen which will be effective for rejecting H_0 given some alternative hypothesis (H_A), e.g. the correlation between log *per capita* change and density, when there is a linear decline with density. The P value is the probability of observing a value of the test statistic as extreme as, or more extreme than, the observed value when H_0 is true. The P value can be interpreted as a measure of the degree of consistency with H_0 (Cox and Snell, 1981). When $P < 0.05$ we say that the null hypothesis is rejected at the 5% level of statistical significance. The type I error is the probability of rejecting H_0 when H_0 is true, i.e. the chance of a false positive. The power of the test is the probability of rejecting H_0 when H_A is true, and is usually calculated in terms of the magnitude of the departure from H_0.

Formulating density independence as a random walk model introduces some simplifying assumptions including (i) a species with single non-overlapping generations; (ii) a closed population; (iii) a random series of uncorrelated environmental effects. These assumptions are secondary to the null hypothesis but are required for the calculation of a P value. It is important that the type I error should be robust to departures from these extra assumptions, and in particular that it should not be inflated leading to spurious rejection of the null hypothesis.

5.4.2 Bias in tests using linear regression

Linear regression has been widely used to test for density dependence in time series data. The method of Morris (1959) tests for unit slope in the regression of log N_{t+1} against log N_t using a t-test. This is statistically equivalent to a test of zero slope in the regression of log N_{t+1}–log N_t against log N_t used by Varley and Gradwell (1960). It is well known that these tests applied to the random walk model are biased and produce an inflated type I error from the effect of serial dependence (Maelzer, 1970; St Amant, 1970). In a plot of log N_{t+1} against log N_t the points are not statistically independent because log N_t is plotted on both axes (except for the first and last values). Large/small values tend to be followed by the smaller/larger values, which reduces the slope of a fitted regression line and leads to spurious density dependence (Table 5.1)

Varley and Gradwell (1968) proposed a two-sided regression test for density dependence which combines the regressions of log N_{t+1} on log N_t and log N_t on log N_{t+1}. Density independence is rejected when both slopes are significantly different from one (using a t-test) and are on the same side of unity. Slade (1977) showed that the test is too conservative, producing only two rejections at the 5% level in 600 series of random walk data.

Slade (1977) also examined tests for density dependence using the slope of the major axis and the standard major axis in the plot of log N_{t+1} against log N_t. Statistical significance was obtained from confidence limits for the slope using a formula based on a random sample from a bivariate normal distribution, which ignores the problem of serial dependence. Simulations using density-independent data showed that the estimated type I error (5% level) varied with series length $n=7, 22, 122$ as (20%, 2%, 2%) and (5%, 1%, 2%) for the major axis and standard major axis, respectively.

Table 5.1 Estimated frequency of detecting density dependence using a one-sided t-test for unit slope in the linear regression of log N_{t+1} on log N_t in series of simulated data from a random walk with no drift; estimates based on 5000 series

Series length	Mean slope	Frequency of detection (%)	
		5% level	1% level
10	0.55	32	7
20	0.76	40	12
50	0.90	50	14

5.4.3 Tests for direct density dependence using serial data

The tests described below are grouped as parametric tests, which assume a random walk with a normal distribution of steps, and distribution-free tests, in which the distribution of steps is not specified. However, both approaches make two important assumptions: (i) successive steps are statistically independent, i.e. there is no autocorrelation in the environmental stochasticity (section 5.4.7); (ii) there is no measurement error in the population density (section 5.4.5)

(a) Parametric tests

In some parametric tests the null hypothesis is a random walk with no expected drift; other tests allow for drift.

H_0: *Normal random walk – no expected drift*
For these methods the sampling distribution of the test statistic on the null hypothesis is independent of the variance of step size and the P value depends only on the length of the series. Reddingius (1971) and Bulmer (1975) independently developed tests based on the Gompertz model, but Bulmer's test is the more widely used. The test statistic is

$$R = \frac{V}{U} = \frac{\sum_{t=1}^{n}(X_t-\bar{X})^2}{\sum_{t=1}^{n-1}(X_{t+1}-X_t)^2} \tag{5.10}$$

where X_t denotes log density in year t and $\bar{X}$ is the mean log density. The test statistic has intuitive appeal because it measures long-term variation in log density relative to variation in short-term change. It is related to the lag 1 autocorrelation coefficient approximately by $(1-1/n)/[2(1-r_1)]$. The null hypothesis is rejected for small values of R, with the lower 5% point closely approximated by $R_{0.05}=0.25+(n-2)0.0366$. Bulmer's test applied to series with expected drift is conservative, i.e. a type I error less than the nominal value, because drift tends to increase R.

Vickery and Nudds (1984) proposed the Pearson correlation coefficient between change in log density and previous log density (r_{dX}). The P value was estimated using simulated data from a normal random walk model with mean and variance chosen to match those of the observed series. Mountford (1988) provides percentage points of r_{dX} for series of various lengths.

H_0: *Normal random walk – with expected drift*
For the random walk with drift, the distribution of the test statistic on the null hypothesis depends on the drift and variance. However, the P value

can be estimated using the parametric bootstrap method (Efron and Tibshirani, 1993). Sets of simulated data (bootstrap samples) are generated from a random walk model with estimated drift and variance, and the value of the test statistic is calculated for each sample. The bootstrap P value is the proportion of the values in the simulated series which are more extreme than the observed value.

Dennis and Taper (1994) developed the parametric bootstrap likelihood ratio test for the stochastic Ricker model. The test statistic is equivalent to the Pearson correlation coefficient between change in log density and density (r_{dN}). Pollard *et al.* (1987) suggested a similar approach using the Pearson correlation coefficient between change in log density and previous log density (r_{dX}). For series with no expected drift, the tests are slightly biased: from 5000 series of simulated data with $n=20$, estimated type I errors (level 5%) were 6.0% (S.E.=0.3%) for Pollard *et al.* and 5.7% (S.E.=0.3%) for Dennis and Taper.

(b) Distribution-free tests

Distribution-free tests use the fact that in a random walk all possible permutations of the observed changes are equally likely, whatever the distribution of steps. The P value can be estimated by repeatedly applying the test to series of random walks generated by using random permutations of the observed changes. Randomization of the observed changes generates population trajectories which all have the same initial and final densities. This restriction on the set of possible outcomes suggests a reduction in power of the test. By fixing the trend, distribution-free tests do not utilize the information provided by absence of trend as evidence of density dependence.

The randomization procedure has been applied with a range of test statistics each motivated by properties of a regulated series. Pollard *et al.* (1987) used the Pearson correlation coefficient between change in log density and previous log density (r_{dX}), based on the Gompertz model. Reddingius (1971) and Den Boer and Reddingius (1989) proposed the observed range in log densities, i.e. $LR=\log(\text{highest density})-\log(\text{lowest density})$, based on the property of fluctuations in density within limits. Crowley (1992) developed a randomized attraction test based on the return tendency. Potential attractors are the ranges between successive values of the densities sorted by increasing order. Changes between generations from a density outside the attractor that fail to move towards the attractor are termed 'violations' The test statistic is the minimum number of violations in the set of potential attractors.

5.4.4 Statistical power

The power of a significance test is the probability of rejecting the null hypothesis when a specified alternative hypothesis is true. Power depends on the significance level of the test, so that there is a trade-off between the type I error and power. Applications of statistical power to tests of density dependence are limited by the uncertainty on the form and strength of density dependence. However, power can be used to compare different tests provided that they have the same significance level.

(a) Strength of density dependence and series length

Table 5.2 gives estimates of the power for tests of Bulmer, Pollard *et al.* and Dennis and Taper, applied to simulated data from (a) the Gompertz model and (b) the Ricker model. For example, using the Gompertz model with a series of length $n=20$, $\beta=0.80$ and initial density equal to the mean, the percentage power of Bulmer's test is only 19%. The length of series to achieve a power of 90% for $\beta=0.5$ is estimated as $n=30$ (Solow and Steele, 1990). Dennis and Taper showed that the power of their test increases with the intrinsic growth rate (r) but does not depend on the slope (b) in the linear relationship between change in log density and density. For the above models, power was greatest for Bulmer's test followed by Dennis and Taper and then Pollard *et al.*

These results apply to models of density dependence for which the test statistic has been chosen to have high power, and are therefore likely to overestimate the power for real populations where the models are at best an approximation (section 5.4.9). The results emphasize, however, that even in favourable situations, relatively long series may be required to have a good chance of detecting density dependence.

(b) Initial population density and transient effects

A trend in the time series caused by an initial displacement from the mean can markedly affect the power of the test. Pollard *et al.* (1987) showed that the power of Bulmer's test for the Gompertz model ($n=10$, $\beta=0.8$ and $\sigma=0.10$) was reduced to near zero when the initial log density was displaced from the mean (by six standard deviations of log density) but the power of the Pollard *et al.* test increased to about 38%. Dennis and Taper (1994) reported a similar effect for the Ricker model ($n=10$, $r=0.3$, 1.2 and $\sigma=0.05$), except for low initial densities and small growth rates where it was difficult to distinguish the series from exponential growth. Table 5.3 shows how random variation in the initial density reduces the power of Bulmer's test but increases the power of the tests of Pollard *et al.* and Dennis and Taper. The increase in power is explained by the greater range of population density provided by the initial displacement. For the Gompertz series in Table 5.2a, Bulmer's test has greater power provided that the initial displacement is less that two standard deviations. In general, the effect will depend on the magnitude of the displacement, the intrinsic rate of increase, the amount of random variation and the length of the series. Other types of trends can reduce power (section 5.4.6).

Table 5.2a Estimated power (%) of three tests for density dependence (5% level) applied to simulated data from Gompertz model. Estimates based on 1000 series each with 500 randomizations (Pollard *et al.*, 1987) and 500 bootstrap samples (Dennis and Taper, 1994) with initial densities set to mean population density. Figures in parenthesis based on random initial densities generated from 50 iterations of the model starting at the mean

n	β	*Bulmer* *R*	*Pollard* *et al.*	*Dennis* *and Taper*
10	0.8	9(8)	5(7)	5(7)
20		19(16)	6(9)	9(12)
30		31(23)	11(12)	15(15)
40		42(38)	15(19)	22(22)
50		55(53)	23(28)	28(31)
10	0.5	23(22)	7(11)	8(11)
20		63(55)	29(29)	32(32)
30		88(88)	59(64)	60(64)
40		98(98)	83(85)	81(82)
50		100(100)	96(96)	92(93)
10	0.2	48(41)	15(20)	19(24)
20		95(53)	68(73)	68(71)
30		100(100)	96(97)	93(95)
40		100(100)	100(100)	100(100)
50		100(100	100(100)	100(100)
10	0.0	66(56)	31(34)	31(35)
20		98(99)	87(89)	87(88)
30		100(100)	99(100)	99(100)
40		100(100)	100(100)	100(100)
50		100(100)	100(100)	100(100)

(c) Environmental stochasticity

The power of Bulmer's test is approximately independent of the variance of the stochastic component in the Gompertz model. Dennis and Taper (1994) showed that the power of their test increased with the standard deviation of population change, but that the effect was slight for values less than the intrinsic growth rate. These results appear to be counter-intuitive in the context of regression analysis where increased random variation tends to obscure relationships and reduce statistical power. However, in serial data an increase in variance of population change increases the range of population density, so that power for detecting a correlation between *per capita* change and density is maintained, or even

Table 5.2b Estimated power (%) of three tests for density dependence (5% level) applied to simulated data from the Ricker model. Estimates based on 1000 series each with 500 randomizations (Pollard *et al.*, 1987) and 500 bootstrap samples (Dennis and Taper, 1994) with initial densities set to mean population density. Figures in parenthesis based on random initial densities generated from 50 iterations of the model starting at the mean

n	r	Bulmer R	Pollard et al.	Dennis and Taper
10	0.2	11(8)	5(6)	5(7)
20		21(14)	8(10)	13(14)
30		35(26)	14(15)	22(23)
40		43(34)	20(19)	32(32)
50		50(49)	25(29)	40(42)
10	0.5	27(20)	9(11)	16(15)
20		58(55)	30(32)	41(48)
30		84(81)	56(59)	72(75)
40		94(92)	77(77)	89(90)
50		98(97)	90(90)	97(96)
10	0.8	47(41)	18(23)	28(34)
20		92(88)	68(70)	81(83)
30		99(99)	94(93)	97(98)
40		100(99)	99(99)	99(100)
50		100(99)	100(100)	100(100
10	1.0	66(58)	33(37)	45(50)
20		97(97)	85(88)	91(93)
30		100(100)	98(99)	99(100)
40		100(100)	100(100)	100(100)
50		100(100)	100(100)	100(100)

increased. The neutral effect of environmental stochasticity needs to be qualified because it refers to a particular form of random variation in the intrinsic growth rate. The effect of other types of stochastic variation could be different. For example, following a 'bad' year a population may be displaced far below its mean, so increasing the power for detecting density dependence during the subsequent increase phase (section 5.4.4b)

(d) Power comparisons using simulated data

Holyoak (1993a) reported an extensive simulation study which compared the tests of Bulmer, Pollard *et al.*, Reddingius and Den Boer, and Crowley. Data were simulated using three density-dependent models:

Table 5.3 Estimated type I error (%) of three tests for density dependence (5% level) applied to simulated data from a random walk with drift r, standard deviation of step length $\sigma=0.50$, and random measurement error with standard deviation σ_U. Estimates based on 1000 series each with 500 randomizations (Pollard *et al.*, 1987) and 500 bootstrap samples (Dennis and Taper, 1994)

		Test procedure					
		Bulmer (R)		Pollard et al.		Dennis and Taper	
r	σ_U	$n=10$	$n=20$	$n=10$	$n=20$	$n=10$	$n=20$
0	0	5	5	5	4	5	5
	0.05	5	6	5	6	6	8
	0.10	6	7	6	6	7	7
	0.20	7	10	6	7	6	8
	0.30	13	15	8	12	9	14
	0.40	16	20	10	13	13	15
	0.50	22	33	12	21	18	21
	0.70	30	49	16	32	17	36
	1.00	39	67	23	48	27	55
0.10	0	4	4	5	7	5	7
	0.05	5	5	5	4	6	5
	0.10	6	5	6	5	7	7
	0.20	8	6	7	5	7	6
	0.30	10	11	7	8	8	10
	0.40	12	16	7	12	11	16
	0.50	20	24	10	16	12	20
	0.70	25	35	15	24	18	27
	1.00	28	55	18	39	22	32
0.20	0	3	1	6	4	6	5
	0.05	2	1	5	5	4	6
	0.10	3	2	6	4	7	5
	0.20	5	2	5	4	6	6
	0.30	7	4	6	3	6	7
	0.40	8	7	6	8	6	9
	0.50	12	9	10	7	11	10
	0.70	20	19	14	15	15	16
	1.00	28	32	17	24	20	21

(1) the Ricker model (R), $N_{t+1}=N_t\exp[r(1-\alpha_t N_t)]$
(2) the multiplicative logistic (ML), $N_{t+1}=N_t[1+r(1-\alpha_t N_t)]$
(3) the model of Maynard Smith and Slatkin (P), $N_{t+1}=RN_t/[1+(aN_t)^{b_t}]$.
   The parameters α_t and b_t were allowed to vary in time as normal variates. In model R, the stochastic variation affects the slope of the log *per*

capita change with density, but has little effect at low density. Model ML is unrealistic because negative densities could occur by chance. Model P has the unusual statistical property of no stochastic effect at the critical density of $1/a$.

The study covered a wide range of parameter values chosen to encompass the range of natural variation in population density, and used 25 time series of 20 generations to estimate the power of each test. Average percentage power (5% level) for the models (R, ML, P) was as follows: Bulmer (57%, 27%, 42%); Pollard *et al.* (61%, 52%, 51%); Reddingius and Den Boer (46%, 15%, 38%); Crowley (20%, 61%, 42%). The relatively high power of the Pollard *et al.* test differs from the results in Table 5.2, emphasizing the point that no single simulation study can provide definitive conclusions about the relative performance of different tests. The statistical power will depend on the choice of model, so that care is needed in comparing tests.

5.4.5 Measurement error

Random measurement error in the estimated densities induces a negative correlation between *per capita* change and density. This tends to inflate the type I error of tests for density dependence, leading to spurious detection. Table 5.3 illustrates the effect for tests of Bulmer, Pollard *et al.*, and Dennis and Taper. The bias increases with the ratio of the standard deviation of the measurement error to the standard deviation of the change in log density, but is relatively small for a ratio less than half. Note that the bias increases with series length.

These results differ from those reported by Dennis and Taper (1994), who claimed that the type I error of their test was hardly influenced by massive amounts of sampling error. This was based on a Poisson model (CV proportional to $1/(N)^{1/2}$) applied to an increasing series, which gave a CV less than 7%. Their claim of increased power without an associated increase in type I error is therefore limited.

(a) Bulmer's second test

To allow for measurement error, Bulmer (1975) proposed the test statistic $R^* = W/V$, where

$$W = \sum_{t=1}^{n-2} (X_{t+2} - X_{t+1})(X_t - \bar{X}), \quad V = \sum_{t=1}^{n} (X_t - \bar{X})^2$$

and where X_t denotes log density. The random walk null hypothesis is rejected for small values of R^*. However, the test lacks power, e.g. for $n=25$, $\beta=0.50$ and no measurement error, the power is only 26% compared with 74% for Bulmer's R.

Reddingius and Den Boer (1989) showed that R can be adjusted to allow for measurement error by using $R^c=[V-ns_{U}^2]/[U-2(n-1)s_{U}^2]$ where s_{U}^2 is an estimate of the measurement error variance. Applying the method to the data in Fig. 5.2 using a measurement error CV=14% gives $R=0.617$, $R^c=0.619$, so the conclusion is not affected. An alternative approach is to examine the effect on R by computer simulation using estimates of the drift (r) and standard deviation (σ) assuming the random walk null hypothesis.

(b) Problems with low counts

When counts are low, estimates of population density will have relatively large sampling error, and there may be problems with zeros. A zero could mean extinction followed by recolonization in which case the assumption of a closed population fails. Or, a zero may be a chance observation of a low density. To apply tests which use the logarithm of density, some authors have replaced zeros by 0.5 or worked with $X=\log(N+0.5)$. The type I error of Bulmer's test was estimated by simulating data from a random walk model with Poisson counts. This showed that tests applied to series with low counts can be seriously biased (Table 5.4).

5.4.6 Population trends

Temporal trends can be broadly grouped as density-independent, i.e. exponential growth/decline, or density-dependent reflecting (i) a system-

Table 5.4 Estimated type I error (%) of Bulmer's test (5% level) applied to series of length $n=20$ from a random walk model with observed number of individuals as a Poisson variable. N_0, expected number of individuals at start; σ, standard deviation of change in log density

σ	N_0	Type I error
0.2	5	74
	10	56
	30	30
	50	19
0.5	5	35
	10	26
	30	12
	50	10

atic change in the mean density or (ii) a return to a mean level following some perturbation. A particular series could contain a complex mix of the different types of trend which may be difficult to distinguish. Density-independent trends inflate Bulmer's test statistic and lead to a conservative test, with type I error less than the nominal value, but randomization tests and the parametric bootstrap tests allow for density-independent trends in calculating the *P* value. Trends caused by an initial displacement from the mean reduce the power of Bulmer's test but increase the power of tests of Pollard *et al.* and Dennis and Taper (section 5.4.4b). Trends in the mean, however, tend to obscure density-dependent relationships and thereby reduce power.

Testing for density dependence with temporal trend in mean density

Holyoak (1996) proposed a test for density dependence which is effective when there is a linear trend in mean log density. The method applies the test of Pollard *et al.* to the detrended series of log densities $X_t - \bar{d}(t-1)$, where $\bar{d}$ is the mean change in log density. The power of the test for a stochastic Ricker model with exponential trends ranging between $(-0.02, 0.08)$ was on average less than 1% before detrending, compared with 64% after detrending. For 60 sets of bird time-series data spanning 31 years, detection rates increased from 17 to 45%, even though in many cases the trends were not statistically significant. Holyoak concluded that even weak trends in mean density can hinder the detection of density dependence. This contrasts with the improved detection when the trend results from a return to the mean following an initial displacement (section 5.4.4b). Holyoak's method therefore represents a useful development in tests for density dependence.

5.4.7 Patterns of environmental variation

The models and methods for detecting density dependence in time-series data assume that environmental stochasticity can be represented as a series of random perturbations, from a normal distribution in the case of parametric methods. Ideally, tests should be robust and little affected by autocorrelated environmental variation or non-normality in the distribution of year-to-year changes. In some cases, the effect of environmental variation on the population can be related to exogenous variables such as weather, but this information is not utilized. This section presents some results on how patterns of environmental variation affect the type I error of the tests and how they can be used to increase statistical power.

(a) Autocorrelated environmental stochasticity

Solow (1990) examined the effect of autocorrelated stochastic variation on Bulmer's test by using a Gompertz model in which the random term follows a first-order autoregressive scheme, i.e. $Z_t = \theta\, Z_{t-1} + \epsilon_t$, where $-1 < \theta < 1$ and the ϵ_t are independent normal variates with mean zero and variance σ_ϵ^2. The type I error (5% level) was estimated for density-independent series, length $n=25$ and values of $\theta = (-0.9, -0.5, 0.0, 0.5, 0.9)$ as (86%, 39%, 5.2%, 0.3%, 0.0%), respectively. Negative autocorrelation leads to spurious detection, whereas for positive values the test is conservative (see also Maelzer, 1970; Reddingius, 1990; Solow, 1991).

Incorporating autocorrelation into the Gompertz model gives log densities in successive generations related by

$$X_{t+1} = \mu + (\beta + \theta)(X_t - \mu) - \beta\theta(X_{t-1} - \mu) + \epsilon_t \qquad (5.11)$$

i.e. autocorrelation in the stochastic component modifies the effect of direct density dependence and induces delayed density dependence (Royama, 1981). When there is no density dependence ($\beta = 1$), the model produces unbounded fluctuations with variance of log density in generation t approximately equal to $\sigma_\epsilon^2 t/(1-\theta)$. In finite series, the sample lag 1 autocorrelation coefficient tends to increase with θ, which explains the increased rejection rate of the random walk null hypothesis using Bulmer's test when $\theta < 0$. This also illustrates the general point that a statistically significant test for density dependence does not necessarily imply bounded fluctuations.

Fox and Ridsdill-Smith (1995) note that there is a problem of indeterminacy in the above model. For example, models with ($\beta = 0.9$, $\theta = -0.8$) and ($\beta = -0.8$, $\theta = 0.9$) have the same dynamics, even though the effect of the density dependence in the underlying Gompertz model is much stronger in the latter. However, examination of the sample autocorrelation function and partial autocorrelation function may help interpretation. In particular, the lag 1 autocorrelation coefficient is given by $\rho_1 = (\beta + \theta)/(1 + \beta\theta)$, with lag two partial autocorrelation $\pi_2 = -\beta\theta$ and zero for lag three or more ($\rho_1 = 0.36$ and $\pi_2 = 0.72$ in the above cases). A sample lag 2 partial autocorrelation significantly greater than zero may therefore indicate spurious detection using Bulmer's test. Note that this contrasts with (i) a zero value expected for the first-order Gompertz model; (ii) a negative value associated with quasi-cycles (sections 5.3.3a, 5.4.8a).

(b) Non-normal stochastic variation

The effect of non-normality was examined for tests of Bulmer, Pollard *et al*, and Dennis and Taper using a random walk model with step size sampled from an extreme value distribution (i) skewed to the left, i.e. a greater chance of unusually small values; (ii) skewed to the right; (iii) symmetrical, with values chosen from (i) and (ii) with equal probability.

Estimated type I errors (5% level) using 5000 sets of simulated data for series of length $n=10, 20, 30$ were as follows (left skew, right skew, symmetrical): Bulmer (4.7%, 4.7%, 4.9%); Pollard *et al.* (5.2%, 5.1%, 5.1%); Dennis and Taper (9.0%, 3.3%, 6.1%). Values for Pollard *et al.* are close to the nominal level as expected because the test is distribution-free, and Bulmer's test is slightly conservative. However, the test of Dennis and Taper has a positive bias when the distribution of steps is skewed to left, and is conservative when skewed to the right.

(c) Weather effects

In some situations, incorporating weather effects can lead to more powerful methods for detecting density dependence. A simple empirical model for analysis is

$$X_{t+1} = a + \beta X_t + \gamma W_t + Z_t \qquad (5.12)$$

where X_t denotes log density in year t, W_t is a weather variable correlated with the change in density between t and $t+1$, and Z_t is a random effect. A test for density dependence uses the t ratio for density in a least-squares fit of the model with density and weather. To allow for the problem of serial dependence (section 5.4.2), the P value is estimated by bootstrap sampling from the fitted null model with $\beta=1$ (Rothery *et al.*, 1997). The method was applied to indices of abundance for three species of butterfly in which annual changes were correlated with summer temperature. This showed stronger evidence against the null hypothesis than the tests of Bulmer and Pollard *et al.* (Table 5.5). In principle, the approach could be applied using other variables related to population change, such as food or the density of a predator (e.g. Dempster *et al.*, 1995; Elkington *et al.*, 1996).

Table 5.5 Illustration of effect of allowing for temperature in testing for density dependence. Data are indices of annual abundance for three butterfly species, 1977–94, taken from the United Kingdom Butterfly Monitoring Scheme. r_{dT}, correlation between *per capita* change in log index and temperature. P values estimated from 5000 series of simulated data with 500 randomizations (Pollard *et al.*, 1987) and 500 bootstrap samples (see text)

Species	Temperature month (year)	r_{dT}	P value Bulmer (R)	P value Pollard et al.	P value Allowing for temperature
Brimstone *Gonepteryx rhamni*	August ($t-1$)	0.59	0.07	0.20	0.002
Peacock *Inachis io*	August ($t-1$)	0.49	0.03	0.06	0.003
Meadow Brown *Maniola jurtina*	July (t)	0.32	0.35	0.21	0.017

5.4.8 Delayed density dependence

The methods described in section 5.4.3 are designed to have high power for detecting direct density dependence when the *per capita* change from generation t to $t+1$ depends on the density in generation t. Delayed density dependence occurs when the *per capita* change depends on density in generation $t-1$ or earlier, after allowing for any intermediate effect of direct density dependence.

(a) Tests for delayed density dependence

The sample autocorrelation function (ACF) and the partial autocorrelation function (PACF) are useful diagnostics of delayed density dependence. In particular, time lags can cause quasi-cycles and a characteristic damped periodic ACF (section 5.3.3). An associated test uses the PACF and the result that for the pth order autoregressive scheme, the coefficients from lag $(p+1)$ onwards are approximately independently and normally distributed with mean $-1/n$ and variance $1/n$ (Quenouille, 1949). Turchin (1990) proposed a regression test based on the coefficient in the lag 2 Ricker model: Dennis and Taper (1994) showed that the type I error of the one-sided test (5% level) is inflated (5.7–9.1%) but that the two-tailed test gives reasonable results (3.1–5.4%). Holyoak (1994a) proposed a randomization test using the correlation coefficient between change in log density and lag 2 log density after partialling out the effect of previous density. The method tests the null hypothesis of a random walk, with power for detecting delayed density dependence. Recently, Takashi *et al.* (1997) extended the bootstrap approach of Dennis and Taper using both the Ricker and Gompertz models. An intriguing approach was developed by Manly (1990) based on the anticlockwise spiral in a plot of log *per capita* change against log previous density.

Turchin (1990) gives some examples of where tests for direct density dependence can fail to detect any effect, even though there may be strong delayed density dependence. Analysis of time-series data from 14 forest insect populations using the regression test for the lag 2 Ricker model showed clear evidence of delayed density dependence in eight cases, but direct density dependence was detected in only one of these using the lag 1 Ricker model. Note that this result does not imply that delayed density dependence necessarily reduces the power of tests designed to detect direct density dependence. For example, the power of Bulmer's test was estimated using 1000 sets of simulated data from a second-order linear autoregressive model (section 5.3.3a) starting at the population mean with $n=20$, $\beta_2=(0.0, -0.5, -1.0)$ and different values of β_1 as follows: $\beta_1=0.5$ (63%, 86%, 98%); $\beta_1=0.9$ (10%, 18%, 43%); $\beta_1=1$ (6%, 12%, 20%). Power increases with the strength of the delayed density dependence, but is low when direct density dependence is weak, as in most of Turchin's examples, or absent (i.e. $\beta_1=1$).

(b) Application to simulated data

Holyoak (1994a) examined Turchin's regression test, Pollard *et al.*'s modified test and the lag 2 sample partial autocorrelation coefficient using simulated data from three models: (1) a lag 1 Ricker model; (2) a lag 2 Ricker model containing no direct density dependence with intrinsic growth rate *r* sampled from normal distribution; (3) the Nicholson–Bailey host–parasitoid model with a negative binomial distribution of parasitoid encounters, and a log normal distribution for the host intrinsic growth rate. Detection rates of delayed density dependence were similar for Turchin's test and the modification of Pollard *et al.* However, the type I error of Turchin's test was too high (approximately 9% at the 5% level). The tests of Bulmer and Pollard *et al.* rejected the null hypothesis even when only delayed density dependence was present, showing that a significant result should not necessarily be interpreted as evidence of direct density dependence.

(c) Time scales for detecting delayed density dependence

Holyoak (1994b) showed that delayed density dependence could be masked when tests are applied to annual counts based on totals over two or more generations per year. Turchin's tests for delayed and direct density dependence were applied to simulated data from two stochastic models: (1) the Nicholson–Bailey model with parasitoid aggregation and a random host intrinsic growth rate; (2) the lag 2 Ricker model with no direct density dependence and random variation in the equilibrium. Tests were carried out using the total number of individuals at different sampling frequencies of 1, 2, 3, 4, 5 or 10 consecutive generations. The results showed that the detection of delayed density dependence declined with sampling frequency, but that the detection rate of direct density dependence increased for the host–parasitoid model, and increased and then declined for the lag 2 Ricker model.

(d) Autocorrelated environmental stochasticity

Section 5.4.7a illustrates how autocorrelation in the random component of the Gompertz model can lead to spurious delayed density dependence. Williams and Liebhold (1995) demonstrated the effect applying Turchin's test and the lag 2 partial autocorrelation coefficient to simulated data from a lag 1 Ricker model, with stochastic variation in the equilibrium. The authors argued for more information on both endogenous and exogenous factors to understand and predict population dynamics.

5.4.9 Applications to field data

Tests of density dependence have been applied to series of data on mammals, birds and insects. Often the assumption of single, non-overlapping generations has been ignored by including long-lived species or species with several generations per year. A notable exception is the work of Den Boer and Reddingius (1989), who analysed 16 series of more than 10 years from six different species of insect and found detection rates (5% level) as follows: Bulmer (3/16), Reddingius and Den Boer (1/16) Pollard *et al.* (0/16). Most other applications involve relatively few species or data sets, and use different tests, which gives a rather piecemeal picture and hinders a systematic comparison of methods. However, Bulmer's test has generally performed well compared with other methods. Three recent extensive studies of insect populations illustrate some of the issues discussed in this section and some of the problems of interpretation.

Woiwod and Hanksi (1992) examined 5715 time series of annual counts of 263 species of moths in light traps and 94 species of aphids in suction traps at sites throughout Britain in the Rothamsted Insect Survey (Woiwod and Harrington, 1994). The detection rates for Bulmer's test were 79% in moths and 88% in aphids in series of longer than 20 years, with corresponding values of 57% and 84% for the test of Pollard *et al.* Further analysis showed higher detection rates in longer series and lower in series showing a temporal trend. The incidence of delayed density dependence (Turchin's test) was approximately that expected by chance, but this could have been masked, or manifested as direct density dependence for species with several generations per year (section 5.4.8). The high detection rates for aphids are consistent with the strong density-dependent migration and recruitment reported in studies of deciduous tree-dwelling species (Dixon and Kindlmann, Chapter 9 in this volume). However, time series were included for analysis provided that the mean count was five or more individuals, so the detection rate may have been inflated by the effect of sampling error (section 5.4.5b). Also, the suction traps included winged migrants from many local aphid populations, which casts doubt on the assumption of a closed population.

Holyoak (1993b) analysed density dependence in 171 time series of 10 or more generations from six insect orders and found the following overall detection rates: Bulmer (38%), Pollard *et al.* (28%), Dennis and Taper (26%), Reddingius and Den Boer (22%), Crowley (13%), Turchin lag 2 (12%). The detection rates for the different orders using Bulmer's test were as follows: Odonata (85%), Hymenoptera (43%), Lepidoptera (40%), Coleoptera (34%), Diptera (23%), and Hemiptera (9%). The relatively high detection rate for Odonata was consistent across tests, but other differences between orders varied according to test and the reasons for differences were not clear. Further analysis showed effects of series length and trends on the detection rates.

Wolda and Dennis (1993) analysed over 1700 series of insect data from a wide range of studies on pest and non-pest species sampled using light, pitfall or suction traps. The test of Dennis and Taper was applied to uninterrupted series of at least 20 generations, with zeros replaced by 0.5. No other tests were used. Detection rates generally increased with series length, although in some individual series the pattern was reversed. The results were similar for pest and non-pest species (although delayed density dependence was not examined) and there was no detectable difference between univoltine and bi/polyvoltine species (although the test is not designed for species with more than one generation per year). Higher rates of detection were found in series with more sampling error (judged to be least for pitfall traps and most in suction traps), which was interpreted in terms of statistical power rather than a bias in the type I error (section 5.4.5). Wolda and Dennis emphasized that tests of density dependence are designed to detect statistical density dependence (SDD), defined as a 'statistical return tendency', and that a statistically significant test result does not automatically imply a causal ecological explanation. They illustrated this by demonstrating SDD in rainfall data and in a seasonal migrant moth, *Autographa gamma*, in which individuals present in one year were unlikely to be descendents of those in the previous year. Hanski *et al.* (1993a) accepted the concept of SDD and the limitations of statistical tests, but argued that in the absence of plausible ecological mechanisms which would generate a random series, biological density dependence is the most likely explanation for SDD in such a large number of species (see also Holyoak and Lawton, 1993; Wolda *et al.*, 1994). It is also possible that tests may indeed be correctly detecting density dependence but for the wrong reasons, because of large measurement errors or uncertainty in the relationship between population density and the trap indices.

5.5 DESCRIBING DENSITY DEPENDENCE IN TIME-SERIES DATA

Tests for density dependence may reject the random walk null hypothesis, but they tell us little if anything about the strength or form of density dependence, or its effect on population dynamics. These questions can be addressed by fitting descriptive models for categorizing and comparing the dynamics of different species and extrapolating long-term behaviour (section 5.3; Hassell *et al.*, 1976). A particular aspect of this approach has been the use of models for detecting chaotic dynamics. The associated statistical problems are those of nonlinear regression analysis with the added complication of autocorrelated responses and measurement error in the explanatory variable (Seber and Wild, 1989). This section contains a few general remarks on potential difficulties.

5.5.1 Flexible empirical models

Turchin and Taylor (1992) have suggested a general framework for analysing population time-series data in which densities in successive generations are related by

$$N_t = F(N_{t-1}, N_{t-2}, ..., N_{t-p}, Z_t) \qquad (5.13)$$

where F is a nonlinear function of density and Z_t is a random effect of environmental stochasticity or dynamic noise. The term p is the order of the process, i.e. the maximum time lag beyond which density has an effect. A mathematical justification lies in the representation of the dynamics of a system of p interacting populations as a single-species model of order p (section 5.3.3c; Takens, 1981). Identification of the order p of delayed density dependence may provide clues to underlying mechanisms, although of course many different processes could result in a given time lag.

Turchin and Taylor suggest a flexible family of empirical models for F taken from response surface methodology (Box and Draper, 1987). In particular, for second-order model ($p=2$)

$$N_t = N_{t-1} \exp\left[Q(N_{t-1}^{\theta_1}, N_{t-2}^{\theta_2}) + Z_t \right] \qquad (5.14)$$

were Q is a quadratic form in the power-transformed densities including the cross-product term (i.e. a total of eight model parameters). Perry *et al.* (1993) proposed a similar model which is a quadratic form in log densities, i.e. a second-order linear autoregressive scheme augmented to include quadratic terms. Non-parametric smoothing using neural networks and thin-plate splines has recently been applied by Ellner and Turchin (1995) to provide a flexible empirical framework for detecting chaos in noisy data.

5.5.2 Problems of model selection and fitting

Model development requires the specification of the form and order of the density dependence, and representation of the random component. Ideally, the structure should reflect ecological mechanisms, but often the choices are made on empirical grounds. Preliminary analysis includes time plots to check for trends or other signs of non-stationarity, and calculating the ACF and PACF to suggest appropriate time lags. Models are usually nonlinear in at least one of their parameters (i.e. nonlinear in the statistical sense) and therefore must be fitted using an iterative search algorithm, often by nonlinear regression of $r_t = \log(N_t/N_{t-1})$ on $(N_{t-1},$

$N_{t-2},...,N_{t-p}$). Note that the model proposed by Perry *et al.* is linear in its parameters and therefore can be fitted by ordinary linear regression, although the sampling properties of the estimators are different to those for the linear regression model.

(a) Model misspecification

The wrong choice of the form or order of density dependence can have a large effect on the dynamics of the fitted model. In particular, over-simple models which omit a time lag or a strong nonlinearity may be biased towards more stable dynamics (Turchin and Taylor, 1992). Model selection can be particularly difficult with noisy data when models with different dynamic behaviour may fit the data almost equally well (Morris, 1990; Berryman, 1992). Recently, cross-validation methods have been used as a basis for model selection (e.g. Ellner and Turchin, 1995).

(b) Overparameterization

This occurs when the model contains more parameters than can be reliably estimated from the data, perhaps because of a restricted range or sparse coverage of densities. Typically, the fitted model contains some parameter estimates which are highly correlated and have large standard errors. Sometimes, the iterative fitting procedure may fail to converge, or home in on a local minimum. In other cases, the fitted model may reflect the random vagaries of the data and produce meaningless results as shown by Turchin and Taylor (1992), or be sensitive to the inclusion of further data (Perry *et al.*, 1993). Flexible phenomenological models can produce unrealistic behaviour when extrapolated beyond the range of the fitted model. Such models may be more suited to the analysis of the local dynamics rather than for categorizing global dynamics.

5.5.3 Measurement error

The above discussion applies to densities measured without error. Measurement error leads to additional biases in the estimated model parameters and further problems in model selection. The statistical topic of analysing relationships in which the explanatory variable is measured with error is referred to as 'measurement error models' or 'errors-in-variables'. For simple linear regression the estimated slope is biased towards zero (the so-called attenuation effect) by a factor $V_{xx}/(V_{xx}+V_{uu})$ where V_{xx} is the variance of the explanatory variable measured without error and V_{uu} is the measurement error variance (Fuller, 1987; Carroll *et al.*, 1995). In linear autoregressive models the effect of measurement error is to dampen the autocorrelation function by a similar factor, with variances referring to log density. For more complex models, computer simulation is usually needed to estimate the magnitude of the effect. Table 5.6 gives some numerical results for the Ricker model and illustrates how weak effects of density dependence are overestimated while strong effects are underestimated (Walters and Ludwig, 1987).

When the distribution of measurement errors can be reliably estimated, the method of simulation extrapolation (SIMEX) may be used to reduce bias (Cook and Stefanksi, 1995; Carroll *et al.*, 1995). SIMEX involves simulating the distribution of the parameter estimates obtained by adding increasing amounts of measurement error to the explanatory variables, and then extrapolating back to the unobserved case of zero error.

In some cases, it may be preferable to apply different model-fitting procedures such as weighted least squares or maximum likelihood. Carpenter *et al.* (1994) illustrate the approach using stochastic simulations of predator–prey models and fitting models to plankton time series. Bias

Table 5.6 Bias in the estimate of r for the Ricker model fitted to time-series data with measurement error. Estimates based on 1000 series of simulated data of length n, $b=-r/100$, $\sigma=0.5$ and measurement error in log density with standard deviation σ_u

| r | n | σ_u | Estimated value of r | | Frequency by range (%)[a] | | | |
			Mean	Accuracy (%)[b]	0–1	1–2	2–2.69	>2.69
0.5	10	0.0	0.78	95	74	25	0	0
		0.2	0.83	95	70	29	0	0
		0.5	0.94	91	59	40	0	0
	20	0.0	0.62	96	95	5	0	0
		0.2	0.66	93	90	10	0	0
		0.5	0.75	92	84	16	0	0
	30	0.0	0.59	95	98	2	0	0
		0.2	0.62	93	96	4	0	0
		0.5	0.73	91	91	9	0	0
1.5	10	0.0	1.53	96	9	82	9	0
		0.2	1.47	97	12	79	9	0
		0.5	1.31	97	23	72	5	0
	20	0.0	1.51	95	3	95	2	0
		0.2	1.42	95	4	94	2	0
		0.5	1.20	86	27	72	1	0
	30	0.0	1.50	95	1	99	0	0
		0.2	1.43	93	3	97	0	0
		0.5	1.19	76	24	76	0	0

[a] Deterministic dynamics: (0–1), monotonic damping; (1–2), damped oscillations; (2–2.69), limit cycles; (>2.69), chaos.
[b] Percentage coverage of nominal 95% confidence interval for r.

in the parameter estimates could be reduced by allowing for measurement error in the fitting procedure, but even relatively small errors (CV=10%) caused substantial difficulties in model selection. The authors argued for experimental manipulations to resolve the problem.

5.6 KEY FACTOR ANALYSIS AND DENSITY DEPENDENCE

Key factor analysis is a widely used method for analysing insect life-table data (Varley and Gradwell, 1960). Estimates of the densities of individuals entering different developmental stages are used to estimate the survival between stages. The aims are (i) to identify the survival rates which contribute most to variation in population density between generations; and (ii) to identify mortality factors which are density dependent. The basic method uses a model in which densities in successive generations are related by

$$N_{t+1}=N_t E s_{1t}\, s_{2t} \ldots s_{Lt} \tag{5.15}$$

where E refers to the maximum egg output and $s_{1t}, s_{2t},\ldots,s_{Lt}$ are the survival rates in the different stages in generation t. The k factor for the ith stage is defined as $k_{it}=-\log s_{it}=X_{it}-X_{(i+1)t}$, where $X_{it} = \log N_{it}$, the logarithm of the density on which the factor operates, and total $K_t=k_{1t}+k_{2t}+\ldots+k_{Lt}$. The standard test for density dependence of a k factor uses the linear regression of k_{it} on X_{it}. For serial data, type I errors may be inflated by lack of statistical independence (section 5.3.2), but the effect is complicated by the different contributions of the k factors to total K_t. Measurement error presents problems for both serial and locational data.

5.6.1 Vickery's randomization test for a density-dependent k factor

Vickery (1991) proposed a randomization test of density dependence in key factor analysis for serial data assuming no measurement error. This tests the null hypothesis of density independence by randomizing the set of k factors over different generations, while preserving the pattern of the k factors within a generation. Application to nine sets of published data which had previously shown density dependence in at least one k factor (5% level) produced randomization P values between 2 and 44%, suggesting that some of the early analyses were biased towards detecting density dependence. However, the randomization test indicated at least one density-dependent k factor in eight out of the nine populations. Further analysis using simulated density-independent data showed that the bias in the regression test was associated with both high mean mortality and high variance. Thus, tests involving key factors (i.e. those which account for a large percentage of total K) are prone to spurious detection of density dependence.

5.6.2 Measurement error

In key factor analysis, measurement error in density biases the regression of k_{it} on X_{it} (Kuno, 1971; Ito, 1972). If the relationship is linear with slope γ the bias is approximately equal to $(1-\gamma)V_{uu}/(V_{XX}+V_{uu})$, where V_{XX} is the variance of the X_{it} without measurement error and V_{uu} is the measurement error variance, assumed constant. In general, the magnitude of the bias depends on the ratio of the measurement error variance to the variance in log density, but weak density dependence ($\gamma<1$) is overestimated and strong density dependence ($\gamma>1$) underestimated (section 5.5.3). The regression test for zero slope is biased towards detecting density dependence.

5.6.3 Manly's simulation model for key-factor analysis

Manly (1977, 1990) proposed a model for key-factor analysis which incorporates the order in which mortality occurs through the stages, and a sequential pattern of density dependence. Using Manly's notation, the k factor for stage i is related to the log density entering stage i by

$$k_{it}=\tau_i+\delta_i R_{it}+\epsilon_{it} \tag{5.16}$$

where R_{it} denotes the log density of individuals entering stage i in generation t, τ_i and δ_i are parameters, and the ϵ_{it} are independent random variates with mean zero and variance σ_i^2. The model can be used to express the between-generation variation in log density of any stage in terms of the density-dependent effects (δ_i) and the variances of the random density-independent effects (σ_i^2).

Estimates of the parameters are obtained by fitting the model by least squares. The method is then repeatedly applied to simulated data from fitted models (i) with no density dependence in a given k value to test for an effect, and (ii) with density-dependence to estimate biases and standard errors in the estimated parameters. The approach allows for the complex pattern of serial dependence in the series. It can also be used to examine biases arising from measurement error in densities when an estimate of the measurement error variance is available.

The method was illustrated for the winter moth *Operophtera brumata* in Wytham Wood, near Oxford, England, using life-table data with estimated mortalities for six stages (Varley *et al.*, 1973). Ordinary regression analysis identified over-winter loss (k_1) as the key factor accounting for most of the variation in total K, and detected density dependence in pupal predation (k_5). The simulation method showed that the regression analysis was biased towards detection of spurious density dependence for over-winter loss, but confirmed that pupal predation was density dependent with a stabilizing effect (section 5.6.4). Repeating the analysis adding random measurement errors in $\log_{10}$ density (S.D.$=0.05$) did not affect the conclusions. Note that density dependence is not detected by the tests of Bulmer, Pollard *et al.*, and Dennis and Taper applied to the densities of each life stage in successive generations ($P>0.05$).

Manly's model is linear in log density, but in principle the approach applies to other formulations. For example, survival probabilities could be related to density and other explanatory variables on a scale which guarantees values in the range (0, 1), e.g. $\log(s/(1-s))$ or $\log[-\log(s)]$ are widely used in survival analysis (McCullagh and Nelder, 1983; Cox and Oakes, 1984; Lebreton *et al.* 1992).

5.6.4 Role of *k* factors in population dynamics

A *k* factor may be density dependent, but this does not necessarily imply that it has a large effect on the population dynamics. It might be unable to account for the observed mean population density, or the magnitude or pattern of the fluctuations in density between generations, its effects being compensated by other factors. One approach to examining the role of a *k* factor is to use simulated data from models in which the strength of density-dependent and density-independent effects are allowed to vary. Simulation models of the winter moth data suggest that pupal predation (k_5) is the main factor regulating population density (Varley *et al.*, 1973; Manly, 1977). However, Den Boer (1986c) has shown that simulated changes in population density using the observed values of the *k* factors, but with k_5 fixed at its mean value, reduces *LR* for both larvae and adults. A similar effect occurred by simulating changes using randomization of the observed values of k_5. Den Boer's null model has been criticized because it constrains all simulated populations trajectories to have the same initial and final values (Latto and Hassell, 1988). However, further simulations (Manly, 1977) show that a reduction in *LR* for adults is unlikely for the fitted model with direct density-dependent pupal predation, supporting the doubts expressed by Den Boer. Note that the interpretation of dynamics of the winter moth is further complicated because densities of larvae and pupae in successive generations show evidence of delayed density dependence (Turchin's test: larvae, $t_{14}=-2.79$, $P<0.05$; pupae, $t_{14}=-2.87$, $P<0.05$; also see Roland, Chapter 13 in this volume). For further discussion, including problems of formulating null models for assessing the effects of a density-dependent *k* factor, see Latto and Berstein (1990a, 1990b) and Den Boer (1990c).

5.7 DENSITY-DEPENDENT POPULATIONS IN SPACE AND TIME

The methods for analysing density dependence (sections 5.3 to 5.6) are based on models in which the *per capita* change is a function of the average population density from generation to generation, i.e. they assume a spatially homogeneous population. More realistically, populations are spatially heterogeneous with processes operating on a range of different scales in subpopulations linked by dispersal. Mortality and dispersal may show spatial density dependence across patches, reflecting competition within patches or aggregation of predators or parasites in response to patch density. This complex spatial structure raises important issues for identifying density dependence, including (i) how spatial density dependence within a generation translates into temporal density dependence between generations, and whether the effect is sufficient to regulate the population; (ii) the effect of spatial heterogeneity on methods for analysing density dependence; (iii) the design of population studies and the use of spatio-temporal sampling schemes.

5.7.1 Heterogeneity, stochasticity and density dependence

When analysing population change between generations using average densities aggregated over different spatial scales, the question arises whether the spatial heterogeneity and stochasticity can obscure the detection of temporal density dependence. Hassell (1986a, 1987) examined the issue using a simple model for a univoltine insect herbivore (De Jong, 1979). In the model, adults disperse and deposit eggs on plants which form subpopulations. The eggs hatch into larvae which compete for food and develop to form the next generation of dispersing adults. The pattern of dispersal is described by a negative binomial distribution, and the survival of eggs on an individual plant is density dependent with an exponential mortality factor. The expected density of adults in generation $t+1$ is given by

$$E[N_{t+1}]=M\Sigma Fn\exp(-\mu Fn)P_n \qquad (5.17)$$

where M is the number of plants, F is the number of eggs laid per adult and P_n is the probability of n adults on a plant. This expression can be reduced to $N_{t+1}=N_t F\exp(-\mu F)/(1+aN_t)^{k+1}$, where k is the index of dispersion of the negative binomial distribution and $a=[1-\exp(-\mu F)]/Mk$. The model illustrates how the spatial distribution can affect the form of the between-generation density dependence. Simulations showed that random variation in fecundity failed to obscure density dependence between generations, but with random variation in the dispersion parameter, density dependence was less clear.

Mountford (1988) examined the effect of heterogeneity and stochasticity in a herbivore patch model with contest competition within patches. The power of the test of Pollard *et al.* (modified for a normal random walk model with no drift) was estimated in relation to the probability of survival of an individual larva (λ), the strength of density dependence (μ) and the degree of clumping ($1/k$). For fixed λ and μ, power decreased with $1/k$, i.e. spatial heterogeneity obscured the detection of density dependence. However, for fixed λ and $1/k$ the power decreased with the strength of density dependence μ. The seeming paradox is explained by the predominant effect of population size on power, with higher power in a larger population. For a given mean population size, spatial hetero-geneity enhanced the detection of density dependence. Further simula-tions showed that the effect of stochastic variation in the model parameters was neutral provided that variation did not change the mean population size. The study illustrates the subtle and complex effects of heterogeneity, stochasticity and nonlinear interactions (May, 1989).

5.7.2 Spatio-temporal sampling schemes

In a spatio-temporal scheme, densities are recorded (i) in time, over sev-eral successive generations; (ii) in space, according to some stratification to allow for spatial heterogeneity within a generation. Ideally, densities are recorded on a spatial scale at which density-dependent processes are operating. Note that the analysis of patterns of density dependence across spatial sampling units avoids the complication of serial correlation in time-series data. A strength of spatio-temporal data is that it can pro-vide a basis for developing mechanistic models of population dynamics. Some examples are as follows.

Hassell *et al.* (1987) illustrate the advantages of spatio-temporal sam-pling in a population study of the viburnum whitefly *Aleurotrachelus jelinekii*. A key-factor analysis on 16 generations of data failed to detect density dependence with the potential to regulate the population. However, an analysis of the numbers from egg stage to adult on 30 labelled leaves detected density dependence in eight out of nine genera-tions. The results were used to develop a model in which survival from egg to adult was density dependent and the number of eggs per leaf fol-lowed a negative binomial distribution (section 5.7.1). The model fitted the data reasonably well, predicting the observed pattern of population increase to an equilibrium level, and showing how the population could be regulated by within-generation density dependence.

Liebhold and Elkington (1989) describe a spatio-temporal sampling scheme for the Gypsy moth *Lymantria dispar* which involves constructing a life table over a spatial matrix of sampling points. These so-called 'mul-tidimensional' life tables may be used to detect density dependence on a fine spatial scale which would be missed by analysis of densities in suc-cessive generations.

Dempster *et al.* (1995; Chapter 4 in this volume) studied 10 species of phytophagous insects and seven parasitoids inhabiting the flower heads in patches of two herbaceous plants, black knapweed (*Centaurea nigra*) and lesser burdock (*Arctium minus*). Populations were monitored in more than 50 patches of each plant species over 2 years. A test of density dependence related the *per capita* change in numbers from one year to the next (i.e. between generations) to the initial patch density measured as numbers per resource (i.e. flower heads for hosts, or per host for parasitoids). Density dependence was detected in nine out of 16 species analysed. Further analysis failed to detect spatial density dependence in parasitism, suggesting that resources were limiting population growth through intraspecific competition.

Several large-scale monitoring schemes have a spatio-temporal structure with records at a fixed number of sites over several years (e.g. Rothamsted Insect Survey, Woiwod and Harrington, 1994; Butterfly Monitoring Scheme, Pollard and Yates, 1993). These schemes offer scope for (i) analysing temporal density dependence by combining data from several series; (ii) exploring variations in dynamic behaviour on a large scale; and (iii) analysing population synchrony (Pollard, 1991; Hanski and Woiwod, 1993).

5.7.3 Choice of spatial scale

Choice of spatial scale is crucial in spatio-temporal studies. Sometimes there may be an appropriate natural sampling unit, such as a leaf or patch, but in other cases the choice may be less clear, especially for relatively mobile species. Heads and Lawton (1983) illustrated the importance of spatial scale for detecting density dependence in a study of mortality in the holly leaf miner *Phytomyza ilicis* caused by the larval parasitoid *Chrysocharis gemma*. The relationship between mortality and host density was examined using nested quadrats varying in size from 0.03–1 m^2. Density-dependent mortality was strongest at the smallest sampling scale, became weaker as the data were aggregated into larger sampling units, and was not detectable at the largest scale.

Ray and Hastings (1996) examined the effect of spatial scale on the detection of density dependence reported in 79 insect population studies. Spatial scale was measured in terms of movement relative to size of area over which density was measured as high, if more than 10% of the population moved across the boundary, and low otherwise. Average detection rates for the mobile stages (small and large instars and adults) were 69% for high movement ($n=16$, i.e. one small instar, two large instars and 13 adults), compared with 23% for low movement ($n=66$, i.e. 31 small instars, 34 large instars and one adult). However, it is difficult to draw any sound conclusions from this analysis because (i) the results of individual studies are taken at face value; (ii) most studies used ordinary regression to test for density dependence; (iii) the method of measuring spatial scale appears rather arbitrary; (iv) the comparison of detection rates in high and low groups is largely confounded with the difference between small and large instars and adults. More generally, methods of analysing density dependence usually assume a closed population for which it is reasonable to relate densities in successive generations or stages in the life cycle. This is unrealistic when there is movement in and out of the population, especially over a small area. Such movement could add a random component to observed density, akin to measurement error, and lead to spurious detection of density dependence. The study of density dependence on a spatial scale that is small relative to the scale of movement of the insect raises difficult problems of interpretation.

5.8 DISCUSSION

The problems associated with the identification of density dependence in population data depend on the nature of the study, the type of data, and the scope and methods of analysis. In general, there are fewer problems with experimental studies involving relatively homogeneous populations in controlled environments, where densities can be manipulated and measured with relatively small error, than in observational field studies on heterogeneous populations in variable environments where density may be difficult or impossible to manipulate and measure accurately.

Tests of density dependence using time-series data apply a significance test of the null hypothesis of a random walk. A significant result indicates a fluctuation with a return tendency but does not show the pattern or strength of density dependence, nor does it identify an ecological mechanism. Tests using ordinary regression are badly biased, leading to spurious detection of density dependence, but these tests have been superseded by methods which provide a reliable P value under the random walk null hypothesis. There is no such thing as a universal optimal test of density dependence because statistical power depends on the assumptions made about the form of the data. However, for density-dependent populations fluctuating within the normal range of variation, Bulmer's test is relatively powerful for detecting density dependence. If there is a marked initial displacement from the mean, then this can be exploited by using the test of Pollard *et al.* or Dennis and Taper. Detection of density dependence is hindered by temporal trends in mean density, but Holyoak's modification of the test of Pollard *et al.* provides a solution to this problem when trends are approximately linear on a log scale.

The statistical power of tests for density dependence depends on the length of the series. Even in a favourable situation, a test may fail by chance to reject the random walk null hypothesis in a series of 20 generations of density-dependent data. Furthermore, patterns of density dependence may be obscured by variability caused by environmental changes. The point is emphasized by the fact that the none of the above tests for direct or delayed density dependence detected an effect when applied to densities of the garden chafer spanning 29 years (Milne, 1984). In some cases, it may be possible to include exogenous variables, such as food or weather, to increase power for detecting density dependence.

In general, the application of tests for density dependence to time-series data should be part of a broader statistical analysis which examines (i) time plots for trends or other types of non-stationarity; (ii) the autocorrelation function and partial autocorrelation function for delayed density dependence; (iii) phase plots for patterns of density dependence; (iv) quantile plots of log density and changes in log density for non-normal data and outliers. For routine application to large data sets, where it may be impractical to examine each series in detail, a sensible strategy to avoid missing interesting effects would be to screen the data by applying a small battery of tests chosen to cope with different patterns of density dependence (e.g. Bulmer, Pollard *et al.*, including Holyoak's modification for temporal trend, Dennis and Taper, and Turchin's test for delayed density dependence). The results of all the tests should, of course, be reported.

Measurement error affects all tests for density dependence leading to an inflated type I error and spurious detection. The only satisfactory way to allow for this is to obtain data on the distribution of the measurement errors. Computer simulation can then be used to estimate the magnitude of the bias. Measurement error is a general problem for the identification of density dependence, which also arises in fitting models to time series, k factor analysis of life-table data and the analysis of spatial density dependence. Computer simulation is a powerful tool for estimating the magnitude of the effect for different types of data and methods of analysis, but it cannot compensate for densities subject to unknown measurement error.

Long-term time series of population densities are fundamental to the study of population dynamics and density dependence, but progress in understanding requires more detailed data, other types of field studies and manipulative experiments.

The most widely used approach for insect populations is to collect life-table data over several successive generations. Densities and mortalities are estimated at different stages in the life cycle so that the effect of a density-dependent factor may be related to the density on which that factor operates, thereby increasing the chances of detecting density dependence. Such studies aim to identify causes of the mortality and underlying density-dependent mechanisms. Thus, the approach can provide a basis for developing mechanistic population models. The usual method of k factor analysis is affected by problems of serial dependence and measurement error, and the problem is complicated by the different contributions of the different k factors to total change. Computer simulation of data from fitted models is a recommended way of examining the biases and precision of parameters in estimated density-dependent relationships, and for checking the sensitivity of the analysis. Simulation model experiments can be used to assess the role of a density-dependent k factor in population dynamics, but care is needed in formulating appropriate null models.

Conventional life-table data can be relatively ineffective for detecting temporal density dependence in populations where regulation results from density-dependent processes operating within generations across different subpopulations or patches. Spatio-temporal sampling schemes at an appropriate spatial scale can be used to examine patterns of density dependence arising from spatial heterogeneity in density. They are also required for studies on the effect of density on movement, an important and difficult topic which has not been addressed in this chapter. Spatio-temporal sampling schemes will usually require a relatively large outlay of sampling effort, but this may be offset by a more rapid progress towards understanding, generating testable hypotheses, and developing population models. Spatial sampling schemes are intended to complement rather than replace long-term life-table data which are essential for estimating temporal variation in the pattern of spatial density dependence, testing model predictions, and detecting temporal density dependence between generations.

ACKNOWLEDGEMENTS

I thank Dr Jack Dempster, Dr David Elston, Dr Marcel Holyoak, Dr Ian Newton, Dr Ernie Pollard, Tim Sparks, and two referees for helpful comments on the manuscript.

6

Host–parasitoid dynamics

H. C. J. Godfray and C. B. Müller

6.1 INTRODUCTION

The dynamic interactions between populations of parasitoids and their hosts have been closely studied by ecologists for nearly a century. There are perhaps three main reasons why they have received so much attention. Firstly, parasitoids are exceptionally numerous, if often overlooked, members of virtually all terrestrial habitats. Quite how many species of parasitoids there are on earth is by no means clear, but the most likely figure is somewhere between 0.5 and 2 million (Godfray, 1994). The second reason is that parasitoids can cause high mortality to their hosts and thus are potential allies in man's war against insect pests (Waage and Hassell, 1982; May and Hassell, 1988; Godfray and Waage, 1991; Mills and Getz, 1996; Murdoch and Briggs, 1996). The study of parasitoid population dynamics was initiated by applied entomologists, and the consequences of the release of parasitoids as biological control agents provide some of the most powerful testimony for the ability of parasitoids to control host densities. The final reason why parasitoids have attracted study is that their relatively simple life cycles make them attractive model systems for the theoretical and experimental study of many aspects of producer–consumer interactions (Hassell, 1978; Hassell and Godfray, 1992). In comparison to predators, where relating the amount of food consumed to the numbers entering the next generation is often difficult, host location and oviposition by the female parasitoid are linked in a very straightforward manner to recruitment.

A brief summary of the history of host–parasitoid population dynamics might go something as follows. The scientific foundations of the subject were laid chiefly by applied entomologists, such as Howard, Fiske,

Insect Populations, Edited by J. P. Dempster and I. F. G. McLean. Published in 1998 by Kluwer Academic Publishers, Dordrecht. ISBN 0 412 83260 7.

Thompson, and especially Nicholson. Nicholson, whose centenary has just been celebrated, was immensely influential in setting a research agenda that is still very active today. He considered a monophagous parasitoid attacking a host that suffered no other density-dependent mortality, in a situation where both hosts and parasitoids have discrete and synchronized generations. In collaboration with the mathematician Bailey, he showed that the simplest possible model of a randomly searching parasitoid (giving rise to a Poisson distribution of encounters per host) predicted an unstable interaction with steadily diverging population oscillations (Nicholson and Bailey, 1935). What Nicholson and Bailey had identified was the inherent propensity of most consumer–resource systems to oscillate; the same propensity that causes the neutral cycles of the Lotka–Volterra predator–prey model, but magnified by the time delay implicit in a discrete generation framework.

Nicholson and Bailey posed what we shall call the classical problem of host–parasitoid population dynamics: how can a coupled host–parasitoid system persist? The challenge of solving this problem was first taken up by Varley in the 1960s, but was largely developed by Hassell and his colleagues in the following two decades (see below). Parallel to this, and spurred by the advent of high-powered computers and the development of 'systems ecology', other ecologists attempted to understand host–parasitoid population dynamics through the construction of large and detail-rich simulation models, normally tied to particular pest systems (e.g. Barlow and Dixon, 1980; Gutierrez and Baumgärtner, 1984). To this day, the two approaches have remained largely separate, to the detriment of both (for recent examples and an introduction into this literature see Ravlin and Haynes, 1987; Axelsen, 1994; Gutierrez *et al.*, 1994; Flinn and Hagstrum, 1995; Barlow *et al.*, 1996; Mills and Gutierrez, 1996; Van Lenteren *et al.*, 1996; Van Roermund *et al.*, 1997a, b).

By the mid-1980s, the perceived concentration of intellectual and other resources on the classical problem had led to a reaction. Perhaps the model of a monophagous parasitoid regulating its host was not the best paradigm to understand the ecology of many parasitoids in the field? Hosts attacked by parasitoids need not invariably be regulated by them, while local host–parasitoid populations may actually be unstable after all, with persistence occurring over larger spatial scales (Murdoch *et al.*, 1985; Price, 1991c). But while some called for the complete abandonment of the classical approach, most workers today adopt a pluralist approach, acknowledging the diversity of issues and problems facing parasitoid ecologists.

Our aims in this chapter are twofold. Firstly, to provide a concise account of what we believe are the most exciting problems facing population ecologists working with parasitoids, and an entry into the more technical literature. Secondly, we have selected one problem that our research has been directed at, to discuss in more detail. This is the question of how to study the population dynamics of complex communities of hosts and their parasitoids.

6.2 A TAXONOMY OF ISSUES

Population ecologists want to understand the factors regulating and controlling the numbers of their study organisms. At a very crude level of analysis, parasitoid densities are likely to be determined by interactions with their host, as envisaged in the classical problem; competition within their own trophic level; or a combination of the two. The first model of parasitoid population ecology is called 'top-down' by insect–plant people, as it implies that herbivores are regulated by their parasitoids, while the second model is called 'bottom-up', because now the herbivores are regulated by competition for food, while regulation of the parasitoid population is not intimately linked with host dynamics. An important unanswered question is the extent to which different parasitoid species are best understood using one or the other model, or a combination of both. Can we discern patterns, such as leaf miners tending to be regulated by their parasitoids, while butterflies and their parasitoids better fit the bottom-up model? Are natural and agricultural systems likely to be different? Is it worth seeking these general patterns at all, or should we just proceed case by case? Each paradigm has its own set of unsolved issues. While much progress has been made on the classical problem, we still do not have a good understanding of how coupled host–parasitoid interactions may persist in the field. The issue of parasitoid regulation in bottom-up regulation has received much less attention and there are many unresolved issues, in particular: what stops parasitoid populations increasing in density until they begin to influence host abundances?

An orthogonal set of issues concerns the scale of investigation required to understand parasitoid population dynamics. There are two main hierarchies of scale to be considered. The first is spatial scale. To what extent can one study local populations in isolation from other populations of same species? The second hierarchy is referred to as community scale. To what extent can individual host–parasitoid interactions be studied without considering other parasitoids that attack the same host or other hosts of the parasitoid, and beyond that, does one need to consider plants, competitors and other natural enemies?

There are two further issues that we do not have space to discuss here. In addition to spatial and community scales, there is another hierarchy of scale, of behavioural detail. Parasitoid population dynamicists frequently summarize the behaviour of the host and parasitoid in terms of two constant parameters: host fecundity and parasitoid searching efficiency. But the last 30 years have seen an enormous amount of research into parasitoid behaviour, and we now have exquisite understanding of the mechanistic and evolutionary basis of many parasitoid behaviours (see review by Godfray, 1994). How much of this detail should we consider in trying to explain parasitoid population dynamics? Can an explicitly evolutionary approach to parasitoid behaviour contribute to understanding population dynamics? Aspects of these questions are discussed by Hassell and May (1985); Bernstein *et al.* (1988, 1991); Visser *et al.* (1992); Godfray (1994); Driessen *et al.* (1995); Ives (1995); Driessen and Visser (1997); and Weisser *et al.* (1997).

Finally, how best can population ecologists assist in the design and implementation of biological control programmes that use parasitoids? Is there a real dichotomy between population ecologists trying to discern general patterns in nature and using relative simple strategic models and thinking, and applied biologists concerned with particular pests using complex simulation models? Is the role (if any) of population ecologists to produce broad guidelines about which species are likely to be most efficacious, or can they contribute to the operational decisions faced on a daily basis by pest managers? Mills and Getz (1996); Murdoch and Briggs (1996) and Murdoch *et al.* (Chapter 7 in this volume) provide recent discussions of these problems.

It is easier to pose than to answer these questions, and we make no pretence that in this chapter we shall provide any devastating new insights. Yet we hope to highlight some issues that we believe have received less attention than they deserve, and to convey some of the excitement in current parasitoid ecology.

6.3 BOTTOM-UP OR TOP-DOWN?

The vast majority of insect herbivores are apparently too rare to be limited by their food supply and are subject to often quite high levels of parasitoid attack. But are they rare because their populations are regulated at a comparatively low density by their parasitoids (top-down regulation), or are they rare because of a more subtle interaction with their food supply (bottom-up regulation)? Perhaps herbivore intrinsic rates of increase are relatively modest so that only occasionally are they subject to competition for food, or perhaps the acres of verdant foliage are misleading and the hosts compete for limited high-quality resources. Both scenarios raise interesting questions. If top-down regulation occurs, how do host–parasitoid systems persist in the face of a tendency for diverging oscillations, identified by Nicholson? This is the classical problem discussed in the next section. But if bottom-up regulation occurs, what factors regulate parasitoid densities and prevent them from influencing their host densities?

Let us consider first the experimental evidence that can be mustered by the proponents of top-down regulation. Firstly, laboratory experiments conclusively demonstrate persistent regulation of host density some way below carrying capacity in a variety of different systems (e.g. Utida, 1957; Bellows and Hassell, 1988; Tuda and Shimada, 1995; Bonsall and Hassell, 1997). Secondly, biological control works – not always, but with sufficient frequency to persuade hard-headed programme managers to invest relatively large amounts of money in biological control programmes (Beddington *et al.*, 1978; Waage and Greathead, 1988). Thirdly, there are a few modern ecological field studies that point to parasitoids as the main factor regulating a host population (e.g. Luck and Podoler, 1985; Gould *et al.*, 1990; Murdoch *et al.*, 1996b; references in Berryman, 1997). Finally, many ecologists find it extremely hard to envisage how the nearly 300 species of herbivore found on a tree such as the English oak, *Quercus robur* (Southwood, 1961) could each be regulated by resource competition.

A bottom-up regulator might set forth the case as follows. If you survey the careful life-table studies of insect herbivores that are attacked by parasitoids, it is rare that parasitoids display the type of temporal density dependence required of a regulatory agent (Dempster, 1983; Stiling, 1987). Moreover, the tendency for diverging oscillations shown by coupled host–parasitoid systems suggests that top-down regulation is unlikely, and a model in which parasitoid dynamics are decoupled from that of their hosts is more likely to explain the continued presence of parasitoids in nature. Finally, resource depletion is seen in nature, and may be very much more subtle than "but the earth is green" arguments imply.

The bottom-up ecologist would go on to criticize the top-down position. Yes, parasitoids can regulate their hosts in the laboratory, but these are artificial situations (almost analogue computers) set up to prove a point. Moreover, some artificial systems show diverging oscillations and no regulation. Biological control, by its very nature, usually occurs in artificial habitats, and its success may be overestimated by the lack of controls. The density of an introduced pest that has reached epidemic level might have declined anyway, irrespective of the release of a parasitoid. The few field studies where parasitoids have been shown to be important must be balanced against the many that show the opposite, while hunches about oak tree herbivores are no substitute for data.

The top-down ecologist would also have something to say about hunches, but would start by attacking the claim of bottom-up's proponents that the survey of field studies was a good argument against parasitoid regulation. Firstly, the species surveyed are not an unbiased sample of insect herbivores. They have been the subject of intensive ecological investigation and more likely than not selected for study because of their convenience or obviousness. An ecologist planning a long-term field study is more likely to choose a common species, especially if it shows interesting behaviour, such as resource depletion, than a rare species that exists at such low density that sampling is time-consuming and difficult.

Moreover, detection of parasitoid regulation requires that studies are carried out at the correct temporal and spatial scale (Walde and Murdoch, 1988; Hassell *et al.*, 1989a).

So where do we stand? Few would argue that either model is invariably correct and the question is of degree rather than kind (e.g. Hunter and Price, 1992). More field studies are required of particular host–parasitoid systems, and ideally these studies should be conducted in parallel with the parameterization of specific quantitative models. It is a shame that some of the best recent field studies of particular systems consist of series of experiments and hypotheses without an overlying quantitative framework within which different processes can be assessed. We are also optimistic about what biological control programmes can tell us about broader issues, especially if carried out in conjunction with ecological studies.

6.4 THE CLASSICAL PROBLEM

How might coupled host–parasitoid interactions persist in nature? There has been an enormous amount of research on this problem, and we can do no more than touch on some of the major contemporary issues in this field. For more extensive reviews and discussion see Hassell (1978, 1986b, 1998); Hassell and Waage (1984).

The problem identified by Nicholson and Bailey is that in the simplest model parasitoids tend to overexploit their hosts, which drop to low population densities. This drop is followed by a decrease in parasitoid densities to such a level that hosts suffer very little mortality and their populations begin to rocket, followed by an even bigger increase in parasitoid densities. These successive rounds of host overexploitation and parasitoid recovery end with the extinction of the parasitoid or of both host and parasitoid. Such diverging oscillations have been demonstrated in artificial laboratory experiments (Burnett, 1958).

The Nicholson–Bailey model assumes that the risk of being parasitized is a linear function of the number of parasitoids (and independent of host density). This assumption is unrealistic when host densities are sufficiently high that parasitoid attack is limited by handling time (the time taken to attack a host and lay eggs) or by egg supply. The relationship between the number of hosts attacked and host density is normally thought to increase with declining slope to an asymptote (type 2 functional response, Holling, 1965). The inability of parasitoids to control high densities of hosts implied by a type 2 functional response is a factor that contributes towards the instability of the interaction (Hassell and May, 1973; Getz and Mills, 1996). A type 3 functional response, which is sigmoid in shape, can help stabilize an interaction as parasitism initially increases super-linearly with rising host densities. However, the stabilizing influence is not strong and cannot by itself stabilize the Nicholson–Bailey model (Hassell and Comins, 1978).

For both the type 2 and type 3 functional response, at high host densities the risk of parasitism becomes a function of the ratio of parasitoid to host density, called ratio dependence. In recent years, some authors have advocated the general replacement of models based on linear risks (e.g. Nicholson–Bailey) with those based on ratio dependence (Akcakaya *et al.*, 1995; for an opposing view see Abrams, 1994). While there is a strong case for adopting ratio-dependent models in some areas of population dynamics, in parasitoid studies it seems a retrograde step, as (i) we already have a well understood mechanistic model that encompasses both the linear and ratio-dependent cases, and (ii) a purely ratio-dependent model has the unhappy property that as host densities approach zero the risk of parasitism approaches infinity.

There are two main ways of stopping the diverging oscillations identified by Nicholson from leading to the breakdown of the interaction. The first is if the efficiency of the parasitoid declines at high parasitoid density. Such a decline might occur if two females disrupt each other's searching or oviposition when they meet, something that is more likely to occur at high parasitoid densities. Behavioural studies have shown that, in some species, females fight if they encounter each other, although a more common behaviour is for females to disperse more readily after encounters with a conspecific (reviewed by Hassell, 1978; Godfray, 1994). It was shown in the 1960s that interference, as this occurrence is known, was capable of stabilizing the Nicholson–Bailey model by preventing high densities of parasitoids from destroying the host population (Hassell and Varley, 1969; Hassell and May, 1973). Laboratory studies indicated that interference might be strong enough to allow host–parasitoid persistence, but most workers today doubt whether levels of interference can be high enough in the field to allow a stable interaction.

Several related phenomena lead to a decline in parasitoid efficiency at high host densities. The evolutionary theory of local mate competition predicts that the proportion of female eggs laid by parasitoids with the appropriate natural history should decline with increasing population density (Hassell *et al.*, 1983; Comins and Wellings, 1985). Alternatively, if parasitoids do not avoid superparasitism, and if the progeny of more than one female compete in an overcompensating manner on or in the body of the host, then there is a resulting decrease in parasitoid efficiency with density (Taylor, 1997). But while these special arguments may apply to particular systems, they are unlikely to have wide generality.

The second way of stopping diverging oscillations is to give the hosts some protection from high densities of parasitoids. There are several equivalent ways of looking at this. Firstly, one can talk about the host having a refuge from the parasitoid, although this is apt to be confusing as the refuge may be probabilistic, while the word refuge is used in a more particular sense to describe a physical protection from parasitism. A more accurate, though more technical, description is to say that the host population experiences heterogeneity of risk of parasitism. Individuals with a comparatively low risk of parasitism are sheltered at times of high parasitoid density. An advantage of thinking about heterogeneity of risk is the remarkable result obtained by Pacala *et al.* (1990) (see also Pacala and Hassell, 1991; Hassell *et al.*, 1991b; Jones *et al.*, 1993) that for a very wide variety of models based on the Nicholson–Bailey framework, stability occurs whenever the coefficient of variation of risk exceeds one. Finally, one of the consequences of heterogeneity of risk is that the *per capita* efficiency of the parasitoid declines as parasitoid densities increase, because more and more parasitoids are chasing a diminished pool of potential hosts. As a result, models of this type resemble those for interference, which has led to the present phenomenon being called pseudo-interference, although we believe this cumbersome and not very informative term is best abandoned.

So much for the description of the process, but what biological factors may be responsible? Clearly, a physical refuge can give rise to heterogeneity of risk and help stabilize an interaction (Hassell and May, 1973; Hassell, 1978; Holt and Hassell, 1993; Hochberg and Holt, 1995). Physical refuges have been invoked more frequently in recent years to account for the persistence of host–parasitoid systems (Hawkins *et al.*, 1993; Begon *et al.*, 1995; Murdoch *et al.*, 1995, 1996b). The easiest refuges to conceive of are those involving hosts that live in concealed sites, such as galls, internally within shoots and wood, or within a deep substrate (Price and Clancy, 1986b; Lampo, 1994; Freese, 1995). Parasitoids searching on the surface of the plant or substrate can attack only those hosts within range of their ovipositors. A good laboratory example of physical refuges is provided by stored-product moths whose larvae feed in bran or other substrates, and who can escape parasitism by burrowing deep below the surface (Begon *et al.*, 1995; Sait *et al.*, 1995). The dynamics of laboratory populations depend on the degree to which the substrate is arranged to allow the host refuge from parasitism. However, it is much harder, though not impossible (Murdoch *et al.*, 1995, 1996b), to think how external feeders, and hosts inhabiting leaf rolls or leaf mines, can achieve physical refuges.

A second straightforward source of heterogeneity of risk is the temporal refuge that occurs if host and parasitoid phenologies do not precisely match (Münster-Svendsen and Nachman, 1978; Godfray *et al.*, 1994). Some hosts may complete development before or after the main peak of parasitoid activity. Such a mismatch may occur by chance in certain years, or may occur every year if, for example, hosts and parasitoids have different thermal or climatic optima.

The process that has received by far the most attention is spatial heterogeneity of risk. We consider here within-population spatial heterogeneity, that is, variation in risk of parasitoid attack within a population that mixes homogeneously each generation (we consider larger spatial scales later on). The classic approach is to consider a population of hosts divided into patches across which parasitoids distribute themselves at the beginning of the generation (Hassell and May, 1973, 1974). If parasitoids tend to aggregate to certain patches rather than others, and if they stay there throughout the generation, then hosts in infrequently visited patches are in a probabilistic refuge and the host–parasitoid interaction is more likely to persist. Parasitoids may aggregate towards certain patches because they contain more hosts, or for reasons unconnected with host density (May, 1978; Chesson and Murdoch, 1986), but stability occurs as long as some hosts have a sufficiently low risk of parasitoid attack (aggregation of parasitoids to patches with low host density can thus result in host–parasitoid persistence, although under more stringent conditions; Hassell, 1984).

The role of local spatial heterogeneity, and in particular parasitoid aggregation to patches of high host density, has been reappraised in recent years (Murdoch and Stewart-Oaten, 1989; Godfray and Pacala, 1992; Ives, 1992b; Murdoch *et al.*, 1992a; Taylor, 1993; Rohani *et al.*, 1994). It has become apparent that the assumption of a single episode of parasitoid distribution across patches is quite critical. Consider a parasitoid that homes in on host aggregations. If, instead of remaining in the patch throughout its life, it disperses after it has partially exploited the patch and looks for other host aggregations, then the host heterogeneity of risk will be reduced, possibly to a level that leads to instability of the system. For parasitoids that do not respond to host density, within-generation dispersal will cause a reduction in the heterogeneity of risk if successive rounds of parasitoid movement mean that most patches are visited, but will have no effect if certain patches are consistently avoided. Thus, local spatial heterogeneity may or may not contribute to the persistence of the system, depending on the detailed natural history of the system. There is a need for more studies of parasitoid movement patterns in the field (Waage, 1983; Casas, 1989; Driessen and Hemerik, 1992; Jones *et al.*, 1996).

Hosts can sometimes survive parasitoid attack, especially by endoparasitoids, through mounting an immune attack. The most common type of immunological response is encapsulation, in which a layer of cells is laid down around a foreign body; these cells subsequently melanize to form a dark capsule, killing the parasitoid it contains, in a way that is not fully understood. Within-species variability in host resistance to parasitoids has often been observed, and several authors have shown that this can be a source of heterogeneity of risk sufficient to stabilize the host–parasitoid interaction (Hassell and Anderson, 1984; Godfray and Hassell, 1990). However, the analysis of the population dynamic consequences of the variability in host resistance assumes that heterogeneity of risk remains constant across generations. This poses a problem, because there is now considerable evidence for a strong genetic underpinning of differences in host resistance (e.g. Carton *et al.*, 1992; Henter and Via, 1995; Kraaijeveld *et al.* 1998). Natural selection will thus tend to select for an optimum genotype and hence a reduction in heterogeneity of risk. The picture is complicated if we assume that there are costs to increased host resistance (Kraaijeveld and Godfray, 1997), and if natural selection operates simultaneously on parasitoid virulence. There is a need for both further modelling and experiments to explore this issue, though some preliminary results of ours, using joint population genetic and population dynamic models, cast doubt on host resistance as a factor allowing persistent interactions (A. Sasaki and H. C. J. Godfray, unpublished data).

Building on Nicholson and Bailey's work, most theoretical studies of host–parasitoid population dynamics have used models phrased as coupled difference equations. But while appropriate for temperate species with synchronized generations, this modelling framework fails to capture some of the population interactions that occur in systems with overlapping generations. This was first noted by Murdoch *et al.* (1987) who showed that if a host possesses a developmental stage that is both relatively long and immune from parasitoid attack, then this can lead to a persistent host–parasitoid interaction. The long-lived invulnerable stages act as a refuge that allows the host population to survive temporary periods of high parasitoid abundance.

The model developed by Murdoch *et al.* (1987) incorporated host and parasitoid age structure, making use of techniques pioneered by Gurney and Nisbet (Gurney *et al.*, 1983) in studies of resource competition in single populations. This technique is known as the lumped-age-class approach and is particularly suitable for studying insects whose life cycle is divided into discrete stages. The resulting models are phrased as differential equations, but incorporating time lags. Godfray and Hassell (1989) (see also a prior simulation study, Godfray and Hassell, 1987; and subsequent work by Godfray and Chan, 1990; Gordon *et al.*, 1991) used this technique to show that a host–parasitoid interaction in a relatively nonseasonal environment could show persistent cycles with a period of approximately one host generation, if the developmental period of the parasitoid was approximately 0.5 or 1.5 times that of the host. The cycles occur because of a phenomenon akin to resonance between the two time lags in the system; a more biological interpretation is given in the legend to Fig. 6.1. Cycles with a period of approximately one host generation are widespread in tropical plantation pests, and this may be due to interactions with their parasitoids (Godfray and Hassell, 1987, 1989) or pathogens (Briggs and Godfray, 1995). Reeve *et al.* (1994) have shown that this mechanism is likely to be responsible for generation cycles in a subtropical planthopper system.

An explicit consideration of age structure is also important in interpreting the dynamic consequences of host feeding, and certain parasitoid

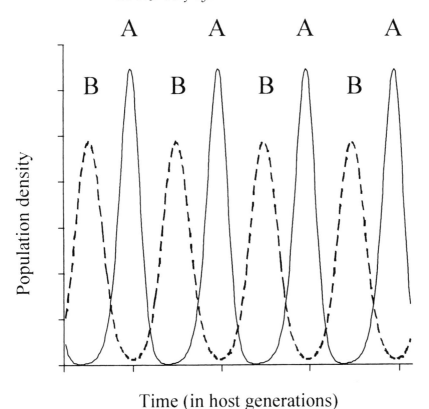

Time (in host generations)

Fig. 6.1 Generation cycles in host–parasitoid models. In continuous time, hosts (solid lines; host stage susceptible to parasitoid attack) and parasitoids (broken lines; searching adults) can both show persistent cycles with a period approximately equal to the generation time of the host, and with parasitoid maxima coinciding roughly with host minima. For this to occur, the parasitoid generation time must be 0.5 or 1.5 times that of the host. The cycles persist because most parasitism occurs when hosts are abundant (*A*), even though at this time parasitoids are relatively rare. The adult parasitoids that result from parasitism at time (*A*) emerge at time (*B*) where they reduce the density of hosts that survive to the next generation. Cycles also require that the adult host and parasitoid are relatively short lived compared to the developmental lags.

sex ratio and clutch-size strategies. Many parasitoids use small hosts for food and larger hosts for oviposition sites (Jervis and Kidd, 1986; Kidd and Jervis, 1991); lay male eggs on small hosts and female eggs on large (Charnov *et al.*, 1981); produce clutch sizes that increase with the size of their host; or produce offspring with greater fecundity or longevity on larger hosts (Godfray, 1994). The resulting age-structured dynamics can be very complex, with patterns of stable equilibria, persistent cycles and divergent oscillations depending quite critically on the biology of the system (Bellows and Hassell, 1988; Murdoch *et al.*, 1992b; 1997; Briggs *et al.*, 1995; Shea *et al.*, 1996). However, a broad pattern can be discerned (Murdoch *et al.*, 1997). What is common to all these behaviours is that parasitoid attack on young hosts contributes fewer female recruits to the next generation than does an attack on old hosts. Moreover, if young hosts are killed, they are clearly not available for more profitable (from the population point of view) parasitism later on. As the value of young hosts is reduced, the effect is initially to contribute to the stabilization of the dynamics because a large parasitoid population, which destroys a large proportion of the stock of young hosts, has less opportunity to propagate itself on the remaining old hosts. However, if the value of young hosts is further reduced, this density dependence becomes over-compensating and the result is cyclic population dynamics with a period of one to two generations. Murdoch *et al.* (1997) modelled age-structured populations in which the host has a long, invulnerable and stabilizing adult stage (as in scale insects), but it seems likely that the same processes will influence the dynamics of more typical host–parasitoid interactions.

6.5 PARASITOID DYNAMICS WITHOUT HOST REGULATION

Let us now turn to the factors that regulate parasitoid densities when the host itself is not regulated by that parasitoid. We need to identify the factors that prevent the parasitoid from increasing in density to a level at which host populations are severely depleted. As mentioned earlier, this scenario has received less attention than the classical model, and we attempt a simple classification of the ways parasitoid populations may be regulated in circumstances when the host itself is not regulated by the parasitoid. Our list is not exhaustive, and alternative schemes could be developed (Hochberg, 1996b). Note that a regulatory factor tends to return the population towards a stationary probability distribution of densities, not necessarily towards a fixed equilibrium point. We begin by concentrating on local populations of monophagous parasitoids.

6.5.1 Parasitoids regulated by interference competition

A few species defend patches of hosts from conspecifics, even though they are unable to parasitize all the hosts contained in a patch (for example, some scelionid egg parasitoids). Such direct competition, related to the behavioural interference discussed above, could regulate parasitoid numbers at a level below that at which they influenced host recruitment. However, as with the behavioural interference discussed above, there are few cases of strong direct competition in the field.

6.5.2 Parasitoids regulated by exploitation competition for hosts not in refuge

Suppose that the host population is largely protected from parasitism in a physical or temporal refuge. We have already discussed such refuges in the previous section, but now we assume that a sufficiently large fraction of the host population is protected that its logarithmic growth rate would exceed zero even if all unprotected hosts were parasitized (i.e. parasitoid regulation is impossible). In these circumstances, parasitoid densities would be regulated by indirect (exploitation) competition for hosts, with host dynamics decoupled from those of the parasitoid. Hawkins (1994) has argued that few hosts are regulated by their parasitoids, and that physical refuges from host attack are pervasive among herbivores.

Suppose now that the size of the refuge varies from generation to generation, but that again parasitoid regulation is not possible (the long-term geometric-mean growth rate is greater than zero even when all hosts outside the refuge are attacked). Exploitation competition will 'bite' most strongly in those years when the host refuge is relatively large. In such populations, it might appear as if the parasitoid is able to attack most hosts in most years. The mechanism of parasitoid population regulation would thus be unclear (especially if parasitoid population growth rates are relatively low) unless a rare year of intense competition is observed.

A related situation occurs when parasitoids have a much narrower phenological range than their host. Parasitoids may be regulated by exploitation competition for those hosts in the correct stage at the right time of year, but be unable to regulate the hosts, most of which are able to avoid the parasitoid temporally. Again, if the temporal overlap of hosts and parasitoids varies from year to year, high levels of exploitation competition may be observed only occasionally. Some of the parasitoids that attack the sawfly galls studied by Price and colleagues (Chapter 14 in this volume) may be regulated in this manner.

6.5.3 Parasitoids regulated by exploitation competition for hosts; coupled dynamics with bottom-up host regulation

Consider now the case where a host is subject both to intraspecific regulation through, for example, density-dependent competition for food, and to parasitoid attack. When intraspecific competition is weak and parasitoid attack strong, the former has relatively little effect on the host's population dynamics and we are again in the realm of the classical model discussed in the previous section. Because, ultimately, all hosts must be subject to some sort of intraspecific competition for resources, an implicit assumption of the classical model is that parasitoids are able to regulate their hosts at a density sufficiently low that intraspecific competition is not important.

As intraspecific competition becomes relatively stronger, both host density dependence and parasitoid attack can combine together to stabilize the host–parasitoid interaction (Beddington *et al.*, 1978; Hassell, 1978). Exactly if and how this occurs can depend quite critically on the biology of the system. In particular, the order in the life cycle of intraspecific competition and parasitoid attack, and whether the host density dependence is under- or overcompensating, is significant (Wang and Gutierrez, 1980; May *et al.*, 1981; Bernstein, 1986; Hochberg and Lawton, 1990). To take an extreme case, consider a host–parasitoid interaction described by the original Nicholson–Bailey model, when the host is subject to a ceiling form of density dependence with strong overcompensation. The resulting dynamics are highly cyclic or even chaotic. Note that if the density dependence acting on the host also increases parasitoid mortality, perhaps because parasitized individuals are more susceptible to stress, then the ability of the parasitoid to depress or regulate the host population density will be reduced (May *et al.*, 1981; Bernstein, 1986). Ives and Settle (1996) have recently observed particularly high parasitoid mortality in dense aphid populations and have suggested that this may explain their inability to control pest levels of aphids.

Now take this trend one stage further, and consider a parasitoid with a very low searching efficiency. The parasitoid can persist only if it is able, on average, to produce one female offspring per generation to replace itself. By analogy with the epidemiological literature we can call the density of hosts at which this occurs the host threshold. If the host threshold exceeds the host's carrying capacity, then clearly the parasitoid cannot persist. If the host threshold is near but below the carrying capacity, then the parasitoid can persist without being responsible for regulating the host population. The parasitoid does have some influence on host population levels, however: when parasitoid densities rise then the host population falls, leading to a drop in parasitoid reproductive success. The parasitoid is regulated via exploitation competition for hosts. But if host densities rise, the equilibrium is restored by the intraspecific density dependence of the host rather than by the density dependence in the host–parasitoid interaction.

6.5.4 Parasitoids regulated by exploitation competition for hosts; coupled dynamics with top-down host regulation

This scenario is similar to the previous one, except that now the host is not bottom-up regulated but is top-down regulated by another natural enemy (the case where this is another parasitoid is discussed below). It is possible for the parasitoid to persist in the system, but without being the major factor responsible for host regulation.

The conditions for a persistent three-species interaction to occur are more stringent than those for the two-species interaction just discussed. In the simplest case, the equilibrium host thresholds for parasitoids and predators (see section 6.5.3 above) must be identical, otherwise the species able to survive at the lowest host threshold excludes the other. However, various factors may allow co-existence of natural enemies, a subject we shall return to below when we discuss parasitoid communities.

6.5.5 Parasitoids regulated by exploitation competition for non-host resources

Some parasitoids require food, such as honeydew or nectar, as adults. Although we cannot provide an example, it is at least conceivable that parasitoids are limited by exploitation competition for these non-host resources.

6.5.6 Parasitoids regulated by their natural enemies

Parasitoids are attacked by their own natural enemies, of which the most important from a dynamic perspective are probably hyperparasitoids. We can distinguish three main ways in which hyperparasitoids may influence their hosts. Firstly, they may reduce their numbers to a sufficiently low level that the primary parasitoid no longer has a major influence on host dynamics. How the hyperparasitoid maintains primary parasitoids at this low level is the classical problem discussed above. Secondly, there may be a dynamic three-species interaction between the hyperparasitoid, parasitoid and host. Such interactions have been studied within a Nicholson–Bailey framework by Beddington and Hammond (1977) and May and Hassell (1981). In general, it is harder for a three-species system to persist than a two-species system. Finally, the hyperparasitoid may have little influence on its host. How the hyperparasitoid persists can then be explained by the same arguments that we are currently deploying to explain the persistence of a non-regulating primary parasitoid.

6.5.7 Parasitoid regulated on a larger spatial or community scale

Thus far we have considered the local persistence of a monophagous parasitoid that does not regulate its host. But a locally unstable interaction might still persist at a regional level as part of a metapopulation with continual colonizations balancing extinctions. The density of a polyphagous parasitoid may be regulated by its coupled interactions with one host, yet still cause mortality of other hosts with which its dynamics are not coupled. It is to these processes, operating on larger spatial and community scales, that we now turn.

6.6 THE SPATIAL DIMENSION

Our discussion so far has concentrated on local population dynamics; the only spatial heterogeneity mentioned has been small scale, within the ambit of a single parasitoid. More technically, we have assumed the population is homogenized relatively frequently so that the system has little spatial memory. Starting locally is a sensible research strategy, but in the past 10 years ecologists have asked whether the answer to the classical problem of local instability might lie in processes operating on a wider spatial scale.

Research has focused on two issues: the study of continuous populations of hosts and parasitoids occupying a one- or two-dimensional habitat; and the study of loosely connected networks of local host–parasitoid systems/host–parasitoid metapopulations. The first result that emerges is that if a local host–parasitoid interaction is stable, then embedding it in a spatial framework is unlikely to alter this fact. The second result is more interesting: host–parasitoid interactions that are unstable locally may persist globally. For this to occur, local populations must fluctuate out of synchrony, so that if one is destined to go extinct it can be 'rescued' via immigration from another.

Consider first a host–parasitoid interaction in continuous space. Most work has retained the assumption of discrete and synchronized host and parasitoid generations and used a variety of mathematical techniques – for example, cellular automata, coupled map lattice, or integrodifference equations – to incorporate the spatial dimension. The most robust patterns that emerge are of spiral waves of hosts and parasitoids propagating through space (Hassell *et al.*, 1991a; 1994; Comins *et al.*, 1992; Boerlijst *et al.*, 1993; Rohani and Miramontes, 1995; Wilson *et al.*, 1998). Due to dispersal, the host population moves through space as a travelling wave which is 'chased' by a second travelling wave of parasitoids. Just as in time parasitoid population peaks follow those of hosts in the Nicholson–Bailey model, so the parasitoid wave follows the host wave in space. Indeed, there is a mathematical symmetry of time and space in these models. If the spatial arena in which the interaction is taking place is large enough, then hosts and parasitoids are able to persist, even though one or both would go extinct in a local interaction. The wavelength and coherence of the spirally propagating waves depend on the detailed assumptions of the model, especially the average dispersal distances of the two species. For some parameter values, the spirals disappear completely and one is left with a seemingly patternless array of host and parasitoid densities that has been called spatial chaos. An interaction described by this type of dynamics would appear to have relatively stable densities of hosts and parasitoids if numbers were averaged over a broad spatial scale, but would show dramatic fluctuation in density if sampling occurred at just a single locality.

Classical host–parasitoid metapopulation models do not explicitly consider spatial structure, and hence their predictions are more straightforward. The crucial parameter is the frequency of host and parasitoid dispersal between patches. For a locally unstable interaction to persist globally, dispersal must be neither too little nor too much. Too little and the rescue effect is too weak and colonization cannot compensate for extinction; too great and the ensemble of populations become dynamically linked and oscillate in unison to oblivion (Murdoch *et al.*, 1984; 1985; 1992a; Reeve, 1988; Taylor, 1988; Adler, 1993).

Until recently, arguments about the importance of spatial processes have had to rely on a rather poor empirical base, typically, informal observations of the results of biological control releases. However, the past year has seen a number of important experimental studies published. The spatially most extensive has been Roland and Taylor's (1997) work on *Malacosoma disstria* in Canada. *M. disstria* is an outbreak pest of trees in aspen parkland. A large grid of 127 sampling points within a 25×25 km study area was set up, and the density of larvae and their percentage parasitism estimated at each. The pattern of parasitism was linked to the local extent of woodland, as estimated from satellite imagery. The authors found that the four species of parasitoids responded differently both to host density and to local woodland fragmentation. Moreover, body size explained some of the differences. Large species showed positive density-dependent parasitism and their distribution was correlated with large-scale measures of forest fragmentation. The smallest species showed negative density-dependent parasitism with its distribution correlated with small-scale measures of forest fragmentation.

The results reported so far are a snapshot of the dynamics from a single generation. But this scale of study offers the prospect of testing some of the theory described above. One interesting result to emerge already is the spatial density dependence shown by the larger parasitoid. While it was argued above that local spatial density dependence may actually tend to destabilize an interaction if parasitoids visit many patches within a generation, large-scale density of the type reported here is likely to be a strongly regulating factor, as the distances involved would preclude many episodes of redistribution (see also Gould *et al.*, 1990; Ferguson *et al.*, 1994).

While metapopulations are something of the 'flavour of the month' in population dynamics, there are actually rather few well documented examples of ensembles of populations in a dynamic equilibrium of extinctions and colonization. Perhaps the best entomological example is the population of Glanville Fritillaries (*Melitaea cinxia*) in the Åland archipelago of western Finland (Hanski *et al.* 1995a; Chapter 1 in this volume). An ensemble of small populations show frequent local extinctions, but also local recolonizations. Extinctions have been shown to be correlated with population size, and most recently with the presence of a locally specific parasitoid, *Cotesia melitaearum*. Cases of parasitoid, but not host, extinction, and of parasitoid colonization of host patches, have also been documented (Lei and Hanski, 1997). The picture is complicated by the presence of other parasitoids in the system, including another primary parasitoid that does best in patches where *C. melitaearum* is absent, and a polyphagous hyperparasitoid which shows a strong density-dependent response to *C. melitaearum* cocoons.

Finally, Harrison (1997) and colleagues have identified a curious spatial pattern in a *Lupinus* herbivore, the moth *Orgyia vetusta*, that may result from a host–parasitoid interaction. While lupin is relatively widespread along a coastal strip of northern California, persistent outbreaks of the herbivore occur only on certain patches of its host plant, and these may persist for many years without spreading. Experiments have excluded differences in host-plant quality, leading Brodmann *et al.* (1997) to ask whether parasitoids dispersing from the outbreak area might cause a halo of intense parasitism in the surrounding bushes, that prevents the spread of an outbreak. Experimental studies involving placing larvae in the field at varying distances from the outbreak show that indeed parasitism is highest in the areas surrounding the outbreak. However, questions remain: in particular why should parasitoids disperse away from outbreaks and so presumably lower their reproductive fitness, and why do not parasitoids reduce population densities within the outbreak? Interference between searching parasitoids is a possible answer, and one amenable to experimental investigation.

6.7 COMMUNITY PATTERNS

The second major axis of scale is community complexity. Isolated host–parasitoid interactions are a relative rarity in the field, especially in natural as opposed to agricultural habitats; the vast majority of hosts are attacked by more than one parasitoid and the vast majority of parasitoids can develop on more than one host. To what extent is it necessary to consider community complexity in our attempts to understand the population dynamics of parasitoids? There have been two main approaches to extending host–parasitoid studies to more than two species. The first is to build upwards incrementally through the addition of extra species of hosts and parasitoids. The second is to try to consider directly the dynamic properties of complex communities.

Consider, firstly, communities of three species: one host, one parasitoid and a third component. If the third component is another host, we have a situation that was first studied in a more general context by Holt (1977) and more specifically for host–parasitoid dynamics by Holt and Lawton (1993, 1994). Let us suppose that the interaction between the parasitoid and either of its hosts is stable when considered in isolation, the stability arising through any of the processes discussed previously. When the three-species community is assembled we expect it to be non-persistent, with the host that can survive at the lower equilibrium density being triumphant. At this equilibrium, the parasitoid is sufficiently common that the other species cannot invade: the parasitoid exerts too strong a mortality on it. This result has many resonances in other fields of ecology. For example, in competition between two species of plant, the species that can survive at the lower equilibrium level of the limiting resource wins out (Tilman, 1982).

This result is important in understanding community structure. One species of host that feeds on a different resource from a second, which it may never encounter in the field, can still be responsible for the exclusion of the latter because of indirect interactions mediated via natural enemies. This type of indirect effect has been called apparent competition, because many of its community consequences are similar to those of direct competition. Because of apparent competition, parasitoids may appear to be specialists, not because of any physiological or ecological constraint, but purely because of the dynamics, what Holt and Lawton (1993) have called dynamic monophagy. Similarly, hosts may evolve to avoid the depredations of parasitoids attacking other hosts, what Jeffries and Lawton (1984) called evolution into enemy-free space.

Dynamic monophagy is not the only outcome when one parasitoid attacks two hosts, otherwise the complexity of many natural communities would be hard to explain (Holt and Lawton, 1993). However, for two hosts to persist, both must have mechanisms that allow them to increase in density when rare. This is a general criterion for the persistence of two competitors, and much work on ecological coexistence can be applied to parasitoids. Briefly, mechanisms that allow coexistence include (i) refuges from parasitoid attack for both species; (ii) switching, in which the parasitoid concentrates its attack on the more common species of host (Murdoch, 1969); (iii) environmental variation that affects both host species in different manners, so that each host has higher growth rates under different conditions (see Chesson and Huntly, 1989, for discussion in another context; this possibility has received little study in the parasitoid literature), and (iv) local instability but persistence in a metapopulation (Hassell *et al.*, 1994; Comins and Hassell, 1996).

Although apparent competition involving parasitoids has often been invoked in discussions of their community structure, there have been few studies that have attempted to assess its importance (reviewed by Godfray, 1994). Bonsall and Hassell (1997) worked with a laboratory system consisting of two stored-product moths (*Plodia interpunctella* and *Ephestia kuehniella*) confined to separate arenas but linked by a parasitoid, *Venturia canescens*, which moves between arenas and attacks both hosts. The individual interactions between the parasitoid and either of the two hosts were stable, but the three-species interaction did not persist, with *E. kuehniella* always going extinct. In the field, Settle and Wilson (1990) have argued that the reduction in the abundance of the grape leafhopper, *Erythroneura elegantula*, that occurred after the spread of the congeneric *E. variabilis*, was caused not by interspecific competition, but by *E. variabilis* acting as a reservoir for an egg parasitoid, *Anagrus epos*, that severely attacks *E. elegantula*. Evans and England (1996) have shown that the presence of pea aphids (*Acyrthosiphon pisum*) increases the rate at which alfalfa weevils (*Hypera postica*) are parasitized by the ichneumonid *Bathyplectes curculionis*. Although *B. curculionis* does not parasitize aphids, it is attracted by, and feeds on, aphid honeydew.

The second simple three-species community consists of two parasitoids attacking a single host. If the two parasitoids have identical biologies, then the three-species system cannot persist, and the parasitoid that can maintain itself at the lowest equilibrium host density survives. For persistence to occur, interspecific competition must be less than intraspecific competition so that either parasitoid species has an advantage when rare. One way that this can be obtained is by assuming that both species have a clumped distribution across host patches, but that the two distributions are not, or are only weakly, correlated (in other words the host experiences independent heterogeneity of risk of attack by the two parasitoids). Such a mechanism underlies (often implicitly) the predicted coexistence of parasitoids in a number of models (Hassell and Varley, 1969; May and Hassell, 1981; Hogarth and Diamond, 1984; Kakehashi *et al.*, 1984; Godfray and Waage, 1991; Briggs, 1993; Briggs *et al.*, 1993). Interspecific competition is also less than intraspecific competition if two parasitoids have subtly different niches, attacking the host in different microhabitats (Kakehasi *et al.*, 1984).

A different, though related, mechanism that allows coexistence is called the competition–colonization trade-off. If the host has a patchy population structure with new patches continually being colonized, then a species of parasitoid that is an inferior competitor but a superior colonizer can persist by virtue of its ability to find patches not yet discovered by the better competitor. It is interesting that soon after the competition–colonization trade-off was being introduced into mainstream population ecology (Hutchinson, 1951; Skellam, 1951), applied entomologists were independently developing the same ideas to explain patterns in parasitoid communities (Pschorn-Walcher and Zwölfer, 1968; Zwölfer, 1971). Finally, two parasitoids may coexist in a variable environment if each species is favoured under certain conditions. The variability may be caused by environmental variation, or in certain circumstances by fluctuations in host population densities that arise as part of the dynamics of the host–parasitoid interaction (Armstrong and McGehee, 1980; Briggs, 1993; Briggs *et al.*, 1993).

A different type of one-host/two-parasitoid interaction occurs if one of the parasitoid species is sufficiently a generalist that its dynamics can be considered as uncoupled from those of its host. Hassell and May (1986) analysed such a system, where they assumed that the generalist had a relatively rapid numerical response. The results were quite complex, depending on the details of the natural history, such as the order of specialist and generalist attack during the host's cycle. An interesting result was the prediction of multiple stable states for some parameter values. Berryman (1997) has recently argued that the frequent occurrence of population cycles in forest insects are caused by parasitoids rather than by pathogens or interactions with resource supply, as has normally been assumed. He suggests that the persistent cycles may arise from direct (first-order) density dependence caused by generalist parasitoids combined with delayed (second-order) density dependence due to specialist parasitoids. The important role of generalist parasitoids has been demonstrated by experiments with gypsy moths (*Lymantria dispar*). Gould *et al.* (1990) created mini-outbreaks of gypsy moth by placing huge numbers of larvae in the field, and found strong density-dependent parasitism by generalist parasitoids. Interestingly, similar experiments with another forest lepidopteran, *Malacosoma disstria*, showed only weak density-dependent parasitism but a strong density-dependent mortality from bird predation (Parry *et al.*, 1997).

Finally, parasitoid communities can also be complicated by adding hyperparasitoids, a subject discussed above, and by the addition of other natural enemies such as pathogens (Hochberg *et al.*, 1990; Begon *et al.*, 1996a, b).

Only a few studies have attempted to encompass more than three species. Wilson *et al.* (1996) considered a community of two hosts attacked by three species of parasitoid; two 'specialists' and one 'generalist', the latter being the competitive dominant. The whole five-species community could exist provided that the heterogeneity of risk for both of the two hosts, and for each of the two parasitoids that attacked them, was sufficiently large and uncorrelated. For intermediate values of the heterogeneity of risk, different subcommunities could persist, depending on which species could survive on the lowest-density host population (which in turn is largely dependent on relative searching efficiency). However, for quite a wide range of parameter values, a new type of dynamic behaviour was observed. In this region, no fixed community of species could persist indefinitely. For example, a community consisting of the two hosts and the generalist might be susceptible to invasion by one of the specialists, with the consequent demise of the generalist. This would pave the way for the invasion of the second specialist to form a community which itself was not persistent, because the generalist could invade, so leading to the loss of the specialists. If such a community were to exist in the field, what would be observed would depend on the relative probabilities of immigration of the different species; and the whole ensemble might persist as a metapopulation.

Hochberg and Hawkins (1992, 1993) have taken a different approach to modelling parasitoid communities. They were motivated by Hawkins and Lawton's (1987, see also Hawkins 1994) observation that the number of species of parasitoid per host varied depending on the host's feeding niche. If hosts are ranked in approximate order of concealment: root feeder, stem borer, gall former, leaf miner, leaf roller/spinner, external feeder, then parasitoid diversity peaks for intermediate levels of concealment (leaf miners have most parasitoids). They suggested that the extent of concealment was correlated with the fraction of the host population in a refuge from parasitism, and they built a model to predict the number of parasitoids that could co-exist on host populations that varied in the proportion of individuals fully protected from parasitism. Two types of parasitoid were considered: generalists with decoupled dynamics, and specialists that potentially could regulate the host because of sufficient heterogeneity of risk. Their model predicted maximum parasitoid diversity for intermediate levels of the refuge which they considered consistent with the data. We have expressed concerns about the conclusions of this model (Godfray, 1994; p. 326), both in the assumption of a strong correlation between host concealment and refuges from parasitism (we doubt that leaf miners enjoy more refuges than rollers or external feeders), and also in some of the technical details of the model (for example, generalists were counted if their density in the environment was greater than a threshold, while the data are normally collected by rearing a fixed number of hosts). Nevertheless, this work is an important first step in developing a dynamic theory of large communities of hosts and parasitoids.

6.8 THE STRUCTURE OF HOST–PARASITOID COMMUNITIES

The major problem with building a population dynamic theory of parasitoid communities is the vast number of ways in which different combinations of hosts and parasitoids can interact. Whereas with simple communities of 2–5 species a full analysis is possible, for larger and more realistic communities it seems to us that a general theory is impossible and that models must be built around patterns observed in the field. This is what Hochberg and Hawkins attempted, in the model described at the end of the previous section, although the data they had available were patterns in the number of parasitoid species attacking a wide variety of different hosts, rather than descriptions of particular communities.

We have argued that the essential building blocks for a quantitative theory of parasitoid communities are food webs describing the trophic interactions that occur in nature (Memmott and Godfray, 1992, 1994). We refer to such food webs as parasitoid webs and divide them into three classes: connectance webs which simply display the associations recorded with no quantitative information; semi-quantitative webs where the relative numbers of each parasitoid species per host are included; and quantitative webs in which all hosts, parasitoids, and host–parasitoid associations are given in the same units. Quantitative parasitoid webs are by far the most valuable ecologically, although also the hardest to collect. However, because of the relative ease with which host–parasitoid trophic relationships can be established and quantified, the construction of quantitative parasitoid webs is considerably easier than the construction of quantitative food webs composed of predators and their prey. Indeed, although we shall not pursue this argument here, we believe that a comparative study of parasitoid webs may make a valuable contribution to exploring the general issue of community ecology and food web topology.

Working with colleagues, over the past few years we have constructed two quantitative parasitoid webs describing the communities centred on leaf miners in tropical dry forest in Costa Rica (Memmott et al., 1994) and temperate deciduous woodland in the UK (Rott, A. and Godfray, H. C. J., unpublished data). We have also initiated what we hope will be a long-term study of an aphid–parasitoid–hyperparasitoid community centred on an abandoned field in the UK. In this study, quantitative food webs are constructed monthly during the season when aphids and their parasitoids are active. As we write, we are in the middle of collecting the data for the fourth field season, although only the first 2 years' data are fully analysed (Müller et al., in press), and the larger project of using the data as a basis for modelling complex communities is still embryonic. We conclude this chapter by describing the structure of the host–parasitoid community that we have found so far, and sketching our working hypotheses about the important dynamic factors influencing the community.

Our study site is an abandoned damp field called Rush Meadow at Silwood Park in the south of England. The field, 18 000 m² in size, is largely surrounded by woodland and has a rich flora of grasses and herbs, with a few woody shrubs and climbers, such as bramble (*Rubus* sp.) and broom (*Cytisus scoparius*). Each year all the plants growing in the field are quantitatively sampled using transect methods. On a monthly basis, all plants are scanned for breeding aphids and those infested are quantitatively sampled to estimate the densities of aphids and aphid mummies (parasitized aphids). The combination of information on aphids and mummies per plant unit, and of the absolute density of the plant units, allows us to express the density of host and parasitoid in units of per metre squared. The composition of the parasitoid community is determined by large-scale rearings of aphid mummies. Subterranean aphids, which tend not to be parasitized or to have a specialized parasitoid fauna, were not sampled

In the first 2 years of the study we recorded 25 species of aphid in the field site. Because of difficulties in identifying parasitized mummies of two pairs of species, we present food webs containing 23 taxa of hosts (although we continue to refer to aphid species). Of the 23 species of aphid, we recorded no parasitoids from eight species. While the mean abundance of hosts from which we reared parasitoids was greater than that of hosts without parasitoids, the difference was not significant. The 15 aphid species were parasitized by 18 species of primary parasitoid, the majority members of the braconid subfamily Aphidiinae, which are specialist aphid primary parasitoids, the others members of the genus *Aphelinus* (Aphelinidae, Chalcidoidea). All aphid primary parasitoids lay their eggs inside the living aphid, the larvae eventually causing their host to attach itself to the plant, to form what is called a mummy. In their turn, the primary parasitoids were attacked by no less than 28 species of secondary parasitoids. Aphid secondary parasitoids fall into two distinct classes. The first, which we call hyperparasitoids, attack the still-active aphid prior to mummification, laying their eggs within the body of the primary parasitoid which itself lies inside the body of the aphid. The second group, which we call mummy parasitoids, attack the aphid after mummification, so killing the primary parasitoid and also the hyperparasitoid, if present. We recorded 18 species of hyperparasitoid, all belonging the Alloxystinae (Cynipoidea, Charipidae), and 10 species of mummy parasitoid of more varied taxonomy (Megaspilidae, Ceraphronoidea; Pteromalidae, Encyrtidae, Chalcidoidea). We note in passing that there are currently only 16 species of Alloxystinae on the British checklist. We believe that the species diversity of this group has been severely underestimated in the past, a conclusion supported by recent molecular studies of this group (F. van Veen, unpublished data).

Aphids and parasitoids are active from late April to September or October, with peak diversity of aphids and parasitoids in June and of hyperparasitoids in July. Figure 6.2 shows an example of a quantitative parasitoid web for 1 month, and Fig. 6.3 a summary web obtained by combining the monthly data for the 11 webs constructed in 1994 and 1995. Although we can usually associate secondary parasitoids with particular species of primary parasitoid, this is not always possible unambiguously and we present the raw data, with secondary parasitoids linked with aphids.

One of the main questions we wanted to ask of the web was the extent to which aphids and primary parasitoids might interact through shared natural enemies; in other words, what is the scope for apparent competition? Parasitoid webs, like food webs, are static snapshots of dynamic communities and hence inferences about processes must be made with

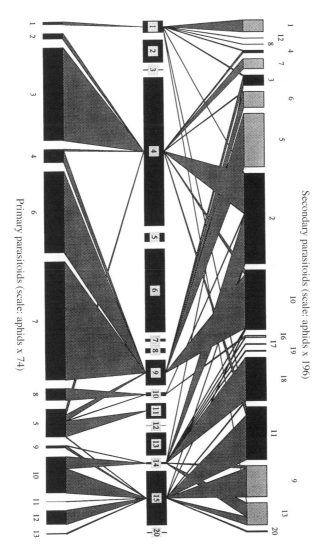

Fig. 6.2 Quantitative parasitoid web for June 1994. The middle set of bars represents the relative abundances of different aphid species in June 1994. The total aphid density is 197/m²; the numbers are species codes. Note that some aphids are not parasitized. The bottom range of bars represent primary parasitoids, the numbers again being species codes. The scale for primary parasitoid density is 74 times that for aphid densities. The widths of the links between aphids and primary parasitoids represent the relative importance of different hosts. The top range of bars represents the equivalent data for secondary parasitoids. The secondary parasitoids are associated with aphids rather than primary parasitoids, for reasons given in the text. Black bars represent mummy parasitoids and grey bars hyperparasitoids (see text).

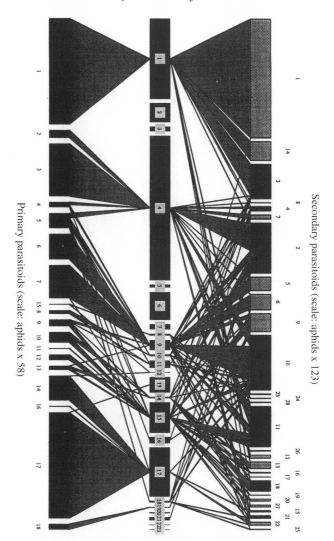

Fig. 6.3 Combined quantitative parasitoid web. Combined food web for 1994 and 1995. The total aphid density is 746/m². The joint densities were obtained by adding together the host and parasitoid densities in each month. Drawing conventions as in Fig. 6.2.

caution (Paine, 1988). Nevertheless, the presence of shared natural enemies is a necessary (but not sufficient) condition for apparent competition to occur (Cohen, 1990). The potential for apparent competition can be illustrated by constructing what in food web theory is called a predator overlap graph (Cohen, 1990), hence we shall refer to it as a parasitoid overlap graph. Figure 6.4 shows each of the 15 species of aphids from which parasitoids were reared as a vertex linked by an edge if they share a particular class of natural enemy. Note that at this stage we are using only the qualitative information in the parasitoid web (i.e. treating the web as if it were a connectance graph). Two conclusions emerge from these graphs. Firstly, mummy parasitoids are much more important than either primary parasitoids or hyperparasitoids in linking together the community. Secondly, the number of links per aphid species is overdispersed (this can be shown by a formal statistical test), with species differing markedly in their number of connections with the rest of the community.

Because we have quantitative data we can pursue further the assessment of the scope for apparent competition. However, because quantitative food webs have seldom been collected, there are no quantitative versions of predator overlap graphs in the literature that we can employ, and so we have had to develop our own (Müller *et al.*, in press). Consider a species of parasitoid that attacks more than one host in a community. If the parasitoid population mixes each generation, then we can ask what fraction of the parasitoids attacking species i are likely to have developed as parasitoids of species j. We call this quantity d_{ij} and make three points. Firstly, d_{ii} exists and is the fraction of parasitoids that would have developed on the same species of host; d_{ii} may approach 1 for a species largely attacked by monophagous parasitoids. Secondly,

$$\sum_j d_{ij} = 1$$

– all parasitoids of species i must have developed on some host or another. Finally, d_{ij} will normally not be identical to d_{ji}. We suggest that the quantity d_{ij} is a possible measure of the influence of species j on species i via shared natural enemies, although we acknowledge that alternative metrics could be advanced.

Figure 6.5 shows the quantitative parasitoid overlap diagrams that are equivalent to the connectance parasitoid overlap graphs in Fig. 6.4. These diagrams also show the difference in connectivity due to the three categories of parasitoids, but they illustrate some new features. Firstly, the magnitude of the d_{ii} terms is higher for primary parasitoids and hyperparasitoids than for mummy parasitoids. Most parasitoids in the first two categories tend to be host specialists, while mummy parasitoids tend to be more generalist. Secondly, indirect interactions involving primary parasitoids and hyperparasitoids tend to be asymmetric, and involve common species having a potentially major influence on rare species in the community. Lastly, indirect interactions involving mummy parasitoids tend to be stronger and more symmetric.

Before we describe our working hypothesis about the dynamics of the aphid–parasitoid community, we need to say a few words about other

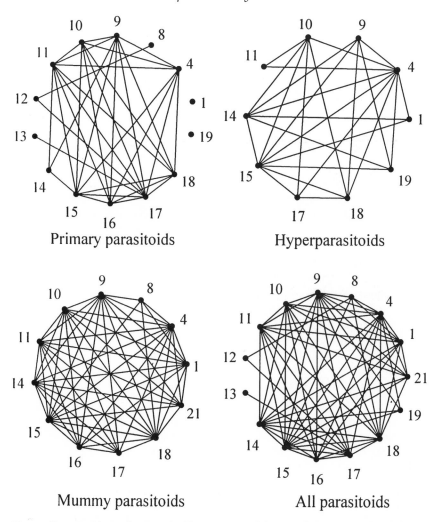

Fig. 6.4 Parasitoid overlap graph. The vertices of the graph represent the different aphids (indexed by their numeric codes) from which at least one parasitoid of the different categories was reared. If two aphid species share a common parasitoid, their respective vertices are linked by an edge. The graph shows the potential for apparent competition between different species.

natural enemies. A clear shortcoming of parasitoid webs is that they contain no information about other organisms that prey on the same species of host. In our community, aphid predators, especially, Coccinellidae, Syrphidae and Cecidomyiidae, are important and can cause severe aphid mortality. A predator quantitative food web has been constructed (R.

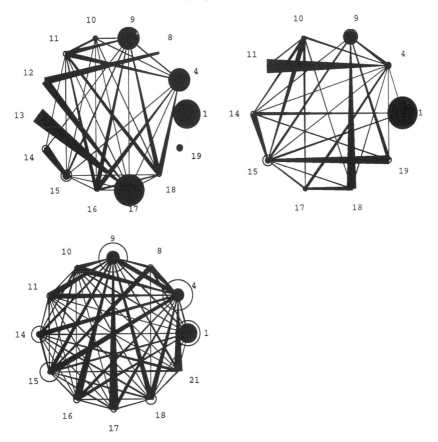

Fig. 6.5 Quantitative parasitoid overlap diagram. As Fig. 6.4, but now (i) the size of the vertex represents the relative abundance of the different species, (ii) the extent to which the vertex is coloured black represents the degree to which that species acts as a source of its own parasitoids (i.e. the fraction shaded black equals d_{ii} – see text), and (iii) the width of the link between aphids i and j at i represents the importance of j as a source of parasitoids attacking i (d_{ij} – see text).

Cooke, unpublished data) which shows widespread polyphagy. Experimental studies have demonstrated short-term apparent competition between aphid species feeding on different host plants, mediated by shared predation (Müller and Godfray, 1997). Aphid fungal pathogens are also present in our system but have not proved to be an important source of mortality, at least so far. During preliminary observations in the year before the study started, a year with heavy rainfall, quite extensive fungal mortality was observed.

Our working hypothesis is illustrated in Fig. 6.6. We hypothesize that the community consists of a series of relatively tightly linked interactions

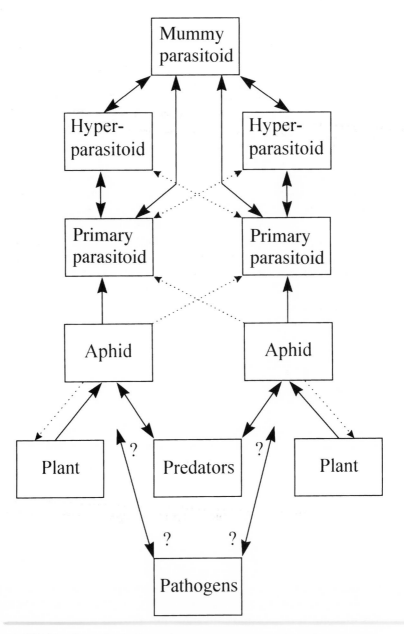

Fig. 6.6 Working hypothesis about the important links in the aphid/natural enemy food web. Two aphid species are shown for simplicity; solid arrows mark hypothesized important dynamic influences, broken lines less important influences. Note that predators and mummy parasitoids link together aphid–parasitoid–hyperparasitoid groups at different trophic levels.

between aphids, primary parasitoids and hyperparasitoids, centred on different host-plant species (or closely related groups of plant species), which we shall call an APH (aphid–parasitoid–hyperparasitoid) group. Sometimes, more than one aphid, primary parasitoid or hyperparasitoid may be present, but links with other APH groups are weak. The primary parasitoids and hyperparasitoids in the APH groups are attacked by mummy parasitoids which are far more catholic in their host preferences and provide a dynamic linkage between different APH groups. Considering the low parasitism rates attained by most primary parasitoids, and the high rate of hyperparasitism, we believe it likely that secondary parasitoids regulate primary parasitoids, but that primary parasitoids do not regulate aphids (see also Mackauer and Völkl, 1993). We have argued that different APH groups are linked together at the top of the trophic chain by mummy parasitoids; we also believe that shared predators link the APH groups at the aphid level (Müller and Godfray, 1997). An important question that can be answered by experimentation is whether mummy parasitoids respond to an increase in the abundance of primary parasitoids in an APH group by density-dependent attack (switching – Murdoch, 1969). Such a response could be an important factor stabilizing the host–parasitoid dynamics within the individual APH groups.

7

Biological control of insects: implications for theory in population ecology

William W. Murdoch, Cheryl J. Briggs and Timothy R. Collier

7.1 INTRODUCTION

For decades, biological control of insect pests has been a rich source of concepts and theory for population ecology. Researchers interested in biological control introduced both the basic idea of density dependence (Howard and Fiske, 1911) and early parasitoid–host models (Nicholson and Bailey, 1935). Even metapopulation theory had its origins as a potential source of insight into biological control (Levins, 1969). Here we look to biological control for data that might shed light on theory in population ecology.

We have chosen to organize this chapter around a few issues that have arisen in biological control, rather than laying out topics in population ecology theory and asking what biological control tells us about each. In any case, questions central to biological control relate to the major theoretical issues of interest to population (and even community) ecologists. In addition, biological control issues sharpen some theoretical questions and give them a useful focus. For most of the chapter, we ignore the distinction between classical biological control, in which alien enemies are introduced to control alien pests, and control of resident, or alien, pests using native enemies.

Population dynamics theory typically focuses on the dynamical properties of the steady state – whether it is stable, what types of cycles occur, etc. Stability can of course be relevant to biological control, since we want

Insect Populations, Edited by J. P. Dempster and I. F. G. McLean. Published in 1998 by Kluwer Academic Publishers, Dordrecht. ISBN 0 412 83260 7.

the pest population to stay below an economic threshold over long periods. But above all, low pest density is the goal, and in many models the forces that suppress the pest more strongly do so at the cost of stability (Murdoch, 1990). We largely ignore this trade-off in the rest of this chapter, and mention it here to alert the reader that it remains an issue of interest in connecting successful biological control to theory.

We look mainly at problems that we have found interesting, rather than attempting a thorough review in a short space. We begin by considering spatial processes, currently a major topic in theoretical ecology. We then move to generalist enemies, and particularly their potential for driving prey extinct, which again implies that space is important. This topic relates to a long-standing controversy over the relative efficacy of specialist versus generalist enemies, and is an area in which theory is relatively sparse. We next explore the ecological issue of competing and co-existing enemy species, reflecting the hoary question in biological control of whether one should release one or multiple enemy species. Finally, we discuss a quite new body of theory related to size-structured interactions between parasitoid and host populations, and comment on its relevance to ideas about density dependence and to biological control.

In each section below, we begin by presenting relevant theory and then discuss evidence. In some cases the theory has been developed before the evidence became available, but in others the theory is a response to evidence. Thus examples of biological control have been both a source of tests and a stimulus for new theory.

A comment on terminology is needed. When discussing models we have used 'steady state' instead of the more common 'equilibrium', as the latter has a particular meaning in this book. Thus a steady state is the condition at which the rate of change is zero. It is stable if perturbations from the steady state die away with time, and unstable if the perturbations increase with time. We often simply use 'density' when it will be clear that steady-state density is implied.

7.2 ROLE OF SPATIAL PROCESSES

There is now a large body of theory investigating how spatial processes might affect population dynamics. The theory covers three main areas: metapopulations or ensembles of subpopulations, refuges, and aggregation by the natural enemy. We deal with these separately in the next three sections.

7.2.1 Metapopulations and ensembles

Theory

Two approaches can be distinguished.

(a) Current usage typically defines metapopulations as a system of subpopulations in patches in which persistence of the metapopulation occurs through a balance between extinction and colonization of subpopulations; little or no attention is paid to population trajectories within patches (Hanski and Gilpin, 1997). 'Classical' metapopulation theory (Harrison and Taylor, 1997), based on Levins (1969), assumes all patches have the same probabilities of extinction and colonization. Other possible metapopulation configurations include one or more patches whose populations are effectively safe from extinction over the period under consideration and which act as a key source of colonists for other patches. These safer populations thus act like spatial refuges.

(b) The alternative framework, which we label ensemble dynamics (Murdoch, 1990), follows the numbers of individuals in local subpopulations. In these models we might never observe local extinction. Instead, movement among local populations serves to reduce local (and ensemble) fluctuations, and render the populations more stable. The theory ranges from linked predator–prey subpopulations (Maynard Smith, 1974; Murdoch and Oaten, 1975; Reeve, 1988) to spatially explicit models of connected populations (Comins *et al.*, 1992) or of individual organisms (Wilson *et al.*, 1995). In all cases, limits on dispersal ability play a crucial role in regional dynamics, as can spatial heterogeneity.

We have distinguished these approaches for the convenience of the reader; but we believe they are simply points on a continuum. Consider, for example, a recent model by Ives and Settle (1997) for temporary crops in which local extinction is crucial. The model assumes a spatially distributed set of fields in which the crop is harvested and replanted with varying degrees of synchrony. The pest is assumed to go extinct at harvesting. However, there is some carry-over of the specialist predator between croppings and, apart from recurrent pest extinction, the model is of the ensemble type. The effects of pest rate of increase and various assumptions about dispersal among fields are examined. A potentially useful piece of insight is that asynchronous planting allows predators to move on to the crop early in the pest's population development on a field, and hence enhances that predator's ability to prevent high pest density.

Evidence

We do not separate evidence into metapopulations and ensembles, as they seem to merge in real systems, however we begin with examples close to the former and finish with evidence about the latter. We are mainly concerned, at first, with whether local extinction and recolonization are sometimes important, rather than whether the system fits a precise definition of metapopulation.

Murdoch *et al.* (1985) presented evidence that, in some cases, local pest populations might be unstable – indeed, in a few cases the pest appeared to be driven locally extinct – and that persistence of the interacting populations was to be sought at a larger spatial scale. One example, control of mosquitoes in small water bodies by the back-swimming bug *Notonecta*, seems to illustrate the second type of metapopulation dynamics mentioned above: mosquitoes colonize stock tanks with *Notonecta*, which then drives them extinct, but mosquitoes also exist in source tanks or pools that appear to lack *Notonecta* permanently.

In a second example from that paper, winter moth (*Operophtera brumata*) in Nova Scotia appeared to have been driven locally extinct in oak woodlands by introduced and native natural enemies. Recent data suggest the moth is indeed absent in some oak woodlands, but persists in orchards and urban shade trees (Roland and Embree, 1995). The winter moth recently invaded Vancouver Island, where control was again successful (Roland, 1994). In this case, the moth was still present in oak woodland in 1990, but was declining exponentially, as can be seen if the data of Bonsall and Hassell (1995, Fig. 2) are replotted on a semi-log scale. From 1983–85 the exponential rate of decline was 1.04 per year; from 1985–90 it was 0.16 per year. So over the latter 5-year period the population was still halving in density about every 4 years, and this population may also be heading to extinction.

Walde (1991, 1994) provided a good demonstration of local pest extinction in the successful control of apple mites by predatory mites. She showed that the predatory mite drives the prey mite locally extinct on several spatial scales: individual trees, groups of trees, and probably even occasionally whole orchards. Extinction at the orchard level suggests that the prey mite may even exist as a classical metapopulation, though this would be extremely difficult to establish. Walde and Nachman (1998) summarize evidence that control of mites in greenhouses by predatory mites involves local extinction, but prolonged interaction over a large area.

Local extinction and colonization are clearly essential features of pest control in some temporary crops, especially where the pests and natural enemies die off between seasons (because of harvest) and immigrate on to the crop in each new season. For example, the silverleaf whitefly (*Bemisia argentifoli*) and its parasitoids colonize new cotton and row-crop fields from native plants, weeds and established fields (Coudriet *et al.*, 1986; Byrne *et al.*, 1990). Often entire populations of whiteflies disperse from one crop to another as fields are rotated during the season. Whitefly dynamics on both a regional and local scale appear to be determined by a combination of the spatial distribution of different crops, differential colonization by whiteflies and parasitoids, and the timing of crop rotation (e.g. Byrne *et al.*, 1990; Riley *et al.*, 1996; D. N. Byrne, personal communication). Pesticide use, however, is also critical and the silverleaf whitefly is not effectively controlled by native or introduced parasitoids.

Settle *et al.* (1996) provide evidence that spatially out-of-phase dynamics may be important in the control of pests of rice by predators in

Indonesia. Predators are more common, especially early in the crop's development, in areas planted asynchronously than in those planted synchronously, as the theory developed for this situation suggests should be the case (Ives and Settle, 1997). They also provide experimental evidence that a greater abundance of predators early in the development of the crop leads to better pest control.

Murdoch *et al.* (1996b) looked for ensemble dynamics, and tested for its contribution to stability, in a stable biological control system in a long-lived crop. They experimentally isolated single grapefruit trees containing subpopulations of California red scale (*Aonidiella aurantii*) and its highly successful controlling parasitoid *Aphytis melinus*. The evidence was strongly against a significant role for ensemble dynamics: isolated populations were no more temporally variable than were those connected to the regional ensemble.

In summary, examples from biological control support a role for metapopulation dynamics broadly defined; natural enemies can drive pests locally extinct, and dispersal and colonization appear to be key in several systems. One experimental test shows ensemble dynamics are not important in one system. This is simply one example, however, and the prevalence of ensemble dynamics in successful biological control is still an open question.

7.2.2 Refuges

Theory

Ecologists have developed theory for spatial refuges over a long period and, with rare exceptions (e.g. McNair, 1986), it tells us that a refuge from the natural enemy will not only increase the pest density but will tend to stabilize the interaction. Hawkins *et al.* (1993) also proposed that degree of control is inversely related to the presence of refuges. They define the degree of 'refuge', however, as 1–(maximum fraction parasitized), and we feel this definition is perhaps too broad. For example, a parasitoid will cause little parasitism in the pest population if it is itself limited by another factor such as a hyperparasitoid, which would have dynamic consequences quite different from those caused by a refuge. In what follows, therefore, we have in mind real refuges. The work of Hawkins *et al.* does, however, further stress the point that real refuges can be expected to reduce control, as can features such as invulnerable stages that in some instances are usefully thought of as temporal refuges.

Theory also tells us that temporal refuges, in which there is a seasonal mismatch between the pest and the natural enemy, are also stabilizing (Griffith and Holling, 1969; Godfray *et al.*, 1994). It seems likely that mismatches large or small are common in the field, but we do not know of a case where they have been shown to be important in biological control.

Evidence

It is likely that spatial refuges play an important role in thwarting pest control in some seasonal crops. Cultural practices, such as removing non-crop plants that harbour the pest, have often been a response to the presence of refuges, though there is sometimes a tension between removing sites for the pest and sites that also harbour natural enemies. (For example, pruning trees in vineyards help support over-wintering populations of an egg parasitoid of the grape leafhopper, *Erythroneura elegantula* (Corbett and Rosenheim, 1996; Murphy *et al.*, 1996)).

As far as we know, the only field study testing the dynamic effects of a refuge is experimental work on red scale, *A. aurantii* (Murdoch *et al.*, 1996b). Scales on the bark of the trunk and structural branches in the tree's interior are parasitized only rarely by *Aphytis*, and this interior population can constitute 90% of the adult female scales (Murdoch *et al.*, 1995). When Murdoch *et al.* (1996b) removed the interior scales, the population on the exterior decreased by about 60% compared with controls. Since exterior scales are the economic pests (on fruit), this result confirms the theoretical expectation that a refuge increases pest density. However, surprisingly, removal of the refuge did not reduce the stability of the exterior scale population, which is contrary to theoretical expectations. It is quite possible, of course, that refuges will be important for stability in other systems.

7.3 AGGREGATION IN SINGLE POPULATIONS AND METAPOPULA-TIONS

Theory

A recent review can be found in Murdoch *et al.* (1997). Although models typically are discussed in terms of spatial aggregation, the differences in relative vulnerability among pest individuals, which are key to the behaviour of models, can arise from non-spatial processes. We begin by considering aggregation at a within-population scale.

The effect of aggregation is best understood in terms of the degree to which relative risk is strongly aggregated among pest individuals (Chesson and Murdoch, 1986). Such aggregation arises naturally from aggregation that is unrelated to local pest density. Early models in the Nicholson–Bailey framework suggested aggregation is a powerful source of stability (Bailey *et al.*, 1962; Hassell and May, 1973; May, 1978), its effects now summarized in the 'CV^2 rule' (Pacala *et al.*, 1990), while those in continuous time suggested it is not (Murdoch and Oaten, 1975).

Murdoch and Stewart-Oaten (1989) showed that aggregation by the enemy in response to local pest density, where the enemy responds to changes in local pest density within the generation, would probably not be stabilizing. It might well be destabilizing but, by increasing efficiency, would lead to a reduced pest density. Murdoch and Stewart-Oaten suggested that stability should be lost in Nicholson–Bailey models that allow the enemy to respond to changes in pest density within the generation, and this was confirmed by Rohani *et al.* (1994).

The Murdoch and Stewart-Oaten model also showed that aggregation independent of local pest density would not be stabilizing if the distribution of attacks varied rapidly through time. Rohani *et al.* concluded, however, that aggregation that is independent of pest density is stabilizing. This result depends, we suggest, on the assumption that enemy attacks are distributed, in a fixed fashion throughout the generation, across patches in which the pest occurs. The key to their result is that the relative risk run by each pest individual does not change throughout the vulnerable period of its life. The results of Reeve *et al.* (1994) in a continuous-time model support this view. We expect that attacks that are independent of local pest density, but that also change fast enough relative to the time scale of the system, will remove the stabilizing effect.

The effect of aggregation independent of local pest density is thus likely to depend on, among other things, the relative time scales in the interaction. Aggregation in space is most likely to be stabilizing, and to interfere with pest suppression, when it requires the enemy to press attacks on a dwindling portion of the pest population, for example when (i) the pest is immobile during the vulnerable (usually immature) stage; (ii) the enemy distributes its attacks in response to some fixed spatial aspect of the environment unrelated to pest density; and (iii) the enemy does not modify this behaviour, as repeated attacks on the more vulnerable portions of the pest population reduce pest density in these locations. Otherwise, aggregation is not likely to induce stability, but should enhance pest suppression.

Aggregation has been explored in the context of an ensemble of subpopulations by Ives (1992a), Godfray and Pacala (1992) and Murdoch *et al.* (1992a). These models were concerned with stability rather than pest suppression, and the overall messages are that aggregation by the enemy is probably secondary in its effects on stability in comparison with the effect of asynchronous dynamics, and that its effect on stability is not easily predicted.

The Ives and Settle (1997) models are focused on pest suppression. They show that predator movement among asynchronously planted fields tends to reduce pest density. They also show that aggregation to pest density (i.e. the predators remain longer in a field when the pest is more dense) tends to increase ultimate pest density. This occurs because, in these models, pests are suppressed more, all else being equal, if the predator attacks early in the pest population trajectory. Fields that have high pest density are late in the pest trajectory, predators then have less effect on maximum pest density achieved, and these fields retain predators that could do more good by moving to more recently infested fields.

Evidence

Rate of parasitism increases with local pest density in some situations, and in some (but not all) cases such patterns may be induced by local parasitoid aggregation (Walde and Murdoch, 1988). However, there is little direct evidence for aggregation of parasitoids and predators to locally higher pest densities. Gould *et al.* (1990) increased larval gypsy moth (*Lymantria dispar*) densities on 1-ha plots and showed that a generalist parasitoid, *Compsilura concinnata*, caused greater parasitism in the higher-density plots; they speculated that the parasitoid immigrated into these plots. Parry *et al.* (1997) showed that birds reduced small-scale increases in tent caterpillar (*Malacosoma disstria*) density and, again, it seems likely that local movement was involved.

In none of these cases can we disentangle the dynamical effects of spatial density dependence in the attack rate from other factors such as predatory and parasitoid switching, or overall numerical response. As noted, aggregation might lead to more effective control, but if so it is likely to be at the detriment of stability.

A rare example of an experimental study of the effect of aggregation in reducing prey outbreaks (although not in a biological control setting) comes from Kareiva's (1987) study of ladybugs attacking aphids on goldenrod. He showed that ladybugs aggregated to dense colonies of aphids; and that an experiment reducing ladybug movement led to a tenfold increase in aphid density.

7.4 THE ROLE OF GENERALIST ENEMIES

Theory

Most entomologists have thought that specialist enemies make the best biological control agents. This belief probably rests on the notion that specialists are likely to have evolved a higher efficiency on the pest and are more likely to have a rapid numerical response. As a consequence, most cases of classical biological control have involved the introduction of specialist predators and parasitoids. There have, however, been supporters of generalist predators (e.g. Ehler, 1977), and occasionally generalists have been introduced, e.g. coccinellids for the control of aphids. Also, in many cases of non-classical control, generalist predators simply are part of the natural enemy complex attacking the pest, especially in temporary crops.

There is in fact a dearth of theory for general enemies that takes into account their biological details to the same extent that they are incorporated in specialist–parasitoid–host theory (Murdoch *et al.*, 1998).

Generalist predators are especially difficult since their diets, behaviour, physiology and often habitat change with size. The theory mentioned below is therefore unavoidably simplistic.

We begin with (a) a model involving local pest extinction. We then turn to two other topics: (b) apparent competition and (c) intraguild predation (or omnivory). In all three cases the theory can predict pest extinction, so space must be implicit if the interaction is to persist.

(a) 'Resident' generalists

A model by Murdoch *et al.* (1985), written with biological control in mind, relates indirectly to the spatial issues discussed above. It illustrates that generalist enemies can control a pest by consistently driving it extinct. The key feature of the model is that the enemy is always present in the crop at or above some minimal density (P_{min}), which is maintained by alternative prey. In this model the 'other' prey species are only implicit, implying that the enemy may have little effect on their densities, which is probably common if several to many prey species are involved. The model is:

$$H_{t+1}=H_t f(H_t, P_t)$$
$$P_{t+1}=g(H_t, P_t) \tag{7.1}$$

where H_t, and P_t are the pest and enemy densities, respectively, f is the fraction of hosts that survive attack, and

$$g(H_t, P_t)>P_{min}=g(0, P_{min}) \tag{7.2}$$

Equation 7.2 states that the enemy density will always $>P_{min}$ if the pest is present, and will remain at P_{min} if the pest is absent. Thus, the enemy is assumed to respond to the pest numerically. This model makes sense only if there is 'somewhere else' from which the pest recurrently reinvades the crop. It is thus an implicitly spatial model.

The model has the steady state ($0, P_{min}$) at which the pest is extinct and the enemy persists because it can feed on other prey. The steady state is stable if $f(0, P_{min})<1$. That is, the pest will be driven extinct if the rate of change of the pest population, f, is negative when the pest population is small (i.e. when H_t is approximately 0) and the enemy population density is P_{min}. The inequality $f(0, P_{min})>1$ is thus an invasion criterion for the pest.

The implications of the model can be made more transparent by assuming a specific f, for example the standard Nicholson–Bailey form:

$$f=\lambda \exp(-aP_{min}) \tag{7.3}$$

where λ is the pest's finite rate of increase in the absence of the enemy and a is the per head attack rate of the enemy.

In this situation, the pest can be driven to extinction ($f<1$) if $aP_{min}>\ln\lambda$. Thus we see that the chances of control are increased by increasing either the enemy's attack rate or its minimal density, either of which needs to be higher as the pest's potential rate of increase is higher.

This model relates to those developed recently by Ives and Settle (1997) for spatially distributed seasonal crops. Although the predators in their models attack only the pest species, the authors have in mind a system (rice) in which the predators are generalists. The key to control in the models is early attack on the pest by predators either invading the crop, or being carried over from a previous crop through feeding on alternative prey. A model combining these two approaches would therefore explore factors that increase P_{min} early in the pest season.

(b) Apparent competition

The above model can be thought of as an example of apparent competition (Holt, 1977), although the pest's 'competitor' is the array of implicit prey species rather than a single species. Holt and Lawton (1993) have recently explored in detail several models of apparent competition between insect hosts via shared natural enemies. Their main point was that, provided mechanisms fostering co-existence are not operating, one host species will inevitably be driven extinct as a result of apparent competition. Again, their model is implicitly spatial since it assumes the pest re-invades from somewhere else.

The apparent competition model suggests that an alternative host can induce extinction of a pest via a shared enemy, if it possesses one or more of the following features. (i) Its rate of increase in the absence of the parasitoid is higher than the pest's. This seems unlikely, given that pest insects tend to have high capacity for increase in the crop environment. (ii) The parasitoid is more effective on the pest species for any one of several reasons, for example if the pest is preferred or more accessible. This seems a more likely feature.

Generalist enemies might of course adequately control a pest species without driving it extinct. Many mechanisms potentially allow co-existence of pests that share a natural enemy, e.g. a refuge for the host or switching behaviour by the natural enemy (Murdoch, 1969; Roughgarden and Feldman, 1975; Holt and Lawton, 1993). Thus generalist enemies might adequately control several or even many pest species without driving them extinct. Except for the simple three-species case, however, there is little theory for this complex situation.

(c) Intraguild predation

Generalists often attack species in more than one trophic level. For example, generalist insect predators often consume smaller, competing predators as well as pest herbivores; facultative hyperparasitoids devour both hosts and the immatures of competing parasitoids. Indeed, this may be a common phenomenon in both biological control (Rosenheim *et al.*, 1995) and natural systems (Polis *et al.*, 1989). Theory for such intraguild predators has been synthesized recently by Holt and Polis (1997).

Although they reach several conclusions about the population dynamic consequences of intraguild predation, we focus on one major result: co-existence of an intraguild predator and its competitor/prey requires that the latter be a more efficient enemy. Consequently, intraguild predation or the introduction of an intraguild predator (or facultative hyperparasitoid) is predicted to disrupt biological control by raising pest density. This result is directly analogous to the effect of introducing a parasitoid that wins in multiparasitism (below); multiparasitism is effectively like intraguild predation in that usurpation of the host leads to the death of a heterospecific competitor (Briggs, 1993).

Evidence

(a) 'Resident' generalists

Advocates of biological control and integrated pest management have long used generalist enemies in practical schemes, for example involving intercropping (e.g. Andow, 1990). In fact this and other practices, for example maintenance of over-wintering places for predatory insects (Thomas *et al.*, 1991), can be viewed as an attempt to maximize P_{min} in Equation 7.2. Indirect evidence that generalists control some potential pests comes from situations where insecticide spraying has killed off generalist predators and turned previously innocuous herbivores into new pests.

Some good examples now exist to illustrate the above theory, a couple of which we mentioned above under the role of spatial processes (mosquitoes and mites). The mite example is especially illustrative. Walde (1994) showed that the predatory mite drives the pest mite extinct on several spatial scales. The interaction between the prey and predatory mite can be ongoing, both because the pest mite recurrently invades from areas where it is not driven extinct, and also because the predatory mite remains in the apple orchards at all times, and can do so because it has alternative prey and can subsist on plant material. As noted earlier, Walde demonstrated that prey mite densities in experimental plots were converging towards zero, i.e. zero prey is a stable steady state.

As noted above, Gould *et al.* (1990) created experimental populations of gypsy moth (*L. dispar*) larvae at different densities on 1 ha plots. Gypsy moths were not present on these plots before the experiment, yet the populations were reduced to extremely low densities (though not to zero) by a polyphagous parasitoid that was present in the area and is maintained there by alternative hosts.

Generalist predators also appear to be able to stabilize pests, at acceptably low density, in some cases of successful biological control. Roland and Embree (1995) conclude that parasitoids probably caused the initial decline in winter moth density in various areas in Canada, but that generalist pupal predators (probably beetle larvae) are key to maintaining the pest at low density thereafter, via intense and directly density-dependent predation (but see Bonsall and Hassell, 1995). Birds have long been suspected of providing such density-dependent mortality on forest insects, including spruce budworm (Ludwig *et al.*, 1978) and perhaps tent caterpillars (Parry *et al.*, 1997). It is not clear from such studies, however, what the processes are – for example, switching among prey species, or numerical or spatial responses – that lead to such effective predation.

As noted above, Settle *et al.* (1996) provide a nice example of the crucial role of generalist predators in controlling pests of rice. Here the importance of alternative prey in maintaining the predators was explored explicitly. There appears to be a nice seasonal fit between alternative prey, which peaks early in the crop's development and serves to produce a high density of generalist predators (our P_{min}), and the later build-up of pest herbivores, which can then be suppressed by the predators.

(b) Apparent competition

Settle and Wilson's (1990) study of leafhoppers in California is a potential case of apparent competition leading to the reduction and possible eventual extinction of a native leafhopper pest species (see also Holt and Lawton, 1993). Settle and Wilson showed that the native grape leafhopper (*Erythroneura elegantula*) declined after the invasion of the variegated leafhopper (*Erythroneura variabilis*) in 1980. Exclusion appears not to have been mediated by resource competition or direct interference, but through the effects of a shared parasitoid, *Anagrus epos*, which develops more successfully on the native species.

Another potential (though anecdotal) example of apparent competition in a biological control system is provided by the citricola scale (*Coccus pseudomagnoliarum*). It was introduced to California early this century, and came under immediate control in southern coastal citrus by a complex of exotic and possibly native parasitoids of two other scales. In Central Valley citrus, where one of the hosts, and the generalist parasitoid *Metaphycus helvolus*, are missing, the citricola scale is still a pest. Early biological control workers suggested that the key to control in the south was the presence of alternative hosts for *Metaphycus* (Flanders, 1942; Bartlett, 1953). This system illustrates what Holt and Lawton (1993) describe as a puzzle for theory: multiple hosts co-existing with several (up to five) shared generalist parasitoids.

(c) Intraguild predation

Rosenheim *et al.* (1995) review the evidence for intraguild predation in biological control. They point out that the phenomenon is quite common, and many studies document it. However, few studies address the prediction that intraguild predation disrupts control. In perhaps the most instructive study, Rosenheim *et al.* (1993) found that predation of lacewing larvae by hemipteran predators reduced the within-season suppression of cotton aphids (*Aphis gossypii*). We know of no cases in which the introduction of an intraguild predator or facultative hyperparasitoid has disrupted classical biological control. However, long-term quantitative studies of the effects of these enemies are not available.

7.5 HOW MANY ENEMY SPECIES?

Theory

Whether it is better to release only one, several, or many natural enemy species has been a controversial issue for three decades (reviewed by Briggs, 1993; Hochberg, 1996a). It leads us into competition theory, and even aspects of community structure and dynamics.

Species can compete in a variety of ways. Theory has been developed for some of this variety, and the messages from different models can be not only different but conflicting with respect to biological control (Table 7.1). Simple theory of exploitative competition (e.g. MacArthur and Levins, 1967), which implicitly views the pest as an indivisible resource, predicts that the most effective enemy will reduce the pest to a steady-state level too low for any other enemy species to persist, and that therefore we need not fear releasing multiple species: the winner is also by definition the best control agent.

This simple picture is lost, however, when we take into account the different ways in which species compete and the complexity of pest life history. For example, competing parasitoid species can also engage in interference competition in the form of multiple parasitism, larval competition and hyperparasitism. These are essentially forms of intraguild competition. Parasitoids that are superior at interference competition may displace species that are superior in exploiting the pest, and hence reduce the degree of control (Briggs, 1993). Now theory tells us we should release only the most effective pest-exploiter, not the winner.

Several models of multiple parasitoids attacking a single host species have included self-limitation in the parasitoid attack rate, representing mutual (within-species) interference or certain types of non-random search by the parasitoids. In these models, the parasitoid rate of increase is limited not only by the density of the host, but also by its own density. If this within-species density dependence is stronger than the competi-

Table 7.1 Implications of competition theory for biological control. Theory based on different mechanisms of competition or co-existence yield different advice on the release of multiple natural enemies. Details on models in each class of theory are given in the text. In the second case of enemies that are self-limiting, co-existence is less likely when intraspecific limitation<interspecific limitation

Theory	Result	Release
Simple models	Best competitor gives most control	All species: best wins
Enemies interfere	Co-existence can increase pest density	Best agent, not winner
Enemies are self-limiting:		
Intraspecific > interspecific limitation	Co-existence decreases pest density	Many species
Intraspecific < interspecific limitation	Added species may increase pest density	Best agent, not winner
Stage-structured pest	Winner may increase key pest stage	Best agent, not winner

tion between the species, then more than one parasitoid species can co-exist, and additional self-limited species can improve control (e.g. May and Hassell, 1981; Hogarth and Diamond, 1984; Godfray and Waage, 1991). These models argue for the release of multiple species. In contrast, if intraspecific density dependence is lacking, or is weaker than the inter-specific competition, then co-existence is less likely and additional para-sitoids can lessen the degree of control (e.g. Kakehashi *et al.*, 1984; Hochberg, 1996a). These models argue for the release of only the single most effective parasitoid species.

The presence of different life stages in the pest can also complicate matters. Verbal theory has sometimes argued that control will be improved by releasing complementary species, for example those that attack different stages or in different seasons (Huffaker and Kennett, 1966; Takagi and Hirose, 1994). Briggs (1993), however, showed in a stage-structured model that attacking different juvenile host stages did not promote parasitoid co-existence, and that the parasitoid that won in com-petition might not be best for pest control: it induced the lowest density of the host stage attacked by its competitor, but not necessarily the lowest density of the most damaging stage, or of the total host population.

The theory summarized in Table 7.1 thus suggests that "how many species should we release?" is not a useful general question. We need, instead, to ask "what kind(s) of species, and how many, should we release given the modes of competition and co-existence among them?"

Evidence

Evidence on the efficacy of releasing one versus many parasitoid species is as mixed as the theoretical predictions. Co-existence of several natural enemies in biological control is probably the rule. Many, if not most, sys-tems involve more than a single natural enemy (e.g. Clausen, 1978).

However, the mechanisms underlying co-existence and the relationship between the number of enemies and the degree of control remain unclear.

Frequently, several natural enemies contribute to pest mortality. For example in mulberry orchards in Europe, each of three introduced parasitoids contributes to overall parasitism of the peach scale (*Pseudaulacaspis pentagona*) during a given season (Pedata *et al.*, 1995), and many predatory species seem to be involved in control of rice pests in Indonesia (Settle *et al.*, 1996).

Several species of natural enemies can act in complementary ways to control the pest. The parasitoid *Aphytis paramaculicornis* was introduced first to control olive scale, *Parlatoria oleae*, in California. However, it attacks only the larger stages and does poorly in summer. Control was incomplete and outbreaks still occurred. Another parasitoid, *Coccophagoides utilis*, which attacks smaller stages especially in summer, was then introduced, and the scale was brought under control (Huffaker and Kennett, 1966). Theory that explains the success of this combination does not exist. Takagi and Hirose (1994) suggest that a combination of two parasitoids is able to control arrowhead scale because one species (*Aphytis anonensis*) is effective at driving down the scale from high densities, while the other species (*Coccobius fulvus*) can regulate the population at low densities (see Roland and Embree, 1995, for an analogous situation).

There are also cases, however, in which one species seems to be the primary source of control even though many species have been released or established. Over 50 species of enemies have been introduced into southern California to control red scale on citrus, but *Aphytis melinus* seems to be the controlling agent, even though at least one other parasitoid and a predatory beetle can usually be found in a given grove (Murdoch *et al.*, 1995). Seventeen enemy species were released against black scale, *Saissetia oleae*, in Israel, but successful control was attributed mainly to the action of a single parasitoid species (*Metaphycus bartletti*) (Argov and Rossler, 1993). In these cases control was achieved only after these particular parasitoids were released, and the presumption is that they alone lead to control, though this is difficult to establish.

Finally, biological control also offers some of the clearest examples of competitive exclusion of one species by another. The parasitoid *Aphytis lingnanensis* was moderately successful at controlling red scale on citrus in southern California before the introduction of *A. melinus*. Within a few years of its release, *A. melinus* displaced *A. lingnanensis* over most of the area and gave improved control (Luck and Podoler, 1985). Also on citrus, the bayberry whitefly (*Parabemisia myricae*) became a pest in the late 1970s. Two parasitoid species from Japan were released, but a short time later a native or accidentally released species, *Eretmocerus debachi*, appeared. Within a few years it had excluded the released parasitoids and depressed the whitefly to extremely low levels (Rose and DeBach, 1992).

Although there are good examples illustrating the potential outcomes of competition among natural enemies, unresolved issues include the mechanisms of co-existence, and the effect of competition among enemies on pest suppression. The critical experiment of removing one or more natural enemies may be logistically impossible, and the release of new enemies is driven by practical needs rather than a desire to test theory. Encouragingly, however, there are some cases where models have been successful in explaining the outcome of competition in real systems. A study of potential control agents for control of mango mealybug (*Rastrococcus invadens*) by Godfray and Waage (1991) illustrates how realistic stage-structured models might be used to predict the success of biological control agents. Murdoch *et al.* (1996a) were able to explain why *A. melinus* was able to displace *A. lingnanensis* rapidly and achieve better control of red scale: *A. melinus* is able to use slightly smaller scale for the production of female offspring.

Successful biological control seems to us an especially attractive arena for investigating the mechanisms and effects of co-existing natural enemies. Here, we know that the shared resource for the natural enemies is scarce – that is the definition of success. So the problem of co-existence in the face of competition is acute. Furthermore, many theoretical mechanisms allowing co-existence result in weaker pest control. How are control and co-existence related?

7.6 DYNAMICAL IMPLICATIONS OF SIZE- AND STAGE-STRUCTURED INTERACTIONS

Theory

The response of a parasitoid to an encountered host individual almost universally depends on the size or stage of the host. Theory that takes into account this fundamental fact has been developed over the past decade. Recent reviews can be found in Murdoch and Briggs (1996) and Briggs *et al.* (1998), and Murdoch *et al.* (1997) integrate several aspects into a single framework.

Parasitoid responses to host age/size include the following. Adult female parasitoids tend to feed on younger (smaller) and parasitize older (larger) hosts, and on older hosts they tend to lay more eggs, lay more female-biased clutches, survive better, develop faster, and turn into more fecund adults.

A summary of theoretical results is as follows.

(i) Invulnerable host stages tend to stabilize an otherwise unstable interaction, especially if the adult is invulnerable (Murdoch *et al.*, 1987).

(ii) Cycles with a period equal to the host development time tend to occur when the adult host is short-lived and there is density dependence in the per head attack rate of the parasitoid (Godfray and Hassell, 1989).

(iii) The various age-related responses listed above all lead to a larger gain for the future female parasitoid population as the age of the host attacked increases. This leads to a kind of delayed density dependence in the parasitoid population. If the density dependence is not too strong or too delayed, it tends to stabilize the interaction (by suppressing the inherent Lotka–Volterra-like cycles). Otherwise it induces 'delayed feedback' cycles with a period equal to one to three times the host development time (Murdoch *et al.*, 1997; Briggs *et al.*, unpublished data). The delayed density-dependent effect is reduced when we take into account the effect on the future parasitoid population of the production of host meals or male parasitoids from smaller hosts (Briggs *et al.*, unpublished data). Murdoch and Briggs (1996) and Briggs *et al.* (1998) explore how pest steady-state density is influenced both by parasitoid behaviour that depends on the stage of the pest attacked, and by properties of the pest.

Density dependence in higher-dimensional models

One of us has already discussed the classic controversy over the role of density dependence versus density independence in population dynamics (Murdoch, 1994) and we will not cover that same ground here. Suffice it to say that the Andrewartha–Nicholson controversy seems, to us, to be resolved now that spatial processes are an integral part of theory on population dynamics. We add just one point.

It is clear that simple models must have one or more vital rates dependent on density, or densities will be unbounded. For example, in the absence of such dependence, the simplest single-species population in a stochastic environment will randomly walk to extinction. Again, even deterministic predator–prey populations that incorporate any delay and lack such density dependence will fluctuate with ever-increasing amplitude. The notion of density dependence therefore continues to be useful.

The picture is less clear, however, in higher-dimension models (e.g. models with several or many species, or several or many age classes in each species). There, populations can be well regulated (i.e. the steady state is stable, or there is a stable limit cycle or other bounded attractor) with no simple dependence of vital rates on density, and with no obvious source of density dependence in general. For example, the Murdoch *et al.* (1987) parasitoid–host model is in every way equivalent to the neutrally stable Lotka–Volterra model (all vital rates are fixed and independent of density), except that it distinguishes between immatures and adults. If the adult host stage is invulnerable and sufficiently long-lived, the model's steady state is stable.

It may be possible to plot the output from such a model to illustrate that there is a complex delayed relationship between the rate of change of the population and density at some previous time(s). But this is a result of the model, not an explanation for its dynamics, and the notion of density dependence turns out to be not especially useful for understanding how an invulnerable adult stage might help stabilize a real population. There are some circumstances, therefore, where the concept of density dependence will not be helpful.

Evidence

Many studies have documented stage- or size-dependent sex allocation, host feeding and oviposition in a variety of parasitoids, including parasitoids used in biological control (reviews in Godfray, 1994; Heimpel and Collier, 1996; van Alphen and Jervis, 1996; Murdoch *et al.*, 1997). However, these studies have largely been done in the context of testing behavioural models, and the theoretically possible range of dynamics has yet to be demonstrated in natural or biological control systems.

A difficulty in exploring the relevance of this type of theory is that it may be impossible to test the models directly. It will usually be impossible to alter the stage-dependent behaviour of the parasitoid experimentally. We will need to rely, therefore, on estimating parameters as accurately as possible and then asking whether field dynamics match model predictions. We comment briefly on the above numbered points.

(i) Approximate parameter values for the model with invulnerable adults were estimated for the interaction between *Aphytis melinus* and red scale in citrus groves in California. The results suggested that, while the invulnerable adult stage might contribute to the observed stability, it could not alone account for it (Murdoch *et al.*, 1987). Murdoch and colleagues are developing a much more detailed model of this system, with more accurate parameter values, and this conclusion might yet change.

(ii) Godfray and Hassell (1989) suggested that cycles with a period close to the duration of the immature pest stages, seen in several tropical pest species whose adult stage is short-lived, can be explained by the stage-structured model discussed above. There is no evidence from these systems whether the crucial mechanism that produces the cycles (density dependence in the parasitoid attack rate) exists, and this type of system would repay more detailed study.

(iii) A rather simple model of increasing gain with age of the pest attacked was developed for the *Aphytis melinus*–red scale interaction, and predicted damped cycles with a period of the order of twice the red scale development time (Murdoch *et al.*, 1992b). Although there is evidence that semi-discrete cohorts of red scale occur with approximately this period, they may also occur in the absence of parasitoids (R. Luck, personal communication), and there is, as yet, no evidence that such cycles occur in real systems, though no search has been made for them.

A broader question is whether such stage-dependent interactions can explain the dynamics of appropriate biological control systems; i.e. those whose protagonists have the potential for overlapping generations. The models can in principle explain, for example, why a system is stable, even though (as mentioned above) there may be no simple density-dependent relationships. The apparent success of such models in giving insight into the relative success of competing enemies (see section 7.5) suggests they are worth examining in real systems.

The red scale example mentioned above bears on the issue of density dependence in higher-dimensional models. Reeve and Murdoch (1986) and Murdoch *et al.* (1995, 1996b) analysed three sets of data from red scale populations, each covering several generations of the scale and at least six of the parasitoid. They failed to find any evidence that either of the two parasitoids, *Aphytis melinus* or *Encarsia perniciosi*, caused density-dependent parasitism. Yet the system is remarkably stable (Murdoch *et al.*, 1995). The explanation might be that these studies were not long enough, that density dependence is obscured by seasonal changes, or that there is so little variation in pest density that we cannot detect the density dependence. But if the explanation for stability lies in the stage-structured interactions, we may never detect density-dependent processes.

An alternative explanation is that these parasitoids are not the source of stability in a direct sense. They might, instead, lower the pest density to a level at which other, perhaps rather weak, density-dependent factors can stabilize the system. There is, for example, a beetle predator present at low density in our study groves, and it or some other factor may induce simple density dependence, though we also failed to find evidence that predation losses were density dependent (Murdoch *et al.*, 1995).

However, even a parasitoid that causes purely density-independent mortality, and is not directly stabilizing via its size-dependent responses, may yet be crucial to stability. Model pest–enemy systems are often more likely to be stable if the pest's rate of increase is reduced. This rate of increase usually refers to the pest's ability to increase in the absence of an enemy that induces density-dependent mortality. Thus a parasitoid that substantially reduces the pest's potential for increase could easily shift the steady state from an unstable to a stable region of parameter space. Hence, although the literature is replete with the conclusion that a given parasitoid is not important to control because it does not induce density-dependent mortality, the pest population might well be uncontrolled and unstable in the absence of such a parasitoid.

7.7 DISCUSSION

Our review shows that biological control systems have provided useful evidence relating to, or stimulating, a broad array of theory in population ecology. They illustrate various spatial processes, including the role of metapopulations, refuges and natural enemy aggregation; provide some of the best evidence on competitive displacement; strongly suggest that generalist predators can be a source of both stability and prey extinction; and have provided the stimulus for an increasingly complete theory of size-structured host–parasitoid interactions. The message from biological control, however, is also in many cases mixed – for example, on the issue of multiple enemies – and unresolved theoretical issues certainly remain.

Biological control has for many decades stimulated developments in theory for the dynamics of hosts and their specialist parasitoids in the context of a single, if sometimes spatially heterogeneous, population. The time would seem ripe now for it to stimulate the fusion of two other aspects of population ecology: spatial processes and generalist predators. While theory for spatially distributed populations has been developing apace, that on generalist predators has been relatively dormant. A natural arena in which these two forces come together, illustrated by the models of Ives and Settle (1997), is biological control in spatially distributed, seasonal (i.e. short-lived) crops. Here spatial processes may be overwhelmingly important, generalist predators seem frequently to be the key potential control agents, and their movements across the habitat must often be a crucial process.

Finally, we comment on the need for better information about biological control in practice. While we have plenty of theory, we have remarkably few well studied examples. In spite of pleas from many of its practitioners, those who practise biological control are not usually funded to carry out detailed studies that get at mechanisms. We also suggest that population ecologists have much to gain by studying biological control. Biological control is effectively a large-scale population experiment that can yield insight into consumer–resource and competitive interactions in a community that is often mercifully less complex than many in which ecologists work. The phenomenon of interest is usually unambiguously in operation: the enemy controls the pest at low density. Examples of biological control have stimulated much predator–prey theory and, properly exploited, they can also provide its testing ground.

ACKNOWLEDGEMENTS

This work was supported by grants Nos DEB94-20286, DEB96-29136 and DEB93-19301 from NSF, and No. 96-35302-3753 from the NRI Competitive Grants Program, USDA.

8

The effects of qualitative changes of individuals in the population dynamics of insects

S. R. Leather and C. S. Awmack

8.1 INTRODUCTION

In 1957, W. G. Wellington highlighted the fact that ecologists had neglected the role of the individual in the population dynamics of animals. He pointed out "populations are composed of individuals, and individuals differ" (Wellington, 1957). More recently, Lomnicki (1988), in a series of elegant and wide-ranging models, showed that individual variation was an important influence on the population dynamics of a species, whether plant or animal. Interestingly, evolutionary biologists consider that the early population biologists treated populations as individual organisms (Cronin, 1991). This chapter discusses some of the factors that can cause differences between the individual members of an insect population (the quality of each insect) and relates changes in insect quality to the dynamics of the populations to which these individuals belong.

Population dynamics is "the study of changes in the number of organisms in populations and the factors influencing these changes" (Solomon, 1969). It is not possible to understand insect populations in terms of numbers alone: it is essential to have knowledge of the factors that affect the quality of individuals to understand numerical changes in insect populations. As Lomnicki (1988) points out, ecological space is heterogeneous, individuals vary, and resource partitioning is never equal. These interactions shape the dynamics of populations at the individual level.

Insect Populations, Edited by J. P. Dempster and I. F. G. McLean. Published in 1998 by Kluwer Academic Publishers, Dordrecht. ISBN 0 412 83260 7.

Here, insect quality will be considered in terms of the major factors that influence population dynamics: fecundity, mortality and movement. All the effects on insect quality discussed below will be related to these first two factors, but the role of dispersal will not be ignored. It could be argued that at the population level, as defined in this volume, dispersal is only another way of expressing recruitment or mortality. An insect that shows high effective fecundity (by which we mean a large percentage of its gametes survive in subsequent generations) will be deemed to be of better quality than an insect which shows low fecundity. Equally, although mortality is inevitable for all living things, insects which produce high numbers of viable offspring will be defined as being of better quality than those which die before reproduction or before they have produced their maximum potential number of offspring.

The timescale over which quality is measured is also of importance to insect population dynamics: an insect that is poorly adapted to one set of environmental conditions may be well adapted to a different environment. Therefore, in an attempt to discuss the long-term implications of insect quality for population dynamics, changes in insect quality will be related not only to the generation being investigated but also to subsequent generations. Individual insects are influenced by a number of factors, both biotic and abiotic. It is often not under the control of the individual either to avoid or to modify these, for example, offspring cannot influence parental effects. Other factors, such as host quality (suitability of the host as a food source and refuge (Leather, 1994), cannot be modified to a great extent by an individual insect, although exceptions include changes in sink strength, gall formation, and the induction of plant defences.

It is important to understand which factors affect the characteristics of the individual before considering the impacts that qualitative changes in individuals have on the dynamics of a population. It is also important to distinguish between obligate changes in individual quality (such as those engendered by seasonality in response to predictable cues such as day length) and adaptive changes in response to cues such as host quality or crowding.

In Fig. 8.1a, only the immediate effects of the environment (i.e. all of the factors that may influence insect quality) affect the individual insect. This is not ecologically realistic, however, as no insect can escape the influence of the quality of its parents (and often its grandparents) (Fig. 8.1b). The interactions between different insect generations all have potential to affect the quality of individuals.

This chapter will discuss the effects of changes in insect quality on insect population dynamics: firstly, by considering how insect quality may be affected by the abiotic environment; secondly, by examining the effects of nutrition on insect quality; and finally, by describing the effects

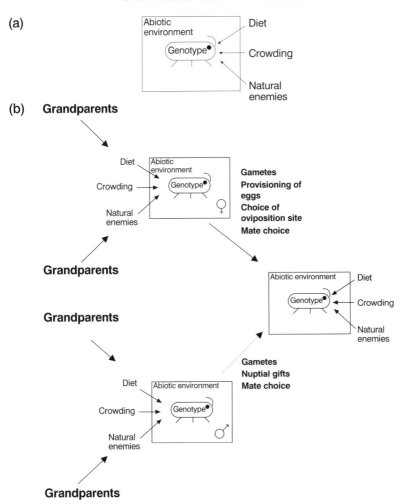

Fig. 8.1 Environmental factors affecting the quality of insects (a) over one generation, and (b) over more than one generation. Male parent's contribution is represented by a dotted line as not all insects have a male parent.

that other organisms can have on the quality of the individual insects within a population.

Size and age are very important factors of insect quality that impinge across all areas of an insect's life, as they have marked influences on fecundity, fertility, dispersal and survival. It is thus appropriate to discuss them separately before moving on to other factors that influence individual insect quality.

8.1.1 Size

It has long been perceived that there is a strong positive link between female size in insects and their fecundity, be it achieved or potential (Honek, 1986), although the universality of this relationship has been called into question (Leather, 1988; Klingenberg and Spence, 1997). The correlation is particularly weak in those insects where longevity is short in the field, or where further progeny are developed after the adult moult.

Size in insects is correlated with a number of life-history parameters in addition to fecundity, e.g. adult life span, mating success, and fertility (Leather, 1995). Large individuals tend to be better able to defend themselves against predators, although they may be more apparent (Leather and Walsh, 1993). They have more reserves for dispersal and also tend to live longer than smaller individuals of the same species (Leather, 1988). Insects that live longer tend to produce more offspring than those that live for a shorter period of time (Leather, 1994). Size also affects the fertility of insects, for example, large codling moths lay a greater number of fertile eggs than smaller ones (Deseo, 1971).

8.1.2 Age

Female insects do not produce their offspring at a constant rate over their lifespan, and in nature insects rarely achieve their full reproductive potential (Leather, 1988). There is generally a rapid burst of reproduction at the beginning of the insect's reproductive life, e.g. aphids generally produce over half their offspring in the first quarter of their adult life (Dixon, 1985), and then fecundity gradually declines. In addition, as the offspring produced at the beginning of the reproductive period are intrinsically more valuable than those produced later, the first offspring produced tend to be larger and fitter than those produced later (Leather, 1995), e.g. egg size of some Lepidoptera decreases with increasing adult age (Leather and Burnand, 1987), and so it may not be appropriate simply to consider total fecundity in insect population models. Fertility also tends to decrease with increasing age of the female (Leather, 1995), and thus a constant daily reproductive rate cannot always be assumed. This may have implications for predator–prey models which assume both constant host quality and a constant daily reproductive rate.

8.2 DIRECT EFFECTS OF ABIOTIC FACTORS ON THE QUALITY OF INDIVIDUALS AND THE POPULATION DYNAMICS OF INSECTS

The three most important abiotic factors that an individual insect faces during its lifetime are temperature, photoperiod and humidity. These can affect individual performance either by direct effects on the physiology of an insect, or via their effects on the quality of its food. These two different effects of the abiotic environment are considered in turn, and related to changes which are genetically fixed (for example, insects entering diapause as the days get shorter in the autumn (Leather *et al.*, 1993)), and to effects which are not genetically fixed (e.g. the increase in insect reproductive rates as temperature increases to a threshold; Gilbert and Raworth, 1996). Although abiotic cues may have effects that are important for the population dynamics of a species, e.g. the diapause preparation response to shortening day lengths of the braconid *Cotesia rubecula* (Nealis, 1985), these are seasonally and geographically determined, and are inextricably linked with particular stages of the insect life cycle. They are thus not of the same order of effect as, for example, the temperature-induced change in colour of an aphid (Dixon, 1972) or the increased reproductive output of insects as temperature increases (Gohari and Hawlitzky, 1986; Leather, 1995).

8.2.1 Direct effects of temperature, photoperiod and humidity on insect quality

As insects are poikilothermic, temperature is one of the most important abiotic factors affecting all their activities. It affects most life-history parameters, particularly developmental period, pre-reproductive delay, fecundity, fertility and flight, along with other behaviours (Gilbert and Raworth, 1996).

Insects all show a similar response to temperature between certain upper and lower limits (Fig. 8.2), although these thresholds are not the same for all insect species. There exists an upper and a lower threshold, above and below which development and reproduction do not occur (Leather, 1995). The developmental threshold is usually lower than the reproductive threshold (Gilbert and Raworth, 1996). Rates of fecundity or development increase between these two thresholds, usually linearly, to an optimal point, which often matches the normal climatic distribution of the insect (Leather, 1995). There is normally an inverse relationship within a species between development rate and adult size: the higher the temperature, the smaller the individual (Gilbert and Raworth, 1996). This may have implications for the regulation of populations by natural enemies. The apparent contradiction is readily explained by remembering that the size–fecundity relationship is valid within a temperature band, but that the size–temperature relationship operates across the temperature bands. This can, of course, have profound implications for population development, as size and fecundity are generally well correlated.

A second implication of changes in temperature is that they may result in changes in the genetic structure of insect populations. If the change in an environmental cue is predictable, for example, if winter temperatures were to increase predictably, then there may be the potential for the insect species to adapt to this change: those individuals which emerge at

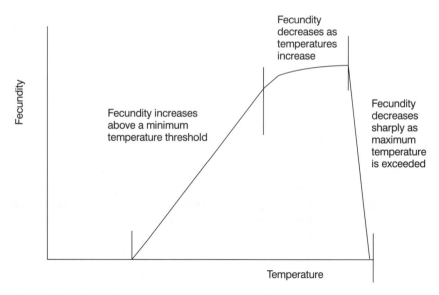

Fig. 8.2 The effects of temperature on insect growth and reproduction.

the right time will be selected for, while those that emerge at the wrong time will not. This can occur only if insects are able to anticipate their environment accurately. This is not out of the question – timing of sexual production in the autumn in the bird cherry aphid, *Rhopalosiphum padi*, is based on the temperatures experienced by individuals in August (Ward *et al.*, 1984). Random changes in the environment, whether or not they result in the same average change over a long period of time, will not have this effect, as insects will not be able to predict the direction of change from one generation to the next. Unpredictable changes in the environment could lead to greater variability within the population, at least when it comes to this particular trait, with the result that the population would not be as well adapted to its environment as it was before. Also, the stability of the population could be significantly influenced by the genetic diversity available (Den Boer, Chapter 3 in this volume).

In seasonal environments, photoperiod is one of the most reliable cues available for insects to distinguish seasonal changes. Photoperiod affects the morph of the individual produced in aphids, e.g. decreasing photoperiods stimulate the production of sexual forms (Kenten, 1955; Dixon and Glen, 1971) and can also affect emergence in spring from overwintering forms. Photoperiod also has marked effects on the reproductive behaviour of insects; for example, the moth *Plutella maculipennis* is more fecund under long-day conditions than under short-day conditions (Harcourt and Cass, 1966). However, although these changes do have impacts on population dynamics, they are obligate responses and do not vary between generations. Although there is variation between individuals in the same species, it is less likely to be a significant factor in population control than abiotic factors such as temperature.

Humidity can have marked effects on insect fecundity and fertility (Leather, 1995). Thus, oviposition in bollworms (*H. zea*) increases with increasing humidity and the percentage hatch of eggs is greater at higher than at lower relative humidities (Ellington and El-Sokkari, 1986). These effects do not affect the subsequent quality of the offspring, only the number that hatch successfully, so humidity will not be considered further. It should be noted, however, that an increase in humidity greatly increases the susceptibility of insects to fungal pathogens as spore germination increases (Oduor *et al.*, 1996) and this can lead to epizootics (outbreaks of disease), individual deaths and decreases in population size.

Temperature and photoperiod can also have qualitative effects on individual insects by increasing their fecundity or stimulating the production of dispersive forms, as in aphids (Dixon, 1985). As an example, consider an insect which uses a mixture of cues to determine the time of emergence from an egg/diapause (such as a combination of temperature and photoperiod), one of which is fixed (day length) and one of which is variable (temperature). If the variable cue were to change so that it was out of the range previously experienced by the insect (e.g. if winter temperature were to increase), the potential exists for the insect to become out of synchrony with its host plant. This would have important effects on insect quality, because the diet that the emerging insect needs might not be available, and there might be an increased vulnerability to extreme events such as late frosts. This could also affect the synchrony between the insect and its parasitoids, which might rely on a different combination of environmental cues for their emergence (Cammell and Knight, 1992).

8.2.2 Indirect effects of temperature, photoperiod, humidity and gaseous air pollutants

For herbivorous insects, abiotic factors can have a profound effect on the quality of their host plants. Abiotic factors such as air pollution and drought stress can affect the quality of insects and their subsequent fecundity and survival via their effects on plant quality. Drought stress is well known to be associated with insect outbreaks, and this is thought to be due primarily to increased mobility of nitrogen within the plant, leading to increased insect reproduction and larger populations (White, 1984) particularly when the quality of the host plant is generally poor, as in many non-arable environments. However, Watt (1994) has proposed that insect performance is favoured by the high temperatures often associated with drought.

Atmospheric pollutants such as nitrogen oxides (Dohmen *et al.*, 1984), ozone (V.C. Brown, 1995) and carbon dioxide (Watt *et al.*, 1995) also influence insect performance via their effects on the host plant. This is thought to be associated with a response to stress by the plant, leading to changes in nutritional quality which can influence insect population growth. The majority of these pollutants enhance the performance of sap-feeding herbivores such as aphids, whilst chewing herbivores such as Lepidoptera and Coleoptera tend to be adversely affected. Thus, insect populations which have different feeding strategies (i.e. suckers or chewers) will be differently affected by changes in host-plant quality caused by environmental stress. This suggests that changes in insect quality can be determined by the feeding strategy of an individual insect, and that it is not possible to make predictions about population dynamics for all insects: different guilds must be considered separately. It has even been argued that abiotic factors, often acting via their effects on food quality, can have qualitative effects on insects which are then passed down through the generations, thus influencing their population dynamics (Rossiter, 1994, 1995).

8.3 THE EFFECTS OF BIOTIC FACTORS ON THE QUALITY OF INDIVIDUALS AND THEIR POPULATION DYNAMICS

8.3.1 Nutritional factors affecting insect quality and population dynamics

Nutrition can be defined as the quality and quantity of food eaten by an insect during its larval and/or adult life (Leather, 1995). Nutritional factors can be divided into two groups: diet quality and diet species. The nutrition of an insect, as either an adult or larva, has marked effects on its fecundity and fertility. Some insects are relatively unaffected by adult nutrition, for example the Lepidoptera, whereas others, such as aphids, are strongly affected by both adult and larval nutrition and also that of their grandmothers (Dixon, 1985).

8.3.2 Composition of the diet: insect nutritional requirements

In order to grow and reproduce, insects must have access to certain essential nutrients for at least part of their development. Although many adult insects do not feed, in order to develop all larvae must have access to essential nutrients, the most important of which is nitrogen. One of the main limiting factors for insect quality is the availability of nitrogen in their diet (Strong *et al.*, 1984). The so-called 'nutritional hurdle' faced by herbivorous insects occurs because plants contain very low concentrations of nitrogen (for example, phloem sap contains only between 0.03%

and 1% amino acids; Klingauf, 1987). A second consideration is that only a small percentage of the nitrogen ingested by an insect is in a form which it is able to use directly. An insect can thus be assimilating a very low percentage of the total nitrogen in its diet, and must ingest large amounts of food to obtain adequate nutrients. One of the implications for insects of the generally low quality of their diets is that very small changes in composition (for example, the addition of a nitrogen fertilizer) can have very large effects on insect quality, in terms of both size and subsequent fecundity (Markkula and Tiittanen, 1969; Shaw and Little, 1972; Hartley and Gardner, 1995). As an example, consider the changes in performance of the sycamore aphid, *Drepanosiphum platanoidis* as it feeds on sycamore (*Acer pseudoplatanus*) throughout the summer (Dixon, 1987a). The quality of the phloem sap (in terms of the concentration of amino acids) changes throughout the summer and influences aphid populations: when the amino acid concentration of the phloem is low, aphid populations are small, whereas when amino acid concentrations increase in the autumn (as the trees transport nitrogen away from the leaves before abscision), aphid populations increase (see Dixon and Kindlmann, Chapter 9 in this volume).

Insects also have other nutritional requirements (carbohydrates, vitamins, minerals, lipids, sterols, etc.) apart from nitrogen. However, these nutrients do not appear to influence insect quality as much as nitrogen, and so will not be discussed here. The changes in C:N ratio observed when plants are grown at elevated CO_2 are thought to adversely affect insect quality, because they dilute the nitrogen available to the insect and not because of any intrinsic detrimental effect of carbohydrates on insect metabolism (Lincoln *et al.*, 1993). As well as having direct effects on insect quality via changes in fecundity, size, morph, etc., nutritional factors can also influence behavioural traits of insects. When the aphid *Aulacorthum solani* was reared at elevated CO_2 for several generations, it lost the ability to produce or respond to alarm pheromone (Awmack *et al.*, 1997a). This of course could have profound effects on the population dynamics of that particular species.

(a) Effect of nutrition on larvae

Regardless of whether the insect feeds as an adult or not, larval nutrition is the most important factor determining potential changes in individual quality. The size an adult insect achieves, within genetic and environmental constraints, is dependent on the nutrition experienced by that individual as a larva. Larvae of *H. zea* fed on corn produced females that were extremely fecund and produced a large percentage of fertile eggs, but those larvae fed on cotton square produced infertile females (Lukefahr and Martin, 1964). Pine sawflies are also strongly affected by the quality of the plant on which they were reared, even when feeding on host plants of the same species but of different clonal origins. Host plants allowing good larval growth and survival resulted in large, highly fecund females (Auger *et al.*, 1990). In species which show a different response to host-plant quality it may not be sensible to use an average measure of performance in population models.

In general, insects reared under conditions of poor larval nutrition, caused by crowding or poor quality food, develop into small, low-quality adults with low fecundity and reduced longevity, seen e.g. in predators such as the stonefly *Megarcys signata* (Peckarovsky and Cowan, 1991) or herbivores such as the willow-feeding chrysomelid *Chrysomela knabi* (Horton, 1989). However, some of the adverse effects of poor insect quality may be offset by the decrease in palatability of these insects to predators (Rowell-Rahier and Pasteels, 1986). Changes in insect quality caused by nutritional factors may, therefore, have unexpected effects on population dynamics.

(b) Effect of nutrition on adults

The fecundity of insects is linked to the amount of reproductive investment; for example, the number of ovarioles within the reproductive tract is positively correlated with fecundity in many species of insect (Wellings *et al.*, 1980; Bennettova and Fraenkel, 1981). The number of ovarioles can be determined genetically: some insects show no variability in ovariole number, e.g. most Lepidoptera (Leather, 1995); or the number of ovarioles may be nutritionally modified, e.g. poorly fed flies, *Sarcophaga bullata*, develop into small adults with a low number of ovarioles (Bennettova and Fraenkel, 1981). On the other hand, fecundity can be determined by an interaction between genetic and environmental factors. In aphids, for example, many species exhibit individual variation in ovariole number (Wellings *et al.*, 1980; Leather *et al.*, 1988), and although the distribution of individuals produced with different ovariole numbers is genetically controlled, the number of surviving individuals in each ovariole class is strongly influenced by larval nutrition (Ward *et al.*, 1982). In adverse environments, selection favours those insects with a low ovariole number. The fecundity of subsequent generations may thus be determined by the frequency of alleles for ovariole number surviving from the previous generation. Aphids with high reproductive investments, i.e. many ovarioles, are less likely to survive to adulthood unless they are in a high quality environment, but they have a higher realized fecundity if they do survive.

The pre-reproductive delay and/or the time an insect takes to reach the adult moult is also strongly influenced by larval nutrition. On a good quality host, an insect develops more quickly and reaches a larger size than it does on a poor quality host (Leather, 1994). If an individual reaches adulthood more quickly and is able to begin its reproductive life sooner, then this has significant effects on the population dynamics of those insects with short seasons of occurrence. So, for example, the pine beauty moth (*Panolis flammea*) develops more quickly on certain provenances of lodgepole pine (*Pinus contorta*) from the south coastal region of the USA, than it does on those from Alaska (Leather, 1987), and in turn has a shorter pre-oviposition period on these same host plants (Leather, 1985). The result is that in the field, outbreaks of *P. flammea* occur regularly in lodgepole pine plantations made up of south coastal provenances, but never in plantations composed of Alaskan lodgepole pine (Watt *et al.*, 1991; Leather and Knight, 1997). This is a widespread phenomenon in insects: aphids reared on good quality hosts show faster development, begin to reproduce more quickly and produce more offspring than aphids from the same clone reared on poor quality hosts (Leather and Dixon, 1982; Awmack *et al.*, 1996, 1997b). Nutritional factors which affect insect quality greatly affect the population dynamics of such insect species.

8.3.3 The effects of different host species on insect quality

The choice of host is another factor that may have a great influence on the quality of insects and their subsequent fecundity. Specialist insects are generally assumed to perform better on a host than are generalist insects on the same host, although this has not often been quantitatively demonstrated. The two types of generalist recognized by Bernays and Minkenberg (1997) distinguish between insects which are able to feed on a wide range of hosts, and those which specialize individually on a wide range of hosts. These two different strategies have implications for insect quality: an insect that can perform averagely well on any host will be of a different quality to one that performs exceptionally well on one host (in effect as a specialist) but poorly on other hosts (Fig. 8.3). Insect quality in the population (fecundity, suitability as hosts for predators and parasitoids, etc.) will be much more variable in a population made up of a wide range of specialists, than one whose individuals do averagely well on all hosts. It has also been pointed out that generalists need to allocate a much larger proportion of their resources to information processing than do specialists (Dall and Cuthill, 1997).

Host acceptance is mediated by a wide range of chemical, visual and physical cues. Plants that are well defended against herbivorous insects often lead to poor insect quality, as the herbivore must expend metabolic energy on detoxification of secondary metabolites which could otherwise be allocated to growth and reproduction. Insects either detoxify these compounds and escape from competition from other insects, e.g. milkweed and ragwort specialists (Oyeyele and Zalucki, 1990), or they avoid them. An interesting effect of toxic secondary metabolites on insect quali-

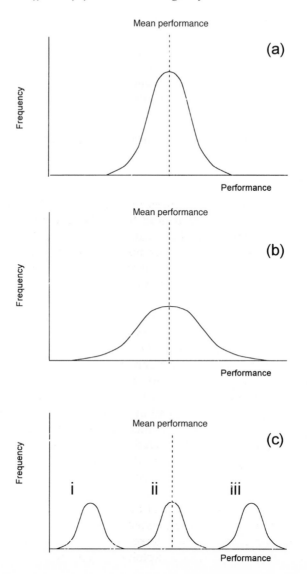

Fig. 8.3 Possible differences in performance between (a) specialist and (b, c) two types of generalist herbivores on the same species of host plant. (a) Specialist has low variability with all individuals showing similar performance; (b) generalist shows high variability with some individuals performing better than specialists but some performing worse – note that the mean performance of the population may be the same; (c) generalist population split into 'host races' which each feed on separate host-plant species. The individuals which feed on the same host plant as in (i) and (ii) show similar characteristics to the specialist herbivore described in (a).

ty is their effect on the subsequent risk of attack from predators and parasitoids. It has even been argued that some individual insects may 'accept' a reduction in quality and hence performance, and feed on poor quality/toxic hosts in order to escape from natural enemies (Rowell-Rahier and Pasteels, 1990). It is, however, difficult to propose a mechanism for an individual insect to make this kind of decision.

8.3.4 The effects of other insects on insect quality and population dynamics

(a) Crowding

Crowding has important influences on the quality of insects and can directly affect their development and fecundity. In some lepidopterans, and in aphids, a certain level of crowding can enhance the quality of the diet by causing the plant to divert nutrients to an area that is being depleted by herbivores. This means that the diet for these insects is of a better quality than if they were feeding singly. Thus, some insects perform better when they are reared in groups (Ghent, 1960; Dixon and Wratten, 1971). Crowding may also be beneficial for individual insects, as they may be able to consume more plant material before induced defences are produced by the plant (Denno and Benrey, 1997).

In locusts, increasing population density engenders a shift from sedentary, highly fecund individuals to more mobile, less fecund forms (Uvarov, 1961). The size of the insect population can also influence the quality of the individual insects within it, and *vice versa*. There are, however, trade-offs involved. Insects which are crowded can deplete their resources before they have completed development. In this case some insects, e.g. the nymphalid *Chlosyne janais* (Denno and Benrey, 1997), disperse during the fourth instar and feed as solitary individuals. In aphids, crowding leads to production of winged morphs which are able to disperse to new host plants, but only at the cost of decreased reproduction, because of the metabolic energy needed to produce and maintain wings (Dixon, 1985). Thus, a temporary decrease in insect population size may occur as individuals disperse and exploit new resources.

(b) Parental influences on reproduction and quality of offspring

The quality of an insect can have a large effect on the quality of its progeny. Good quality females tend to produce good quality offspring which develop more quickly and have greater fecundity, leading to increases in population size. Insects can influence the quality of their offspring in several ways, each of which have slightly different effects on the resulting quality of the next generation.

(c) Maternal influences on the quality of the next generation

The effect of the female on the fitness of individual insects was long over-looked. However, work on aphids (Dixon and Wratten, 1971; Dixon and Dharma, 1980; Dixon et al., 1982; Leather, 1989) indicated that females influenced the fitness of their offspring in just as marked a way as mammals, birds and other groups exhibiting parental care (Rossiter, 1994, 1995).

Females have more influence on the quality of the subsequent generation than males. Like males, a female insect can be fit herself, although there is evidence that fitness in one environment is not a guarantee of fitness in all environments. She can chose a good quality mate (although again 'good' will be to some extent environmentally defined). A female can also vary allocation of resources. Faced with a resource of a given quality and a finite amount of metabolic energy available for reproduction, a female insect can allocate her resources to maximize the quality of her offspring and thus the performance of the population. A female may therefore produce a few, high quality offspring, or several poorer quality offspring depending on the environment (see below).

Resource allocation by the female may be affected by the nature of the host on which she is laying her eggs. Insects that are restricted to discrete sources of food, such as seeds, galls or carcasses, and which depend intimately on these resources for survival, might be expected to be more likely to invest in resource allocation, and thus the quality of their offspring, than those which are able to utilize widespread resources. On a widespread resource, a female insect needs only to assess its quality in terms of nutrient availability, secondary metabolites, and so on. A discrete resource may not support a female's offspring to maturity and so it may pay her to reduce the number of offspring produced on it. Parasitic Hymenoptera are also able to vary the sex of their offspring according to their perception of host quality (Charnov *et al.*, 1981; Charnov, 1982) and assess the presence/absence of superparasites (Field *et al.*, 1997).

8.4 PHENOTYPIC PLASTICITY AND INSECT QUALITY

8.4.1 Egg provisioning, clutch size and host quality

Provisioning of eggs (and a choice of a good quality host plant for initial oviposition) can lead to better quality offspring (whether bigger or better nourished) which in turn leads to females which have either (i) more resources to provision their eggs; (ii) greater fecundity; or (iii) both. All of this can lead to better quality insects which are fitter (in terms of fecundity) even in sexually reproducing insects. There can thus be improvements in insect quality without any evidence of selection. For example, the gypsy moth, *Lymantria dispar*, does not allocate equal amounts of vitellogin to its eggs. The host plant on which the female is reared significantly affects the amount of yolk provisioning, so that females arising from *Populus tremuloides* invest more in their eggs than those arising from *Quercus rubra, Quercus prinus* and *Pinus rigida* (Rossiter *et al.*, 1993).

Individual females do not lay equally sized clutches. Many insects have the capacity to adjust their clutch size in relation to host quality, so that on poor quality hosts many species of insects will lay fewer but larger eggs. For example, the pine beauty moth, *P. flammea*, lays fewer, larger eggs on Alaskan lodgepole pine than it does on south coastal lodgepole pine (Leather *et al.*, 1985; Leather and Burnand, 1987). The black bean aphid, *Aphis fabae*, produces larger nymphs on poorer quality host plants (Leather *et al.*, 1983). The larger offspring are better able to cope with the poorer quality hosts (Leather, 1994).

8.4.2 Nuptial gifts, female fitness and fecundity

Female quality, and hence offspring quality, can also be improved by the donation of nuptial gifts. Nuptial gifts are defined as potentially nutritious substances given to the female by the male in conjunction with mating and subsequently used by the female to aid longevity and/or reproduction (Boggs, 1995). Nuptial gifts vary from gifts of food, to simply passing nutritive accessory gland fluid into the female together with the sperm (Boggs and Gilbert, 1979), to allowing the female to eat part or all of the male during mating. Whatever the form of the gift, the effect is to increase the fitness of the female, either in the short term by increasing egg production over the next few days, as seen in some orthopteran species (Gwynne, 1988), or even perhaps increasing lifetime fecundity (Boggs, 1995). Male nuptial gifts may affect juvenile development time: larger donations from the male will result from longer developmental times (Boggs, 1981). Survival may also be enhanced, as in some butterfly and moth species, where repeated matings increase female survival (Svärd and Wiklund, 1988; Royer and McNeil, 1993).

More importantly, male nuptial gifts can affect individual dispersal. Firstly, nutrient donations can allow females to disperse or migrate. Female monarch butterflies obtain lipids from males, and it has been shown that multiple matings in this species are necessary for long-term persistence of the population (Wells *et al.*, 1993). The alternative scenario is that females can perceive the males as a nutrient resource and refrain from dispersal, remaining instead in the vicinity of the males. This has also been shown for monarch butterflies, but in this case in Australia where overwintering aggregations are not formed and the butterflies breed all year round (Zalucki, 1993). As the composition of a population is affected by its age structure, dispersal habits of individuals, and population size, all of which are influenced by female fitness, nuptial gifts may play an important role in the population dynamics of some insects.

The effective population size is a measure of the number of individuals that contribute offspring to the next generation, as opposed to the actual number of individuals present in a population at any one time (Boggs, 1995). Although a population will have a characteristic mean number of matings per individual, some individuals will mate more often than others. Male nuptial gifts can affect the refractory period of the males, i.e. a large donation may prevent the male from mating for a longer time than a male that only makes a small donation to the female (Sakaluk, 1985; Burpee and Sakaluk, 1993). This of course has potential implications for population genetics.

8.5 INDIVIDUAL CHARACTERISTICS

The tent caterpillar, *Malacosoma pluviale*, a forest pest in North America, exhibits a behavioural polymorphism. If larvae from a single egg mass are analysed, most are less active and show no detectable response to light, but a few individuals respond to light and are more active (Sullivan and Wellington, 1953). The active larvae are able to locate and establish on food plants, whereas the inactive (sluggish) larvae need to follow a silk trail to find food sources, and have lower consumption rates and slower development times than the active larvae (Wellington, 1957). The sluggish larvae are also less resistant to disease than those that are active. However, the sluggish larvae are favoured in high-density populations because the active larvae may pick up viral and fungal diseases more readily, and they are also more susceptible to predators due to their greater apparency. The occurrence of the active larvae leads to greater dispersion of the population and the development of outbreaks, so that these two different larval qualities can have major impacts on the population dynamics of this species.

This concept of variable fitness was later embodied in the 'polymorphic fitness hypothesis' of Baltensweiler (1993), working on the larch bud moth *Zeiraphera diniana*. *Zeiraphera* possesses two colour morphs depending on its food plant – mainly black on larch, and a lighter orange on pine and spruce. The frequency of colour morphs on larch changes from an intermediate colour to black on a regular basis. Defoliation events occur at 8–10-year intervals in the Upper Engadine in Switzerland, and have done so since 1850. Larch (*Larix decidua*) flushes 2 weeks earlier than cembran pine (*Pinus cembra*), and the outbreaks of the moth on larch occur when the black form of its larvae is in the ascendancy. The reason for this is that the coincidence of egg hatch with bud break of the host plant is very important in determining rates of survival in *Z. diniana*, and the dark larch forms hatch earlier than the 'hybrid' crosses between the two races.

Both of these studies have their roots in the hypothesis put forward by Chitty (1955) who suggested that changes in the vitality (fitness) of individuals in successive generations, caused by changes in population density, could lead to greater population stability. Basically, his argument was that since population increase will lead to survival of many weaker genotypes, any harmful environmental factor will have a proportionally greater effect at population peaks than troughs.

Some populations of insects have different forms which respond differently to the same abiotic cues. For example, the sycamore aphid, *D. platanoidis*, has two morphs, one red and one green. The red form, which is produced in response to high temperatures, is more active and develops more quickly than the green form which is produced at the same time (Dixon, 1972). This red polymorphism persists until autumn (offspring of red viviparous mothers are always red), but eggs that are laid by red oviparae produce green fundatrices. The red form is better able to withstand high temperatures and overcrowding. This has also been documented in the grain aphid, *Sitobion avenae*, which again produces a brown and green morph. The green morph is able to produce winged offspring from winged adults (Watt and Dixon, 1981) in response to crowding, while the brown form cannot. As winged aphids have a lower reproductive rate than non-winged aphids, this can influence the growth of a population in that population growth after colonization of a habitat may be slower than that of the population left behind.

In addition, recent work (Losey *et al.*, 1997) has shown that the red form of the pea aphid, *Acyrthosiphon pisum*, is susceptible to attack by predators, whilst the green form is attacked more frequently by parasitoids. Colour is therefore an important quality that can have marked effects on the population dynamics of a species, leading in some cases to dramatic population fluctuations.

Another example of balancing selection is the phenomenon of resistance to chemical insecticides exhibited by, for example, the aphid *Myzus persicae*. The four different resistance levels shown by this aphid species to insecticides range from susceptible (with no resistance) to highly resistant. The balance between these four genotypes is maintained because, despite heavy selection for insecticide resistance in the summer months, highly resistant aphids do not survive the winter as well as less resistant ones. The polymorphism in the population is thus maintained by balancing selection (Foster *et al.*, 1996).

8.6 CONCLUSIONS

It has long been recognized that the quality of individual insects within a population varies greatly both within and between generations. What has not always been realized is that these changes can be directional and persistent, and can have marked effects on the population dynamics of some insect species. It is tempting to approach the study of the population dynamics of an insect species using the concept of a standard insect as the unit to be modelled. However, high levels of variability can have major influences on insect population dynamics. Variability may be caused by the environment, such as the timing of emergence from the over-wintering stage; or the influences of different host plants on the ability of populations to increase to outbreak levels; or by genetic factors such as ovariole number in aphids; or by an interaction between genetic and environmental factors.

This variability may influence the dynamics of insect populations by affecting dispersal (via the formation of winged or more active morphs); vulnerability to natural enemies such as predators and pathogens; and the ability of herbivorous insects to colonize new host plants and detoxify secondary metabolites. The variation of the quality of individuals within populations is too great to ignore safely, and should wherever possible be incorporated into studies of their population dynamics.

Part Two

9

Population dynamics of aphids

A. F. G. Dixon and P. Kindlmann

9.1 INTRODUCTION

Most aphid species reproduce both asexually and sexually, with several overlapping parthenogenetic generations between each period of sexual reproduction. This is known as cyclical parthenogenesis and, in temperate regions, sexual reproduction occurs in autumn and results in the production of over-wintering eggs, which hatch the following spring and initiate another cycle. For their size, aphids have very short developmental times and potentially prodigious rates of increase and, as a consequence, are typically little affected by the activity of insect natural enemies (Kindlmann and Dixon, 1989; Dixon, 1992; Dixon *et al.*, 1995, 1997). That is, aphids show very complex and rapidly changing within-year dynamics, with each clone going through several overlapping generations during the vegetative season, and potentially being made up of many individuals which can be widely scattered spatially. The production and survival rates of the eggs determine the numbers of aphids present the following spring.

The study of the population dynamics of aphids living on wild herbaceous plants is difficult because their host plants vary in abundance and distribution from year to year. Tree-living aphids, in addition to being very host specific, live in a habitat that is relatively stable both spatially and temporally. Therefore, it is not surprising that most long-term population studies on aphids have been on tree-dwelling species. The studies on deciduous tree-dwelling species (Dixon, 1963, 1966, 1969, 1970, 1971, 1975, 1979, 1990; Dixon and Barlow, 1979; Barlow and Dixon, 1980; Dixon and Mercer, 1983; Chambers *et al.*, 1985, Wellings *et al.*, 1985; Dixon *et al.*, 1993, 1996) have revealed strong direct density-dependent recruitment

Insect Populations, Edited by J. P. Dempster and I. F. G. McLean. Published in 1998 by Kluwer Academic Publishers, Dordrecht. ISBN 0 412 83260 7.

and dispersal, and an inverse relationship between the numbers of aphids hatching from eggs in spring and the numbers present several generations later in autumn (Dixon, 1970, 1971); this delayed response has been referred to as the seesaw effect.

This account will be mainly concerned with the analysis of two data sets. That for the sycamore aphid, *Drepanosiphum platanoidis*, was collected in Glasgow from 1960–73 and the Turkey oak aphid, *Myzocallis boerneri*, in Norwich from 1975–95. The role of intraspecific competition operating via strong direct density-dependent migration and recruitment and delayed density-dependent recruitment in regulating the abundance of these aphids is well documented. However, the studies on these species have revealed that they show very different within-year dynamics, with two similarly sized peaks in abundance in the sycamore aphid, one in late spring/early summer and the other in autumn; and predominantly only one peak in abundance in the Turkey oak aphid, which occurs in late spring/early summer (Fig. 9.1). In addition, the between-year dynamics in the two species differ greatly, with relatively constant yearly totals in the sycamore aphid on all the eight trees studied in Glasgow, and a marked cyclical pattern in the yearly totals in the Turkey oak aphid on the two trees studied in Norwich (Fig. 9.2). The relative rarity of sycamore aphids in 1973 resulted from the still autumn of 1972 when the aphid achieved a very high level of abundance, which resulted in it being abundant the following spring. As in 1961, the aphid declined dramatically in abundance in 1973 following the very high population in spring, giving a low overall abundance in 1973 (Dixon, 1979). The relative stability in sycamore aphid numbers from year to year is strikingly different from the cyclical pattern previously reported in suction trap catches (Dixon, 1990; Figs 9.2 and 9.3), and which proved so attractive to theoreticians (Turchin, 1990; Turchin and Taylor, 1992). The objective of this chapter is to understand why the patterns in the population dynamics of these species differ, and to account for the apparent discrepancy in the between-year dynamics of sycamore aphid numbers in the air and on sycamore. That is, the explanation will be constrained by reality.

9.2 MATERIALS AND METHODS

Census data for both the sycamore and Turkey oak aphids were collected weekly from bud burst to leaf fall by sampling 80 leaves at random from the lower canopy of eight sycamore and two Turkey oak trees located in the vicinity of Glasgow and Norwich, respectively (Dixon, 1963, 1966; Sequeira and Dixon, 1997). The re-analysis of the census data for only two of the eight sycamore trees is presented here, namely for the tree that was studied for longest and in greatest detail, and the one closest to the suction trap used in this study. In addition, the numbers of aphids

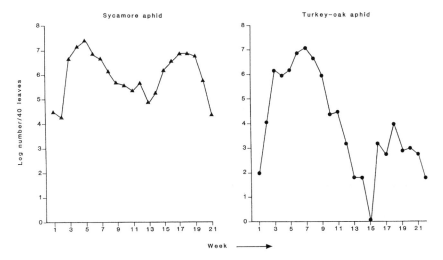

Fig. 9.1 Within-year trends in the weekly numbers of aphids on a sycamore and a Turkey oak tree in one year.

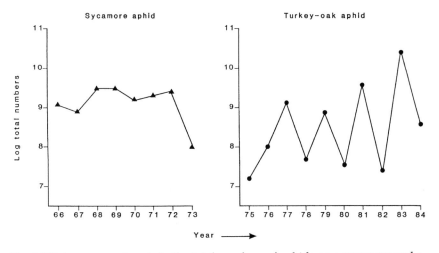

Fig. 9.2 Between-year trends in the total numbers of aphids on a sycamore and a Turkey oak tree in Glasgow and Norwich, respectively.

were recorded weekly on a Turkey oak in Glasgow over a period of 3 years (1971–73) and on four caged Turkey oak saplings (1 m high) over the course of a year in Norwich.

The reproductive rates of both the sycamore and Turkey oak aphids were obtained by clip-caging adults individually on leaves and monitoring the number of offspring they produced daily. This was done over a period of 3 years for the Turkey oak aphid and 6 years for the sycamore

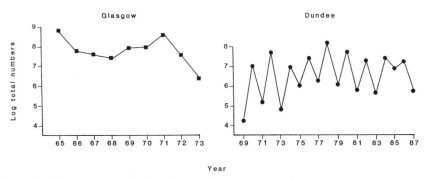

Fig. 9.3 Between-year trends in the total numbers of sycamore aphids caught each year in a suction trap positioned close to two of the trees sampled in Glasgow, and in the Rothamsted Insect Survey trap located at Dundee.

aphid, both in Glasgow. The soluble nitrogen content of the leaves of a Turkey oak, collected weekly in Norwich throughout a year, was extracted, and the concentration determined using a nitrogen analyser. The nitrogen content for sycamore was obtained from Dixon (1963).

The suction-trap catches of sycamore aphids were from two sources. One trap was located 10 m south of the canopy of one of the sycamore trees sampled for aphids in Glasgow. The opening of this trap was positioned 2 m above soil level and it was run for 9 years (1965–1973). The other trap was the Rothamsted Insect Survey trap at Dundee, which is 106 km north-east of Glasgow. The numbers of sycamore aphids caught by this trap were supplied by Dr R. Harrington.

The within-year dynamics of both species are largely determined by seasonal changes in host quality. Aphids do best when amino acids are actively translocated in the phloem. In spring, the leaves of trees grow and import amino acids via the phloem; in summer, leaves are mature and export mainly sugars; and in autumn, the leaves senesce and export amino acids and other nutrients. Thus the leaves of trees are most suitable for aphids in spring and autumn. This seasonality confounds population analyses, which may be complicated further by the length of the vegetative season varying from year to year. To overcome this problem and to address the differences in the population profiles shown by the two species, each year was divided into a spring increase, a summer decline and an autumn increase (Fig 9.4). The latter is very much weaker in the Turkey oak aphid than in the sycamore aphid (Fig. 9.1). The rates of increase or decline in each of these periods were determined by regression. In order to overcome measuring errors, the census data were smoothed by calculating moving averages (length three sampling intervals, equal weight) prior to analysis.

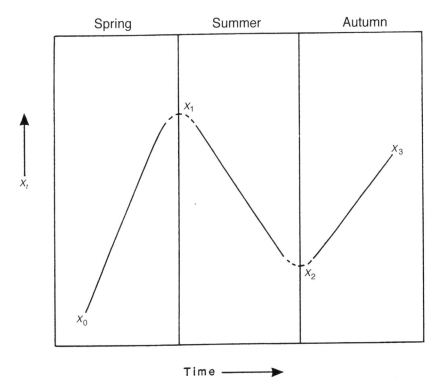

Fig. 9.4 Diagrammatic representation of the within-year population dynamics shown by deciduous tree-dwelling aphids.

Spring rates of increase in the adult population each year were calculated as the slopes of the relationships between the natural logarithms of the numbers of adults and time. The first point was always that when adults were first recorded in a year. The last point was that which gave the longest period of exponential growth for calculation of the growth rate and a regression coefficient significantly different from zero ($P<0.05$). This method was adopted because, as the population approached the summer peak (x_1), the population growth becomes sigmoidal rather than exponential. The rates of the summer declines were calculated as the slopes of the relationships between the natural logarithms of the numbers of adults and time. The first value was the first measurement following the summer peak (x_1) and the length of the series was again that which resulted in a regression coefficient significantly different from zero. This was necessary because population change in the summer trough in abundance was sigmoidal rather than negative exponential. The rates of increase in autumn were calculated in the same way as the spring rates of increase. The slope of the relation between the numbers in autumn and those present the following spring was also determined by regression.

9.3 RESULTS

9.3.1 Within-year dynamics

(a) Rates of increase

The spring rates of increase were always negatively correlated with initial numbers of adults, and the summer rates of decline were also negatively correlated with the numbers of adults at the instant when summer migration starts, i.e. when the population curve becomes sigmoidal rather than exponential. In addition, the autumnal rates of increase were negatively correlated with the minimum number of adults in summer. These results were independent of species and site (Table 9.1, Fig. 9.5).

A model of the within-year dynamics utilizing the above relationships was derived as follows: if $x(t)$ is the natural logarithm of the population density at time t; x_0 is the natural logarithm of the population density in spring; s_1, s_2 and s_3 are the rates of increase before the summer peak is reached, of the subsequent decline, and of the autumnal increase, respectively; x_1 is the natural logarithm of the population density when migration begins, which here for simplicity coincides with the summer peak population density; and x_2 is the population density at the end of summer, then

$$x(t)=x_0+s_1t \text{ before the peak is reached,}$$
$$x(t)=x_1+s_2t \text{ during summer, and}$$
$$x(t)=x_2+s_3t \text{ in autumn}$$

where t denotes time from the beginning of the corresponding period. Because of the inverse relationships between the spring rates of increase, summer rates of decline and autumn rates of increase, and initial numbers $s_1=k_1x_0+q_1$, $s_2=k_2x_1+q_2$ and $s_3=k_3x_2+q_3$, where k_1, k_2 and k_3 are the slopes, and q_1, q_2 and q_3 the intercepts of the three rate relationships (Table 9.1). Then it follows:

$$x_1=x_0+\{k_1x_0+q_1\}T_1=q_1T_1+(1+k_1T_1)x_0 \qquad (9.1)$$
$$x_2=x_1+\{k_2x_1+q_2\}T_2=q_2T_2+(1+k_2T_2)x_1 \qquad (9.2)$$
$$x_3=x_2+\{k_3x_2+q_3\}T_3=q_3T_3+(1+k_3T_3)x_2 \qquad (9.3)$$
$$x_{0,t+1}=k_4x_3(t) \qquad (9.4)$$

where k_4 is the slope of relationship between the numbers next spring and this autumn; T_1, T_2 and T_3 are the durations of the periods of the spring increase, summer decline and autumn increase, respectively; and $x_{i,t+1}$ means x_i in year $t+1$ ($i=0, 1, 2, 3$). In both species, k_4 is slightly larger than 1, the value used in this study.

Table 9.1 Slopes (k_i), intercepts (q_i) and correlation coefficients (r_i) of the relationships between natural logarithm of numbers of adults at the beginning of spring, summer and autumn, and the subsequent rate of increase or decline

	Site	Spring increase				Summer decline				Autumn increase			
		k_1	q_1	r_1	P	k_2	q_2	r_2	P	k_3	q_3	r_3	P
Sycamore	1	-0.50	3.85	-0.92	<0.001	-0.17	-0.05	-0.67	<0.05				
aphid	2	-0.43	3.14	-0.96	<0.001	-0.29	1.56	-0.70	<0.05	-0.34	2.68	-0.97	<0.01
Turkey oak	1	-0.10	0.95	-0.61	<0.01	-0.15	0.02	-0.45	<0.05	-0.19	0.8	-0.17	NS
aphid	2	-0.11	0.90	-0.64	<0.01	-0.10	-0.32	-0.42	<0.05	-0.01	0.71	-0.03	NS

NS, not significant

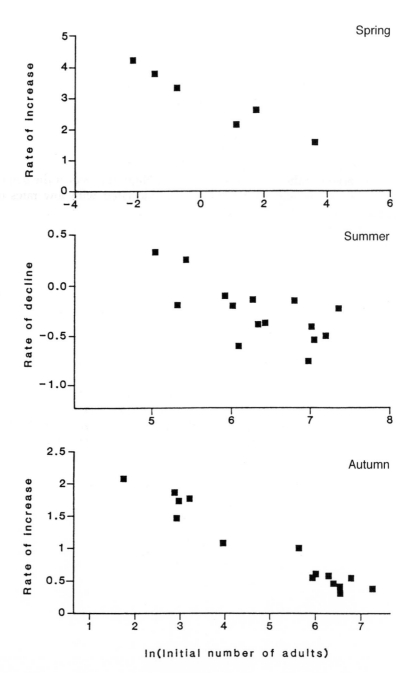

Fig. 9.5 The rates of increase and decline achieved on two trees relative to the initial numbers of sycamore aphid adults present at the beginning of spring, summer and autumn, respectively.

From this it follows:

$$x_{i,t+1}=C_i+k_4(1+k_3T_3)(1+k_2T_2)(1+k_1T_1)x_{i,t} \tag{9.5}$$

where C_i is a constant and $i=0, 1, 2, 3$. The behaviour of Equation 9.5 depends on the product of the three brackets and k_4 on the right hand side. If we write $B_i=1+k_iT_i$, then Equation 9.5 simplifies to

$$x_{i,t+1}=C_i+k_4B_3B_2B_1x_{i,t} \tag{9.6}$$

Using the parameters derived for sycamore and Turkey oak aphids cited in Table 9.2, the above system describes the within-year population trends observed in the two species (Fig. 9.6). High numbers at the beginning of spring, summer and autumn are associated with low rates of increase and high rates of decline, and *vice versa*. The net effect of this is that the sycamore aphid shows a strong seesaw in abundance between spring and autumn.

(b) One or two peaks in abundance

In both species of aphid there is a tendency for numbers to increase in autumn. The autumnal peak in abundance in the Turkey oak aphid is always at least one and often two orders of magnitude smaller than the summer peak, which is very different from that observed in the sycamore aphid where both peaks are similar in size (Fig. 9.1). This could be due to the nutritional quality of the host, reflected in the reproductive rate of the aphid in autumn, differing on sycamore and Turkey oak, and/or the extent to which the populations of the two species decline in summer.

The trends in soluble nitrogen in the leaves of sycamore and Turkey oak (an indicator of nutritive quality) are similar in that it is high in

Table 9.2 Parameters for the sycamore and Turkey oak aphids used in the population model

	Aphid	
Parameters	Sycamore	Turkey oak
k_1	−0.4	−0.11
q_1	3.5	0.9
T_1	2	8
k_2	−0.3	−0.1
q_2	1.5	0
T_2	2	8
k_3	−0.3	−0.1
q_3	2.7	0.8
T_3	8	8
k_4	1	1

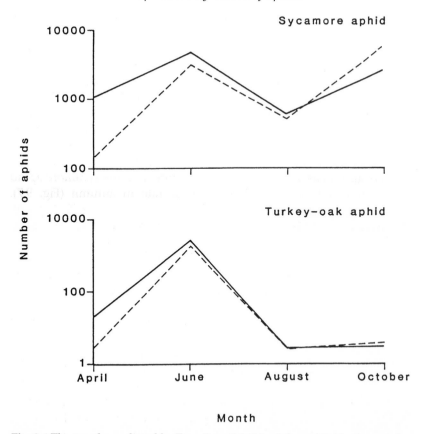

Fig. 9.6 The trends predicted by Equations 9.1, 9.2a, 9.3a and 9.4 in the numbers of sycamore and Turkey oak aphids in years starting with high numbers (solid line) and low numbers (dashed line).

spring when the leaves are actively growing, falls to a low level in summer, and then increases in autumn prior to leaf fall (Dixon, 1963, 1971). Accepting that the concentration of soluble nitrogen in the leaves is a good indicator of changes in host quality, then Turkey oak should be a better host for aphids than sycamore. However, on sycamore there is another species of *Drepanosiphum*, *D. acerinum*, which continues to reproduce all through summer when *D. platanoidis* is in reproductive diapause. Thus, although soluble nitrogen levels give an indication of changes in host quality in time, they cannot be used as an absolute indicator of host quality for a particular species of aphid.

A more direct measure of host quality is given by the reproductive rate achieved by the aphid. This has been measured for both species by clip-caging adult aphids on the leaves of their host trees throughout the course of several years in Glasgow. In addition, the Glasgow census data on the two species includes a record of the numbers of small nymphs and adults present each week. The ratio of the number of small nymphs in a week to the number of adults in the preceding week is correlated with the numbers of offspring produced per adult per day over the same period, for both sycamore and Turkey oak aphids ($r=0.79$, $P<0.001$; $r=0.89$, $P<0.001$, respectively). Thus the ratio of small nymphs to adults can be used as a measure of the reproductive rates achieved, and allows comparison of the seasonal trends in reproductive performance of the sycamore aphid in Glasgow and the Turkey oak aphid in Norwich. This indicates that the Turkey oak aphid has a higher reproductive rate than the sycamore aphid in spring and early summer, but a similar rate in autumn (Fig. 9.7). Therefore, the small autumnal peak in abundance of the Turkey oak aphid, relative to that achieved by the sycamore aphid, cannot be attributed to the latter having a higher rate of increase in autumn.

Interestingly, the population trends differ between those on the two Turkey oak trees in Norwich, on the one tree in Glasgow, and the caged saplings (Fig. 9.8). This cannot be accounted for in terms of nymphal recruitment rates measured by the ratio of small nymphs in a week to the number of adults the preceding week. Both the seasonal trends and rates achieved were very similar in all three cases. However, plotting the changes in numbers of adults between one week and the next divided by the number of large nymphs in the first week indicates that, in the middle of the year, the trees in Norwich lost more adults than they gained by recruitment. This is in marked contrast to what was observed in Glasgow and on the caged saplings (Fig. 9.9). Caging clearly reduced the losses due to natural enemy activity and migration. The natural enemy activity on the

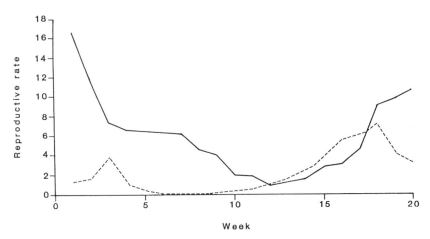

Fig. 9.7 The seasonal changes in the reproductive rates achieved by sycamore aphids (dashed line) and Turkey oak aphids (solid line).

Glasgow tree was greater than on the Norwich trees but, possibly more importantly, the Glasgow tree was enclosed on two sides by a very tall building, the walls of which often took on a yellow colouration from the countless numbers of Turkey oak aphids that settled there during the migratory period in summer. It is likely that many of these aphids subsequently flew back onto the tree. That is, it is likely that the presence of high walls on two sides of the tree severely restricted migration. Therefore, the small autumnal peak in abundance in the Turkey oak aphid in Norwich is mainly determined by the very low level to which the population declines in summer (Fig. 9.1). The extent to which populations of this species decline in summer appears to depend on the net rate of loss of adults, which is likely to depend on the extent to which the host tree is isolated from other trees. As sycamore is far more abundant than Turkey oak, sycamore aphid populations do not decline to such low levels in summer, and consequently can achieve high levels of abundance in autumn.

9.3.2 Between-year dynamics

(a) On trees

The total numbers of aphids counted on sycamore from year to year vary less than those on Turkey oak (Fig. 9.2), which tend to cycle in abundance. This cyclical pattern is seen even more strikingly in the year-to-year changes in the June numbers (Fig. 9.10b), but less so in other months. However, June numbers are not correlated with those in the following year. If the numbers in June are large, the numbers next June are always small. If the numbers in June are small, they can be either small or

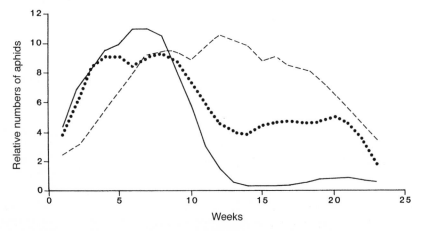

Fig. 9.8 The within-year population profiles of Turkey oak aphids on single trees in Glasgow (dotted line) and in Norwich (solid line), and on caged saplings (dashed line), each scaled so that the peak numbers are similar.

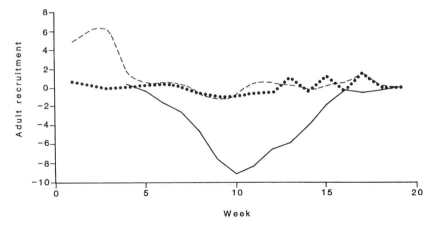

Fig. 9.9 The adult recruitment to populations of Turkey oak aphids on single trees in Glasgow (dotted line) and in Norwich (solid line), and on caged saplings (dashed line).

large in the following June. That is, the relationship between June numbers in one year and those in the following year is triangular (Fig. 9.10c). In addition, as the June numbers make the largest contribution to the yearly totals, the latter also tend to cycle (Fig. 9.10a).

Why do the numbers in June cycle from year to year? The minimum number of aphids present in summer does not depend on the peak numbers present in summer (Fig. 9.11a). If the peak number of aphids present in summer is large, then there is intense competition for resources and the aphids present at the beginning of autumn are small and have a low fecundity (Dixon, 1990), which should affect the population rate of increase in autumn. This was tested by determining the slope of the dependence of ln(density+1) against time in the first 6 weeks following the summer minimum. The relationship between the population rate of increase in autumn and the size of the summer peak is triangular (Fig. 9.11b). When the summer peak is large, the population rate of increase in autumn is never large – the aphids are less fecund, which is in accord with the predictions. If the summer peak is small, then the population rate of increase in autumn can be either small or large. This variability may be a consequence of high predation in some years, as predators attracted to large numbers of aphids in summer may have a marked effect on the few aphids that remain after the summer migration. The population rate of increase in autumn, however, was positively correlated with the size of the peak in the next year (Fig. 9.11c). In summary, the dynamics shown by the Turkey oak aphid are less predictable than those of the sycamore aphid, especially in summer and autumn. This is also

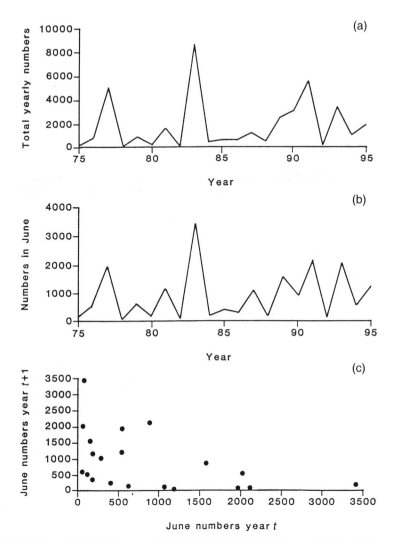

Fig. 9.10 The total numbers of Turkey oak aphids recorded yearly and in June, and the relationship between numbers of aphids in June for that year and the following year, 1975–95.

seen in the lower values of the regression coefficients of the relationships defining the spring, summer and autumn phases of the dynamics in the Turkey oak aphid compared with the sycamore aphid (Table 9.1).

In order to take this stochastic behaviour into account, the between-year population dynamics of the Turkey oak aphid were described by Equations 9.1 to 9.6, but with Equations 9.2 and 9.3 replaced by:

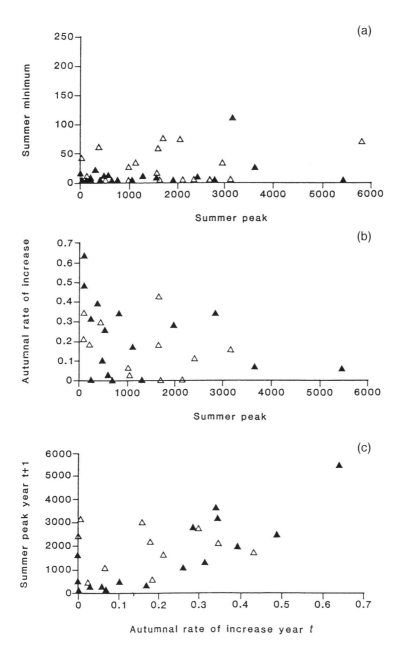

Fig. 9.11 The relationships between (a) the minimum number of Turkey oak aphids in summer and the summer peak numbers; (b) the autumnal rates of increase and the summer peak numbers; and (c) the summer peak numbers for one year and the next, for two trees (closed and open triangles, respectively).

$$x_2=x_1+(k_2x_1+q_2)T_2+r.RND_1=q_2T_2+(1+k_2T_2)x_1+r.RND_1 \qquad (9.2a)$$
$$x_3=x_2+RND_2(k_3x_2+q_3)T_3=RND_2q_3T_3+(1+RND_2k_3T_3)x_2 \qquad (9.3a)$$

where RND_1 is a random number between -0.5 and $+0.5$, and r is a constant ($=10$, as the stochastic behaviour associated with the summer decline in this species is marked), which together simulate the extremely low numbers, the random noise in migration and the sampling error. RND_2 is a random number between 0 and 1 which takes into account that the population may be negatively affected in autumn.

In this system, using the parameters for Turkey oak aphid (Table 9.2), the numbers in June also cycle (Fig. 9.12a) and the plot of numbers in June against numbers the preceding June is triangular (Fig. 9.12b). That is, the cyclical pattern is driven by the inverse relationship between the size of the spring peak and the population rate of increase in autumn. Interestingly, in the absence of noise in the system, the cyclical pattern is more definite but the amplitude of the fluctuations is smaller. In addition, running the system for periods equivalent to 200 years reveals a weak but significant seesaw effect.

The trend in sycamore aphid numbers can be predicted from the above equations, using the parameters for the sycamore aphid (Table 9.2), and making $r=1$. The latter can be justified as, although there is noise in the sycamore aphid system, mainly a consequence of variability between years in the wind speed in autumn (Dixon, 1979), it is considerably less than in the Turkey oak aphid system. The prediction of the model is that there should be little change in sycamore aphid numbers from year to year (see Fig. 9.15 below). This stability is attributable to the two within-year peaks in abundance being similar in size and both making a major contribution to the yearly totals. The inverse relationship between the size of the spring and autumn peaks is such that a change in the size of one peak is compensated by a similar but opposite change in the size of the other.

(b) In the air

The yearly catches of the suction trap positioned close to two of the sycamore trees closely reflected the average abundance of the aphids on those trees and, like the total numbers on the trees (Fig. 9.2), showed relatively little change from year to year. In contrast, the yearly catches recorded by the Rothamsted Insect Survey trap at Dundee fluctuated markedly from year to year (Fig. 9.3).

The size of the annual catch taken by the Dundee trap is mainly determined by the aphids caught in June, October and November (Fig. 9.13), with the numbers taken in June making up only 25% of the total caught in these three months. The sum of the catches in these three months is

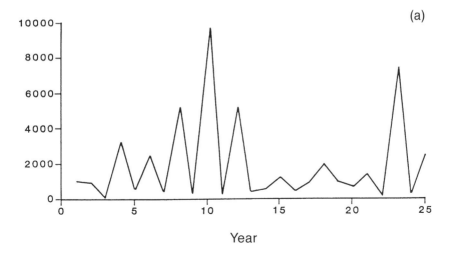

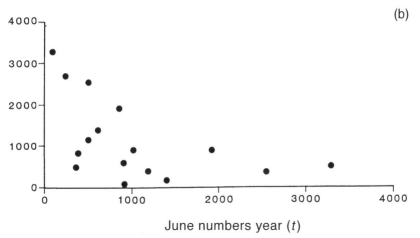

Fig. 9.12 The model's prediction of (a) the numbers of Turkey oak aphids present in June; and (b) the form of the relationship between the numbers in June in one year and the next.

well correlated with the yearly totals ($r=0.95$, $P<0.01$) and show similar dynamics, i.e. they cycle (Fig. 9.13). Therefore the Dundee trap mainly catches aphids late in a year. In addition, the ratio of the catches of the local and the Dundee trap each month did not remain constant. Early in a year, the local trap catches many more aphids relative to the Dundee trap than late in a year (Fig. 9.14). This change in flight behaviour can be represented by a correction factor, c, which increases in value from 0.08 to 1 from spring to autumn. Biologically this makes sense as early in a season not all trees are equally infested with aphids, mainly because of differences in tree phenology (Dixon, 1990); as a consequence, movement locally between trees is likely to be more advantageous, as there are great risks in migrating between clumps of trees. Later in a year all trees in the immediate vicinity are likely to be more equally infested and it is then possibly more advantageous to exploit heterogeneity in the levels of infestation between clumps of trees. The Rothamsted Insect Survey suction traps, because they are positioned well away from woodland and trap insects flying high above the ground, are more likely to take aphids migrating between clumps of trees than between trees within clumps.

Accepting that the dynamics of the sycamore aphid can be represented by Equations 9.1 to 9.6, with the modification in Equations 9.2a and 9.3a, it is possible to predict the relative trends in the yearly sizes of total tree counts and Dundee suction trap catch. The tree count will be proportional to $x_0+x_1+x_2+x_3$, and the trap catch to $c_0x_0+c_1x_1+c_2x_2+x_3$, where c_0, c_1 and c_2 are $<<1$. Using these relationships, the prediction is that the trap catch will oscillate from year to year, but the tree count will not (Fig. 9.15), as observed (c.f. Figs 9.2 and 9.3).

9.4 DISCUSSION

This study confirms that regulation of aphid abundance occurs within rather than between years. By dividing each year into three phases, spring increase, summer decline and autumn increase (Fig. 9.4), it was possible to avoid the confounding effect of seasonality. The population

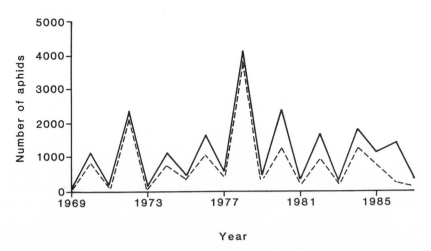

Fig. 9.13 The total numbers (solid line) of sycamore aphids caught each year in the suction trap located at Dundee from 1969–87, and the numbers caught in June, October and November each year (dashed line) over the same period.

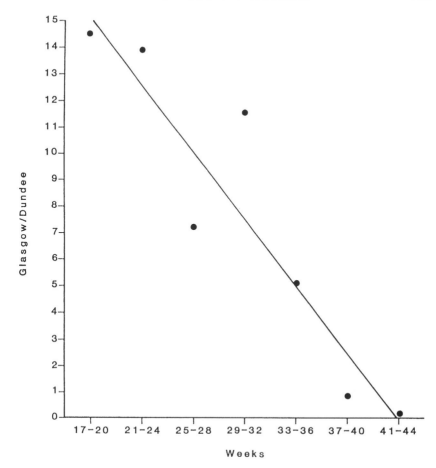

Fig. 9.14 The relationship between the ratio of the numbers of sycamore aphids caught in the Glasgow and Dundee suction traps and time in weeks from the beginning of a year.

will increase, decrease and increase in these three phases, respectively, irrespective of aphid abundance. That is, there is an equilibrium trajectory rather than an equilibrium point. What is important for understanding the population dynamics is how the rates of increase and decline are related to aphid abundance.

Intraspecific competition in each of these phases results in direct density-dependent effects on adult size, recruitment and migration (Dixon, 1963, 1966, 1969, 1970, 1971, 1975, 1979, 1990; Barlow and Dixon, 1980; Dixon and Mercer, 1983; Dixon *et al.*, 1993; Chambers *et al.*, 1985; Wellings *et al.*, 1985), which operate via the rates of increase and decline to regulate aphid abundance. In addition, there is a delayed density-dependent

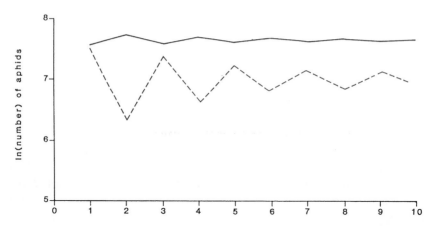

Fig. 9.15 The model's predictions of the year-to-year changes in the total numbers of sycamore aphids on trees (solid line) and caught in the Dundee suction trap (dashed line).

response that operates between phases, with high peaks of abundance in spring/summer reducing the rate of increase in autumn (Dixon, 1975, 1990).

A marked feature of the population dynamics of deciduous tree-dwelling aphids is the seesaw effect. This is the negative correlation between the numbers of aphids in the first generation in spring and the last generation in autumn, which is present in lime aphid (*Eucallipterus tiliae*; Dixon, 1971), pecan aphid (*Monellia caryella*; Liao and Harris, 1985), sycamore and Turkey oak aphids. However, the effect varies in strength between species; in the two species studied in detail here, it is strong in the sycamore aphid and weak in the Turkey oak aphid.

In those species in which the seesaw effect is strong, it appears to be driven mainly by one component in the yearly cycle – the autumnal rate of increase. This could operate in one of several ways: high aphid numbers early in a year adversely affect the quality of (a) the host plant in autumn, (b) the aphid, or (c) both the aphid and the plant. High numbers of the pecan aphid early in a year adversely affect the quality of pecan for the aphid later in the year (Wood *et al.*, 1985; Alverson and English, 1990; Bumroongsook and Harris, 1991). In the sycamore aphid the effect appears to operate directly through the aphid, with high aphid abundance early in a year resulting in the production of very small aphids, which are late in coming into reproduction in autumn and have a low reproductive rate (Dixon, 1975). There is no evidence that a high abundance of this aphid in spring induces changes in sycamore in autumn that are detrimental to the aphid (Dixon, 1975; Wellings and Dixon, 1987).

The Turkey oak aphid shows a weak seesaw response (Dixon, 1990; Dixon *et al.*, 1996; Kindlmann and Dixon, 1998). That is, a general feature of the deciduous tree aphid system is a within-year 'memory' that transfers information on abundance from spring to autumn.

Unexpectedly, the differences in the shapes of the yearly population profiles of the two species studied in detail here cannot be attributed to differences in the nutritional quality of the host trees. In the Turkey oak aphid, the marked and consistent asymmetry in the sizes of the summer and autumnal peaks appears to be determined by the very large migratory losses during summer. It is likely that, in a Turkey oak forest, trees would gain as well as lose aphids, and the net loss in summer is likely to be less than from isolated trees, like those in Norwich. In other aphid–tree systems, an abundance of aphids early in a year renders the host less suitable for aphids later in the year. This appears to be the case in the pecan aphid (Liao and Harris, 1985; Alverson and English, 1990). In other species, such as the lime aphid, there is evidence that the host is similarly affected (Barlow and Dixon, 1980), but much less so than pecan. In systems where aphids have a marked adverse effect on the quality of their host plant, a single peak in abundance is to be expected in those years when aphids are abundant in spring. Furthermore, in such systems one would expect aphid clones that produce their sexual forms early in a year to be at a selective advantage over those that produce them late in a year, which could account for the early production of sexual forms in species such as the lime aphid (Dixon, 1987b). That is, although the three stages in the seasonal development of leaves (growth, maturity and senescence) characteristic of all trees, have a dramatic effect on the reproductive activity of their aphids, it is not the only factor determining the shape of the within-year population profile. A major factor is migration, with isolated mature trees suffering greater losses and showing markedly different population profiles to trees in forests. Moreover, the average abundance of aphids on isolated trees is less than on trees in single-species stands, which supports the theoretical predictions (Dixon and Kindlmann, 1990). Thus, in pondering the reason for the differences in the population profiles and average levels of abundance of aphids on trees, differences in host-plant phenology and isolation appear to be more important than any intrinsic differences in nutritional quality (Dixon, 1990).

This analysis again reveals that the time scale over which most regulation occurs in deciduous tree-dwelling aphids is considerably less than a year. The density-dependent processes operate mainly on individuals during the course of their development, and the combined effect serves to bring the population density back to the equilibrium trajectory. That is, the density-dependent processes operate continuously rather than between generations. Therefore, analyses of data sets made up of yearly totals of the numbers of aphids caught in suction traps will not reveal when aphid abundance is regulated. However, once there is a good understanding of how the numbers of a particular species of aphid on its host plant are regulated, as is the case for the sycamore aphid, then aerial populations can be predicted and the prediction validated by reference to suction-trap catches. The pattern in the yearly suction trap catches of the sycamore aphid gives strong support to the contention that the within-year seesaw in abundance is an important feature of the population dynamics of this aphid.

There are studies on the population dynamics of coniferous tree-dwelling aphids. The longest (12 years) is that of Kidd (1990a, b, c) on the large pine aphid, *Cinara pinea*. The average numbers of this species recorded each year showed a 4–5-year cycle in abundance. To account for this, Kidd suggested that the natural enemies and/or induced plant defences showed a delayed density-dependent response operating between years. A simulation model developed to explore the effect of these two factors predicted a stable 5-year cycle, but only under a very restricted set of conditions. The assumptions regarding the nature and operation of these delayed responses need to be tested. The population dynamics of the Todo-fir aphid, *Cinara todocola*, on young saplings of Todo fir growing in a nursery, were studied for 10 years by Yamaguchi (1976). Unfortunately, the interpretation of the results is confounded by changes in the quality of the host plant for this ant-attended aphid, which first increased and then decreased over the period of the observations. However, the rate of population increase in early summer is density dependent. The study by Furuta (1988) on *Cinara tujafilina* over 3–5 years revealed that, as in the previous species, there is only one peak of abundance each year and that there is a tendency for years with high and low peaks in abundance to alternate. This was attributed to the activity of predators, in particular syrphids. This claim was supported by a predator removal experiment, which resulted in higher peak abundance in 2 out of 3 years. Scheurer's (1964, 1971) short-term (3-year) but detailed studies on several species of *Cinara* lend support to the idea of a seesaw effect, with high peak numbers resulting in very few over-wintering eggs, and *vice versa*. A major factor in this appears to be migration, which is more marked in years when aphids are abundant in spring. The consequence of this, as there is predominantly only one peak in abundance each year, is that these species tend to alternate in abundance from year to year. Similarly, a 7-year study of *Cinara pectinatae* revealed that the numbers of eggs laid in autumn on silver fir, *Abies alba*, is inversely related to peak numbers in summer (Maquelin, 1974). Because of its economic importance, the green spruce aphid, *Elatobium abietinum*, has been extensively studied, but only intensively over relatively short periods at any one site. These studies have revealed direct density dependence acting on recruitment and migration, and cold weather acting during winter as the major disturbing factor (Hussey, 1952; Day and Crute, 1990). In summary, coniferous and deciduous tree-dwelling aphids show similar population dynamics. In particular, there is some evidence for the seesaw effect in coniferous tree-dwelling aphids. The notion that the delayed effects of natural enemies and/or plant defences operate over periods of a year or more needs to be tested experimentally.

There is no doubt that the natural enemies of aphids can reduce their rate of increase, occasionally dramatically, and the use of hymenopterous parasitoids and ladybirds in the biological control of aphids is claimed to have been successful. Indeed the first and outstanding success in biological control was the use of the ladybird *Rodolia cardinalis* to control a coccid, the cottony cushion scale, *Icerya purchasi* (Doutt, 1964). This led to a universal expectation that the abundance of organisms is regulated by their natural enemies. Indeed the array of insect natural enemies associated with the sycamore aphid is awesome (Dixon, 1997). However, as aphidophagous predators develop relatively slowly compared to aphids and avoid ovipositing in patches of aphids already being exploited by conspecifics, they are ineffective at regulating the numbers of aphids (Hemptinne *et al.*, 1992; Dixon *et al.*, 1995, 1997; Doumbia *et al.*, 1997). Hymenopterous parasitoids can mature on one aphid and would appear to be potentially more likely to regulate aphid abundance. However, the Turkey oak aphid is not attacked by parasitoids in the UK. In the case of the sycamore aphid, their effectiveness is reduced by their longer developmental times relative to that of their host; by the action of hyperparasitoids, which in many cases are less specific than the primary parasitoids; and by the vulnerability of parasitoids to aphid predators (Dixon and Russel, 1972; Hamilton, 1973, 1974; Höller *et al.*, 1993; Mackauer and Völkl, 1993). In addition, because of the risk of hyperparasitism, primary parasitoids are likely to cease ovipositing in a patch where many aphids are already parasitized, as high levels of primary parasitism make the patch attractive to hyperparasitoids. By continuing to oviposit in patches of aphids already attacked by conspecifics, both predators and parasitoids reduce their potential fitness (Ayal and Green, 1993; Kindlmann and Dixon, 1993).

Interestingly, coccids, which develop very slowly compared to aphids (Dixon *et al.*, 1997), appear to be more frequently regulated by natural enemies (Taylor, 1935; Doutt, 1964; Murdoch and Nisbett, 1996). This raises the question: why are some phytophagous insects apparently regulated by top-down and others by bottom-up processes? A major factor could be the generation-time ratio of the phytophagous insect and its natural enemies. When the natural enemies develop faster than their host/prey, as is the case for many coccids, then regulation by top-down processes is possible; however, where they develop more slowly, as is the case for aphids, then regulation by bottom-up processes is more likely.

For very good pragmatic reasons, there are very few long-term studies of insects. Such studies need to be encouraged as they provide the reality against which to test theoretical predictions. Population studies at different sites within the range of an aphid should also be encouraged, as there are indications that the within-season dynamics of the sycamore aphid differ in the north and south of the UK (Dixon *et al.*, 1993). In the case of tree-dwelling aphids, population regulation appears to be mainly by means of direct and delayed density-dependent factors operating within a year, with weather and natural enemies perturbing the system. Differences in within-year population profiles appear to be more a consequence of the spatial isolation than nutritional quality of the host plants. Hopefully, this analysis goes some way to bridging the "deep chasm" between reality and theory highlighted by Taylor (1989).

ACKNOWLEDGEMENTS

This study was funded by a grant (No.GR9/01899) from the Natural Environment Research Council and from NATO (No. CRG 920553).

10

The population dynamics of Tephritidae that inhabit flower-heads

N. A. Straw

10.1. INTRODUCTION

The Tephritidae contains about 4500 species of small- to medium-size aca-lypterate flies distributed widely across all zoogeographic regions of the world (Foote, 1984). With few exceptions, the larvae are phytophagous and feed inside specific plant structures, such as flowers, leaves, stems, roots or fruit. The greater number of species can be divided into two main ecological groups: those which inhabit flower-heads of herbaceous Compositae (=Asteraceae), and those which attack the soft fruits of a wider range of plants, including many shrubs and trees. The tephritids that inhabit flower-heads include most species in the subfamilies Tephritinae and Myopitinae, and are dominant in Palaearctic and Nearctic regions, whereas frugivorous tephritids, often called fruit flies, belong primarily to the subfamilies Dacinae and Trypetinae and are most abundant in tropical and subtropical areas (Foote, 1984; Freidberg, 1984; Robinson and Hooper, 1989; White, 1989).

In one sense, the lifestyles of these two groups of tephritids are similar. Their host plants are often highly scattered, and the food resources exploited by the larvae are characteristically packaged into discrete units within which the larvae are confined. This packaging of resources influences competition for food and the population dynamics of the different tephritids in fundamentally similar ways. However, in relation to man the two groups are viewed quite differently. The tephritids that inhabit

Insect Populations, Edited by J. P. Dempster and I. F. G. McLean. Published in 1998 by Kluwer Academic Publishers, Dordrecht. ISBN 0 412 83260 7.

flower-heads are generally considered benign or beneficial. Most are associated with weed or ruderal plant species growing on waste ground, marginal areas or poorly managed pastures, and the flies have no economic significance. Some of these host plants, however, especially carduine thistles (e.g. *Cirsium* and *Carduus* spp.) and knapweeds (*Centaurea* spp.), have been introduced into other parts of the world, particularly from Europe to North America or Australia, where they have become serious weeds of rangelands and other extensive grazing systems. Tephritids, along with other insects that inhabit flower-heads, have been introduced as biological control agents to reduce seed production and plant vigour, with varying degrees of success (Harris, 1989; Julien, 1992).

The frugivorous tephritids, in contrast, include some of the most important pests of commercial fruit orchards, e.g. *Rhagoletis cerasi* on cherries and *Ceratitis capitata* on various fruits in warmer temperate regions; *Dacus oleae* on olive in the Mediterranean area; and *Dacus tryoni*, *Anastrepha suspensa* and related species on a wide range of soft fruits in subtropical and tropical countries (Robinson and Hooper, 1989). Much of the information available on the population dynamics of frugivorous tephritids has come from studies on these key pest species (Fletcher, 1987; Aluja, 1994). However, the population behaviour of these pests is probably influenced by the intense management of the habitat by man. Plant distributions in commercial orchards are usually standardized and resource levels increased artificially, and this may influence the degree and variability of resource exploitation by the tephritids. Many frugivorous species, particularly those in warmer climates, are multivoltine and have a complex population structure that is very different from that of the univoltine species found in temperate regions, especially the species that inhabit flower-heads.

The aim of this review is to describe the main processes that determine the characteristic population dynamics of tephritids, and it concentrates on the species that inhabit flower-heads in semi-natural habitats for which a number of detailed population studies have been completed. It seems most appropriate to seek relationships between tephritids and their host plants and natural enemies in those habitats in which plant distributions and complexes of associated insect species are most likely to resemble those occurring in natural situations, and with which the behaviour and physiological characters of the tephritids have adapted. Certain parallels can be drawn with the population dynamics of frugivorous species attacking non-commercial host plants, but less information is available for these species. For a description of the population dynamics of frugivorous tephritids in managed ecosystems, the reader is referred to Boller and Prokopy (1976); Fletcher (1987); Robinson and Hooper (1989); Aluja (1994); and McPheron and Steck (1995).

10.2. TEPHRITIDS INHABITING FLOWER-HEADS OF COMPOSITAE

Tephritids that inhabit flower-heads in the Palaearctic, with few exceptions, are confined to and basically comprise the sub-families Tephritinae and Myopitinae (Hendel, 1927; Zwölfer, 1965; Freidberg, 1984; White, 1989). They represent about 70% of all tephritids found in the region, and are almost exclusively associated with Compositae. A small number of species in the same or closely related genera mine or form galls in stems or roots of the same or other Compositae species. The tephritid fauna of the Nearctic shows a similar composition (Goeden, 1987; Goeden and Ricker, 1987; Goeden and Teerink, 1993).

The majority of tephritids in flower-heads are monophagous or oligophagous and univoltine, and their seasonal development is closely synchronized with that of their host plants. However, three distinct lifestyles can be recognized depending on whether the species attack flower-heads relatively early or late, and on whether the species does or does not form a gall (Straw, 1989a). Members of the tribe Tephritini (Tephritinae), e.g. *Tephritis, Paroxyna*, lay their eggs into flower-heads at a relatively early stage, and the first- and second-instar larvae feed on the immature achenes (seeds) and soft receptacle tissues before or soon after flowering. Damage to the receptacle may induce the formation of undifferentiated callous tissues which the larvae consume, but true galls are not produced (Straw, 1989b). Most species pupariate in the flower-heads in late summer, emerge in the autumn and over-winter as adults.

Species in the Myopitinae, e.g. *Urophora*, also oviposit into flower-heads at a relatively early stage, but the larvae induce the formation of differentiated, structurally complex galls (Freidberg, 1984). Gall tissue forms around each larva which establishes and the tissues coalesce to form a multilocular gall inside the flower-head, or more rarely in the stem, which is often large and woody (Redfern, 1968; Lalonde and Shorthouse, 1985; Shorthouse, 1988). Fully grown (third-instar) larvae over-winter in the gall cells.

In the third lifestyle, characteristic of the tribe Terelliini (Tephritinae), e.g. *Terellia, Cerajocera, Chaetorellia*, eggs are laid into flower-heads rather later, usually just before flowering, and the early larval instars initially feed inside individual achenes after fertilization, when these are rapidly accumulating cotyledon material (Straw, 1989b). Galls are not produced. The older larvae feed generally on the maturing achenes and receptacle tissues, and they over-winter when fully grown, in the flower-head or on the ground.

It is common for host-plant species with large flower-heads which produce relatively large achenes, particularly Compositae: Cardueae, to be attacked primarily by one early- and one late-attacking tephritid species

(Straw, 1989a). The early-attacking species is usually either a non-galling representative of the Tephritinae or a galling Myopitinae species (e.g. *Urophora*). For example, flower-heads of lesser burdock, *Arctium minus*, are attacked by *Tephritis bardanae* and *Cerajocera tussilaginis*, and those of black knapweed, *Centaurea nigra*, are attacked by *Urophora jaceana* and *Chaetorellia jaceae*, early- and late-attacking pairs of species, respectively. In flower-heads occupied by Myopitinae, particularly of *Centaurea* species, early-attacking Tephritinae species tend to be scarce. The reasons for this are not clear (Straw, 1989a). Flower-heads containing very small achenes, e.g. on host plants belonging to Compositae: Anthemideae and Senecioneae, tend to be attacked only by species of Tephritini.

Tephritid larvae are typically attacked by a number of parasitoid species, some of which are relatively specific. The flower-heads are also inhabited by caterpillars of small moths of which usually two to three species are present. Some of these are monophagous, others oligophagous, and they act as hosts to a variety of generally polyphagous parasitoids. The caterpillars are primarily phytophagous, feeding on achenes and other flower-head tissues, but they also act as facultative predators of the tephritid larvae. Various other dipteran, cynipid and coleopteran larvae occur more or less abundantly, depending on the Compositae species. The insect communities of the flower-heads form rather self-contained ecological systems and they have been used extensively to explore patterns in community structure, competition and species co-existence (e.g. Zwölfer, 1987, 1988, 1994; Romstöck and Arnold, 1987).

10.3. EARLY POPULATION STUDIES: THE ROLE OF NATURAL ENEMIES OPERATING INSIDE FLOWER-HEADS

Tephritids inhabiting flower-heads have been popular subjects for ecological and population studies because most of their life-cycle stages, and those of the insects with which they are associated, are confined within the flower-head and are easily sampled. Collections of flower-heads taken during the summer, or especially at the end of the season just before seed fall, can be dissected and the contents analysed to estimate population numbers of eggs, larvae and sometimes pupae, and numbers dying from different causes (Varley, 1947; Redfern, 1968, 1983). This neat packaging of the flower-head insect community was recognized early by Varley (1947) who monitored three generations of the knapweed gallfly, *U. jaceana*, in the flower-heads of *Centaurea nemoralis* (probably *C. nigra*) from 1934 until 1936, in the first study to attempt to apply to field data the host–parasitoid theories of Nicholson (1933) and Nicholson and Bailey (1935).

Varley (1947) estimated the numbers of tephritid larvae killed by parasitoids and predators, and dying from unknown causes, inside the flower-heads, and also attempted to estimate over-winter losses when most of the galls, which are produced by the larvae, were lying on the ground. Fecundity, and the conditions affecting fecundity and adult behaviour, were investigated in some detail. Four main periods were recognized when particular mortality factors killed larvae or pupae. During July, 20–26% of young larvae died before forming galls and 14–45% newly established larvae were killed by the endoparasitoid *Eurytoma tibialis*. In August and September, 9–21% of the older larvae were parasitized by several endoparasitoid species and 10–12% were eaten by caterpillars. Most galls fell to the ground in the autumn and 61.5% disappeared before the following spring, probably removed by mice. Mice were also found to have opened many of the galls recovered in the spring and to have consumed the contents of about 64% of the gall cells. Also in the spring, 31–60% of the surviving larvae and pupae were parasitized by a small number of polyphagous parasitoid species. In 1936, 44% of overwintering larvae were drowned when the ground became saturated after heavy rain. These data were summarized in a life table which gave the numbers of *U. jaceana* (per m^2) at the start of each generation and surviving each stage, and the numbers and proportion killed by each mortality factor. Overall generation mortality was about 99%.

The pattern of mortality described by Varley (1947) is fairly typical of tephritid–flower-head systems, and similar life tables have been constructed for several other species (Redfern, 1968, 1983; Straw, 1986; Romstöck-Völkl, 1990; Dempster *et al.*, 1995a; Tables 10.1 and 10.2). In Varley's study, the largest number of individuals, 86% of those present at seed fall, died over the winter and, following further parasitism, this left few adults to lay eggs. Parasitoids, particularly *E. tibialis*, were the largest single cause of mortality inside the flower-heads (figures corrected for subsequent predation (Varley *et al.*, 1973) suggested an even higher rate of kill), and Varley (1947) concentrated his attention on whether these natural enemies acted in a density-dependent manner and were responsible for maintaining the gallfly densities observed. He decided that even though *E. tibialis* had the potential to regulate its host population, the substantial density-independent mortality over the winter disrupted the interaction between host and parasitoid, and enabled the gallfly to persist at higher densities than would otherwise have occurred. The only other density-dependent process which could be identified, interference competition between larvae soon after egg hatch, did not appear to be important at the generally low densities observed.

The main criticism of Varley's study has been that it was probably not unsurprising that the theoretical model, parameterized by the field data, should give predicted steady densities of the gallfly similar to the averages observed in the field. The study did not provide an independent test of the ability of the mathematical model to predict host population

Table 10.1 Life table for the population of *Tephritis bardanae* inhabiting Monks Wood National Nature Reserve during 1983–85. Estimated total population numbers, l_x (±95% confidence limits), numbers dying (d_x) and percentage dying at each stage ($q_x=[d_x/l_x]\times100$)

Event	1983 l_x	1983 d_x	1983 q_x	1984 l_x	1984 d_x	1984 q_x	1985 l_x	1985 d_x	1985 q_x
Eggs laid	41 200 (±33 400)			5800 (±2400)			(6900)		
Failed to hatch		–			1600	(27.5%)		–	
Eggs hatched	41 200 (±33 400)			4200 (±1700)			(6900)		
1st and 2nd larvae died		24 200	(58.7%)		2200	(51.8%)		–	
3rd instar larvae established	17 000 (±8000)			2000 (±800)			6900 (±4500)		
Parasitized by *H. albipennis*		11 000	(64.3%)		1200	(58.9%)		2030	(29.4%)
Other parasitism		–			40	(1.8%)		120	(1.8%)
Unidentified causes		180	(1.1%)		120	(5.7%)		320	(4.6%)
Host-plant death		1750	(10.3%)		–			100	(1.4%)
Died at pupation		60	(0.4%)		30	(1.3%)		120	(1.8%)
(Sampled)		(370)	(2.2%)		(30)	(1.6%)		(50)	(0.7%)
Pupae	3700			630			4200		
Killed by caterpillars		180	(4.9%)		–			370	(8.8%)
Other predation		–			10	(1.4%)		–	
Unidentified causes		640	(17.4%)		20	(3.3%)		650	(15.7%)
Adults emerged (autumn)	2900 (±1500)			600 (±230)			3100 (±2000)		

Table 10.2 Life table for the population of *Cerajocera tussilaginis* in Monks Wood National Nature Reserve during 1983–85. (l_x, d_x and q_x as in Table 10.1)

Event	1983 l_x	d_x	q_x	1984 l_x	d_x	q_x	1985 l_x	d_x	q_x
Total adults Egg shortfall	—	—		65 800	7.6×10⁶	(99.0%)	(8000)	(9.1×10⁵)	(98.0%)
Eggs laid Failed to hatch	884 000 (±471 000)	79 000	(8.9%)	79 000 (±19 600)	8300	(10.4%)	19 000 (±9100)	1250	(6.6%)
Eggs hatched 1ˢᵗ and 2ⁿᵈ larvae died	806 000	460 000	(52.1%)	71 400	27 000	(33.8%)	17 800	1660	(9.3%)
3ʳᵈ-instar larvae established	345 000 (±120 000)			44 500 (±8600)			16 100 (±9500)		
Gall failed		400	(0.1%)		400	(0.9%)		250	(1.5%)
Parasitized by *H. albipennis*		78 800	(22.8%)		13 600	(30.7%)		2050	(12.7%)
Other parasitism		9300	(2.7%)		1200	(2.7%)		200	(1.2%)
Predation		21 800	(6.3%)		2500	(5.5%)		230	(1.4%)
Starvation		11 500	(3.3%)		2400	(5.3%)		2500	(15.3%)
Cannibalism		13 100	(3.8%)		3400	(7.5%)		200	(1.2%)
Disease?		13 300	(3.9%)		1300	(2.9%)		620	(3.9%)
Unidentified causes		14 600	(4.2%)		400	(0.9%)		200	(1.2%)
Host-plant death		28 300	(8.2%)		—			700	(4.4%)
(sampled)		(4500)	(1.3%)		(1000)	(2.3%)		(160)	(1.0%)
3ʳᵈ-instar larva alive in September Over-winter loss	150 000 (±50 000)	84 000	(56.0%)	18 300 (±4600)	(10 200)		9000 (±5000)		

size. A further difficulty in interpretation is that the variability in the estimates of parasitism rates or other mortality factors was not given, and perhaps too much confidence was placed on the mean rates which were used to calculate searching efficiencies. However, Varley's study was extremely influential in developing the concept of regulation by density-dependent processes.

10.4. QUANTIFYING FOOD RESOURCES: THE ROLE OF FLUCTUATIONS IN FLOWER-HEAD NUMBERS

Two aspects which Varley (1947) did not consider were how changes in flower-head numbers between years might have affected the mean numbers of gallfly larvae per flower-head, and how typical, or self-contained, was the population that he studied. The study site used by Varley (1947) comprised an area of 30×70 m in which knapweed grew in profusion, and he collected flower-heads and fallen galls from 1 m² plots within this area. Sampling did not extend to areas nearby where knapweed was less abundant or where gallflies were present at low densities. Consequently, Varley's observations on fly behaviour and mortality might relate only to conditions in a large, relatively dense patch of *Centaurea*. Within this area the number of flower-heads fell from 240 per m² in 1935 to 140 per m² in 1936, and infestation of flower-heads declined from 25 to 9%. Changes in flower-head numbers could, therefore, have had an important influence on population densities.

10.4.1 Studies on tephritids inhabiting flower-heads of *Arctium minus*

The food resources available to the larvae of a flower-head-inhabiting tephritid are confined entirely within the flower-head (although gall-forming species may draw nutrients from elsewhere in the plant). Therefore, the total food resources available to the population can be estimated by counting the numbers of flower-heads and assessing flower-head size. Changes in flower-head numbers between years, or changes in total flower-head biomass, can then be compared with estimates of the insect population. This approach was taken by Straw (1986, 1991) who monitored the populations of insects in the flower-heads of *A. minus* in Monks Wood National Nature Reserve, near Huntingdon, Cambridgeshire, UK, during 1983–85, and related changes in total population size and in densities per flower-head to changes in flower-head numbers (Fig. 10.1, Tables 10.1 and 10.2).

Monks Wood covers 157 ha and is surrounded by arable farmland in which *A. minus* was rare. It was considered, therefore, that the populations of the two tephritids inhabiting *A. minus* flower-heads in the wood, *T. bardanae* and *C. tussilaginis*, were relatively isolated and more-or-less

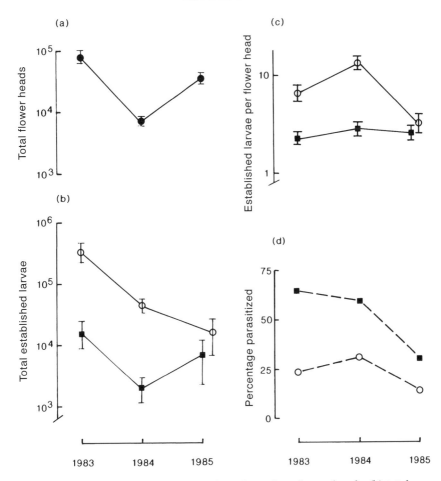

Fig. 10.1 Changes in (a) total number of *Arctium minus* flower-heads; (b) total population size of *Tephritis bardanae* and *Cerajocera tussilaginis*; (c) densities of tephritid larvae per attacked flower-head; (d) rate of parasitism of tephritid larvae, in Monks Wood National Nature Reserve, 1983–85 (after Straw, 1986, 1991). Squares, *T. bardanae*; circles, *C. tussilaginis*.

self-contained. The distribution of *A. minus* plants was mapped each year and total numbers of flowering plants and flower-heads estimated. Plants were largely restricted to paths, rides and cleared areas within the wood, which meant that they could be located easily (Straw, 1991). *A. minus* is a short-lived, monocarpic species, a facultative biennial (Gross *et al.*, 1980; Gross and Werner, 1983). Individual plants grow vegetatively as a rosette for up to 5–6 years, after which, when a large size has been attained and conditions are favourable, they produce a single large flower-stem. Flower-stems reach about 1 m in height and produce, on average, 40–60 spherical flower-heads, each of which contains about 20–30 large achenes. Plants die after flowering. Tephritid numbers in the wood were estimated from samples of flower-heads collected in September (Straw, 1991).

10.4.2 Fluctuation in resources

Average numbers of flower-heads per flower-stem, and flower-head size, were similar each year, but the number of plants which flowered varied widely. 2224 flowering plants were present in 1983, 101 in 1984 and 1710 in 1985. These changes in the number of flowering plants were unrelated to the overall size of the *A. minus* population, but depended on the proportion of rosettes which flowered (Fig. 10.2). As a result of the changes in flower-stem numbers, total flower-head numbers fell by 13-fold between 1983 and 1984, and increased five-fold between 1984 and 1985 (Fig. 10.1a). The populations of the tephritids tracked these changes in resources quite closely, particularly *T. bardanae*, although numbers of *C. tussilaginis* failed to increase in 1985 (Fig. 10.1b). (This was due partly to wet and windy weather which delayed emergence and disrupted the later flight period of *C. tussilaginis* in 1985.)

   Populations of other tephritid species track variation in flower-head numbers in a similar manner, both between years and between patches within the same year (Fig. 10.3; Dempster *et al.*, 1995b). This close relationship between resources and total insect numbers suggests that resource supply has a major influence on changes in the tephritids' total population size. If there was no relationship, then other factors, perhaps mortality caused by natural enemies, would need to be identified as being responsible for holding the tephritid population below, and independently of, the supply of resources.

10.4.3 Population models

Even though the tephritid populations in Monks Wood were correlated with the total number of flower-heads present, this was not in itself conclusive evidence that their populations were resource limited. The question remained as to what determined the numbers of insects relative to the quantity of resources (i.e. what sets the position of the regression line in Fig. 10.3?). Both a ceiling model in which the upper limit to population size is set by the resources (Dempster and Pollard, 1981; Pollard and Rothery, 1994), or an equilibrium model in which the herbivore population is held below that which resources would support by, for instance, mortality caused by natural enemies, would predict a rise and fall in herbivore numbers in line with changes in the amount of resources (= carrying capacity), especially in the latter model if resource availability influenced realized fecundity or the inherent growth rate of the herbivore population (May,

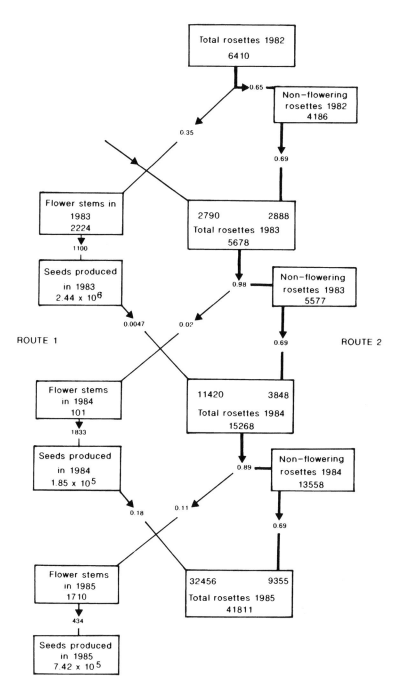

Fig. 10.2 Life table for the population of *Arctium minus* in Monks Wood National Nature Reserve, 1982–85.

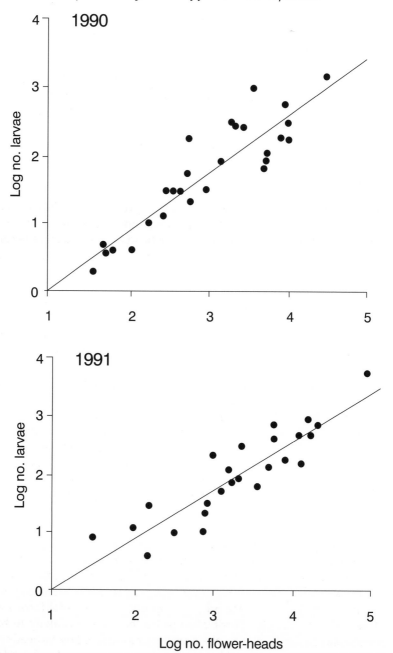

Fig. 10.3 Relationship between patch population size of *Chaetorellia jaceae* and the number of *Centaurea nigra* flower-heads per patch. (Reproduced with permission of Springer-Verlag, from Dempster *et al.*, 1995a, p. 344, Fig. 2.)

1981). Consequently, to understand what determines insect population densities relative to resources, it is necessary to go further and ask what factors adjust densities to the amount of resources present.

In Monks Wood, *T. bardanae* occupied only 11–13%, and *C. tussilaginis* 17–65% of the total flower-heads each year (Straw, 1991). Therefore, although changes in overall population size suggested that the populations were resource-limited, this problem of understanding why all resources were not used needed to be resolved in order to demonstrate that resource limitation was actually operating.

10.4.4 Resource availability

In fact, two processes explained why only a proportion of flower-heads were attacked by the tephritids each year, and both were related to resource availability. Firstly, not all flower-heads were available because the adult flight periods were not exactly synchronized with the periods when flower-heads were suitable for oviposition (Straw, 1991). Females of both tephritid species laid eggs only into a limited size range of flower-heads, well before or just prior to flowering, and flower-heads were only at this developmental stage for 10–11 days (Straw, 1989b). Flower-heads on the same stem developed sequentially, so some flower-heads were suitable for oviposition over a period of about 6 weeks, but adults were active for only part of this period.

This restriction on egg-laying was particularly severe in *T. bardanae* because both adults of this species over-wintered, and numbers on flower stems in the spring, although they appeared early, declined steadily and disappeared before most flower-heads had reached a size acceptable for oviposition. It was estimated that 27–28% of the total flower-heads were available to *T. bardanae* each year. The flight period of *C. tussilaginis* was more closely synchronized with the presence of suitable flower-heads, and for this species 66–97% of the total flower-heads appeared to be available.

The second restriction on exploiting *A. minus* flower-heads arose because of the highly scattered distribution of flowering plants and the small number of flower-heads on each flower-stem that was available for oviposition at any one time. Female tephritids have only a limited time in which to search for flower-heads and, when host plants are highly scattered, they are likely to locate only a proportion of the total (Straw and Ludlow, 1994). Hence the distribution of the host plant is likely to have imposed a spatial limitation on the number of available flower-heads that could have been attacked, and this probably explained why *T. bardanae* used only 39–48% and *C. tussilaginis* 19–67% of the available flower-heads (Straw, 1991).

10.4.5 Adult competition, dispersal and egg shortfall

The question remained, however, whether higher numbers of adults could have increased the number of available flower-heads attacked, or larval densities inside the flower-heads, and whether this might, in the absence of some of the other mortalities operating during the life cycle, have mitigated some of the population changes caused by fluctuation in flower-head numbers. For both tephritids, the answer appeared to be that any increase in adult numbers in the spring would have had only a limited effect. In 1984, when flower-head numbers had fallen, there appeared to be an excess of adults in Monks Wood, and even in 1985, the numbers of adults present could, given their known fecundity (140–320 eggs per female; Straw, 1986), potentially have attacked a much larger number of flower-heads.

The numbers of *C. tussilaginis* adults observed on *A. minus* flower-stems in Monks Wood in 1984 were seven times higher than in 1983, because of the carry-over of a large population, but numbers of eggs laid increased only two-fold. In fact, adult numbers were probably higher than this. Approximately 56% of over-wintering larvae were estimated to have died on the ground over the 1983–84 winter, but even this would have left many more adults in Monks Wood than were recorded on the flower-stems. The difference between the number of eggs that could potentially have been laid and the number actually laid indicated an egg shortfall of 99% (Table 10.3). Many females appeared to have dispersed or were prevented from laying eggs, and any increase in the numbers of females surviving to the spring would have been unlikely to have increased egg densities. The percentage of flower-heads attacked by *C. tussilaginis* remained approximately the same between 1983 (65%) and 1984 (57%), which suggested that increases in adult densities were counterbalanced by greater difficulty in finding flower-heads, or that there was much greater overlap in the flower-heads located when flower-heads were scarce (Straw, 1991).

In 1985, few *C. tussilaginis* adults appeared on the flower-stems, partly because of the smaller numbers produced in 1984, but also because of poor weather. Nevertheless, even in this year, assuming similar over-winter losses, egg shortfall amounted to 98% and most females laid few or no eggs.

The relationship between adult numbers and eggs laid by *T. bardanae* in Monks Wood was more difficult to decipher, because mortality of adults over the winter could not be measured and counts on flower-stems in the spring underestimated adult numbers. Even so, spring counts indicated the minimum numbers of adults that must have been present which could be compared with the numbers emerging in the previous autumn (Table 10.3). These data indicate an apparent over-winter loss of 99% between 1983 and 1984 and negligible egg shortfall, but the reverse between 1984 and 1985. It is unlikely that egg shortfall in 1984

Table 10.3 Estimates of over-winter loss and egg shortfall for the Monks Wood populations of *T. bardanae* and *C. tussilaginis*.

	1 autumn adult/larval population[a]	2 estimated spring adult population[b]	3 minimum spring adult population[c]	4 total eggs laid[a]	over-winter disappearance[d] (%)	egg shortfall[e] (%)	over-winter loss + egg shortfall[f] (%)
T. bardanae							
1982–83	–	–	2000	41 200	–	88	–
1983–84	2900	–	30	5800	99	0	98
1984–85	600	–	550	(6900)[g]	8	85	87
C. tussilaginis							
1982–83	–	–	6000	884 000	–	(0)	–
1983–84	150 000	65 800	2200	79 700	56	99	99
1984–85	18 300	8000	1700	19 000	(56)[g]	98	99

[a] See Tables 10.1 and 10.2.
[b] Autumn population numbers × estimated over-winter survival rate (0.44 for *C. tussilaginis*, unknown for *T. bardanae* adults).
[c] Peak count of males on flower-stems in spring × 2.
[d] Over-winter disappearance for *T. bardanae* = 1–3; for *C. tussilaginis* = 1–2.
[e] Egg shortfall = [(total potential eggs)–(total eggs laid)]/(total potential eggs). Potential eggs = total number of females × maximum fecundity: for *T. bardanae* calculated as the minimum spring number of adult females (3) × 172 eggs/female; for *C. tussilaginis* calculated as the estimated number of females surviving winter losses (2) × 232 eggs/female.
[f] [(Potential eggs in autumn)–(total eggs laid)]/(potential eggs in autumn).
[g] Numbers in parentheses calculated assuming the same over-winter survival rate as recorded for 1983–84.

and over-winter losses before 1985 were so low, and the data suggest that many *T. bardanae* adults dispersed from Monks Wood before spring 1984, when the number of flower-heads produced was small, and many additional adults moved into the wood before spring 1985 when flower-head numbers increased.

Adult *T. bardanae* move to hibernation sites after they emerge in the autumn and many may have moved into Monks Wood at this time. Adults are active again in the early spring and a greater or lesser number may have dispersed or remained in the wood, depending on the number of *A. minus* flower-heads present. *T. bardanae* appeared, therefore, to be much more mobile than *C. tussilaginis* and, as a result, it was able to track resources in the wood more effectively. *C. tussilaginis* apparently dispersed less readily and the population studied was more self contained. This difference in the dispersal ability of *T. bardanae* and *C. tussilaginis* was recorded by Dempster *et al.* (1995b) (see below).

10.4.6 Wider implications

The Monks Wood study demonstrated that overall tephritid population numbers could vary considerably between years, because of changes in the supply of flower-heads, even though only a proportion of the total flower-heads were attacked and egg and larval densities inside flower-heads changed relatively little (Fig. 10.1). The rates of mortality of eggs and larvae inside the flower-heads were similar to those recorded for *U. jaceana* by Varley (1947) and for other tephritids inhabiting flower-heads (Redfern, 1968; Romstöck-Völkl, 1990; Dempster *et al.*, 1995a), but they were not enough to reduce adult numbers sufficiently, even with the considerable over-winter mortality, to have a detectable influence on the numbers of eggs laid in the following spring. It is not possible, therefore, to judge whether within-flower-head mortalities are able to influence tephritid numbers in the following year without taking into account the relationship between the number of adults present in the spring and the numbers of flower-heads available.

Underuse of resources by *T. bardanae* and *C. tussilaginis* arose largely because of the limited overlap between the periods of adult activity and the period when flower-heads were suitable for oviposition. Other tephritid species also attack only a fraction of the flower-heads present in their habitat, typically less than 50% (e.g. Redfern, 1968; Romstöck, 1984; Redfern and Cameron, 1985; Romstöck-Völkl, 1990), and all species studied lay eggs only into a limited size range of pre-flowering buds that are likely to be available for only short periods (Varley, 1937; Berube, 1980; Romstöck-Völkl and Wissel, 1989; Rivero-Lynch and Jones, 1993). Consequently, under-use of resources because of asynchrony between insect and host-plant phenology, which may vary between sites and between years, is probably a common feature of flower-head-tephritid systems, and low infestation rates cannot be taken as evidence that tephritid populations are not resource limited.

10.5 EXPANDING THE SCALE OF OBSERVATION: ADULT DISPERSAL AND METAPOPULATION DYNAMICS

The observations on *T. bardanae* in Monks Wood indicated that this species dispersed away from the wood when flower-heads were scarce, and remained in large numbers, or congregated in the wood, when flower-heads were more abundant. Dispersal on this scale adjusted numbers of *T. bardanae* relatively quickly to local changes in the amount of resources (Fig. 10.1b). *C. tussilaginis*, on the other hand, dispersed less readily and its population became concentrated or diluted amongst the flower-heads locally available as numbers of these fluctuated. The differences in dispersal ability of the tephritids appeared to play a major role, therefore, in how their populations responded to resources.

10.5.1 Spatial population dynamics of tephritids attacking *A. minus* and *C. nigra*

The role of dispersal in tephritid populations has been investigated more extensively by Dempster *et al.* (1995a, b) and Halley and Dempster (1996), who looked at rates of local population extinction, and the roles of immigration and density-dependent processes operating within host-plant patches, in determining local population persistence for the four tephritids inhabiting flower-heads of *A. minus* and *C. nigra*. In this study, a population was defined as the insects occurring on one patch of the host plant. Host-plant patches were defined as either a single plant or a group of plants separated from the next plant or patch by at least 100 m. Patches varied in size from 1 to 146 plants in *A. minus* and from 1 to 429 plants in *C. nigra*.

The study covered an area of 5 km² of agricultural land near Cambridge, UK, over which the host plants were scattered along field edges, on waste ground and in small copses. The position of all flowering individuals was mapped in the spring, numbers of flower-heads per plant measured, and insect populations estimated from samples of flower-heads collected in September (Dempster *et al.*, 1995a). Detailed life-table information, including over-winter losses, was collected for the tephritid populations occurring in four large patches of each plant species. Information on movement of insects between patches and on rates of immigration was obtained by planting out experimental patches of plants at varying distances from natural patches, and by the use of chemical markers (Dempster *et al.*, 1995b).

A total of 78 *A. minus* patches were monitored, although only 38–51 of these produced flower-stems in any one year. Between 15 and 30% of *A. minus* patches disappeared entirely or failed to produce flower-stems in any one year, but these losses were balanced by the appearance of entirely new patches or by flower-stems appearing in patches where they had previously been absent. In contrast, the population of the perennial *C. nigra* was more stable. Only one out of 66 patches disappeared and all patches produced some flower-stems in each year. However, changes in flower-head numbers per patch varied as widely in *C. nigra* as in *A. minus* and across the whole area both species produced similar numbers of flower-heads.

10.5.2 Causes of population change

Population numbers of the tephritids in each patch were closely related to the total numbers of flower-heads present in the patch each year (Fig, 10.3). Regression analysis of changes in population size of the tephritids between 1990 and 1991 against (i) initial density in the patch in 1990 (numbers per flower-head), and (ii) changes in numbers of flower-heads 1990–91, showed that for all four tephritids (and most of the other flower-head insects), changes in patch population size were more strongly related to changes in resources (r^2=0.315–0.604) than with initial densities (r^2=0.160–0.395), and that population numbers tracked resources closely (Fig. 10.3). When the effect of changes in flower-head numbers between years was removed first, the contribution of temporal density dependence within patches to changes in population numbers failed to increase and, especially for *T. bardanae*, tended to become less important (r^2=0.041–0.306).

The tephritids did not occupy all of the flower-heads, and the life tables indicated that total generation mortality was 85–99%, which could have been taken as indicating that the populations were being held below a higher rate of resource exploitation by some of the mortality factors operating during the life cycle. Significant density-dependent mortality factors were difficult to identify from the life-table data because of the small number of generations followed, although mortality occurring after the larvae reached full size, through the adult stage and up to egg laying (k_1) was more closely correlated with initial densities than was mortality between the egg and fully grown larvae inside the flower-heads (k_2). k_1 included over-winter larval and pupal mortalities (except *T. bardanae*), adult mortality, adult dispersal and fecundity. Therefore, as in Monks Wood, the main processes which adjusted larval numbers to the numbers of flower-heads appeared to occur before or at egg laying.

Mortality caused by parasitoids and predators inside flower-heads was included in k_2 and it was concluded that these played a very minor role in determining population numbers in each year (Dempster *et al.*, 1995b). Consequently, regulation of the tephritid populations at a level below that which the available resources would have allowed seemed unlikely. The main density-dependent processes which appeared to have operated were linked to adult dispersal.

10.5.3 Dispersal

The marking experiment showed that the tephritids were able to move considerable distances, although the two species on *C. nigra* moved somewhat less (maximum distances moved: 1.2 and 2.1 km) than the species on *A. minus* (maximum distances 2.8 km) (Dempster *et al.*, 1995b). Similar distances moved have been recorded for *Urophora cardui*, which causes stem galls on *Cirsium arvense* (Schlumprecht, 1989; Eber and Brandl, 1994), and for *Tephritis conura* which lives in flower-heads of *Cirsium heterophyllum* (Romstöck-Völkl, 1990). This long distance dispersal enabled the tephritids to locate new plant patches quickly. Experimental patches were colonized rapidly by all of the tephritid species, even up to 800 m from a source population, and immigration rates into occupied patches were sufficient to allow even small patches, of only one or a few flowering individuals, to be occupied for 50% of the time (Dempster *et al.*, 1995b; Halley and Dempster, 1996). Immigration rates into patches were estimated to be 9–14% for *U. jaceana*, *C. jaceae* and *C. tussilaginis*, and to some extent these rates appeared to be inversely proportional to densities of adults produced per patch, i.e. density-dependent.

For these three tephritids, however, dynamics within patches were still important. Significant relationships between patch population change and initial densities were found for *U. jaceana* and *C. tussilaginis*, even though the temporal density dependence detected had less influence on numbers than changes in resources. Immigration was important because it encouraged persistence by a constant transfer of adults between patches. At this scale, therefore, populations of *U. jaceana*, *C. jaceae* and *C. tussilaginis* appeared to behave as metapopulations, i.e. groups of semi-independent local populations linked by immigration and emigration (Hanski and Gilpin, 1991). Similar population behaviour has been described for *U. cardui* (Eber and Brandl, 1996).

The situation for *T. bardanae* was quite different. Adults of this species dispersed to hibernation sites in the autumn and all host-plant patches had to be recolonized in the spring. Many must have moved considerable distances over the winter period. Only 7% of *T. bardanae* adults produced in a patch and marked there were relocated in the same patch in the following year (Dempster *et al.*, 1995b). Very little temporal density dependence could be detected between mortality rates or patch population change and initial patch densities, and it appeared that densities of *T. bar-*

danae in a particular patch were unrelated to densities in the previous year and had no influence on densities in the following year. The population of *T. bardanae* was apparently extremely mobile, therefore, and operated as a single population over the whole of the study area.

Romstöck-Völkl (1990) observed a similar pattern of population behaviour in *T. conura*. Patches of host thistles, 0.2–1.5 km from source populations, which had been cut down in the previous year to eliminate all flower-head insects, were recolonized immediately in the spring of the following year. The rates of infestation after recolonization were almost the same as those observed before manipulation. A patch on which adult numbers were increased 12-fold in one season returned to pre-manipulation densities the year after. Romstöck-Völkl (1990) concluded that differences in natural rates of infestation between patches were the result of differences in the relative timing of colonization in the spring and plant phenology.

10.5.4 Differences between studies

The overall population of *T. bardanae* in Dempster *et al.*'s (1995a) study was probably fairly stable during the study period, as total numbers of *A. minus* flower-stems and flower-heads were similar each year (346–440 flower-stems in 1990–92; 44 600–51 200 flower-heads in 1989–91). The numbers of flower-stems in individual patches varied considerably, and some patches disappeared entirely, but the fluctuations in individual patch size tended to cancel out at the larger scale, so that the overall amount of resources remained relatively constant. In Monks Wood, numbers of *A. minus* flower-heads rose and fell over the whole site more-or-less simultaneously, and these changes seemed to be determined by variations in the weather. As a result, the population of *T. bardanae* larvae established in the wood, and that of *C. tussilaginis*, fluctuated in parallel with the changes in flower-head numbers.

In one sense, the tephritid populations in Monks Wood operated in a similar manner to populations in the largest patches monitored by Dempster *et al.* (1995a). However, Monks Wood contained 4–5 times as many *A. minus* flower-stems in 1983 and 1985 as were found in the whole 5 km² studied by Dempster and co-workers, and twice the number of flower-heads (Straw, 1991). Consequently, Monks Wood would have produced many more tephritids than a considerable area of the agricultural land surrounding it. For *C. tussilaginis*, immigration into the wood was likely to have been much smaller than emigration, and the wood probably acted as a considerable source population. The dynamics of *C. tussilaginis* numbers inside the wood were probably little influenced by patch populations outside.

In contrast, Monks Wood probably attracted many adult *T. bardanae* in the autumn when these were looking for over-wintering sites, and most will have dispersed again in the spring, the numbers staying depending on the numbers of flower-stems in the wood. The net influx of individuals in 1985 (Table 10.3) suggested that either *A. minus* flower-stems in the wider countryside did not fluctuate in parallel with the numbers in Monks Wood, so that *T. bardanae* from outside could add to numbers in the wood, similar to that observed by Dempster *et al.* (1995a), or, if *A. minus* abundance fluctuated in synchrony inside and outside the wood, *T. bardanae* must have travelled extremely large distances to have congregated in the wood in sufficient numbers.

Total *C. nigra* flower-heads recorded by Dempster and co-workers increased four-fold between 1989 and 1991. Much larger variation was recorded within individual patches, but, as in *A. minus*, patches varied independently, and at the larger scale increases and decreases tended to cancel out. Asynchronous behaviour of individual host-plant patches contributes greatly to the stability of metapopulations, whereas synchronous fluctuation reduces the probability of persistence because rates of immigration and colonization decrease when populations are smallest and local extinction is most likely to happen (Hanski and Gilpin, 1991). The *C. tussilaginis* occupying some of the larger *A. minus* patches in Monks Wood may have operated as quasi-independent sub-populations, but the coincident changes in flower-stem numbers throughout the wood meant that movement between areas was not able to mitigate extreme changes in resource supply. The extent to which different patches of *A. minus* (or of other tephritid host plants) vary in size, in synchrony or independently, at the larger scale and in different habitats, appears crucial to understanding how these tephritid populations persist in the longer term.

10.5.5 Resources provided by other Compositae species

One possible criticism of the studies on the tephritids attacking *A. minus* and *C. nigra* is that these host plants may not be typical of the majority of Compositae species which support tephritid populations. Other host plants may be generally more abundant or fluctuate less between years. In such situations, changes in flower-head numbers might be less important in driving changes in tephritid numbers, and mortality caused by parasitoids and predators inside the flower-heads might be more important.

The data summarized in Table 10.4 show, however, that *A. minus* at least, provides some of the most abundant Compositae flower-head resources for insect populations. The third column in Table 10.4 gives estimates of flower-stem density (numbers per 1000 m^2) for 10 Compositae species (all of the Cardueae and Cichoroideae) present in a 3-km² area of sand-dune habitat in Meyendel, north of The Hague, in the Netherlands (N. A. Straw, unpublished data). Densities were calculated by a plant-centred method which involved counting the numbers of conspecific flower-stems in 1-m-wide, 30-m-long transects laid out in random directions from each of 100 randomly selected flowering plants of each species. The method gave estimates of plant density in the areas where each species grew, and was less susceptible to problems of sampling error compared with density estimates obtained from counts in randomly placed quadrats, even though the total area covered by random quadrats used at the same time was >3 ha. Flower-heads of five of these plant species contained tephritids.

The data show that *A. minus* flower-stems occurred at relatively low density, second only to flowering individuals of *Cirsium vulgare*, but because of their large size they provided relatively high numbers of flower-heads. Further, because its flower-heads were large, on a dry-weight basis *A. minus* produced the largest amounts of flower-head material per unit area, equal to that produced by the relatively abundant perennial *C. arvense*. The overall abundance of all these plant species in the sand dunes at Meyendel was equal to, or greater than, their abundance in neighbouring agricultural areas, and the densities of *A. minus* were not dissimilar to those recorded in Monks Wood. Consequently, many of the Compositae that support populations of flower-head-inhabiting tephritids provide similar, or even less abundant resources than those quantified for *A. minus* and *C. nigra*, and the tephritids exploiting them probably face similar difficulties in finding food.

10.6 PARASITISM AND PREDATION

Parasitoids, especially the relatively specific endoparasitoids, are often responsible for killing the largest number of established tephritid larvae inside flower-heads, although deaths from unknown causes or because of caterpillar predation can sometimes be equally or slightly more important (Redfern, 1968; Dempster, *et al.*, 1995a). Generally, however, parasitoids appear to have little influence on the numbers of tephritid eggs laid in the next generation, because of the large, and probably highly variable, over-winter losses that occur soon afterwards, and because of the restrictions on egg laying imposed by resource availability (see above). Population studies on flower-head-inhabiting tephritids have generally been short but, even without taking into account changes in resources, none has demonstrated significant or consistent temporal or spatial density dependence in parasitism rates at the flower-head, plant or patch scales (Straw, 1986; Romstöck-Völkl, 1990; Redfern *et al.*, 1992; Dempster *et al.*, 1995a). This is probably not surprising given the high variability in

Table 10.4 Plant-centred estimates[a] of flower-stem and flower-head density for co-existing Compositae: Carduaeae and Cichorioideae in the Meyendel sand-dune area of the Netherlands

Plant species	Life-history type	Flower-stems per 1000m²	Flower-heads per 1000m²	Dry weight of flower-heads per 1000m² (g)
Cirsium arvense[b]	Perennial	475–760	3930–6610	715–1470
Sonchus arvensis	Perennial	380–385	890–1640	100–260
Cirsium palustre	Biennial	75–570	3480–26 700	365–3950
Picris hieracioides	Biennial/perennial	165–220	580–930	35–60
Carduus crispus[b]	Biennial	125–130	3960–5150	830–910
Sonchus asper	Annual/biennial	70–180	3360–14 200	280–1260
Carlina vulgaris	Biennial	80–100	370–380	180–250
Hieracium umbellatum[b]	Perennial	85	300	15
Arctium minus[b]	Biennial	35–40	1960–2800	990–1440
Cirsium vulgare[b]	Winter annual/ Biennial	20–25	205–215	290–310

[a] Range of estimates for 1987 and 1988. The density of *H. umbellatum* was recorded only in 1988.
[b] Supported populations of flower-head-inhabiting tephritids.

population estimates that arise because of the highly aggregated distributions of larvae, but it further emphasizes that parasitoids are unlikely to be regulating tephritid populations.

The relationship between parasitism rate and tephritid larval density observed in collections of flower-heads can be misleading. In Monks Wood, *T. bardanae* and *C. tussilaginis* were attacked by the relatively specific endoparasitoid *Habrocytus albipennis*. The population of *H. albipennis* was based, therefore, on the combined larval numbers of the two tephritid species. However, in 1983 and 1984 when the *C. tussilaginis* population in Monks Wood was much larger than that of *T. bardanae*, 85–90% of *H. albipennis* larvae developed on *C. tussilaginis* larvae, and parasitoid numbers depended primarily on the numbers of this host. (Rates of parasitism of *T. bardanae* in 1983 and 1984 were amongst the highest recorded for a flower-head-inhabiting tephritid (Table 10.1.) This arose because *T. bardanae* was more susceptible to attack by *H. albipennis* and this, in combination with a large parasitoid population supported by *C. tussilaginis*, resulted in a much higher rate of parasitism than would have occurred had *T. bardanae* supported the parasitoid population alone.)

Parasitism of *C. tussilaginis* by *H. albipennis* increased from 23% in 1983 to 31% in 1984, when densities of host larvae inside the flower-heads increased, and then fell to 13% in 1985 when densities of *C. tussilaginis* larvae decreased (Table 10.2). This pattern suggested a temporal, and potentially stabilizing, response to changes in host density on the part of *H. albipennis*. However, the high densities of *C. tussilaginis* larvae in 1984 resulted from the carry-over from 1983 of a large population which was concentrated onto the small number of flower-stems present in 1984. The overall size of the *C. tussilaginis* population was smaller in 1984 than in 1983, and seen in this context, parasitism by *H. albipennis* acted in a delayed density-dependent manner, as would be expected of a specific, univoltine parasitoid (Nicholson and Bailey, 1935). Delayed density dependence in a mortality factor is generally considered to be destabilizing (May, 1981).

The regulatory ability of a density-dependent factor needs to be judged in relation to changes in total population size, and not against relative changes in the density of larvae per unit of resources. For the tephritids attacking *A. minus*, changes in total population size were poorly correlated with changes in larval densities inside flower-heads, because of the wide fluctuation in flower-head numbers and the carry-over of a large insect population onto a smaller amount of resources. The same would seem likely in other flower-head systems. The resources provided by other types of plant may fluctuate less, and in these systems, insect densities per unit of plant material may be a better guide to total population numbers, and relationships between mortality factors and insect densities may be a more reasonable indication of potentially regulating interactions. For flower-head-inhabiting tephritids, however, it can be misleading to speculate on the likely regulatory ability of natural enemies from data obtained solely from the dissection of flower-heads.

10.7 THE DIFFICULTY OF DETECTING COMPETITION

10.7.1 Competition between larvae

For insects that are apparently resource limited, it has proved surprisingly difficult to detect competition for food in populations of tephritids living in flower-heads. This is one reason why earlier studies attached greater importance to the easily quantified mortality caused by natural enemies (Varley, 1947). The confinement of flower-head resources into a discrete unit should make competition between larvae relatively easy to detect, and in one sense this is true. The distribution of tephritid eggs and larvae amongst flower-heads is typically highly clumped (Myers and Harris, 1980; Romstöck-Völkl and Wissel, 1989; Straw, 1989c), and collections of flower-heads taken in the same year will always contain some flower-heads in which larval densities exceed flower-head capacity. However, most competition occurs between the first and second instars, between egg hatch and establishment, and is mainly in the form of interference competition for position inside the developing achenes (Straw, 1986). Mortality of larvae at this stage is difficult to detect directly when flower-heads are dissected at a later date, even though the mortality can be considerable (e.g. Tables 10.1 and 10.2). Death of larvae before establishment is usually measured by comparing egg densities with estimates of the number of established larvae.

Being based on a density-dependent process, the competition-induced component of this early larval mortality should be readily identifiable as density dependent (Varley, 1947). However, significant density dependence has rarely been identified, probably because relationships between larval mortality and densities per flower-head contain a large amount of noise caused by variation in flower-head size, and because egg densities are usually estimated from samples collected early in the season and larval densities from samples collected near seed fall, and the combined errors of these estimates obscure any significant relationships.

In comparison, direct competition between older larvae after establishment is easily identified as wounding, cannibalism and crushing, or as reductions in body size in the most crowded flower-heads (Romstöck-Völkl, 1990), but the numbers of older larvae killed because of competition are usually negligible (e.g. 1–8% for the tephritids studied in Monks Wood). Most competitive interactions between larvae are sorted out before establishment.

10.7.2 Competition and population density

The clumped distribution of tephritid eggs amongst flower-heads always results in some flower-heads receiving more eggs than their capacity, except at the lowest rates of infestation. Of the flower-heads attacked by *T. bardanae* and *C. tussilaginis* in Monks Wood in 1983 and 1984, 10–38% received more eggs than they could support, despite the overall under-use of resources. Where food resources are compartmentalized and larvae cannot move between resource units, competition should occur in some units every year, and the frequency of competition should increase gradually as average larval densities rise, as more and more flower-heads receive too many eggs. Competition will occur at all population densities.

This appears to be a common feature of flower-head insect communities, and differs markedly from patterns of competition in insects where resources are not divided into discrete units, or where the larvae are able to disperse. For example, caterpillars of the cinnabar moth, *Tyria jacobaeae*, feed externally on the leaves of ragwort, *Senecio jacobaea*, and are relatively mobile, being able to search for new host plants when food becomes short. Except in occasional years, the caterpillars are able to redistribute themselves and largely avoid competition, but when total larval requirements exceed the supply of plant food, many larvae starve. Dempster (1975, 1982) indicated that competition for food between *T. jacobaeae* caterpillars occurred only above a certain threshold density and that competition was not observed in the majority of years when larval densities were lower. Such a threshold in larval mortality from competition will not occur in the flower-head systems.

10.7.3 Competition between adults

Packaging of food resources into discrete units has a further effect, however, that makes competition harder to detect. When larvae are confined within flower-heads, or other small plant structures, the distribution of larvae depends on the oviposition behaviour of the mobile adult population. If there is direct interference between adult females for oviposition sites, or if adult females can discriminate between attacked and unattacked flower-heads, they can, as individuals, avoid placing their offspring in competitive situations. By doing so, they may increase the efficiency of resource use by the population.

T. bardanae and *Chaetorellia australis*, which attacks flower-heads of *Centaurea solstitialis* and *Centaurea cyanus*, use an oviposition-deterring pheromone (ODP) which deters other females from laying eggs into the same flower-head (Straw, 1989c; Pittara and Katsoyannos, 1990). ODP is well known amongst frugivorous tephritids and appears to be responsible for over-dispersed distributions of egg batches (= ovipositions) at low rates of infestation (Bauer, 1986; Averill and Prokopy, 1989). At higher rates of infestation, egg distributions rapidly become aggregated. ODP does not prevent multiple ovipositions into fruit or flower-heads when oviposition sites are in short supply, as witnessed by the typically clumped distributions of tephritid egg batches amongst flower-heads, but it probably encourages many females to disperse and, by so doing, contributes to the 'invisible' competition between adults which appears to be so important in limiting tephritid numbers to those that resources will support.

Generally, as food resources show greater division into discrete units, the importance of competition between adults will increase and a greater part of competition for food will become invisible to all but detailed investigations. Direct interference between adults on host plants may also increase dispersal and contribute to egg shortfall in tephritids, but this is likely only at very high adult densities (Romstöck-Völkl, 1990).

10.8 THE IMPACT OF TEPHRITIDS ON POPULATIONS OF THEIR HOST PLANTS

A ceiling model of population limitation implies that the insect herbivore has no, or a very limited, effect on the supply of its resources (Dempster and Pollard, 1981; Pollard and Rothery, 1994). An equilibrium model requires that there is a reciprocal relationship either between host and natural enemy populations, or between herbivore and host-plant populations (Caughley and Lawton, 1981; May, 1981). Natural enemies do not appear capable of maintaining equilibrium densities of tephritids that inhabit flower-heads, but by destroying seeds (e.g. Table 10.5) and reducing plant vigour (e.g. Harris, 1980) it is conceivable that tephritids might have an influence on the populations of their host plants.

For two reasons, it is unlikely that the tephritids attacking *A. minus* were able to influence the numbers of flower-heads available to the next generation, and the same may apply to other tephritids inhabiting flower-heads, at least in areas where the species are native. Firstly, the population of *A. minus* in Monks Wood was not seed limited, but depended on the provision of patches of bare ground on which seedlings could establish free of competition from other herbaceous vegetation. Most seeds of *A. minus* fell directly below the flower-stems and germinated within high-density patches in which there was intense self-thinning. Consequently, the tephritids destroyed mainly surplus seeds (Harper, 1977).

In addition, recruitment of new plants into the *A. minus* population in Monks Wood was uncorrelated with total seed production and, instead, depended primarily on climatic conditions which favoured post-emergence seedling survival (Fig. 10.2). Therefore, any influence that seed predation might have had on recruitment through reduced seed production would probably have been masked by the larger variation in seedling establishment which was dictated by the weather.

Table 10.5 Percentage of the total number of *A. minus* seeds produced in Monks Wood destroyed by insect larvae

	1983 (%)	1984 (%)	1985 (%)	
			Rides	Coppice area
C. tussilaginis	24.6	37.5	1.6	6.0
T. bardanae	2.7	4.2	1.4	10.1
Aethes rubigana[a]	3.0	3.4	0.8	1.1
Other insects	0.7	0.6	0.1	–
Total	31.0	45.7	3.9	17.2

[a] Lepidoptera: Cochylidae

The second reason why *T. bardanae* and *C. tussilaginis* could not influence the number of flower-heads was that annual cohorts of *A. minus* flower-stems were produced by alternating, interleaved cohorts of rosettes. The *A. minus* population persisted through time as a large number of rosette plants that showed good survivorship between years (route 2, Fig. 10.2), or through seed production (route 1, Fig. 10.2). Seeds germinated quickly and there was little carry-over of seeds between years. In each year, a variable proportion of rosette plants flowered, and seeds from these individuals germinated over the autumn and winter to produce new rosette plants in the following season. At least one year's growth was required before flowering, so these new rosettes would not produce flower-heads until the next year (Fig. 10.2). Consequently, flower-heads and seeds of *A. minus* were produced by interleaved cohorts of plants, which in a strict biennial would have been separate, alternating cohorts, but which in *A. minus* were linked by rosette survival between years. The tephritid populations, in contrast, jumped from one annual cohort of flower-stems to the next without being able to have any immediate effect on the numbers of flower-stems appearing in the following year.

Between 2 and 35% of the *A. minus* rosettes present in Monks Wood flowered in any one year, and this was determined primarily by climatic conditions at the end of the previous season and over the winter. Therefore, even if the tephritids influenced total rosette numbers on a longer time scale, they still would not have had any immediate control over annual numbers of flower-stems produced. In perennial host plants such as *C. nigra*, variations in flowering also appear to be related to climatic and site conditions, and tephritids attacking the flower-heads are similarly unlikely to have a direct influence on flower-head numbers available to the succeeding generation. The populations of *T. bardanae* and *C. tussilaginis* can be considered as reactive, in that they responded immediately to changes in resource abundance, but non-interactive, in that they had no control over that abundance (Caughley and Lawton, 1981). The same probably applies to many other species of tephritid which attack flower-heads of biennial, facultatively biennial and, in some cases, perennial host plants. Populations of frugivorous tephritids have also been described as reactive but non-interactive in relation to their resources, for similar reasons (Bauer, 1986).

Tephritids used as biological control agents against Compositae weeds have been regarded as having a significant impact on plant populations (Harris, 1989). However, these situations are rather different to those pertaining in the native region, particularly in that the host plant is usually growing at a much higher density over a wide area. Many of the host location problems faced by tephritids in semi-natural habitats in the native region will be absent, and much higher rates of infestation may be achieved (Straw and Sheppard, 1995). In addition, many competing species and natural enemies may be missing (Myers and Harris, 1980; Sheppard and Woodburn, 1996). Consequently, the impact of tephritids inhabiting flower-heads (or stems) on host-plant populations in non-native areas is not necessarily a good guide to impact in their areas of origin.

10.9 CONCLUSIONS

Population studies on tephritids that inhabit flower-heads have, with time, increased the spatial scale of observation, as it has been realized that tephritids are highly mobile and their populations are active, and interact, over distances of at least 1–3 km. The most mobile species, represented by *Tephritis* spp., exist as single, extensive populations that are active over considerable areas, whereas the less-mobile species appear to persist as metapopulations in which immigration and emigration between host-plant patches is important in reducing the probability of local population extinction. Within-patch dynamics are dominated by fluctuations in the numbers of flower-heads. Tephritid numbers track changes in flower-head numbers closely, but total resources are generally under-used and competition for food is difficult to detect. Restrictions on resource use occur because not all flower-heads are available for oviposition, and competition is largely not evident because it occurs mainly between the very young larvae or between the adults. Intraspecific competition between adults results in fewer eggs being laid and greater dispersal. Natural enemies that kill tephritid larvae inside flower-heads appear to have very little influence on tephritid population densities, and there is little evidence that they operate in a density-dependent manner.

11

Population dynamics in the genus *Maculinea* (Lepidoptera: Lycaenidae)

J. A. Thomas, R. T. Clarke, G. W. Elmes and M. E. Hochberg

11.1 INTRODUCTION

Maculinea (Large blue) butterflies are rare and specialized insects. After feeding briefly on the flowers of a specific food plant, the larvae of all five European species live underground for 11 months in *Myrmica* ant nests, where they attain >98% of their final biomass by eating the resources of the ant colony. We have studied aspects of the population ecology of all the European species of *Maculinea* (Table 11.1) and made intensive studies of the population dynamics of two species – *Maculinea arion* and *Maculinea rebeli* – representing alternative strategies for exploiting ant colonies that have evolved in this genus (Thomas *et al.*, 1991).

In the first part of this chapter, we compare aspects of population dynamics within the genus *Maculinea*, to explore whether generalizations can be made for these closely related species. In the second part, we use empirically-based and partly validated models of one *Maculinea* system, in order to examine the effects on populations of heterogeneity in the quality of the underlying habitat, which acts indirectly, bottom-up, on *Maculinea* by altering its main resource, *Myrmica* ants. We aim to show that *Maculinea rebeli* sites may contain different amounts of suitable but heterogeneous habitat that can vary in its spatial arrangement or average quality, with considerable consequences for the populations of both the butterfly and the species with which it interacts. In particular, we ask:

Insect Populations, Edited by J. P. Dempster and I. F. G. McLean. Published in 1998 by Kluwer Academic Publishers, Dordrecht. ISBN 0 412 83260 7.

Table 11.1 Summary of the species with which different species of *Maculinea* interact, their method of feeding and the mean number of pupae found per *Myrmica* nest in the field

Maculinea	*Initial food plant*	*Myrmica host*	*Parasitoid*	*Feeding method*	*Mean no. pupae per Myrmica nest*
M. arion	*Thymus* spp., *Origanum vulgare*	*M. sabuleti*	*Neotypus* sp.	Predator	1.2
M. teleius	*Sanguisorba officinale*	*M. scabrinodis*	–	Predator	1.2
M. nausithous	*Sanguisorba officinale*	*M. rubra*	*Neotypus* sp.	Predator	2.5
M. rebeli	*Gentiana cruciata*	*M. schencki*	*Ichneumon eumerus*	Cuckoo	5.3
M. alcon	*Gentiana pneumonanthe*	*M. rubra* (northern Europe)	*Ichneumon* sp.	Cuckoo	5.9
		M. ruginodis (north-central Europe)			
		M. scabrinodis (southern Europe)			

Can heterogeneity in the quality of habitat affect the local balance of abundance between populations of interacting species in different parts of the same site?

Does the spatial arrangement of suitable but heterogeneous habitat within a site affect the persistence and population size of interacting species?

Can changes in the average quality of habitat within a site counteract either the effects of changing area on the size of butterfly populations, or the effect the butterfly population has, top-down, on the size of *Myrmica* ant colonies?

11.2 BACKGROUND TO STUDIES OF *MACULINEA* SYSTEMS

We described behaviour throughout the life cycle, compiled life tables, and measured variation in natality for 7 and 5 years, respectively, on one site supporting *M. arion* and another with *M. rebeli*. Factors causing mortality or reduced natality were identified and then quantified in each stage of the life cycle in the field, supplemented by laboratory experiments on behaviour, survival and competition in ant nests. The values of most parameters are listed by Thomas (1977, 1995); Elmes and Thomas (1987a, b); Thomas *et al.* (1989, 1991, 1993); Thomas and Wardlaw (1990, 1992); Elmes *et al.* (1991a, b, 1996); Hochberg *et al.* (1992, 1994); Thomas and Elmes (1993). Emigration and immigration by adult butterflies was hard to quantify, although we established that each species lived in predominantly closed populations, with no recorded interchange between sites >3 km apart (unpublished data). *M. rebeli* was studied near its climatic centre of range at *ca* 1500 m in the Spanish Pyrenees and the French Alps. Most studies of *M. arion* were made in the UK near its northern edge of range, with several parameters also measured on sites with warmer climates; predictions were tested on other sites in the UK where the habitat was managed for conservation (Thomas, 1995; Thomas *et al.*, 1997, 1998). Comparative studies of other *Maculinea* were made on various sites throughout Europe (Thomas, 1984a; Thomas *et al.*, 1989; Elmes *et al.*, 1994).

With varying degrees of completeness, measurements were also made of the population dynamics of the initial food plant of both species, the various species of *Myrmica* present (Elmes *et al.*, 1998), and the host-specific ichneumonid parasitoids of *Maculinea* (Thomas and Elmes, 1993). In each case, we attempted to measure the effect of variation in the underlying habitat or weather on each population, as well as the effects of these interacting species on each other. Differences in the habitat were recorded both within and between sites (Thomas, 1995; Elmes *et al.*, 1996). When measuring habitat, we began with Singer's (1972) premise that, narrow though the niches of many butterflies may be (Thomas, 1991), sites in real landscapes that support the larval food do not exist – as many metapopulation models assume – as homogeneous patches of universally suitable (source) or unsuitable (sink) habitat, which change over time only in their number, areas, distribution and the permeability to dispersal of intervening land. Instead, sites may contain a variety of types ranging, among sources, from habitat of optimum quality where r and K are the maximum possible for the species under a given climate, through a continuum of sub-optimal but suitable habitats producing fewer butterflies. Similarly, sink habitat varies from absolute sinks where mortality is total, to sub-areas where the butterfly's intrinsic rate of increase, though <1, contributes some adults to the population, and perhaps increases its probability of persisting through extreme years when normal source types of habitat become unsuitable (Sutcliffe *et al.*, 1997).

Finally, the population and habitat data were combined to construct a range of population models. In the case of *Maculinea rebeli*, six simple predictions were wholly or partly confirmed in the field, giving us the confidence to explore more complex relationships using the models.

11.3 POPULATION ECOLOGY IN THE GENUS *MACULINEA*

11.3.1 Ecology

The main plant, ant and parasitoid species with which each species of *Maculinea* interacts are listed in Table 11.1; a general diagram of population interactions is given in Fig. 11.1. Typical sites of all *Maculinea* species consist of 0.25–5-ha meadows or heath, with clear-cut boundaries defined by the distribution of the early larval food plant, from which the adult butterfly seldom emigrates (Thomas, 1995). Populations of *Maculinea*, *Myrmica*, the initial food plant and parasitoids interact and function at different temporal and spatial scales, and at three trophic levels within any site. The butterfly and parasitoid are univoltine and all food plants are perennial. Uninfested *Myrmica* colonies typically survive for 10 years, but those of any species adopting *Maculinea* larvae (i.e. those within 2 m of a food plant) disband more often, especially colonies of the host *Myrmica* species (Table 11.1) which are more seriously damaged by *Maculinea* caterpillars.

Adult *Maculinea* fly in July and oviposit on the flowers of their initial food plants (Table 11.1). The mean number of eggs laid by *M. arion* in the field (50–75 eggs per female) is similar to that of phytophagous lycaenid butterflies in the UK (Thomas *et al.*, 1991), but that of *M. rebeli* was estimated to be twice as high (Hochberg *et al.*, 1992, 1994); it is unclear whether this difference is inherent or reflects the greater availability of nectar sources and sunnier weather on *M. rebeli* sites in central Europe. Although female *Maculinea* choose particular growth forms of both the flower head and the entire plant when egg laying (Thomas, 1984a), they

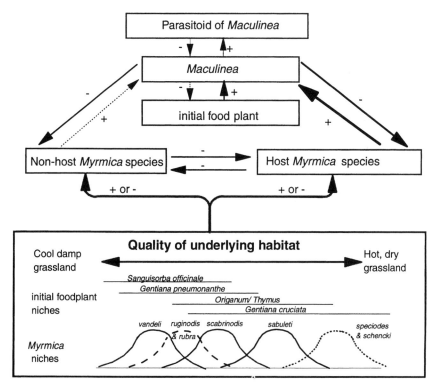

Fig. 11.1 Niche breadths of *Maculinea* food plants and *Myrmica* ants, and interactions between variable habitat and populations of five species in a *Maculinea* 'community module'. The sign and direction of each arrow indicates which species generally have positive or negative direct effects on another's population size. Indirect effects are obtained by the sum of the signs beside two or more arrows (e.g. negative + negative = positive). The boldness of the arrow approximates to the strength of each relationship.

are less selective than most phytophagous butterflies studied (e.g. Thomas, 1991; Dempster, Chapter 4 in this volume). In particular, they oviposit regardless of whether the food plant is growing within the foraging range of any species of *Myrmica* ant, and frequently lay >50% of their eggs near species of *Myrmica* to which their young stages are poorly adapted (Thomas *et al.*, 1989).

Maculinea eggs experience small density-independent mortalities or hatch after a week (Thomas *et al.*, 1991; Hochberg *et al.*, 1992). Caterpillars feed in the flower buds for 2–3 weeks, experiencing density-independent then density-dependent mortalities in the case of *M. rebeli* and *M. alcon*, caused respectively by predation and starvation (Hochberg *et al.*, 1992);

M. arion has similar mortalities acting in the reverse order, with cannibal-ism replacing the starvation of *M. rebeli* and *M. alcon*. On reaching the fourth and final instar, caterpillars fall to the ground and await discovery by *Myrmica* workers, which mistake them for ant larvae and carry them into their nests. Caterpillars cannot disperse more than a few centimetres from the food plant, and die if they are not within the short (*ca* 2 m) for-aging range of a *Myrmica* colony (Elmes *et al.*, 1991a); all ground further away from food plants represents a physical refuge for *Myrmica* (Godfray and Müller, Chapter 6 in this volume).

In established populations of both *M. arion* and *M. rebeli*, the egg and early larval mortalities, including failure at adoption by *Myrmica*, account for 20–40% of typical populations (Thomas, 1977, 1995; Hochberg *et al.*, 1992, 1994). Heavier and more variable mortalities occur inside the ant nests, typically killing 80–90% of the population in an established *Maculinea* population. Most of these mortalities arise from density-inde-pendent incompatibilities with the *Myrmica* species that adopt larvae, and from density-dependent competition for food inside nests.

Workers of all the *Myrmica* species foraging beneath food plants find and adopt *Maculinea* caterpillars with an equal probability, but it is almost entirely in the nests of a single and different species of *Myrmica* that the caterpillars of most *Maculinea* species survive to emerge as adults the fol-lowing summer (Table 11.1; Thomas *et al.*, 1989; Thomas, 1995; Elmes *et al.*, 1991b, 1994, 1998; Hochberg *et al.*, 1992). *M. alcon* differs in using a sep-arate *Myrmica* species in different parts of its range (and may indeed con-sist of three cryptic species of *Maculinea* (Elmes *et al.*, 1994).

A major divergence in behaviour occurs inside the ant nest. The cater-pillars of the two 'cuckoo species' of *Maculinea* (*M. rebeli*, *M. alcon*) mimic the behaviour of ant grubs and are fed directly by the nurse ants with regurgitations, trophic eggs and ant prey (Elmes *et al.*, 1991a, b), but cater-pillars of *M. arion*, *M. teleius* and (almost certainly) *M. nausithous* are carni-vores which prey exclusively on *Myrmica* brood (Thomas and Wardlaw, 1992). Predation is the less efficient method of feeding as the food enter-ing a *Myrmica* colony passes through one extra trophic level (ant larvae) before being assimilated by *Maculinea* caterpillars. On average, we esti-mated that about 350 *Myrmica* workers were needed to produce one adult of a predacious *Maculinea*, whereas 50 workers can rear a cuckoo species (Elmes *et al.*, 1991a; Thomas and Wardlaw, 1992). However, whichever feeding method is used, individual *Myrmica* colonies typically adopt two to three times more caterpillars than they can support in the field, resulting in high intraspecific competition and density-dependent mortality in the host ant nests (Thomas and Wardlaw, 1992; Hochberg *et al.*, 1992, 1994). Under these circumstances, the feeding behaviour again affects survival. Predacious caterpillars in overcrowded *Myrmica* colonies generally eat all the ant brood and then all die of starvation (scramble competition), although they sometimes survive if an offshoot of a neighbouring *Myrmica* colony quickly invades the vacant nest site (Thomas and Wardlaw, 1992). In contrast, the worker ants select particular caterpillars to feed in nests crowded with cuckoo species of *Maculinea*, resulting in the survival of a fixed number of individuals which varies according to the size of the ant nest (contest competition) (Thomas *et al.*, 1993).

Finally, caterpillars may be parasitized by a host-specific ichneumonid parasitoid. Each cuckoo species of *Maculinea* is parasitized by an *Ichneumon* species which selectively enters host *Myrmica* nests to oviposit in its butterfly host: *M. arion* and *M. nausithous* are parasitized by *Neotypus* species, which oviposit in their hosts about 2 weeks earlier on the initial food plant. Both types of parasitoid eventually kill their host 10 months later in the pupal stage (Thomas and Elmes, 1993).

Feeding by *Maculinea* larvae has little discernible impact on any initial food-plant population. Populations of *Thymus* and *Gentiana cruciata* were especially stable in space and time on study sites. The niche of each food plant (and of the adult and early juvenile butterfly) is considerably broader than that of any single ant species (Fig. 11.1). In the case of *G. cruciata*, it may range from cool moist patches with tall vegetation inhabited by *Myrmica vandeli*, *Myrmica rubra*, *Myrmica ruginodis* and *Myrmica scabrinodis*, through intermediate zones where *M. rubra*, *M. scabrinodis*, *Myrmica sabuleti* and *Myrmica schencki* compete, to sparse arid turf where *M. schencki* dominates or competes with more xerophytic ant genera. Each *Myrmica* colony forages over about 11 m^2 of ground and behaves like a separate population; thus a 1-ha meadow, supporting a solitary population of both the butterfly and its food plant, effectively contains a metapopulation of up to 900 *Myrmica* colonies (Clarke *et al.*, 1997). These individual ant colonies are highly dynamic. Small colonies frequently go extinct and vacant nest sites are soon occupied by stepping stone colonization, by offshoots from large neighbouring colonies. *Myrmica* colonies that adopt *Maculinea* larvae are especially prone to extinction, for although individual workers typically live 2 years and, at most, the caterpillars supplant all the vernal brood produced in the colony (leaving the summer-developing ant brood intact as a temporal refuge), the same nests are repeatedly parasitized each year.

Many field experiments have shown that *Myrmica* populations respond rapidly, bottom-up, to changes in the structure of the vegetation, with thermophilous species replaced by cool-loving species within 2–3 years, if the microclimate at the soil surface of a site becomes cooler, for example through less grazing, and *vice versa* (Thomas, 1984b, 1995). Changes in the species of *Myrmica* occupying nest sites are particularly dynamic in regions with cool climates (e.g. UK), where *Myrmica* niches are very narrow (Thomas, 1993; Thomas *et al.*, 1998). But even near its centre of range, the intrinsic annual rate of increase of a *Myrmica* colony varies considerably according to the underlying habitat (Fig. 11.1): in optimum habitat $r_{Myrmica} \approx 2$, similar to r_{max} of *M. arion* and about 10% that of *M. rebeli* in optimum habitat (Hochberg *et al.*, 1994).

In the interactions summarized in Fig. 11.1, the direction and sign on each arrow indicates the effect of one species on the population of another, a plus sign corresponding to a direct beneficial effect and minus to a harmful one. The boldness of the arrow roughly indicates the average relative strength of each interaction. Together, these species constitute a 'community module' (Holt, 1997) – a small group of species which, although living in a wider species-rich community, interact so closely that they are largely insulated from the influence of other species. This arises because the food plants (especially *Gentiana*) are robust perennials, unpalatable to most herbivores and phytophagous insects; *Myrmica* ants are keystone species that dominate their environment, competing mainly between themselves; and *Maculinea* live isolated inside the flowers of their food plant and then underground with *Myrmica* for all but about 3 weeks of the year, during which time mortalities are low. The host-specific parasitoids of *Maculinea* also live for >90% of the year within a *Maculinea* larva inside *Myrmica* nests. Thus each group is likely to exhibit clearer patterns of interaction in the field than other species that are less segregated from their community. Furthermore, sufficient species are present in the *Maculinea* systems for us to seek evidence of indirect (chains of species), as well as direct, interactions between populations (see section 11.4.4 below).

11.3.2 Comparisons between populations of predacious and cuckoo species of *Maculinea*

We cannot make exact comparisons between the population dynamics of each type of *Maculinea* due to differences in location (edge-of-range or central populations), and the different way in which models of each system were constructed. However, every model of each system (Thomas, 1991; Hochberg *et al.*, 1992, 1994, 1996, 1998; Elmes *et al.*, 1996; Clarke *et al.*, 1997, 1998), and all field tests (Thomas, 1995; Elmes *et al.*, 1996) indicate that survival by larvae in *Myrmica* nests is the key variable that directly controls butterfly abundance and persistence. *Myrmica* populations themselves respond bottom-up to changes in the quality of the underlying habitat (Fig. 11.1; Thomas, 1984b, 1995), although in the case of *M. schencki* at least, they may be equally affected, top-down, by parasitization by the butterfly population (section 11.5.2).

Before considering some separate attributes of the dynamics of *M. arion* (section 11.4) and *M. rebeli* (section 11.5), two general predictions about *Maculinea* populations are examined.

(a) Efficient feeding by cuckoo Maculinea species may result in populations of up to seven times the density of those of predacious Maculinea

Random sampling of 192 host *Myrmica* nests containing *Maculinea* pupae, on sites of the five species across Europe, resulted in similar densities of butterflies per nest found to those expected for the two feeding methods (*M. rebeli* + *M. alcon* > *M. arion* + *M. teleius*, $P<0.001$), although the observed difference in carrying capacity in the field was five-fold (Table 11.1) rather than the seven-fold difference predicted from laboratory experiments (Elmes *et al.*, 1991a, b; Thomas and Wardlaw, 1992; Thomas and Elmes, in press). Results for *M. nausithous* show the same productivity per worker ant as *M. arion* and *M. teleius* when account is taken of both the smaller biomass of *M. nausithous* and the larger average size of its host ant colonies (Thomas and Elmes, in press). Since the colony density of all *Myrmica* species, apart from *M. rubra*, are similar in optimum habitat on *Maculinea* sites (Elmes and Wardlaw, 1981, 1982; Thomas, 1980, 1984a; Hochberg *et al.*, 1994; Elmes *et al.*, 1998), figures for the densities of *Maculinea* pupae in *Myrmica* nests (Table 11.1) represent relative differences in the carrying capacity of sites with optimum habitat for butterflies using each feeding method.

(b) The stability of populations of predacious and cuckoo-feeding Maculinea

Theory suggests that populations of cuckoo-feeding *Maculinea* species will fluctuate less than those of predacious species because the main density-dependent mortality in the life of each involves, respectively, contest and scramble competition for food (Hassell, 1976). Time-series data of established *Maculinea* populations exist only for *M. arion* in the UK. In Table 11.2, population CVs (variance/mean) of *M. arion* are compared with those of other butterflies monitored by the UK Butterfly Monitoring Scheme, calculated using the strict criteria for comparing population variabilities described by Thomas *et al.* (1994). The maximum CV of *M. rebeli* was indirectly estimated (see Elmes *et al.*, 1996 and section 11.5.3b below). As predicted, *M. rebeli* appears to have very stable populations compared to *M. arion*, which fluctuates as much as any univoltine phytophagous species with closed populations (Table 11.2). Some of the difference may be attributable to the fact that *M. rebeli* was studied near the centre of its range, whereas *M. arion* was near its northern limit, where populations of phytophagous butterfly species experience erratic 'random-walk' fluctuations over time (Thomas *et al.*, 1994). However, even the variation between range-edge and central population CVs in *Maniola jurtina* – which shows the greatest intraspecific variation of any butterfly monitored – accounts for less than half the difference between the CVs of *M. rebeli* and *M. arion*.

Table 11.2 Population CVs of
phytophagous UK butterfly species
compared with those of *Maculinea arion*
and *M. rebeli*. CVs were estimated using
criteria described by Thomas *et al.* (1994),
except for *M. rebeli* (see text). (I) and (II)
represent annual comparisons of the first
and second generations of bivoltine
species; (c) and (o), species with
predominantly closed and open
populations, respectively (Thomas, 1984a)

Species	CV (%)
Maculinea rebeli **(c)**	**<55**
Anthocaris cardamines (o)	55
Gonepteryx rhamni (o)	60
Polygonia c-album (o)	61
Maniola jurtina (c)	62
Pararge aegeria (c)	64
Aphantopus hyperantus (c)	66
Pieris napi (I) (o)	66
Argynnis paphia (c)	69
Ochlodes venata (c)	69
Thymelicus sylvestris (c)	71
Coenonympha pamphilus (c)	71
Erynnis tages (c)	71
Inachis io (o)	72
Hipparchia semele (c)	74
Pieris rapae (I) (o)	75
Pieris brassicae (I) (o)	75
Pieris napi (II) (o)	75
Lycaena phlaeas (I) (c)	79
Melanargia galathea (c)	83
Polyommatus icarus (I) (c)	84
Lasiommata megera (I) (c)	86
Pieris rapae (II) (o)	86
Boloria selene (c)	88
Callophrys rubi (c)	88
Lysandra coridon (c)	90
Aglais urticae (o)	92
Argynnis aglaja (c)	92
Lycaena phlaeas (II) (c)	92
Maculinea arion **(c)**	**92**
Pieris brassicae (II)	100
Lasiommata megera (II) (c)	113
Aricia agestis (II) (c)	115
Vanessa atalanta (o)	118
Polyommatus icarus (II)	119

Since interspecific comparisons of population CVs of butterflies have not previously been published, it is worth noting that no obvious pattern is detectable among the phytophagous species listed in Table 11.2, except that the annual fluctuations in generation II of bivoltine species are significantly greater than those of the same species in generation I ($P<0.05$) and are also greater than those of univoltine species ($P<0.01$). No significant differences exist between the CVs of univoltine and first-generation bivoltine butterflies, between species with open or closed populations, or between those belonging to different families (Table 11.2).

11.4 THE DYNAMICS OF NORTHERN POPULATIONS OF *M. ARION*

Figure 11.2 illustrates population changes in the egg stage of *M. arion* on four conservation sites in the UK, where the habitat was deliberately managed to ensure that >80% of the flowering *Thymus* coincided with the niche of the host ant, *M. sabuleti* (Thomas, 1995; Thomas *et al.*, 1998). Even in this artificially stable habitat, population fluctuations were erratic (see also Table 11.2; note that periods of exponential growth experienced by new populations were omitted in calculations of species' CVs). In all, 12 annual declines in population size and 20 increases were recorded. At least nine declines occurred when the larval population exceeded the capacity of its host-ant nests (Fig. 11.2, arrows). In two cases (black arrows), an increasing butterfly population overshot the maximum potential carrying capacity of the site; in the other seven cases (grey arrows), local droughts, sometimes exacerbated by overgrazing, temporarily reduced the carrying capacity of individual *M. sabuleti* colonies, resulting in high density-dependent mortalities among *M. arion* caterpillars when the absolute density of the butterfly was sometimes low (Thomas 1995). Another decline was due to a population becoming so small that it experienced additional inverse density-dependent mortalities (Thomas, 1980). This variation in the carrying capacity of a site, coupled with a slow intrinsic growth rate by *M. arion* in the UK ($r_{max}=2$) and its propensity to overshoot, resulted in similar dynamics to those predicted in theory (Dempster, Chapter 4 in this volume) for populations in which growth is ultimately limited by a resource ceiling, and in which there is no lower limit to the level to which populations can fall (indeed, one *M. arion* population became extinct).

11.5 EMPIRICALLY BASED MODELS OF *M. REBELI* POPULATIONS

11.5.1 Models of *M. rebeli* interactions

Data similar to those in Fig. 11.2 do not exist for other species of *Maculinea*. However, empirically based models of the *M. rebeli* community suggest that populations of this cuckoo species are stabilized very differ-

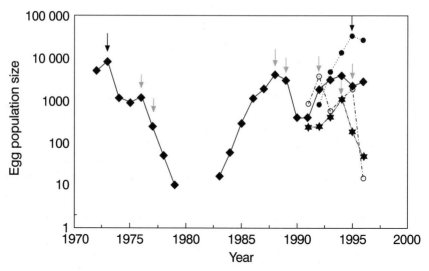

Fig. 11.2 Fluctuations of egg populations on four UK *Maculinea arion* sites. Arrows indicate declines when the carrying capacity of *Myrmica sabuleti* nests was exceeded, because the butterfly population increased above this capacity (black arrows) or because the capacity of ant colonies was temporarily reduced due to drought or overgrazing (grey arrows).

ently from those of *M. arion*, at least near the centre of *M. rebeli*'s range, where *Myrmica* niches are considerably broader than in the UK, and where apparently minor variation in sward structure, or the weather, has less impact on ant-colony sizes (Thomas, 1995; Elmes *et al.*, 1998; Thomas *et al.*, 1998). Various models were constructed to describe population interactions in *M. rebeli*'s community module (Fig. 11.1) (Hochberg *et al.*, 1992, 1994, 1996, 1998; Elmes *et al.*, 1996; Clarke *et al.*, 1997, 1998). All except the first involve minor adjustments to the HCET spatial model of Hochberg *et al.* (1994), which is summarized below.

The HCET model is mechanistic and spatially explicit, incorporating 19 parameters measured on one Spanish site (S1) or in the laboratory. Nine parameters describe mortalities or natality in *M. rebeli*; the other 10 describe the population ecology of *Myrmica*, *G. cruciata* or *Ichneumon eumerus*, or variation in the underlying habitat. The algorithm models a 1-ha square of grassland, subdivided into 900 (30×30) cells (Fig. 11.3a) over which a constant population of gentians is distributed with the clumping recorded at S1. Each cell corresponds to the foraging range of a typical *Myrmica* colony and contains no more than one nest. Micro-variation in habitat (Fig. 11.1) is represented by a linear gradient along the *x* axis of the grid, ranging from cool–moist to hot–dry cells in 30 steps (Fig. 11.3). Each species of *Myrmica* has a different growth rate and nest capacity according to its position across this gradient (Fig. 11.3b, dotted lines), the exact values being indirectly estimated to give the best fit to the relative ant densities measured across the habitat gradient at S1. In the absence of competition from other *Myrmica* species, non-host *Myrmica* persist in cells $x=1$–22 with their optimum habitat at $x=9$; *M. schencki* inhabits cells $x=9$–30, with maximum productivity at $x=22$ (Fig. 11.3b, dotted line).

In parallel to the butterfly's dynamics (section 11.3.1), both within- and between-nest dynamics of the ant colonies were modelled. Only vernal ant brood can be supplanted by *M. rebeli* caterpillars, and the extent to which these reduce the next generation of ants in a nest depends on the number of caterpillars adopted and how long they survive. Hence its

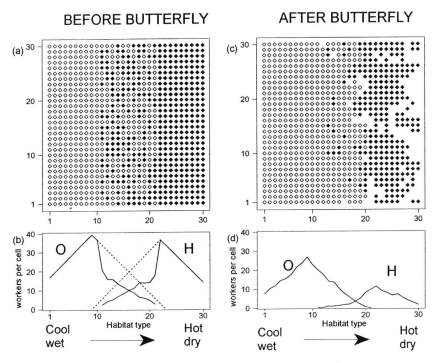

Fig. 11.3 The basic spatial model for *Maculinea rebeli* on site S1, incorporating a gradient in habitat quality ranging from the coolest and wettest ($x=1$) to the hottest and driest ($x=30$) grassland in which *Gentiana cruciata* grows. (a) The distribution of host (black symbols) and other *Myrmica* (open symbols) colonies following 50 years without the butterfly. (b) The density of ant workers per cell across the habitat gradient before the butterfly invades. H, host; O, other *Myrmica*; solid lines show O and H competing, dotted lines are O or H in the absence of its competitor. (c) The distribution of ant colonies in one year >10 years after *M. rebeli* has invaded. (d) Effect of *M. rebeli* on competing *Myrmica* colonies.

impact is usually greater in *M. schencki* nests. There is a probability that each *Myrmica* colony will either die or split to colonize a vacant adjacent cell each year, with both events depending on nest size. We assumed that the first ant colony to arrive in a vacant cell cannot be displaced that year by neighbouring nests, and that caterpillar adoption occurs after the completion of the annual ant-colony dynamics.

To obtain predictions of the population dynamics of species in the *M. rebeli* community module, the spatial model is seeded with ants and run for 50 'years' allowing the ants to reach a dynamic balance. *M. rebeli* is then introduced to interact sequentially with the gentians and ants, experiencing the mortalities summarized above and described by Hochberg *et al.* (1994).

11.5.2 Model predictions of *Myrmica* and *M. rebeli* population dynamics on a site

In the absence of the butterfly, the HCET model predicts that strong competition exists between the two *Myrmica* species over about half the site, in the intermediate type of habitat, with each ant tending to dominate the closer it lives to its optimum habitat (Fig. 11.3b, solid lines). An invading butterfly population has a considerable impact on the ants (Fig. 11.3c, d), both through the direct damage to the colonies it parasitizes and indirectly – because it supplants twice as much vernal brood in *M. schencki* colonies compared to non-host *Myrmica* – by altering the competitive balance in favour of other *Myrmica* over *M. schencki*. Under the habitat conditions simulated at site S1, the model predicts that an invading butterfly population increases rapidly with a reproductive rate of 13–29 adult female offspring per mother, an intrinsic rate of increase roughly 10 times greater than that shown by *M. schencki* (or *Maculinea arion*) in its optimum habitat. After three generations, the butterfly reaches a maximum of *ca* 2500 adults per ha (Fig. 11.4, line a). By then the reduction in the number and size of *M. schencki* colonies is so great that the carrying capacity of the site for the butterfly is reduced to about 20% of its initial value, causing the butterfly population to fall to around 450 adults per ha over the next 10 years. Thereafter, considerable intra- and interspecific turnover occurs between ant colonies in individual nest sites across the intermediate and drier parts of the habitat spectrum, but the overall populations of butterfly and both species of ant on site S1 stabilize at a dynamic equilibrium, with little annual variation in total numbers. Not only is the number of *M. schencki* colonies reduced by 36% on the site (Fig. 11.4, line c), but there is also a 66% reduction in the number of workers per surviving colony (Fig. 11.4, line e). The predicted impact on non-host *Myrmica* is more complex. Due to reduced competition from *M. schencki*, the number of colonies of 'other *Myrmica*' species increases by *ca* 8%; at the same time there is a decrease of *ca* 20% in their total worker numbers, due to the direct damage of the butterfly on their colonies (Hochberg *et al.*, 1994).

Thus we predict that *M. rebeli* quickly reduces the initial carrying capacity of a new site by about 80% through its impact both on its main resource, *M. schencki*, and on *M. schencki*'s competitors, other *Myrmica*. Cycles are just detectable in model simulations between the butterfly and ant populations over time, but the amplitude of both these and stochastic fluctuations is predicted to be very small, due to the high intrinsic rate of increase of the butterfly compared to that of its hosts and to the contest nature of the strong density-dependent mortalities recorded among *M. rebeli* caterpillars in ant nests. There is also a negligible effect on population stability when realistic levels of annual stochasticity in habitat quality are included (Clarke *et al.*, 1997), for example to simulate annual variation in sward structure in the Pyrenees or Alps, or the indirect effects of weather on the growth rates of ants.

Sensitivity tests on each parameter in the HCET model confirm that the size of the butterfly population on a site is determined almost entirely by the collective capacity of the *M. schencki* colonies present to support the butterfly (Hochberg *et al.*, 1994). However, since the butterfly can

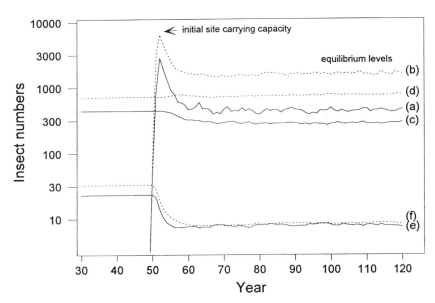

Fig. 11.4 Changes in *Maculinea rebeli* populations (a, b) and the number (c, d) and mean size (in butterfly equivalents) (e, f) of *Myrmica schencki* colonies per ha on model sites, following an invasion by the butterfly in year 50. Solid lines (a, c, e) represent populations on site S1 (Fig. 11.3) containing a gradient of habitat quality. Dotted lines (b, d, e) are predicted population levels in 1 ha of optimum habitat for *M. schencki* ($x=20$–24 on the habitat gradient).

infest only those colonies that have gentians growing within their foraging ranges, the density and clumping of *G. cruciata* also has a considerable indirect influence, both on the initial carrying capacity of the site,
and on the equilibrium levels ultimately attained between the three
insect populations. On established sites, the model predicts that butterfly
and plant populations are positively correlated over low gentian densities, because with more food plants present, more *M. schencki* nests are
exploited. However, above densities of 1200 gentians per ha, the damage
by the butterfly to *M. schencki* is so great that many individual colonies
are driven extinct or are severely diminished in size. Consequently, over
higher densities of gentians, the populations of the butterfly and initial
food plant are predicted to show an inverse correlation (Hochberg *et al.*,
1994, Elmes *et al.*, 1996).

In contrast to the northern populations of *M. arion* (Fig. 11.2), we
therefore predict that *M. rebeli* is regulated more like the Nicholsonian
systems summarized by Godfray and Müller (Chapter 6 in this volume)
and Hassell (Chapter 2), with ant–butterfly populations stabilizing in
equilibrium after the butterfly has reduced the biomass of its host
resource by about 80% of its original value. However, the level at which
host–butterfly populations equilibriate is set by the quality of the underlying habitat, which includes the density of *G. cruciata*. It is a moot point
whether populations of *M. schencki* are regulated bottom-up by habitat
quality, or top-down by the butterfly. Our original simulations (Hochberg
et al., 1994) suggested a simple top-down effect, in which the host ant was
maintained through the impact of the butterfly at around 20% of the
density per ha that its resource could support (Fig. 11.4, lines c and e).
However, in new simulations we predict that a similar 3- to 4-fold difference in both the initial site carrying capacity and the equilibrium levels of
persisting populations can be effected, bottom-up, through altering the
quality of the habitat on a site by changing the original gradient (Fig. 11.4,
solid lines) to more homogeneous habitat of near-optimum quality for *M.
schencki* growth (Fig. 11.4, dotted lines b, d, f). Other predictions of changing habitat quality are described in section 11.5.4c.

The possible top-down effect of *I. eumerus* has been explored only
recently (Hochberg *et al.*, 1996, 1998). Simulations suggest that populations
of *M. rebeli* become slightly smaller and less persistent on sites supporting
the parasitoid, but that these mortalities are unlikely ever to regulate populations of either the butterfly or *Myrmica*, due to the powerful bottom-up
density-dependent interactions that exist between these latter species.

11.5.3 Empirical tests of *M. rebeli* model predictions

Predictions of population dynamics in the *M. rebeli* community have yet
to be tested over time. However, six predictions about the absolute or rel-

ative sizes of populations were tested in single years in the field on other sites. Most were in the French Alps, several hundred kilometres from the original (S1) site in Spain (Elmes *et al.*, 1996; Thomas *et al.*, 1997).

(a) Predicting M. rebeli *population size on other sites*

Seven parameters concerning gentians and their habitat were measured at each of 12 sites in the French Alps and on a 13th in Spain, all of which were known to have supported *M. rebeli* for several years (i.e. ant and butterfly populations could be assumed to be near their equilibrium levels). Ant parameters were not measured because these were predicted by the quality of the habitat, which was rather crudely assessed (Elmes *et al.*, 1996). Actual butterfly population size was recorded in the egg stage, which is conspicuous enough for total *M. rebeli* numbers to be counted without error on most sites. Actual butterfly numbers were then compared with model predictions (Elmes *et al.*, 1996). The match between actual egg densities and model predictions was unexpectedly high (see section 11.5.3b), explaining 86% of the observed variation between sites (Fig. 11.5).

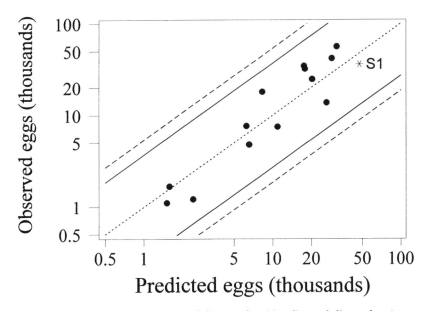

Fig. 11.5 The ability of the HCET model to predict *Maculinea rebeli* egg density per ha on sites other than S1. Dotted line indicates the expected (1:1) relationship if the model were a perfect predictor of *M. rebeli* density. Dashed line represents the 95% confidence limits around the dotted line if *M. rebeli* populations had the same annual fluctuations as the average univoltine British butterfly species (see Table 11.1). Solid line indicates the 95% confidence if *M. rebeli* had the same population CV as the least variable British butterfly species.

(b) M. rebeli *populations are unusually stable*

All models predicted that established populations of *M. rebeli* are exceptionally stable over time, for reasons described in section 11.3.2b. We did not confirm this prediction directly, but obtained indirect corroboration from the close fit between the observed and estimated population sizes of *M. rebeli* (Fig. 11.5). Elmes *et al.* (1996) calculated that, even if the HCET model were a perfect predictor of *M. rebeli* numbers, we would expect a closer fit by chance from single censuses of different sites in different years on <1% of occasions, if each *M. rebeli* population experienced the same average annual fluctuations of UK colonial–univoltine–phytophagous butterflies (Fig. 11.5, dashed line; Table 11.2). In fact, all observed values fell well within the 95% confidence limits expected if *M. rebeli* populations showed the same temporal variation as the most stable of all known butterfly species in the UK (Fig. 11.5, solid line). On this basis, we suggest that *M. rebeli* has a mean CV of <55% on sites in the Alps and Pyrenees (Table 11.2).

(c) Egg and gentian densities are positively correlated over low gentian densities

The parameters included in the HCET model were measured on a site containing 2672 flowering gentian plants per ha, well beyond the threshold above which the model predicts a negative (indirect) impact of the gentian and *M. rebeli* on *M. schencki*, leading to an inverse correlation between butterfly and food plant numbers (Elmes *et al.*, 1996). The other 13 sites examined (Fig. 11.5) supported lower plant densities (38–1645 gentians per ha), spanning the range over which a positive correlation between plant and butterfly populations is predicted. Reassuringly, the relationship (including both the slope and intercept) predicted by the model between low gentian populations and *M. rebeli* egg densities did not differ significantly ($P \geq 0.09$) from that observed in the field in the French Alps.

(d) Apparent competition between gentians and M. schencki: *three predicted effects of* M. rebeli *on ant populations*

The HCET model predicts that *M. rebeli* reduces *M. schencki* populations by considerable and predictable amounts, the precise level being determined mainly by the density of *G. cruciata* and by the nature of the underlying habitat (Fig. 11.4; section 11.5.4c). In addition, due to the stepping-stone manner by which *Myrmica* colonizes vacant nest sites, the model predicts that 'apparent competition' (Holt and Lawton, 1994) should exist between gentians and *M. schencki* colonies, especially at the scale of individual cells. Although no direct interaction is known to exist between gentian and ant populations, fewer, smaller and less persistent colonies of *M. schencki* should live in the vicinity of gentians than in otherwise identical habitat >2 m away.

We confirmed each prediction on up to six *M. rebeli* sites, mainly in the French Alps (Thomas *et al.*, 1997). About half as many *M. schencki* colonies were recorded near a *G. cruciata* plant than in apparently identical habitat 3–4 m away from gentians on the same sites. In contrast, other *Myrmica* species had equal numbers of colonies near and away from gentians; and ants of other genera, which compete with *Myrmica* to some extent but which are not affected by the butterfly, had more colonies situated near gentians, presumably because they could exploit nest sites vacated by *M. schencki*. This last result, although not quite statistically significant, may be an example of 'apparent mutualism' (*sensu* Holt and Lawton, 1994) between gentians and non-*Myrmica* ants. Where a *M. schencki* colony did occur within 2 m of a gentian, it contained, on average, about half the number of workers recorded per colony in apparently identical habitat that had no gentian present. Added to this, over a period of 5 years, the rate of turnover of *M. schencki* colonies in 50 nest sites within 2 m of a gentian was more than twice that of colonies of other species of *Myrmica* in the same nest sites.

11.5.4 Predicted effects of heterogeneity in habitat quality on populations

Our ability to find predicted numbers or patterns among the butterfly, plant and ant populations on all sites studied in the field (section 11.5.3) encouraged us to make further predictions. All concern the effects of variation in the distribution or quality of habitat within sites acting bottom-up through the *Myrmica* ant colonies on the interacting populations (Fig. 11.1). Even in the original model, Hochberg *et al.* (1994) predicted that the mean levels reached between *M. rebeli*, *M. schencki* and other *Myrmica* populations would vary greatly in different sub-areas within the site S1, according to their position within the habitat gradient (Fig. 11.3). Here we consider two new scenarios (sections 11.5.4a and c) and a third (section 11.5.4b) drawn from recent analyses by Clarke *et al.* (1997, 1998).

(a) Effects of heterogenous habitat on apparent competition and apparent mutualism between Myrmica *and* G. cruciata *populations.*

Model simulations suggest that more complex patterns of co-existence between ant and gentian populations may occur locally within sites, compared to the average values predicted for whole sites described in section 11.5.3d. Indeed, both apparent mutualism and apparent competition are predicted to exist between the same pairs of species in different sub-areas within a single site, depending on the nature of the underlying habitat (Fig. 11.6). To our knowledge, this possibility is new to theory. We explain it as follows.

In the coolest ground inhabited by gentians (Fig. 11.6, $x=1$–7 along the habitat gradient), only non-host species of *Myrmica* can ever persist (Fig. 11.3a, b). Thus the sole impact of *M. rebeli* on *Myrmica* is direct and always harmful, resulting in fewer colonies on ground within 2 m of a gentian (i.e. a pattern of apparent competition exists between gentians and other *Myrmica*). In slightly drier habitat (Fig. 11.6, $x=7$–14), non-host *Myrmica* compete with *M. schencki* colonies for nest sites (Fig. 11.3b), but the latter are largely eliminated in this poor habitat (for *M. schencki*) by *M. rebeli*, leaving numerous vacant nest sites available for colonization by other *Myrmica*. Although many 'other *Myrmica*' colonies also experience direct damage by *M. rebeli* in this zone, the comparatively small reduction in their colony size caused by the butterfly is more than compensated for by the greater harm done to its competitor, *M. schencki*. Thus the net effect of these direct negative and indirect negative–negative (= positive) interactions (Fig. 11.1) between *M. rebeli* and non-host ants is an increase in the number of 'other *Myrmica*' colonies living in this zone, resulting in apparent mutualism between these ants and plants. Finally, in suboptimal habitat where 'other *Myrmica*' colonies have a low intrinsic growth rate, and only just persist in competition with *M. schencki* when the butterfly is absent (Fig. 11.3b, $x=15$–21), even the relatively minor damage done by the butterfly makes colonies of 'other *Myrmica*' too small to exploit the many nest sites vacated by *M. schencki*. In other words, the indirect negative–negative impact of *M. rebeli* via *M. schencki* on 'other *Myrmica*' is smaller in this zone than the direct negative effect of the butterfly on these ants, resulting in a return to apparent competition between gentians and 'other *Myrmica*' (Fig. 11.6, $x=15$–21).

A similar pattern of switches between apparent competition and mutualism between *M. schencki* and gentians is predicted over the warmer half of the habitat gradient. However, since the direct damage of the butterfly to this ant's population is always greater, this results in apparent mutualism (i.e. a net increase in *M. schencki* where gentians grow) only in a narrow band of optimal habitat where the values of r and K for this ant were highest (Fig. 11.6, $x=19$–22 along the habitat gradient).

(b) Site ruggedness: the spatial arrangement of heterogeneous habitat within a site

The smooth gradient in habitat quality in the HCET model (Fig. 11.2) was a realistic representation of the terrain recorded on site S1. It is artificial only in the sense that the land was terraced many centuries ago into gentle gradients for agriculture. Several other sub-alpine grasslands supporting *M. rebeli* are terraced; others have more rugged terrain.

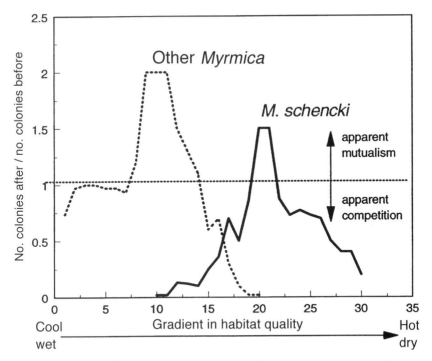

Fig. 11.6 Model predictions of apparent competition and apparent mutualism between *Gentiana cruciata* and two *Myrmica* species in different parts of the habitat gradient on site S1 (see Fig. 11.3), represented as the ratio of the mean number of colonies present in each of 30 types of habitat after and before the butterfly invades a site.

Clarke *et al.* (1997, 1998) explored some consequences on populations of these different terrains by altering the spatial arrangements of heterogenous habitat within a model site. Initially, the same size of site (1 ha) was used as in the HCET model, and we kept exactly the same amounts (30 cells) of each of the 30 slightly different types of habitat within it. However, instead of distributing these as one simple slope from cool–moist to hot–dry ground, we rearranged them to create sites with $k_{(1, 2, 4...64)}$ dry 'hills' set among damper areas (Fig. 11.7), using an algorithm described by Clarke *et al.* (1997). As an extreme case, we also distributed each cell type randomly. The simulations were repeated for sites containing an inverse pattern of damp hollows set among drier ground. Finally, for added realism, we modelled smaller (0.5- and 0.25-ha) sites, and introduced small-scale temporal fluctuations in cell habitat quality across whole sites to mimic annual variation in sward structure or the weather.

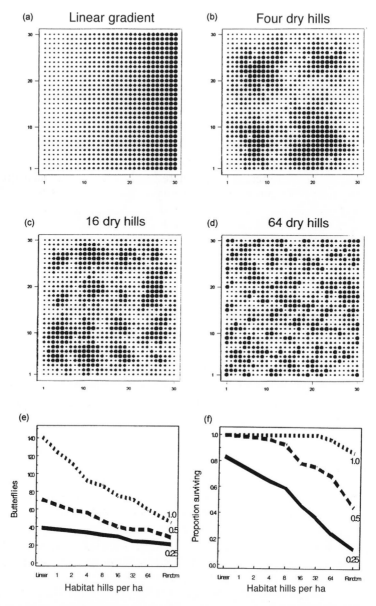

Fig. 11.7 Model predictions of the effect of increasing ruggedness in the distribution of heterogeneous habitat within a *Maculinea rebeli* site. (a) Original linear gradient of habitat quality. (b–d) Exactly the same heterogeneous habitat as in (a) redistributed within the same-sized site in patterns of increased ruggedness. (e) Mean size of surviving *M. rebeli* populations on three sizes (1, 0.5, 0.25 ha) of site with increased ruggedness. (f) Persistence of *M. rebeli* populations on the same sites.

The effect of greater ruggedness is to reduce both the carrying capacity of a site for *M. rebeli* and the butterfly's persistence, especially on small sites (Fig. 11.7). For a given level of ruggedness, very similar results were obtained using damp hollows instead of dry hills; nor did the addition of realistic annual fluctuations to cell habitat quality significantly alter the results (Clarke *et al.*, 1997).

The explanation for these predictions lies in the small spatial scale at which colonies of each 'species' of *Myrmica* functions, and their practice of colonizing vacant nests by stepping-stone dispersal from offshoots of established ant colonies nearby. As each micro-patch of habitat of a certain quality becomes increasingly isolated from others of similar quality in rugged sites, continuous blocks of near-optimum habitat for the host ant become rarer. Under these circumstances, the nest sites of individual colonies of *M. schencki* that are driven extinct by the butterfly are more likely to be colonized by other *Myrmica* species which, once established, are seldom dislodged. Nevertheless, the model predicts that neither ant experiences any significant decline until the butterfly begins to reduce the size and persistence of individual colonies (Clarke *et al.*, 1997). Then, both ants respond like any conventional metapopulation where local extinction exceeds colonization, with particularly severe declines predicted in the case of *M. schencki*. A more interesting prediction is that *M. rebeli* is the greatest loser on rugged sites (Clarke *et al.*, 1997) since, at the spatial scale of the adult butterfly, very little has changed. The site remains the same-sized single patch of habitat of unchanged overall quality, supporting a single population of *M. rebeli*. Nevertheless, it declines because the main resource of its caterpillar (*M. schencki*) functions over a smaller spatial scale, and so its populations are also sensitive to habitat heterogeneity at this scale. These predicted effects perhaps also apply to populations of phytophagous butterflies, whose adult and larval stages also operate on different spatial scales. Indeed, our predictions of reduced persistence in populations on sites with more rugged (but not, in absolute terms, more heterogeneous) habitat may apply to any group of competing populations (especially of plants) where there is strong pre-emptive competition for space, where dispersal is generally very localized (e.g. vegetative), and where the species are differentially prone to external damage, for example from herbivores or disease (Clarke *et al.*, 1997).

(c) Changes in the average quality of habitat within sites

Here we describe model predictions of the effects on *M. rebeli* of altering the size and average quality of the underlying habitat on its sites (Figs 11.4, 11.8, 11.9); Hochberg *et al.* (1998) give additional predictions of the effect of reduced habitat quality on its parasitoid's populations. As in section 11.5.4b, we aim to simulate realistic scenarios, starting with a 1-ha *M.*

rebeli site containing the original gradient of habitat, when traditional agricultural practices are modernized. We first assume that this involves the ploughing or spraying of all the thicker-soiled (moister) ground, corresponding to $x=1–15$ in Fig. 11.3, making it no longer available to *Myrmica*, *M. rebeli* and gentians. This leaves a 0.5-ha site of arid (agriculturally poor) grassland of habitat quality $x=16–30$ on the gradient. In our system, this fragment will now produce slightly more butterflies than the original hectare, if it supported typical densities of gentians (300–500/ha), because the habitat that was lost (quality 1–15) generally acts as a sink to *M. rebeli* (Fig. 11.8c, f). However, at high gentian densities, the effect of the sink areas was to enhance the number of butterflies produced over the drier half of the gradient by reducing the over-exploitation of *M. schencki* colonies there. Thus we estimate that the *M. rebeli* population will fall by 38% if the wetter half of a site possessing a large gentian population is destroyed (Fig. 11.8a).

In practice, in western Europe those fragments of semi-natural grassland that are not 'improved' for agriculture are often abandoned by farmers, and experience vegetational successions that alter the quality of their habitat over time. Some may be obtained as nature reserves, and may be deliberately managed to increase the representation of the optimum habitat (if known) of a desirable species like *M. rebeli*. In Fig. 11.8 we present the results of various simulations in which site size and habitat quality are altered separately or in combination, with the less heterogeneous, but more optimal, habitat distributed either as a gradient on a single site or as four dry hills that have become isolated for *Myrmica* within the original 1-ha 'landscape' (Fig. 11.9).

Figure 11.8 suggests that apparently small changes in habitat quality can have large effects on the carrying capacity of a site for *M. rebeli*, as was also demonstrated in Fig. 11.4 (dotted lines). Depending on the size of the gentian population and the spatial arrangement of the habitat, a change from the original habitat gradient to optimum habitat for *M. schencki* ($x=20–24$) results in a 2.0- to 4.1-fold increase in the density of *M. rebeli* on a 1-ha site. Indeed, 0.6–0.75 ha of the original model site can be lost without any decline in the butterfly population, so long as the surviving area of land changes to optimum quality for the host ant (Figs 11.8 and 11.9). Of course, many insect species experience a dual blow from modern changes in land use, with much fertile land being destroyed and the surviving areas deteriorating in quality for them (Thomas, 1991). We also examined the twin effects of absolute habitat loss and deterioration in habitat quality on *M. rebeli*, and obtained population extinctions in every combination examined (site size reduced to 0.25–0.75 ha; habitat quality changed to $x=10–14$ and 1–15).

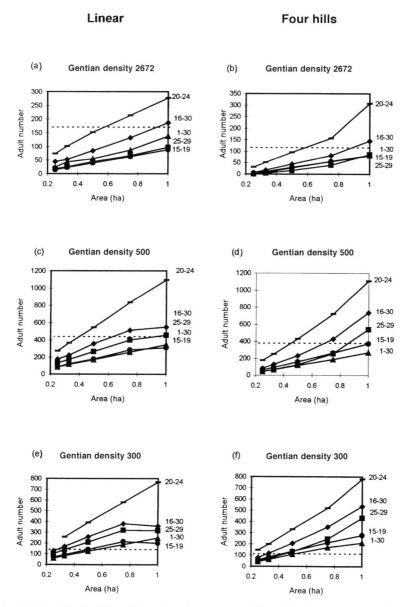

Fig. 11.8 Predicted effects of varying the mean quality of the habitat on sites that vary in size, gentian density and site ruggedness (see Fig. 11.7). Numbers to right of each *y* axis indicate the range of habitat quality from the orginal gradient simulated in each scenario. The number of butterflies supported on site S1 (1 ha) with the original habitat gradient (Fig. 11.3) is indicated by the dotted lines.

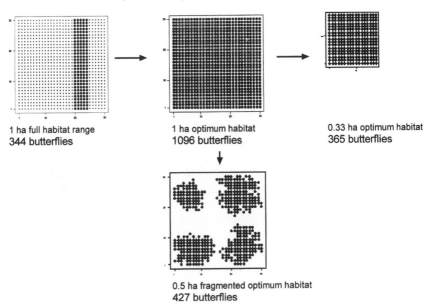

1 ha full habitat range
344 butterflies

1 ha optimum habitat
1096 butterflies

0.33 ha optimum habitat
365 butterflies

0.5 ha fragmented optimum habitat
427 butterflies

Fig. 11.9 Diagrammatic representation of four scenarios of habitat change summarized in Fig. 11.7. Bold symbols indicate optimum quality of habitat for *Myrmica schencki* (*x*=20–24 on the habitat gradient).

11.5 DISCUSSION

We have yet to understand fully the population ecology of the genus *Maculinea*. To date, no predatory species (*M. arion*) has been fully studied near its centre of range nor any cuckoo species (*M. rebeli*) near its edge of range; complex spatial models of interacting populations, as described for *M. rebeli*, have yet to be constructed for a predatory *Maculinea*; certain model predictions about *M. rebeli* have yet to be confirmed (or rejected) experimentally; and knowledge of the population dynamics of all ichneumonid parasitoids is rudimentary. Nevertheless, some firm and other tentative conclusions can be drawn.

In each system of feeding, *Maculinea* caterpillars experience frequent density-dependent competition for food inside *Myrmica* colonies, resulting in mortalities that are strong enough to regulate their populations. In addition, we predicted (Figs 11.3 and 11.4) and confirmed (section 11.5.3; Thomas *et al.*, 1997) that feeding by *M. rebeli* has a major top-down impact on the populations of both its main resource, *M. schencki*, and its resource's main competitors, other species of *Myrmica*. These impacts, in turn, feed back to influence the size of the butterfly's population. There is little doubt that *M. arion* has at least as great an impact on its host-ant populations, because the initial food plant is generally more widespread than *G. cruciata*, and because the predatory caterpillars are more destructive of individual ant colonies, frequently eating the vernal brood not only of the original infested colony (which then deserts its nest site) but also of any new *M. sabuleti* colony that invades that nest (Thomas and Wardlaw, 1992). Finally, regardless of whether some *Maculinea* populations are limited by a ceiling set by their resource, or whether others fluctuate around an equilibrium level, that upper limit or equilibrium level can vary considerably between sites or over time in response to changes in the quality of the underlying habitat, acting bottom-up by altering the values of *r* and *K* of colonies of different competing *Myrmica* species. Near the edges of *Myrmica* and *Maculinea* ranges, small (2–3-cm) differences in the height and density of the sward containing the initial food plant can so alter the microclimate at the soil surface that one species of *Myrmica* may largely replace another even in the absence of *Maculinea* on a site (Thomas, 1984b, 1995); in warmer climates, it takes a change of *ca* 20–50 cm in the mean sward height (or an equivalent change in temperature or humidity) for the same effect (Thomas, 1984a, 1993, unpublished data; Thomas *et al.*, 1998).

Despite these similarities between *Maculinea* species, one difference was found in feeding behaviour which appears to effect their population dynamics. Predatory species exploit ant nests inefficiently, leading to site-carrying capacities of about 20% the value of those for cuckoo species of *Maculinea* (Table 11.1). Also, the density-dependent scramble competition of predatory *Maculinea* theoretically produces less stable populations than the contest competition of cuckoo species. Some empirical corroboration of this prediction was obtained (Table 11.2; Figs 11.2 and 11.5; sections 11.3, 11.5.3b), although another variable may contribute to the observed differences: *M. arion* was studied near its edge of range, where the amplitude of fluctuations in phytophagous butterfly populations is sometimes greater (Thomas *et al.*, 1994); however, it is unlikely that the considerable difference in population CVs (Table 11.2) can be explained by range effects.

Our data suggest – but are inadequate to confirm – that the *M. arion* (predatory) populations studied conform to Milne's (1957) predictions of populations, being limited by a ceiling in their resources. However, observations were made near the edge of *M. arion*'s range: an intraspecific tendency towards this pattern of fluctuations is observed in *Maniola jurtina* and some other phytophagous butterflies towards their northern edges of range (Thomas *et al.*, 1994). In contrast, our models predict that *M. rebeli*, having much reduced its resource, then fluctuates around an equilibrium to which it rapidly returns (Fig. 11.4), due to its high intrinsic rate of increase when its resource is underexploited, and to strong contest competition when the limit of that resource is approached or exceeded. Empirical corroboration rests on the close match found between predicted and observed egg populations on 13 other sites using this model (Fig. 11.5), on the validity of predictions between the relationship between gentian and *M. rebeli* densities on the same sites (Elmes *et al.*, 1996), and on field confirmation that *M. schencki* colonies are depressed in number, size and persistence by similar values to those predicted by the model on sites occupied by *M. rebeli* (Elmes *et al.*, 1996; Thomas *et al.*, 1997). Although phytophagous butterflies such as *M. jurtina* appear to undergo similar (though greater) fluctuations around a mean on sites towards the centres of their range (Thomas *et al.*, 1994), we suspect *M. rebeli* also tends inherently to fluctuate around an equilibrium level due to its feeding behaviour. Nevertheless, we do not consider Nicholson's and Milne's types of population regulation to be incompatible, even in the same species under different environmental conditions (Hochberg, 1996b).

More practically, we believe there are benefits from combining, in one model, various aspects of population dynamics that are usually studied separately. The *M. rebeli* model has four attributes:

it is mechanistic and empirically based, with the mode of action, sequence, value (and usually variation) of each parameter having been measured in field populations or in laboratory experiments;

it contains populations of more than two interacting species, enabling the net effect of both direct and indirect (chains of species) interactions between populations to be explored;

the quality of the underlying habitat was included as a variable, mimicking the conditions found on real sites in two mountain chains;

the model is spatially explicit, enabling population interactions to be studied at different spatial scales, both locally within a site and between whole populations.

A disadvantage of this approach is that it is time-consuming and expensive to collect adequate biological data for realistic models. In addition, any model combining so many variables will be complex; there is a real danger that it will generate spurious predictions. We tried to counter this by testing its ability to predict the population sizes or patterns among species on other sites in another mountain range (Elmes *et al.*, 1996; Thomas *et al.*, 1997). While acknowledging that these tests were far from comprehensive, the results increased our confidence in the untested predictions.

The benefit of this approach is that predictions were made that would not be possible from simpler or more theoretical models. Begon *et al.* (1996b) found that very different predictions may result from studying systems that include more than two interacting populations: ours concerning apparent competition between gentian and ant populations are another example. On the other hand, very few multi-species population models include the quality of the underlying habitat as a variable (if considered at all); by combining these features in a spatial model, we obtained novel predictions that apparent competition and apparent mutualism may exist between two populations in different sub-areas of the same site (Fig. 11.6), and also about effects of the ruggedness of terrain on species' persistence and abundance (Fig. 11.7) and the re-assembly of ant species following the butterfly's extinction (Clarke *et al.*, 1997). Confirmation of either prediction would represent a small step towards explaining patterns in communities through the population dynamics of the constituent species; in the case of site ruggedness, there are also applications to nature conservation (Clarke *et al.*, 1997, 1998).

The pervasive prediction from models of *M. rebeli* is that alterations to the quality of suitable habitat within a site may greatly affect the abundance and persistence of populations. In our simulations we suggest that the spatial arrangement of heterogeneous habitat (site ruggedness) can alter the carrying capacity of a site with a persisting butterfly population by about two-fold (Fig. 11.7); that variation in the mean quality of the habitat can alter this capacity by a further four-fold (Figs.11.4, 11.8, 11.9); and that variation in the density of gentians – which is mainly determined by extrinsic habitat factors – can alter the density of persisting *M. rebeli* populations by about 20-fold (Elmes *et al.*, 1996). Thus large differences are likely to exist in the carrying capacities of different *M. rebeli* sites, even near its centre of range. In the small number of colonies sampled, we recorded a 50-fold difference in the density of *M. rebeli* on isolated but nearby sites (Fig. 11.5). This is similar to the range of densities found between populations of phytophagous butterfly species that were censused in single years in the UK (Thomas, 1984b). Much of that variation was also attributable to differences in the average quality of individual species' habitats on different sites. These between-site differences in butterfly densities are generally an order of magnitude greater than the annual fluctuations experienced by the same species over 7-year periods on individual sites monitored by the UK Butterfly Monitoring Scheme (Thomas, 1984b; Table 11.1).

Finally, since several phytophagous species of butterfly are known to have equally narrow niches to *Myrmica* (Thomas, 1991), it is likely that their populations will also be affected, bottom-up, by similar variations in the quality of their individual habitats, particularly near their edges of range. This is indeed the conclusion of several early autecological studies (Thomas, 1984b, 1991; Warren, 1992). It is therefore disappointing that variation in the quality of a species' (suitable) habitat is seldom included in models of insect metapopulations. We predict that this may be an influential variable, particularly in modern landscapes where the same changes in land use, causing the destruction of much wildlife habitat, often also result in the abandonment of traditional forms of management in the islands of semi-natural biotope that survive.

ACKNOWLEDGEMENTS

We are very grateful to Caroline Skeat for making new model simulations and preparing figures, to D. Moss for calculating population CVs of phytophagous butterflies in the UK, and to C. D. Thomas for many constructive discussions.

12

The dynamics of a herbivore–plant interaction, the cinnabar moth and ragwort

Eddy van der Meijden, Roger M. Nisbet and Mick J. Crawley

12.1 INTRODUCTION

The relationship between the cinnabar moth, *Tyria jacobaeae*, a specialist herbivore, and its food plant ragwort, *Senecio jacobaea*, involves periodic total defoliation (and defloration) of the plant, followed by population crashes of the insect. The insect is not only dependent on ragwort for food, but also for the plant's secondary chemicals – pyrrolizidine alkaloids – that are sequestered by the larvae and probably function as a protection against natural enemies.

Cameron (1935) was the first to describe the phenomenon of defoliation and the following crashes of the population of the cinnabar moth, in a study in Great Britain to evaluate the insect's potential for biological control of ragwort in New Zealand. The accidentally introduced plant became a pest at the beginning of this century, and remains so to the present day, not only in New Zealand but also in several other countries. Many studies followed in Europe, the original area of distribution of both the insect and plant, and elsewhere, stimulated either by the status of ragwort as a noxious weed, or by the intriguing ecological relationship between insect and plant (see Watt, 1987 for a bibliography). Since the late 1980s, the chemical ecology of ragwort and the cinnabar moth has attracted attention. Both aspects of nutrition and chemical defence against herbivores are currently being studied (Soldaat and Vrieling,

Insect Populations, Edited by J. P. Dempster and I. F. G. McLean. Published in 1998 by Kluwer Academic Publishers, Dordrecht. ISBN 0 412 83260 7.

1992; Vrieling *et al.*, 1993; Van Dam *et al.*, 1993; Hartmann, 1995). These aspects, although they may be related to the dynamics of the system, are beyond the scope of this chapter.

Three European population studies (of Dempster, Crawley and van der Meijden and their co-workers) cover long terms, ranging from 9 to more than 20 years. The general conclusion drawn from these studies is that insect population change is dependent on the availability of plant biomass per caterpillar. Population crashes are triggered by food shortage. Plant biomass, in turn, is thought to be dependent on stochastically varying rainfall during the summer months (in which reproduction, early regrowth after herbivory and early seed germination take place). There is no consensus about the effect of herbivory in one year on plant biomass in the next.

In this chapter, we make a new comparison of the available data to seek to understand the dynamics of this plant–herbivore system, and describe a simple model that fits the relationships between the populations of insect and plant in two of the three studies. Finally, we discuss the similarities and differences against the background of the local biotic and abiotic environments.

12.2 OBSERVATIONS AND STUDY SITES

12.2.1 Species

The cinnabar moth is a univoltine insect, with an extremely conspicuous appearance: the larvae have a yellow and black banded colour pattern, the adult is coloured black and cinnabar red. The most accurate population counts of the cinnabar moth are made during the egg stage. Counts of adults, larvae and pupae are time-consuming and considerably less reliable. The eggs are yellow, and laid in conspicuous batches (of 1–150 eggs) on the undersides of the lower leaves of ragwort. Depending on ambient temperature, it takes 5–10 days for the eggs to hatch. The larval period, combining five instars, lasts about 4 weeks.

Apart from an occasional observation (usually of dispersing caterpillars) on *Senecio vulgaris*, the insect is monophagous on ragwort. Females prefer to oviposit on bolting flowering plants. When feeding on flowering plants, larvae consume buds, flowers and the upper leaves before the lower leaves. Buds, flowers and upper leaves contain higher concentrations of both nitrogen and alkaloids. Bioassays have demonstrated that a diet containing buds or flowers results in larger pupae, indicating differences in food quality between different parts of the plant.

Ragwort is an extremely common weed in Europe, especially on sandy soils. It is a so-called biennial weed, but in practice it may live much longer and become perennial. Especially on poor soils or after being damaged, its life cycle may be prolonged. It is generally monocarpic, but may occasionally flower several times. Repeat flowering is most likely when the plants are damaged during flowering, but even undamaged plants are capable of repeat flowering under certain conditions. The plant is famous for its regenerative powers, which determine its weedy character, and also for the secondary metabolites of the pyrrolizidine alkaloid group. These latter substances are toxic to several vertebrates and cause concentration-dependant repellance to some insect species.

Ragwort does not become established in closed swards. Patches of bare ground, such as small disturbances caused by rabbits, are a prerequisite for colonization. Ragwort populations on sand dunes appear to be seed-limited: seed-addition experiments resulted in an increase of the plant's population density. Total defoliation by the cinnabar moth usually leads to the total destruction of seed production in that particular year. Ragwort has a small but long-lived seed bank.

12.2.2 Study sites

Dempster and co-workers (Dempster, 1971b, 1975, 1982; Dempster and Lakhani, 1979) studied the interaction of cinnabar moth and ragwort in the Weeting Heath National Nature Reserve in the UK (Fig. 12.1) from 1966–74. The area is a heavily grazed rabbit enclosure, with a calcareous sandy soil poor in nutrients. The average rainfall is very low for the UK: about 560 mm per year.

The system of Crawley and co-workers is located in Silwood Park, UK (Crawley and Nachapong, 1985; Crawley and Pattrasudhi, 1988; Crawley and Gillman, 1989). The study was started in 1981 and is still continuing. The ragwort population grows in an unfertilized grassland on acid sandy soil (pH 4.0). Growth conditions can be characterized as mesic. The grassland is heavily grazed by rabbits and fenced plots within the study area are mown each August, but the census areas have been unmown since 1990. The productivity of the vegetation at this site is much higher than at the other two sites, as there is little or no bare ground. The average yearly rainfall in Silwood Park amounts to *ca* 660 mm.

Van der Meijden and co-workers studied the interaction in a calcareous sand dune system at Meyendel, near The Hague, the Netherlands, from 1974 (Van der Meijden, 1979; Van der Meijden and van der Waals-Kooi, 1979; Van der Meijden *et al.*, 1991; Van der Meijden and Van der Veen-van Wijk, 1997). Ragwort is growing in local plots (populations) of only a few to several hundred m². Most of these local populations are in unproductive open dune which is grazed by rabbits. Some are found in scrub or forest habitat. The annual rainfall amounts to *ca* 773 mm.

12.2.3 Census methods and population parameters

Census methods differ considerably between these studies. Dempster studied one area of 60×90 m² (0.45 ha) within a much larger area. Each

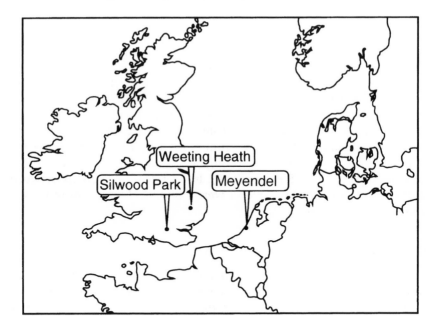

Fig. 12.1 North-western Europe, indicating the study sites Weeting Heath and Silwood Park (UK) and Meyendel (the Netherlands).

season a new set of 150 1-m² quadrats was laid out and used (in a partly destructive way) for collecting the data. Crawley's basic units of study are 10×10-m² plots. He studied a minimum of four of these plots in 1984 and a maximum of 12 in 1985. Since 1990, two permanent 50×20-m² plots (0.2 ha) have been censused each year. Van der Meijden studied about 100 local populations in an area of *ca* 6 km². In each population a permanent plot of 4 m² was marked and recorded.

The population parameters used in this comparative study are not identical for the different sites. Dempster estimated several parameters of the ragwort population. The number of mature plants correlated extremely well with population sample fresh weight over 8 years ($r=0.93$, $n=8$, $P<0.001$), indicating that number of stem plants and biomass are fluctuating synchronously. At Silwood Park, only the number of stem plants was counted yearly over the whole period. For both Weeting Heath and Silwood Park we will therefore use the number of mature ragwort plants (i.e. stem plants) as a parameter to indicate the amount of food for larvae of the cinnabar moth. In the Meyendel populations, only ground cover of ragwort was estimated over the whole period. This parameter correlated well with plant biomass (dry weight; $r=0.92$, $n=25$, $P<0.001$, Van der Meijden, 1979).

Dempster's study is by far the most detailed. He estimated parameters characterizing oviposition, survival and performance of the cinnabar moth, including total number of egg batches and total number of eggs. Crawley measured only the number of egg batches over the whole study period, whereas Van der Meijden measured total number of egg batches and total number of eggs. As there is much variation in egg-batch size among years, we prefer to use the most accurate parameter, i.e. total number of eggs, when available (in Dempster's and van der Meijden's data).

Data on rainfall were obtained from meteorological stations as close as possible to the field sites (for Weeting Heath and Meyendel), or rainfall was measured locally (Silwood Park). In all cases, correlations between rainfall and a biological parameter were based on the amount of rainfall in sets of whole months.

12.3 A COMPARISON OF FIELD DATA

All three ragwort populations suffer frequent total defoliation by the cinnabar moth. At Weeting Heath this took place in five out of nine seasons; at Silwood Park in five out of 16; and at Meyendel in 11 out of 24 seasons. The lower incidence of total defoliation at Silwood is consistent with the lower variance in moth (egg) abundance. The mean amount of biomass per m² per year is highest at Weeting Heath and lowest at Meyendel (Table 12.1; 2 dm² of ragwort in Meyendel corresponds with about 0.75 stem plant). Mean herbivore pressure at the three sites amounts to 8.9, 4.1 and *ca* 18 eggs per stem plant per year, respectively. (It should be kept in mind that stem plant is used here in a relative sense as a unit of biomass. Biomass is made up by stem plants and rosette plants together, and stem plants will vary in size in different soils.) Fluctuations in herbivore load for the three study sites are demonstrated in Fig. 12.2. For Weeting Heath and Meyendel, this figure clearly demonstrates how the insect population grows, overshoots the carrying capacity zone and crashes, whereafter it starts to grow again; the Silwood Park population does not show such a pattern. It is worth mentioning that in some cases, when the cinnabar moth is overshooting the carrying capacity, as in 1968 in Weeting Heath and in 1987 at Meyendel, the number of eggs is 10- to 100-fold higher than the carrying capacity set by its food plant. It is also worth mentioning that, although the carrying capacity is about 10 eggs per stem plant, the mean size of an egg batch is about 35. That means that either a high level of mortality (by predation or adverse weather conditions) is anticipated, or larvae are forced (by their parent) to migrate to other ragwort plants.

Yearly fluctuations in ragwort biomass are much larger at Weeting Heath and Meyendel than at Silwood Park (the coefficients of variation of the yearly change in biomass are almost twice as high). Similarly, the

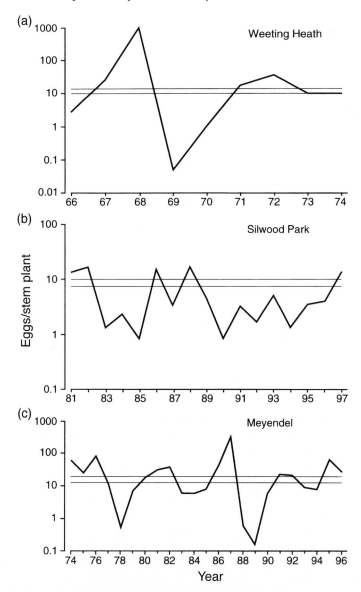

Fig. 12.2 Herbivore load expressed as the number of eggs per stem plant. For Silwood Park, the observations on the number of egg batches per stem plant of ragwort were transformed into the number of eggs per stem plant by taking one egg batch as equal to 35 eggs. For Meyendel, the observations on the number of eggs per dm^2 ragwort were transformed into the number of eggs per stem plant by taking 1 dm^2 ragwort as equal to 0.375 stem plant. Carrying capacity (the number of eggs that will lead to total defoliation of ragwort over the total population) is indicated. (a) Weeting Heath, (b) Silwood Park, (c) Meyendel.

Table 12.1 Comparative statistics on (a) biomass (number of stem plants per m^2 in Weeting Heath and Silwood Park; dm^2 of ragwort ground cover per m^2 at Meyendel); (b) oviposition (number of eggs per m^2); (c) rate of change in 2 successive years in biomass; (d) in oviposition

Site	Mean	Min.	Max.	CV
(a) Biomass				
Weeting Heath	5.6	0.1	18.4	99.8
Silwood Park	4.1	0.1	12.2	78.4
Meyendel	2.0	0.1	6.7	77.2
(b) Oviposition				
Weeting Heath	49.8	0.4	133.7	99.0
Silwood Park	16.4	3.1	44.5	73.9
Meyendel	13.1	0.2	54.2	105.4
(c) Rate of change in biomass				
Weeting Heath	12.4	0.0	84.0	235.1
Silwood Park	1.6	0.2	8.2	130.5
Meyendel	2.6	0.0	26.5	220.3
(d) Rate of change in oviposition				
Weeting Heath	3.2	0.0	8.7	185.4
Silwood Park	1.8	0.2	9.2	123.9
Meyendel	3.0	0.0	21.7	174.0

yearly fluctuations in the number of eggs are largest at Weeting Heath and Meyendel (Fig. 12.3, Table 12.1). The Weeting Heath and Meyendel data show clearly how the insect's (egg) numbers follow fluctuations in plant biomass with a time lag of one season. The pattern of fluctuations at Silwood Park is not so clear.

The Meyendel population of the cinnabar moth shows cyclic dynamics, with a significant 5-year period (see Figs 12.3 and 12.4; Van der Meijden and van der Veen-van Wijk, 1997, Crawley, 1997). The Weeting Heath time series seems to indicate a similar pattern, but it is too short to establish cyclic dynamics. No hint of any cyclic pattern is found in the Silwood time series.

12.4 ELEMENTS OF A SIMPLE MODEL FOR THE CINNABAR MOTH–RAGWORT INTERACTION

Dempster and Lakhani (1979) were the first to develop a population model for the cinnabar moth–ragwort interaction. Their objective was to explain as much of the variation in the Weeting Heath system as possible. Consequently, their model became rather complex. Our purpose is to

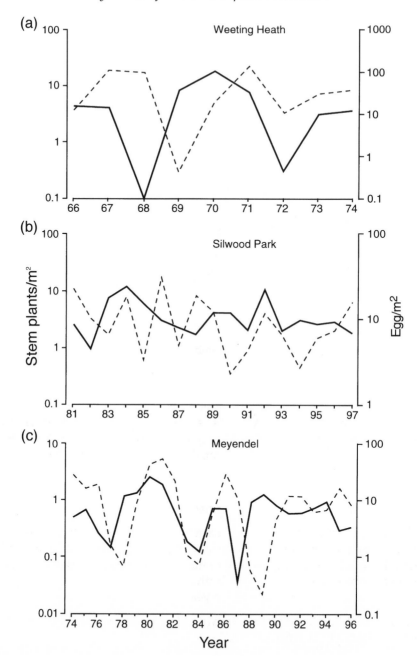

Fig. 12.3 Population dynamics of ragwort (dotted lines) and the cinnabar moth (solid lines). (a) Weeting Heath, (b) Silwood Park, (c) Meyendel. For Silwood Park, one egg batch of the cinnabar moth was taken as equal to 35 eggs. For Meyendel, 1 dm^2 ragwort was taken as equal to 0.375 stem plant.

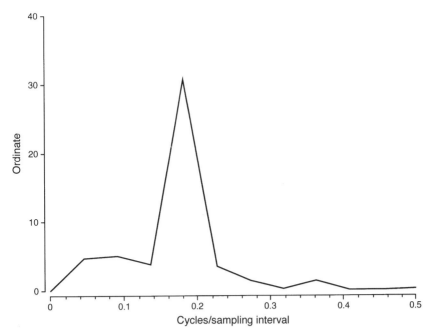

Fig. 12.4 Periodogram of cycles in the number of cinnabar moth eggs at Meyendel.

develop the simplest possible model that allows a comparison of the interaction at the three different sites, which requires a different approach. We restrict ourselves to only a few basic parameters that were measured at each site. Note, however, that these were also basic elements in Dempster and Lakhani's model.

12.4.1 Plant biomass determines cinnabar moth population change

The insect's population dynamics are apparently driven by the availability of food resources. Food shortage leads to crashes in numbers, and food abundance allows the population to increase. If egg production is determined by the amount of food per individual insect or per egg (= ragwort biomass in year n/number of eggs in year n, B_n/E_n), then

$$E_{(n+1)} = \gamma(B_n/E_n)E_n \qquad (12.1)$$

which is equivalent to

$$E_{(n+1)} = \gamma B_n \qquad (12.2)$$

i.e. plant biomass in year n is converted into eggs in year $(n+1)$ with a fixed conversion efficiency γ.

12.4.2 Stochastically varying rainfall determines ragwort production

All three studies mention a relationship between ragwort biomass and rainfall. Biomass at the start of the period of oviposition by the cinnabar moth is affected by rainfall one year earlier (for Weeting Heath and Meyendel during the summer period, when plants may suffer greatly from lack of water in these sandy habitats). This relationship may be formulated as

$$B_{(n+1)} \propto \Phi_n \tag{12.3}$$

in which B is biomass and Φ is a random variable that characterizes rainfall.

12.4.3 Herbivore pressure by the cinnabar moth affects ragwort production

There is a second factor thought to be important for plant production, and that is the amount of herbivory in the previous year. Ragwort plants that are defoliated are able to regrow. As mentioned earlier, the regrowth capacity of ragwort is spectacular (Islam and Crawley, 1983; Van der Meijden *et al.*, 1988). Nevertheless, there is strong evidence that plant production one year after herbivory is reduced. There is no consensus with respect to this effect between the three studies. Dempster and Lakhani (1979) mention that the proportion of plants flowering depends on whether or not they were defoliated in the previous summer. In an experiment at Meyendel, in which the effects of different herbivores were separated, Prins and Nell (1990) established that cinnabar moth herbivory may dramatically reduce ragwort numbers. Van der Meijden *et al.* (1985) similarly reported at an earlier stage of the Meyendel study that herbivory had a negative effect on plant production.

A problem with identifying the effects of herbivory by the cinnabar moth is that the number of eggs or larvae is not a reliable parameter for estimating the amount of herbivory. In some years, egg and larval numbers are so high that they could easily consume the available biomass many times. As a plant can be eaten only once, the data on egg or larval numbers should be truncated, which means effectively that the carrying capacity is the maximum value. The effect of herbivory can than be modelled as

$$B_{(n+1)} = f(E_n / B_n) \tag{12.4}$$

in which B, the plant biomass, is a function of E_n / B_n, the herbivore pressure in year n expressed as the number of eggs or larvae per unit of biomass (B), under the condition that E in this ratio is the smallest number that can totally consume B.

12.5 DATA ANALYSIS

Using the available field data (with or without log transformation) for Weeting Heath, Silwood Park and Meyendel, multiple regressions were used to explore the hypothetical relationships mentioned above.

12.5.1 Plant biomass determines cinnabar moth population change

The correlations between plant biomass in year n and egg numbers in year $(n+1)$ were significant for Weeting Heath (log number of mature plants against log egg numbers) and Meyendel (plant ground cover against egg numbers, untransformed as well as log-transformed; Table 12.2). No relationship could be found between biomass and number of egg batches in Silwood Park, neither for the total census period from 1981–97, nor for the periods with different management regimes (1981–89 with a yearly mowing regime, and 1990–97 when the sampled areas were not mown).

12.5.2 Stochastically varying rainfall, together with herbivore pressure by the cinnabar moth, affect ragwort production

Rainfall and herbivore load in year n correlate significantly with log biomass in year $(n+1)$ for Weeting Heath and Meyendel (Table 12.3). Multiple regression analysis did not indicate important relationships when herbivore load was not truncated as described above and when biomass was not log transformed. The latter result probably reflects a plateau effect when conditions for biomass increase are favourable. For Silwood, it is only for the 1981–89 period (when the site was mown each year) that we find an effect of rainfall on biomass production one year later. In contrast to the other sites, this correlation is negative. Herbivore load had no significant effect in the multiple regression on log biomass ($P=0.22$).

Table 12.2 Correlation coefficients (r) between plant biomass in year n with the number of eggs or egg batches in year $(n+1)$ (1) and log-transformed data of both parameters (2)

Site		r	n	P	Slope	Intercept
Weeting Heath	(1)	0.56	7	NS		
	(2)	0.70	7	0.052	0.74	2.50
Silwood Park	(1)	0.00	15	NS		
	(2)	–0.05	15	NS		
Meyendel	(1)	0.74	21	0.000	6.56	–0.90
	(2)	0.60	21	0.003	0.93	1.42

Table 12.3 Multiple correlation between the amount of rainfall and the level of herbivore load by the cinnabar moth in year n with log biomass of ragwort in year $(n+1)$

Site/parameter	CV	Significance level	r^2
Weeting Heath			
Constant	−0.506	0.65	
Rainfall	0.023	0.02	
Herbivore load	−0.255	0.00	0.724
Silwood Park (1981–97) and (1990–97)			
Constant		NS	
Rainfall		NS	
Herbivore load		NS	0.000
Silwood Park (1981–89)			
Constant	1.904	0.00	
Rainfall	−0.010	0.03	
Herbivore load	0.0015	0.22	0.419
Meyendel			
Constant	5.220	0.000	
Rainfall	0.0086	0.002	
Herbivore load	−0.194	0.001	0.528

Amount of rainfall at Weeting Heath relates to July, August and September and at Silwood Park and Meyendel to June, July and August. Herbivore load is estimated as the number of eggs per mature plant (Weeting Heath and Silwood Park) or the number of eggs per dm^2 of ragwort ground cover (Meyendel).

12.6 THE MODEL

The field data for Weeting Heath and Meyendel (Table 12.2) indicate that the most important interaction between insect and plant involves only two variables (B_n = biomass and E_n = number of eggs), both at the beginning of herbivory in season n, and three scaling factors or parameters (δ, α and γ). Plant biomass production (growth) is reduced by the effect of herbivory one year earlier (Equation 12.4) and the number of insect eggs is determined by the amount of available biomass one year earlier (Equation 12.1):

$$B_{(n+1)} = \delta e^{-\alpha E_n/B_n} \qquad (12.5)$$

An additional variable that affects biomass is stochastic rainfall during the growing season, as described in Equation 12.2. The two effects on biomass can be combined into:

$$B_{(n+1)} = K\Phi_n e^{-\alpha E_n/B_n} \qquad (12.6)$$

in which K is a scaling factor replacing δ and β. Based on the observations at Weeting Heath and Meyendel, Equation 12.5 describes the dynamics of ragwort. For Weeting Heath the dynamics of the cinnabar moth are best described by Equation 12.1 with log-transformed values of $E_{(n+1)}$ and B_n. The cinnabar moth dynamics at Meyendel are also well described by this equation, but untransformed data for $E_{(n+1)}$ and B_n give slightly better results than log-transformed data.

12.7 BEHAVIOUR OF THE HERBIVORE–PLANT MODEL

The parameter K is the amount of biomass in the situation without herbivory; the parameter γ is the conversion factor of biomass into eggs; α is a conversion factor that characterizes the effect of herbivory (expressed in terms of number of eggs per unit of biomass) on biomass one year later. The two conversion factors relate to the mutual effects of herbivores on plants and *vice versa*, and can be combined in one parameter, $\mu=\alpha\gamma$.

Simulations of Equations 12.2 and 12.6 ($\Phi=1$) show how the quantitative dynamics of the model are determined by μ (Fig. 12.5). If $\mu>1$ the system becomes extinct; if $\mu<1$ the model predicts convergence to a stable equilibrium; if $\mu=1$ the model predicts limit cycles (Gurney and Nisbet, 1998) with a period of 6 years. If Φ (the effect of rainfall; $\Phi=\exp\{\Theta_n\}$) is a random variable drawn from a uniform distribution of Θ ($-\frac{1}{2}$, $\frac{1}{2}$), this 6-year period is still found, but obscured (Fig. 12.6).

12.8 DISCUSSION

Much of the variation of the dynamics of the cinnabar moth–ragwort interaction at Weeting Heath and Meyendel can be explained by a simple model with only three variables: ragwort biomass, the number of cinnabar moth eggs, and the amount of rainfall during a crucial period (July–September for Weeting Heath and May–August for Meyendel).

Although arthropod parasitism and predation of eggs and larvae are important factors that may lead to more than 70% mortality (Dempster, 1971b, 1982), food availability determines the carrying capacity of both populations. As shown in Fig. 12.2, cinnabar moth numbers, especially at Weeting Heath and Meyendel, may overshoot this carrying capacity many times leading to mass starvation and reduction of the size of larvae and pupae, and consequently of female fecundity.

Yearly differences in plant biomass production are well explained by rainfall and herbivore load one year earlier in both populations. Considering the coefficients in Table 12.3, rainfall seems to be more important for explaining these differences at Weeting Heath than at Meyendel. This coincides with the much lower yearly rainfall level at the first locality.

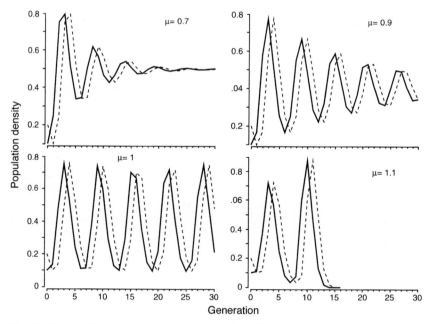

Fig. 12.5 Simulations of quantitative dynamics of Equations 12.1 and 12.5 indicating the interaction between ragwort (solid lines) and cinnabar moth numbers (dotted lines). The composite parameter $\mu = \alpha\gamma$.

The effect of cinnabar moth herbivory on plant production, expressed as herbivore load, appeared to be a crucial element in the model. For both Weeting Heath and Meyendel it provides the element with the highest significance level in the multiple regression on plant biomass. Herbivore load as used in this chapter has a ceiling value, the carrying capacity, i.e. the number of eggs (cinnabar moth larvae) that can survive on one unit of biomass (a stem plant at Weeting Heath or a dm^2 of ragwort at Meyendel).

The model that is developed here, based on ragwort biomass, cinnabar moth egg numbers and stochastically varying rainfall, predicts cycles of abundance of the cinnabar moth with a period of approximately 6 years. The Meyendel data show a significant 5-year cycle for the cinnabar moth. Probably the heterogeneity between the local patches or populations of ragwort may reduce the cycle period. Although there is a high level of synchrony between local populations, there are always some populations completely out of phase (van der Meijden and van der Veen-van Wijk, 1997). Another reason for this discrepancy might be that in Equation 12.2, on insect population dynamics, the information on insect density in year n is obscured by incorporating the biomass per insect ratio in year n in Equation 12.1. This simplification of reality would imply that the insect

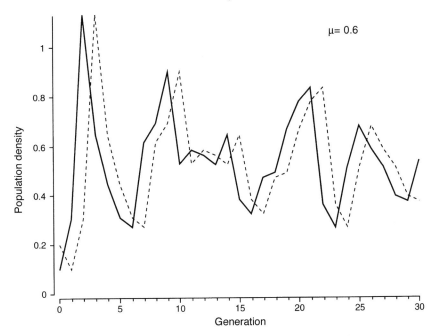

Fig. 12.6 Simulations of quantitative dynamics of Equations 12.1 and 12.6 indicating the interaction between ragwort (solid lines) and cinnabar moth numbers (dotted lines) with the effect of random rainfall.

suffers mortality only due to starvation. It would also imply that the population of the insect could not go extinct. In spite of these problems, it is likely that the model captures the essential mechanisms responsible for the observed cycles at Meyendel and (more speculatively) at Weeting Heath.

Environmental circumstances are fundamentally different at Silwood Park. We could not find any of the above-mentioned correlations for this population, yet in some years the cinnabar moth population may defoliate ragwort to the same extent as elsewhere. A key determinant of ragwort biomass in Silwood Park (and at the other sites as well) is rabbit grazing. Ragwort is eliminated from the plant community within 4 years of the erection of rabbit-proof fences, as a result of intense interspecific competition from perennial grasses (Crawley and Edwards, unpublished results). In rabbit-grazed areas, the next most important factor is the rate of soil disturbance: the open conditions required for recruitment need to be created by animals like moles, rabbits and earthworms, as in the more open Dutch and Weeting Heath communities. Rabbit grazing was not estimated in any of the sites. It seems very likely, however, that fluctuations in the effect of rabbit grazing on ragwort performance have a greater impact at Silwood Park than at the other two sites which are much less productive. This will result in more relaxed competition following an increase in grazing.

Application three times a year of a cocktail of insecticides (a knock-down pyrethroid and a systemic) in Silwood Park is associated with a four-fold increase in ragwort density, but only on plots that are rabbit-grazed and have higher than average rates of soil disturbance. Note, however, that this response to insect exclusion is almost certainly due to reduction in the numbers of root-feeding larvae of the beetle *Longitarsus jacobaeae*, rather than to the exclusion of cinnabar moths. This inference is based on the observation that when cinnabar moth caterpillars are experimentally removed by hand, and presumably *Longitarsus* numbers are not affected, plant numbers actually decrease. This counter-intuitive result reflects the fact that in the Silwood Park grasslands, the main cause of mortality on mature ragwort plants is fruiting. Cinnabar moth feeding reduces (or precludes) fruiting and hence increases mature plant survival. Plants that mature larger crops of seeds are more likely to die over winter (Gillman and Crawley, 1990).

Plant recruitment is not so closely coupled to rainfall and seed production at Silwood as in the more open and drier sites. Seed addition experiments (Johnston, 1992) demonstrate that, contrary to the Dutch situation (van der Meijden and van der Waals-Kooi, 1979), recruitment is not seed-limited in most years and under most management regimes. High-density cohorts of ragwort appear to be produced when there is a coincidence of high disturbance, good seed production in the previous summer (associated with relatively low cinnabar moth impact) and adequate rainfall. There have been two such cohorts during the last 17 years (1984 and 1992). There is no absolute requirement for high rainfall, since high-density cohorts can be created by experimental disturbance even in exceptionally dry years (e.g. in 1995 and 1996; Crawley and Edwards, unpublished results).

Both plant and insect numbers in all populations seem to be clearly affected by population density, judging from the negative slopes of graphs in which population change ($N_{(t+1)}/N_t$) is plotted against population size (N_t) and there is strong direct density dependence (see van der Meijden *et al.*, 1991). The main cause of this effect in the ragwort population appears to be intraspecific competition for limited recruitment opportunities, with microsite availability determined by the interaction between rabbit grazing and the growth of perennial grasses. Moth numbers are limited by intraspecific competition for food (ragwort biomass); the situation in Silwood is far from clear-cut, however, because the negative correlation between population change in cinnabar numbers per unit plant biomass is only marginally significant ($P=0.059$). The Silwood cinnabar population shows no tendency whatever towards regular cycles in abundance.

We cannot explain why, at Silwood, the cinnabar moth population is not affected by the abundance of its food plant, nor why crashes due to overexploitation are not followed by population growth when food is no longer limiting (see Figs 12.2 and 12.3). A tentative suggestion might be that the relatively small sampled insect population is a subpopulation of a much larger unit with much local heterogeneity (Harrison and Thomas, 1991). The importance of such local heterogeneity for differences in local insect density in relation to food resources (in Meyendel) were indicated by van der Meijden and Van der Veen-van Wijk (1997). Shaded areas and plots with predatory ants had a considerably lower egg load than open sandy areas without this predator. These former areas also demonstrate different dynamics of the ragwort population (van der Meijden *et al.*, 1992). Alternatively, it may be that other herbivores, such as rabbits and *Longitarsus* beetles, are the main cause of fluctuations in ragwort abundance at Silwood.

The data presented on the Silwood population (Figs 12.2 and 12.3) indeed indicate a much lower egg load than at the other sites. Mean herbivore load was calculated for Silwood (4.1 eggs/stem plant/year), Weeting Heath (8.9) and Meyendel (*ca* 18). This could be due to a lower fecundity related to food quality, a lower larval survival due to predation or parasitism, or a lower survival because migrating larvae have greater difficulties in finding new food plants in the closed Silwood sward than at the other sites.

There are no indications of any differences in food quality between sites. However, the suggestion is not so easily discarded that the fauna of predatory arthropods present in the rich grassland environment at Silwood has a much greater impact on the dynamics of the cinnabar moth than at the poorer sites at Weeting Heath and Meyendel. This might lead to a reduction in the moth's population growth rate, so that it is unable to respond to rapid changes in food supply. Dempster (1971a) provides evidence for such an effect. The hindering effect of a dense grass vegetation on migrating larvae might have a similar result. The moth's habit of laying eggs in a few large batches usually leads to massive migration by third-, fourth- and fifth-instar larvae. This migration starts before plants are completely defoliated (see Sjerps and Haccou, 1993 for an evolutionary explanation).

Local heterogeneity (metapopulation structure) will affect the chances of survival of the cinnabar moth population (an aspect not dealt with in this chapter), especially when the insect's population greatly overshoots the carrying capacity of its food plant. We refer to other papers on this insect–plant system for discussion of this effect (Dempster, 1982; van der Meijden and Van der Veen-van Wijk, 1997)

This study demonstrates that the population dynamics of a specialist insect herbivore may be resource-driven and resource-limited. At the same time, it illustrates the stochastic effect of weather on the local availability of resources. Finally, it shows how difficult it may be to establish resource limitation, as other mortality factors may temporarily reduce population growth, thus obscuring the ultimate impact of resource limitation.

13

The population dynamics of *Operophtera brumata* (Lepidoptera: Geometridae)

Jens Roland

13.1 INTRODUCTION

Thirty years ago, the debate on the role of density dependence in population dynamics formed a central theme in the Proceedings of the Fourth Symposium of the Royal Entomological Society, which was subsequently published in the book *Insect Abundance*, edited by T.R.E. Southwood (1968). Among the case studies included in that volume, the study on the winter moth by George Varley and George Gradwell has been central in the development of ideas relating to the importance of density dependence in population regulation. Indeed, among studies of herbivorous insect populations, those on the winter moth have had a great impact on the field, perhaps more so than those of any other insect species.

Despite the importance of these studies of winter moth, they have been criticized on several counts: the use of only five trees at Wytham Wood; lack of statistical independence in regressions of mortality versus density (Bulmer, 1975; Royama, 1996); the fact that many of the data are only now being analysed (e.g. Hunter *et al.*, 1997); and that most of Varley and Gradwell's results were published in conference proceedings rather than in the primary literature. However, the impact of these early studies is without question.

In this chapter, I will do three things: (i) briefly review the general features of winter moth dynamics which can be gleaned from the early studies; (ii) present the results of more recent studies, both of native

Insect Populations, Edited by J. P. Dempster and I. F. G. McLean. Published in 1998 by Kluwer Academic Publishers, Dordrecht. ISBN 0 412 83260 7.

populations in Europe and of introduced populations in North America; and (iii) where warranted, present a synthesis of the general dynamics among widely separated winter moth populations. In doing so, I will develop an argument in support of the importance of density-dependent mortality beyond its potential for regulating winter moth populations. I will argue that the greater significance of density dependence for winter moth populations, and possibly for other insect species, is through inter-action between density-dependent mortality factors and other popula-tion processes. I suggest that these interactions produce the two most dramatic patterns of winter moth dynamics: the synchronized cyclicity of native populations in Europe, and the highly successful biological control of introduced populations in North America.

13.2 POPULATION BIOLOGY

Two general features of winter moth dynamics are apparent from the early studies: (i) the importance of synchrony of egg hatch with host-plant bud burst in setting population size in any year; and (ii) the role which natural enemies, in particular predators on pupae, play in popula-tion regulation (Varley and Gradwell, 1968). Timing of bud burst varies considerably between years, and thus its synchrony with winter moth egg hatch also varies. Late bud burst results in poor establishment of hatching first-instar larvae, and in their increased dispersal and thus low density. Early bud burst improves larval survival, reduces dispersal, and results in high larval density. If hatch is significantly later than bud burst, increasing tannin levels in oak foliage results in poor larval performance and low survival (Feeny, 1970). Pupal predation is the only mortality fac-tor which showed any regulatory capability (i.e. temporal density depen-dence), a pattern evident in both unlagged data and data lagged by 1 year (Roland, 1994). The importance of synchrony with bud burst and of predation has been reinforced by many additional studies throughout the 1970s (East, 1974; Holliday, 1975; Kowalski, 1976, 1977; Wint, 1983).

Because these general results are well known, I will emphasize more recent work on winter moth population dynamics, from studies both of native populations in Europe and of introduced populations in North America under biological control (Roland and Embree, 1995; Murdoch *et al.*, Chapter 7 in this volume).

13.3 TEMPORAL DYNAMICS

Long-term data on winter moth abundance at Wytham Wood exhibited a 9- to 10-year pattern of increase and decline with a *ca* 35-fold change in abundance from trough to peak (Fig. 13.1). Longer time series of winter moth abundance from several sites in Germany (Altenkirch, 1991) and in

Norway (Hogstad, 1997) also exhibit a 9- to 10-year periodicity with peaks coincident with those at Wytham Wood (Fig. 13.1). Very long records of outbreaks, from 1862 through to 1968 in forests in Scandinavia (Tenow, 1972), also exhibit a 10-year cycle, with the recent peaks again coincident with those in Britain, Germany and Norway. For introduced populations in North American oak forests, census data either were not collected after population collapse (Embree, 1991), or were collected for only 5 years after the collapse (Roland, 1994). Therefore, no pattern of cyclicity can be assessed. Census data were collected, however, in apple orchards in Nova Scotia from 1966 (the year after the collapse) through to 1980 (MacPhee *et al.*, 1988); these data again show periodic rise and fall, but with a shorter 7–8-year cycle that is not synchronous with cycles in Europe (Fig. 13.1).

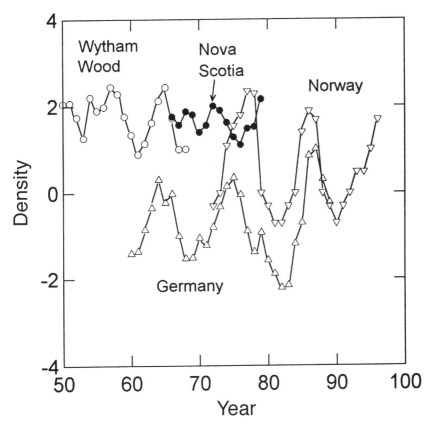

Fig. 13.1 Annual density of winter moth larvae over time, open circles, at Wytham Wood (from Varley *et al.*, 1973); downward open triangles, in Norway (from Hogstad, 1997); and closed circles, in Nova Scotia (from MacPhee *et al.*, 1988); and upward open triangles, of adult moths in Germany (from Altenkirch, 1991).

Long-term, cyclic winter moth dynamics suggest delayed density-dependent regulation (Royama, 1992). Time-series analysis of data from Germany (Fig. 13.2a) and from Wytham Wood (Fig. 13.2b) confirm that winter moth dynamics are cyclic, and are characterized by a lagged density-dependent process (Royama, 1992). General theory of insect population dynamics suggests that mechanisms such as delayed induced plant responses (Haukioja and Hakala, 1975; Rhoades, 1985), or a delayed impact of natural enemies such as predators, parasitoids or disease (Turchin, 1990), can produce cyclic patterns. Are there any such mechanisms apparent from population data collected on winter moth? In the next sections the evidence is considered for each of several possibilities.

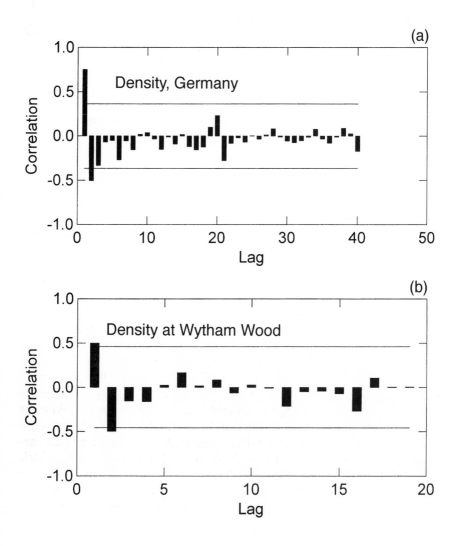

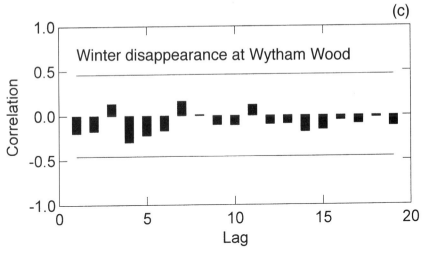

Fig. 13.2 Partial autocorrelation for 16 years' winter moth density data from (a) for 30 years' data from Germany (from Altenkirch, 1991); (b) Wytham Wood showing a significant second-order lagged effect indicative of delayed density-dependent mortality (from Varley *et al.*, 1973); and (c) for 'over-winter disappearance' at Wytham Wood (k_1 from Varley *et al.*, 1973).

13.4 POPULATION PROCESSES

13.4.1 Plant quality

Two principal patterns emerge from early studies of plant quality and winter moth dynamics. The first is the degree to which egg hatch is synchronized with bud burst in setting annual density; the second is the effect of plant quality on winter moth performance. Synchrony with bud burst is thought largely to set the density of larvae on oaks, as described above (Varley and Gradwell, 1968). This was the hypothesized mechanism of 'winter disappearance' (k_1) as a key factor at Wytham Wood (Varley and Gradwell, 1968). Similarly, differences between plant species in timing of bud burst probably explain differences in winter moth abundance; early-flushing species such as apple have higher larval density (Holliday, 1977; Wint, 1983; Kirsten and Topp, 1991) than do late-flushing species such as oak. It is of interest that models of effects of climate change on timing of both egg hatch and bud burst of oak predict little net change in synchrony on oak (Buse and Good, 1996). In contrast, on the unusual host plant Sitka spruce (*Picea sitchensis*), such models predict earlier egg hatch, but little change in bud burst (Hunter *et al.*, 1991), which would result in reduced herbivore density, despite the fact that variation in bud burst of Sitka spruce under present conditions has no effect on population density (Watt and McFarlane, 1991; Dewar and Watt, 1992).

The pattern of 10-year cycles of winter moth could be due to delayed induced plant defences (Haukioja and Hakala, 1975) resulting from winter moth feeding. However, three facts make this unlikely. Firstly, at a given moth density, winter moth pupal weight and fecundity are higher in declining populations for which defoliation of pedunculate oak (*Quercus robur*) would have been high in the previous year, compared to pupal weights in increasing populations for which defoliation would have been low the previous year (Gradwell, 1974). Secondly, although there is a negative within-year effect of defoliation of garry oak (*Quercus garryana*) on winter moth performance, moderate levels of defoliation in the previous year actually result in higher pupal weights compared to light or no defoliation in the previous year (Roland and Myers, 1987). Thirdly, life-table data show no lagged effect of larval density (N_t) on larval mortality (k_{t+1}) (Roland, 1994). These three patterns suggest little potential for a delayed induced defence in oak foliage. There appears to be no strong plant–herbivore mechanism to drive the long-term cyclic dynamics of winter moth. Although the bottom-up effects of plant quality are weak for winter moth, they are much stronger for the tortricid *Choristoneura viridana* even within the same trees (Hunter *et al.*, 1997), again arguing against an induced plant defence driving winter moth dynamics.

If weather were cyclic with a period of 9–10 years, then the effects of bud burst (through weather) could give the appearance of regulated cyclic dynamics. However, bud burst is not cyclic in Fenno-Scandia (Bylund *et al.*, 1995), nor is 'over-winter mortality' (k_1) in Wytham Wood (Fig. 13.2c). Because neither bud burst nor k_1 is itself cyclic, these parameters cannot be expected to directly produce cyclic patterns of moth abundance. However, bud burst may contribute an important but indirect effect in producing cyclic dynamics, as discussed below.

A final effect of plant synchrony on winter moth dynamics may be manifest through the impact of parasitoids. Early bud burst of some plant species (e.g. apple; Wint, 1983), and the concomitant early winter moth larval development, may permit winter moth to escape attack by parasitoids (Roland, 1986a). Similarly, other secondary host plants such as heather (*Calluna vulgaris*; Kerslake *et al.*, 1996) and blueberry (*Vaccinium myrtillus* and *Vaccinium corymbosum*; Horgan, 1993) might provide a refuge from searching parasitoids. Such refugia would, on theoretical grounds, stabilize dynamics on early-flushing apple compared to oak, but would do so at a higher host density. Indeed, despite the presence of parasitoids of winter moth in apple orchards, caterpillar densities are generally higher there than in oak stands (Tenow, 1972; Holliday, 1977; MacPhee *et al.*, 1988; Roland, 1990a; Pearsall and Walde, 1994), but are less variable from year to year (Roland, 1990a).

In general, the effect of host plants on winter moth dynamics seems to be to act as a mild perturbation partially determining density in each year. The relative contribution of plant effects on variation in abundance is low compared to that of natural enemies (Hunter *et al.*, 1997).

13.4.2 Natural enemies

Varley and Gradwell (1963, 1968) and Varley *et al.* (1973) initiated their study of winter moth at Wytham Wood with the specific goal of determining the role of natural enemies on herbivore dynamics. In that population, parasitoids were found to have only a small impact on total mortality, and showed no regulatory ability (direct density dependence, Table 13.1). In contrast, predation on pupae by carabid and staphylinid beetles was strongly dependent on pupal density both in the current year (unlagged density) and in the previous year (lagged density) (Table 13.1). This pattern was confirmed by several additional studies of predation on winter moth in Britain (Frank, 1967; East, 1974; Kowalski, 1976, 1977) which found that density-dependent predation resulted from a combination of both aggregative and numerical responses by beetles (East, 1974; Kowalski, 1976, 1977).

Debate over the importance of density-dependent predation on pupae in Wytham Wood has focused on whether the strength of the density dependence is sufficient to regulate winter moth populations. Models of winter moth populations at Wytham Wood (Den Boer, 1986c, 1988) compared the dynamics of modelled populations either with density-dependent predation or with predation as a random variable (based on observed predation levels among the 19 years of data). In these models,

Table 13.1 Relationship between mortality (*k*-values) and log density for Mount Tolmie, British Columbia (1985–90) and Wytham Wood, Berkshire (1950–68). Mortality was regressed against density in the current year and against density in the previous year (1-year lag). From Roland (1994)

Mortality	*No lag* slope	r^2	*1-year lag* slope	r^2
Mount Tolmie, British Columbia				
Reduced fecundity	0.22*	0.58	0.09	0.39
Larval mortality	0.06	0.00	0.10	0.02
Parasitism	−0.06	0.01	0.59**	0.67
Predation	0.88*	0.59	−0.37	0.10
Wytham Wood, Berkshire				
Reduced fecundity	0.10***	0.43	0.02	0.02
Larval mortality	0.16	0.04	−0.07	0.01
Paraistism	0.01	0.01	0.03*	0.21
Predation	0.32***	0.58	0.31***	0.48

* $P<0.10$; **, $P<0.05$; ***, $P<0.01$.

the dynamics were similar whether predation varied in relation to density, or simply varied. However, this model limited the potential role of density-dependent predation by using the observed values for all the other winter moth mortality factors in each year, producing dynamics similar to those seen in the field. An alternative model (Latto and Hassell, 1987) allowed all mortality factors to vary randomly but only within the range observed in the field, and this resulted in more strongly regulated populations when predation was density-dependent compared to when it was random. Therefore, field and modelled populations of winter moth based on data from Wytham Wood appear to be regulated by generalist predators, but not by parasitoids.

Studies of introduced winter moth populations were conducted in Nova Scotia (Embree, 1966) concurrently with those at Wytham Wood, and later in British Columbia, Canada (Embree and Otvos, 1984; Roland, 1986b, 1990a,b, 1994; Horgan, 1993). In these populations, up to six parasitoid species were introduced from Europe. The dramatic decline of winter moth was attributed to these introduced parasitoids, especially the tachinid fly *Cyzenis albicans* and the ichneumonid wasp *Agrypon flaveolatum* (Embree and Otvos, 1984; Murdoch *et al.*, 1985; Barron, 1989; Embree, 1991). However, during the period of moth decline in both Nova Scotia and British Columbia, mortality of unparasitized pupae in the soil also rose sharply, to over 95% (Fig. 13.3), whereas parasitism typically rose to only 40–60% (Embree, 1965) (Fig. 13.4). Parasitism was weakly density-dependent, but was so only for lagged data (Table 13.1). It is of interest that in populations in Canada, predation is higher on unparasitized winter moth pupae than on parasitized pupae (Roland, 1990b), despite the former being in the soil only half as long (5 versus 10 months). This suggests that in the presence of parasitoids, the impact of predation on winter moth can be particularly severe. No experimental work was done on beetle abundance in either Britain or Nova Scotia, but in British Columbia, removal of beetle larvae by the use of exclusion cages lowered pupal mortality to less than 10% (Roland, 1990b) implicating beetle larvae as the dominant source of mortality from predation.

In all studies of winter moth, predation is strongly spatially density-dependent, both in oakwoods (Varley and Gradwell, 1968; Kowalski, 1977; Roland, 1990a) and in apple orchards (Holliday, 1975; MacPhee *et al.*, 1988; Roland, 1990a; Pearsall and Walde, 1994). Recent studies (1991 and 1992) at Embree's original oak sites in Nova Scotia indicate that predation remains high at about 85% (J. Roland and S.J. Walde, unpublished data) and at 70–80% in apple orchards (MacPhee *et al.*, 1988), much higher than during the period prior to winter moth collapse. One suggestion for this pattern has been that the added parasitism has reduced host density to the level at which predators can take a larger proportion of the prey (Roland, 1990b). Interestingly, winter moth density remains highest in Wytham Wood, lower in British Columbia, and lowest in Nova Scotia

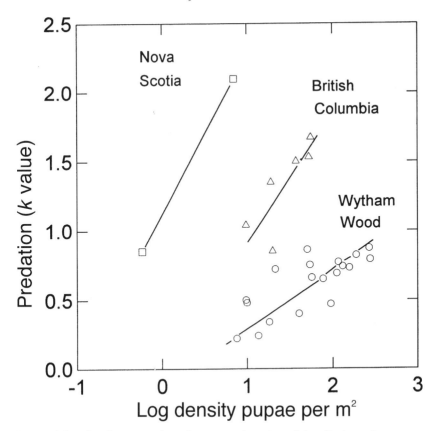

Fig. 13.3 Mortality from pupal predators as a function of density in each year from three populations of winter moth: Wytham Wood (open circles), British Columbia (upward open triangles), and Nova Scotia (open squares). Data for Nova Scotia are for 1966 (higher dot) and for 1992 (lower dot). For the latter year density was not estimated but is known to be very low (Embree, 1991; Roland and Walde, unpublished data). Adapted from Roland (1994), with data for Wytham Wood from Varley *et al.* (1973).

(Fig. 13.3), a pattern which corresponds to weakest density dependence at Wytham Wood, and strongest in Nova Scotia.

Winter moth does have some diseases, including a microsporidian (Canning, 1960; Canning and Barker, 1982), and virus (Wigley, 1976; Cunningham *et al.*, 1981). However, the impact of these is considered to be minor.

13.5 POPULATION STRUCTURE AND SPATIAL SCALE

Winter moth females are flightless; their dispersal is limited to distances they can walk, usually only as far the nearest tree trunk. Even the males,

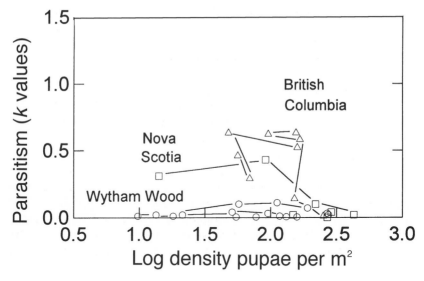

Fig. 13.4 Mortality from parasitoids as a function of density in each year from three populations of winter moth: Wytham Wood (open circles), British Columbia (upward open triangles), and Nova Scotia (open squares). Adapted from Roland (1994), with data for Wytham Wood from Varley *et al.* (1973).

which are capable of flight, disperse only on the order of tens of metres (van Dongen *et al.*, 1996). Ballooning of first-instar larvae is the principal mode of dispersal, which probably occurs to distances of up to 1000 m like that of its congener, *Operophtera bruceata* (Brown, 1962). Recent studies of allozyme variation in winter moth show that (i) there is significant variation among populations only 1–2 km apart (van Dongen *et al.*, 1994), and (ii) that the structure of winter moth populations is more fragmented than that of most other Lepidoptera (van San and Sula, 1993). Because dispersal is so restricted, large-scale synchronization of populations across Europe (Varley and Gradwell, 1968; Tenow, 1972; Altenkirch, 1991; Hogstad, 1997; Fig. 13.1) are probably not due to any synchronizing effect of dispersal.

Dispersal of parasitoids attacking larvae on trees, and of beetles attacking pupae in the soil, is probably greater than that of the winter moth itself. Indeed, the introduced parasitoids in Canada spread at a rate of tens of kilometres per generation (Embree, 1966; Embree and Otvos, 1984). Broadly dispersing parasitoids may leave high-density patches of hosts where they developed as larvae, and then disperse into areas where hosts are less abundant. This process would tend to produce weakly density-dependent or density-independent parasitism (Brodmann *et al.*, 1997). In fact, spatial density dependence of parasitism of winter moth is rare (Embree, 1965; Hassell, 1968; Roland, 1990a; Horgan, 1993; Pearsall and Walde, 1994). Spatial variation in parasitism may reflect more strongly the spatial distribution of host-plant species which vary in their attractiveness to the dominant parasitoid *C. albicans* (Hassell, 1968; Roland, 1986b; Horgan, 1993).

Ground beetles, on the other hand, move and forage at the scale of a few hectares (Den Boer, Chapter 3 in this volume), which is the same scale at which the sample trees are located both at Wytham Wood (M. P. Hassell, personal communication) and at Mount Tolmie on Vancouver Island (Roland, 1986a). If the scale at which predators search is the same as that over which there is large variation in prey density and to which predators aggregate, then detecting the effects of aggregation in response to prey abundance (e.g. Heads and Lawton, 1983) is more probable. The difference in scale at which parasitoids and predators of winter moth search, combined with the limited dispersal of the host, may explain why density dependence in this system is common for predation, but rare for parasitism.

13.5.1 Importance of the density-dependent structure of winter moth populations

Two patterns of winter moth dynamics suggest alternative and additional importance of density-dependent predation of winter moth: (i) synchronous cyclic outbreaks over large geographic areas; and (ii) the dramatic success of biological control in North America. The indirect impact of density-dependent structure on each of these may be greater than its direct effect of simply adding more mortality at high moth density.

13.5.2 Cyclic dynamics

The impact of pervasive and strong density-dependent mortality is to impose an underlying density-dependent structure (Royama, 1992) on winter moth populations. The fact that the primary cause of density-dependent mortality is generalist predators means that predator species may differ between Britain (Frank, 1967; East, 1974; Kowalski, 1976, 1977), Vancouver Island (Roland, 1986a), Nova Scotia (Pearsall, 1992) and other sites, but they produce a similar density-dependent structure in all populations. As proposed by Moran (1953b), and more recently championed by Royama (1992), similar patterns of density-dependent structure can result in geographically synchronous patterns of cyclicity if there are minor perturbations which are correlated over large geographic areas. The most probable perturbation to operate at a scale as large as western Europe is weather. One way in which weather is known to affect all winter moth populations is through its effect on timing of bud burst. For winter moth populations, therefore, the 'Moran effect' of producing synchronous cyclic dynamics across widely scattered populations incorporates the importance of synchrony of hatch with bud burst and density-dependent pupal mortality. Both synchrony with bud burst and density-dependent predation on pupae are very important elements of every winter moth population studied. This mechanism would explain synchrony of outbreak and collapse among European populations (Fig. 13.1), which would reasonably have similar weather, and the lack of synchrony between these populations and those in Nova Scotia (Fig 13.1), where weather would be poorly correlated with that in Europe.

13.5.3 Biological control

Successful biological control of winter moth in North America was achieved by the introduction of parasitoids from Europe to both the east (Embree, 1966) and west (Embree and Otvos, 1984) coasts of Canada. The dramatic success of these introductions has been attributed both to the impact of the parasitoids in adding mortality (Embree, 1966, 1991), and to the impact of generalist predators providing a strong density-dependent mortality (Horgan, 1993; Pearsall and Walde, 1994; Roland, 1994; Roland and Embree, 1995). Again, the widespread and strongly density-dependent predation of winter moth pupae can act to hold populations at a low level once they are suppressed by the addition of mortality from introduced parasitoids. The rate of population change, $R_t[\log(N_{t+1}/N_t)]$, of winter moth in British Columbia is strongly density dependent at low (post-collapse) density (Fig. 13.5), caused by strongly density-dependent pupal predation (Roland, 1994). The point at which density is regulated ($R_t=0.0$) in this population is little affected by variation in parasitism among sites (Roland, 1994). Reduction in parasitism because of poor synchrony of host and parasite development, or because of parasite mortality from hyperparasitoids, for example (Humble, 1985), has little effect on dynamics or on equilibrium density (Roland, 1994). The interaction between strongly density-dependent predation and weakly density-dependent parasitism would, therefore, explain the uniform pattern of biological control of winter moth, despite variation in parasitism between sites (Embree, 1966; Roland, 1994).

13.6 SYNTHESIS

Is there a general pattern of winter moth population dynamics? Or are the dynamics of each population substantially different from each other? Where there are sufficiently long records, densities exhibit periodic rise and fall. Each year, population level is partly determined by the synchrony of bud burst and egg hatch which probably acts as a large-scale

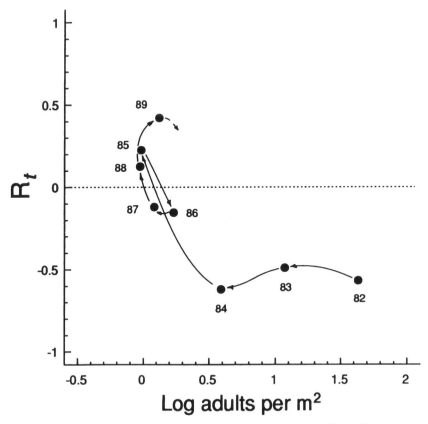

Fig. 13.5 Population change, R_t, from one year to the next ($\log[N_{t+1}/N_t]$) versus log density at Mount Tolmie Park, British Columbia (from Roland, 1994).

perturbation on winter moth populations. These perturbations, acting on populations all of which have a strong density-dependent structure, produce cyclic, synchronized outbreaks across large geographic scales. The presence of this density-dependent predation by generalist predators in all winter moth populations has contributed to its successful biological control; introduced parasitoids reduce moth density to the level where pupal predators can regulate the population. Wherever winter moth populations have been studied, they are characterized by being rife with density-dependent processes, both direct and lagged. Winter moth has been the subject of early, long-term population studies which have fostered many recent studies in both Europe and in North America. Studies of winter moth continue to be at the forefront of arguments in support of density dependence as a key element in insect population dynamics, just as they were 30 years ago.

14

Population ecology of a gall-inducing sawfly, *Euura lasiolepis*, and relatives

Peter W. Price, Timothy P. Craig and Mark D. Hunter

14.1. INTRODUCTION

Thirty years ago the Royal Entomological Society convened its fourth symposium, this time on insect abundance. The published volume (Southwood, 1968) included papers by many of the major figures of the day, providing a valuable historical record of how population studies were approached and what kinds of conclusions were reached. We see that variation in food quality for insect herbivores was hardly studied, and leaders of the day regarded "regulation by food shortage as relatively uncommon" (Wilson, 1968, p. 154). There was also a general absence of comparative approaches and experimental studies in long-term research on population dynamics, a heavy reliance on correlational analysis, and a shortage of hypothesis development and testing. If 'theories' were discussed, they were generally and actually hypotheses developed from studies on mammals. Emphasis was on mortality factors, particularly carnivores as top-down influences on herbivore populations. To what extent have our approaches and conclusions changed over the past 30 years?

Our view is that we still need stronger theoretical developments, more experimental studies, and an emphasis on comparative studies. We also need a more balanced approach to evaluating trophic interactions, both bottom-up, from plants to herbivores; and top-down, from carnivores to herbivores; and lateral effects of competition, facilitation, and other effects. We need more hypothesis testing, although there is no shortage

Insect Populations, Edited by J. P. Dempster and I. F. G. McLean. Published in 1998 by Kluwer Academic Publishers, Dordrecht. ISBN 0 412 83260 7.

of hypotheses, probably the majority these days generated by ecologists working on insects (c.f. Price, 1997). Our own research program over the past 20 years has attempted to contribute these elements to further the field of population biology and dynamics.

14.2. THEORY AND HYPOTHESES

In the development of theory, we take it as axiomatic that theory must be founded on empirical facts and that it must be mechanistic in its formulation, as in the theory of evolution, or the theory of plate tectonics. Our definition of scientific theory is the empirically and factually based mechanistic explanation of patterns in nature. Thus, empiricists such as those who study populations in the field have a basic role to play in the development of theory. After all, Charles Darwin was an empiricist first and developed his theory on empirical facts. As a result, there can be no rift between so-called theoreticians and empiricists.

In fact, we have advocated a Darwinian methodology for the study of insect populations and the development of theory (Price, 1991a, 1996). Collect empirical data, discover broad patterns in nature, and develop an understanding of the mechanisms driving the pattern. Then broaden the comparative approach to extend the theory as widely as possible. The broadest kinds of patterns, and the simplest to detect, are probably phylogenetically based. Thus, a comparative macroevolutionary study of population dynamics within a phylogenetically related group of species provides an obvious conceptual theme with which to construct broad theory. A strict phylogeny of a group used in comparative studies, advocated by Harvey and Pagel (1991), although desirable, is not mandatory as long as the group of species being compared clearly has a common ancestor. Then we have argued that common phylogenetic constraints set much of the ecology and behaviour of the species; an adaptive syndrome of behaviours, life-history attributes, and physiology compensate for these constraints, and these evolved characters set the evolutionary basis for population dynamics, which are emergent properties (Price *et al.*, 1990). This scenario we call the phylogenetic constraints hypothesis (Price, 1994). After considering one sawfly species, we will employ this hypothesis to broaden in a comparative manner the thesis developed on population dynamics, considering relatives of the focal species.

Mechanistic studies toward theory should encompass both proximate and ultimate factors: factors involved with physiological mechanisms such as control of phenology, choice of food, or death through plant resistance; and factors explaining the selective forces that shape the time of events or the advantages of a certain behaviour (Mayr, 1961; Orians, 1962). Knowing that carnivores are important regulators of a herbivore population provides a proximate mechanism, but why a species ultimately remains vulnerable to predation after a long evolutionary history needs another kind of answer.

We also advocate a balanced approach in the development of theory and hypotheses, such that equal opportunity is available for discovering bottom-up effects, top-down effects, and lateral effects (Hunter and Price, 1992). Then their relative importance can be evaluated (e.g. Harrison and Cappuccino, 1995). Equal emphasis on natality and mortality may well change perceptions on population regulation based on life-table analysis, for life tables actually portray a schedule of death much better than birth (Price *et al.*, 1990). Such 'death tables' overemphasize the roles of predators, parasitoids and other mortality factors, because they have generally failed to measure directly the process of birth, female ovipositional behaviour, and the role of variation in plant quality in a female's decisions.

If theory and hypothesis are to develop realistically, surely the basic ingredient must be food supply, which influences 100% of individuals in every population in every generation. Once this is understood in detail, then probably weaker effects can be evaluated up the trophic system. Knowing an insect's perception of food supply, food quality variation, and the carrying capacity of the environment set by food, is fundamental to a mechanistic explanation of population distribution, abundance and dynamics.

Indeed, there is a rapidly growing literature showing that food supply, in terms of quantity, quality and abiotic effects on food for herbivores and other organisms, is probably responsible for limitation of populations well below levels at which the availability of plant tissue would be reduced by defoliation and death. Some examples are provided in Table 14.1. Research on the gall-inducing sawfly, reported in this chapter, contributes to an understanding of how food supply and its perception by insect herbivores results in limitation from below in the trophic system.

14.3. SAWFLY TROPHIC-LEVEL ECOLOGY

To achieve a mechanistic understanding of population trends over many years, the details of interactions up and down the trophic system need investigation involving field studies and experiments. For the focal sawfly species these are largely reported in the literature, so coverage here is brief. Experimental studies on the focal species are listed in Table 14.2.

Euura lasiolepis is a gall-inducing sawfly (Hymenoptera: Tenthredinidae: Nematinae), causing stem galls on its only host-plant species, arroyo willow (*Salix lasiolepis*, Salicaceae), in the western USA. Our research group has studied this and related species in Flagstaff, Arizona and elsewhere since 1979. Briefly, the life cycle involves emergence of adults and mating in the spring. Then the female selects a shoot and injects a fluid that induces gall development (Price and Craig, 1984). Depending on shoot

Table 14.1 Examples showing limitation of insect herbivore populations from the plant-food resources and the interaction of plant and insect. Exogenous factors relate to physical variables and endogenous factors are those involving the plant–herbivore interaction. Only six examples are provided on each type of factor although many others are documented

Insect species	Plant species	Driving variable	Reference
Exogenous factors			
Lygaeus equestris seed-feeding bug	*Vincetoxicum hirundinaria*	Weather on food supply	Solbreck, 1995
Choristoneura occidentalis western spruce budworm	*Pseudotsuga menziesii*	Precipitation	Swetnam and Lynch, 1993; Swetnam et al., 1997
Hyphantria cunea fall webworm	Hardwood trees	Temperature	Morris, 1969; summary in Price, 1997
Schistocerca gregaria & others desert locust	Grasses, herbs and crops	Rainfall	Rainey, 1982; Showler, 1995
Dociostaurus maroccanus Moroccan locust	Grasses and herbs	Weather on food supply, and open ground	Dempster, 1957
Dendroctonus etc. bark beetles	Coniferous trees	Tree stress	Berryman, 1982
Endogenous factors			
Epilachna niponica herbivorous lady beetle	*Circium kagamontanum*	Resource supply and ovipositional restraints	Ohgushi, 1995
Eurosta solidaginis goldenrod gall fly	*Solidago altissima*	Plant resistance	Cappuccino, 1992
Flower-head herbivores 10 species	*Centaurea, Arctium*	Food supply	Dempster et al., 1995
Pristiphora erichsonii larch sawfly	*Larix laricina*	Herbivory reduces availability of oviposition sites	Jardon et al., 1994; Tailleux and Cloutier, 1993
Choristoneura pinus Jack pine budworm	*Pinus banksiana*	Food supply reduced by defoliation	Nealis and Lomic, 1994
Tortrix viridana green tortrix	*Quercus robur*	Food supply	Hunter et al., 1997

quality, the female may or may not also lay an egg (Preszler and Price, 1988). Larvae hatch from eggs in about 14 days, and feed in the parenchymatous tissue of the gall, tunnelling until much of the undifferentiated tissue is consumed. Larvae form cocoons in the galls, usually in November, and pupate in the early spring. Adults emerge in early May and females oviposit in rapidly growing shoots soon after shoot growth has been initiated.

The saw-like ovipositor, an ancestral trait in the sawflies, and the gall-inducing habit can be considered as phylogenetic constraints which are ancestral in the group. The ovipositor is adapted for placing eggs into plant tissues, and such tissues are usually in an early stage of development. This lessens wear on the 'saw' and results in placing the egg in rapidly developing, nutritious tissues with high water content. Such ovipositor structure and oviposition behaviour predispose a lineage to

Table 14.2 Experimental studies on the sawfly, *Euura lasiolepis*, and its host plant, *Salix lasiolepis*

Type	Result	Reference
Bottom-up effects		
Water treatments on plants	High water: higher gall densities and higher survival	Price and Clancy, 1986a
Water treatments on plants	Preference–performance linkage	Craig *et al.*, 1989
Water treatments on plants	High water: more galls, more eggs in galls and higher survival	Preszler and Price, 1988; Price 1990
Fertilizer and water treatments	Shoot growth best predictor of sawfly preference and performance	Waring and Price, 1988
Clonal phenotypic effects of interspecific competition	All species increase as shoot length increases	Fritz and Price, 1990; Fritz *et al.*, 1986
Oviposition stimulant	Phenolic glucoside, tremulacin, only effective stimulant	Roininen *et al.*, 1988
Oviposition deterrent	Oviposition scars become deterrent to subsequent females	Craig *et al.*, 1988
Pruning to simulate browsing herbivory	Shoot length and gall density increase	Hjältén and Price, 1996 1997
Preference–performance	Strongest linkage recorded to date in literature	Craig *et al.*, 1989
Population perturbation	Rapid decline to background levels	Price *et al.*, 1995
Lateral effects		
Competition for oviposition sites	Oviposition scars become repellent to subsequent females	Craig *et al.*, 1988, 1990a
Facilitation	Gall initiation facilities subsequent shoot growth	Craig *et al.*, 1990a
Dispersal	Females highly philopatric to natal site	Stein *et al.*, 1994
Top-down effects		
Parasitoid exclusion	No change in host density in three generations	Woodman, 1990; Price 1990
Window of vulnerability to parasitoid	Narrow window imposed by gall toughening	Craig *et al.*, 1990b
Gall size and parasitoid attack	Larger galls reduce probability of attack	Price and Clancy, 1986b

egg-laying accidents, followed by strong selection, which eventually result in gall formation on rapidly developing plant modules early in the growing season. Ovipositor structure and gall formation constrain life-cycle evolution and behaviour, and the kinds of modules utilized as sites for oviposition.

14.3.1 Strong ovipositional preference and its linkage with larval performance

A strong ovipositional preference exists for the rare, longest shoots in a population of shoot modules (Fig. 14.1, Craig *et al.*, 1986, 1989; Preszler and Price, 1988). This preference is clearly related to high larval performance in terms of survival on the longest shoots (Fig. 14.2, Preszler and

Price, 1988; Craig *et al.*, 1989). Such a strong preference–performance link-age had not been documented in the literature before, although the pref-erence–performance hypothesis that 'mother knows best' has been tested many times (Thompson, 1988; Courtney and Kibota, 1990; Price, 1997). The most vigorous shoots, which become the longest, grow as young ramets from ground level, and as ramets age, shoot vigour and length decline (Craig *et al.*, 1986). This strong relationship between plant growth rate and ovipositional preference motivated development of the plant vigour hypothesis (Price, 1991b), which proposes that many herbivores prefer vigorous plant modules over modules on stressed plants, contrast-ing with the plant-stress hypothesis (White, 1969, 1974).

The notable feature of this relationship, both in the natural field situa-tion and in experiments, is that differences in density of galls and larvae are determined by the female's behaviour in relation to plant module variation and quality. When water supply to plants is relatively high, after heavy winter precipitation in the field, or with high-water treat-ments in experiments, shoot modules grow long. Females prefer these modules for gall initiation, and lay eggs in most of the stems. The result is that in a cohort of 100 eggs in searching females in wet sites, 72.5% are

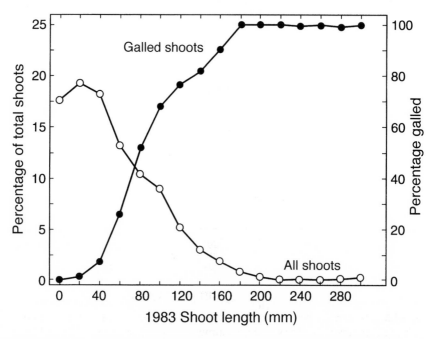

Fig. 14.1 The availability of shoots on 10 willow clones given as the percentage of all shoots in each 20-mm shoot-length category, and the percentage of galled shoots in each category. (Reproduced with permission from Craig *et al.* (1986), courtesy of the Ecological Society of America.)

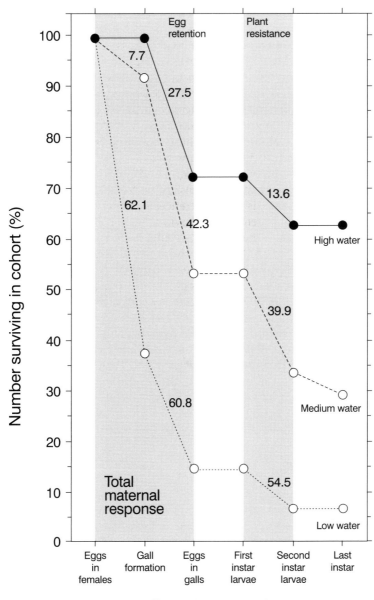

Fig. 14.2 Survivorship curves for three cohorts of sawflies on experimental plants with high-, medium- and low-water treatments. The water treatments result in plants with many long shoots, medium-length shoots and generally short shoots, respectively. Stages in the generation are spaced equally, while in reality oviposition to second instar covers a period of about 40 days. Based on data of Preszler and Price, 1988.

placed in galls (Fig. 14.2). In contrast to these female decisions, in dry sites or low-water treatments, females actually withhold eggs when exposed to short shoot modules, and only 14.9% of eggs in females are deposited in galls. Most of the differences in gall and larval density on high-, medium- and low-water treatment plants is accounted for by female behaviour and natality. Percentage losses of the cohort in each treatment resulting from female decisions in response to host-plant quality were, respectively, high water, 27.5%; medium water, 46.7%; and low water, 85.1%. After oviposition, mortality resulting from host-plant resistance accounted for, respectively, high water, 9.9%; medium water, 22.3%; and low water, 8.4%. However, as a percentage of the cohort surviving to the first instar, there was an increasing loss as water stress increased: high water, 13.6%; medium water, 43.2%; and low water, 58.4% (Preszler and Price, 1988).

Clearly, the ultimate mechanism selecting for female behavioural decisions is larval survival. Mother does know best! Hence, as precipitation declines in any winter, water supply to host plants is reduced, and shoot module size declines, we can anticipate an immediate decline in population in the subsequent new generation $(t+1)$ because females will withhold eggs. Then, there will be a delayed decline $(t+2)$ in population because of low survival in the new generation. This $t+1$ and $t+2$ decline has been documented (Price and Clancy, 1986a). The reverse effects will result from higher-than-normal precipitation. Population density from year to year is driven more by plant-quality effects on natality than on mortality, although plant effects remain strong on larval survival. With such strong bottom-up influences, we must conclude that the possible role of top-down factors cannot be very strong under these conditions.

14.3.2 Proximate factors

Two proximate factors are involved with a mechanistic explanation for the strong preference–performance linkage between oviposition and larval survival. How do females select long-shoot modules, and why do larvae die in the shorter shoots? The oviposition stimulant for *Euura lasiolepis* is the phenolic glucoside, tremulacin (Roininen *et al.*, 1998), and tremulacin concentrations increase in a linear fashion with shoot length (Price *et al.*, 1989). Tremulacin shows the strongest increase in concentration with shoot length of all the six phenolic glucosides present in this willow species. Females walk over young stems and leaves antennating the surface, and eventually probe with the ovipositor, thereby presumably gaining information on phenolic glucoside levels. Evidently, the higher levels in the longer shoots act as a strong oviposition stimulant for sawflies.

Two factors contribute to high larval mortality in the shorter shoot-length categories. One is the natural abscission of shorter shoots, reaching 80% in the shortest categories, but as little as 10% in the longer shoot-length classes. When abscission occurred in the fall, galls dropped to the ground and usually became covered in snow, but no sawflies emerged from galls on these abscised shoots (Craig *et al.*, 1989). Thus abscission of shoots acts as a strong selective factor on female preference for longer shoots. Larvae also die very soon after hatching in the gall as first-instar larvae, with increasing probability as shoot length declines (Fig. 14.2). In experiments in which gall protein and phenolic glucoside levels were manipulated with water and fertilizer treatments, the best predictor of survival was shoot length, not nutrient or chemical defence (Waring and Price, 1988). Therefore, we suspected water balance in the gall as a proximate factor. We erected the osmotic potential hypothesis that reduced water supply to plants results in shorter shoots with higher osmotic potential in galls than on well-watered plants, resulting in an osmotic imbalance and larval death in the shorter shoots. Recent unpublished results are consistent with this hypothesis, but the actual mechanism causing death remains unknown. Hence, this important plant-resistance factor requires more study.

14.3.3 Competition and facilitation

Competition among females for oviposition sites occurred at all densities in natural populations and in experimental settings (Craig *et al.*, 1990a). Females lay eggs with their saw-like ovipositor at a leaf node, through the very young petiole of the leaf and into the stem. A scar is left at the base of the petiole and the exposed phenolics oxidize, presumably to quinones; these become repellent to females about 2 h after the scar is formed (Craig *et al.*, 1988). The scar on the plant acts as if it is a territorial defence against subsequent ovipositions. Because long shoots on poor-quality clones are very rare, competition for the few shoots is inevitable and females either select lower-quality shoots, or withhold eggs, or disperse from the site.

However, gall initiation also stimulates shoot growth distal to the gall, facilitating attack by subsequent females a few days later (Craig *et al.*, 1990a). In an experimental test, previously attacked willow plants received 1.67 times more ovipositions than unattacked plants.

The interaction between competition for oviposition sites and facilitation evidently resulted in even spacing of galls on shoots, and little or no exploitation competition among larvae (Craig *et al.*, 1989, 1990a). At high gall densities in the field, where up to nine may occur on a single shoot, there was no significant decline in survival per gall over the first six galls on a shoot, remaining at about 50% survival. Then survival dropped to about 30% at the ninth position.

The frequency of more than six galls per shoot is very low in natural populations. When sawfly populations were relatively low, as in 1989, only one shoot in 15 000 shoots sampled had six galls present, and no higher number of galls per shoot was found. When populations were relatively high in 1987, 14 shoots in 15 000 had more than six galls (10 with seven galls, four with eight galls), indicating a very low probability of larval competition that results in larval death. We have not quantified sublethal effects of competition on larvae.

14.3.4 Carnivores

Top-down effects in the trophic system are generally weak. On vigorous plants and shoots, galls grow relatively large and the largest gall-diameter classes provide a refugium from small pteromalid parasitoids (Price and Clancy, 1986b; Price, 1988). Thus, where sawfly populations are largest on vigorous plants, attack by small parasitoids is lowest, and a negative density-dependent relationship results. For larger ichneumonid parasitoids with larger ovipositors, galls toughen rapidly and create a narrow window of opportunity for attack per gall (Craig *et al.*, 1990b). Mortality ranged from 2–10% in 4 years of rearings of several hundred galls per year. In general, mortality caused by parasitoids was unrelated to sawfly density in space and time, or inversely density-dependent (Hunter and Price, 1998).

Larval predators include mountain chickadees and grasshoppers, the latter eating gall tissue and so killing sawfly larvae. Both show sporadic attack, and chickadees concentrate attacks only in some years in the highest densities, in the highest-quality sites which are most resilient to mortality factors. These predators do not account for any of the general pattern of population change (Hunter and Price, 1998).

Overall, carnivores are opportunistic in their attacks, and have virtually no influence on the patterns of population change in time or in space.

14.4. PHYLOGENETIC CONSTRAINTS AND THE ADAPTIVE SYNDROME

The phylogenetic constraints imposed by a saw-like ovipositor and the gall-inducing habit are ameliorated by a set of adaptations that maximize persistence in an environment heterogeneous in space and variable in time. This suite of adaptations, the adaptive syndrome, acts as a "coordinated set of characteristics associated with an adaptation or adaptations of overriding importance, e.g. the manner of resource utilization" (Eckhardt, 1979, p. 130). The preference–performance linkage is of major importance, with the associated proximate adaptations enabling detection of long shoots and avoidance of larval death. Spacing of eggs through repellence of oviposition scars improves economy in egg utilization and reduces larval competition. In addition, the sawflies have a haplo-diploid sex determination system. Females lay eggs singly; they can determine the sex of each progeny according to shoot quality, allocating females disproportionately to higher-quality plant modules (Craig *et al.*, 1992).

These evolved characteristics in the phylogenetic constraints and adaptive syndrome, involving life history, behaviour, chemical ecology and sex allocation, largely define the manner in which populations respond to a variable environment. Therefore, we call the population dynamics emergent properties (Price *et al.*, 1990; Price, 1994). They are set by the evolved responses of the sawfly species to variable plant resources. In this context we can examine the way in which populations change in time and space.

14.5. POPULATION DYNAMICS

The trophic-level ecology of the sawfly has established an overarching influence of the host plant and its variation in time and space. This bottom-up scenario can be summarized in a flow diagram of effects, starting with high winter precipitation and its influences up the trophic system (Fig. 14.3). The same diagram could be used to contrast spatial variation among populations in relatively wet sites, such as permanent springs, compared to those in drier sites along temporary drainages. Likewise, the diagram could effectively contrast the population dynamics on young, rapidly growing willow clones or ramets, or those damaged by fire or heavy browsing, relative to the population dynamics on older, less vigorous clones or ramets. Each of these interactions has been examined and established as fact (e.g. Hunter and Price, 1998). Hence we have an empirically based, factual, mechanistic explanation of population increase. Of course, reduced precipitation or water availability has the opposite effects, resulting in population decline.

Our 16-year record of population change in relatively dry sites along a temporary stream, and a 14-year record for wetter sites, such as by a permanent spring, show that populations are relatively stable, varying within two orders of magnitude (Fig. 14.4). Trends are based on three willow clones in the rare wet sites, and 12 clones in the common drier sites. This magnitude of variation is low relative to outbreak or eruptive species which commonly show three to five orders of magnitude change in density (Price *et al.*, 1990). Therefore, we have suggested the term latent population dynamics for these sawflies, as opposed to eruptive population dynamics for outbreak species.

Clearly, water in the form of winter precipitation is the driving variable in population change. Water provides for the needs of the willows during rapid growth in June, the driest month of the year, when sawfly attack is ending, and galls are expanding and eggs are hatching. We have used weather records for Flagstaff to calculate total precipitation in the

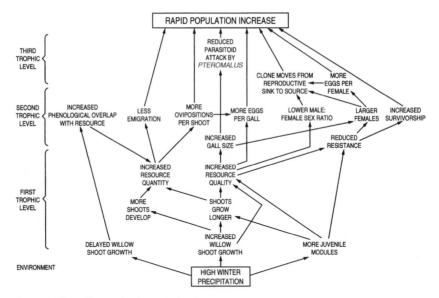

Fig. 14.3 The effects of relatively high winter precipitation on the three trophic levels of willow plants, sawflies and parasitoids.

form of rain and snow for the months October–May. Precipitation during this period influences willow growth, which in turn influences female oviposition behaviour and larval survival in the subsequent generation.

Periodicity in precipitation accounts for the majority of population change. For example, in wet sites precipitation at time t accounts for 52% of the variance in sawfly density in the next generation ($t+1$). (Combined $t-1$ and $t-2$ effects on population size in year t have no additional explanatory power in wet sites.) In dry sites, precipitation in t accounts for 43% of the variation in sawfly density in $t+1$ and 38% of the variance in $t+2$ (Fig. 14.5) (Hunter and Price, 1998). The combined effects on generation t of precipitation in $t-1$ and $t-2$ explain 69.4% of the variance in gall density among years in dry sites. Obviously, this is an overriding influence on population dynamics. The lagged effect of precipitation on sawfly density is illustrated clearly in Fig. 14.6. The periodic fluctuation of precipitation results in periodicity in sawfly populations.

Precipitation is a density-independent factor, and yet its periodicity provides alternating positive and negative feedback on population change through its action on willow module production (Fig. 14.3). While time lags identified in time-series analyses are commonly interpreted to be driven by delayed density-dependent factors such as food, competitors or natural enemies (e.g. Turchin, 1990; Berryman, 1994), it is clear that weather can drive cycles without density dependence, as Williams and Leibhold (1995, 1997) have argued.

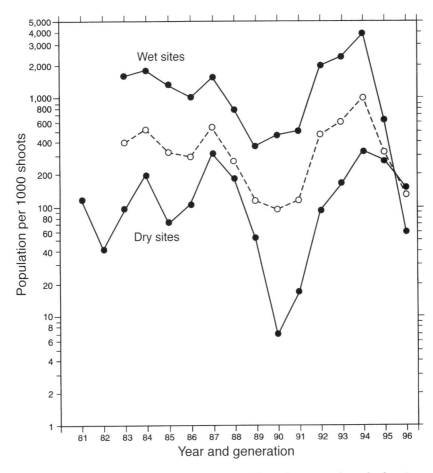

Fig. 14.4 Population trends for 16 years on 12 willow clones in relatively dry sites, and for 14 years on clones in relatively wet sites. The mean population for all clones is shown by open circles and a dashed line.

The ardent density-dependent regulationist will no doubt wish to invoke the role of natural enemies or competition as essential features in this system. We understand the role of carnivores in a mechanistic way much better than in most systems, and we can reject as a hypothesis any important role in population dynamics. Top-down impact on sawfly population dynamics is minor or negligible. Competition is more difficult to discount entirely, because we have established that important interactions among females exist, in section 14.3.3 on competition and facilitation. Knowing that competition occurs at all densities in natural populations and that gall induction facilitates acceptance of a shoot by subsequent females suggests a relatively complex interaction which is yet

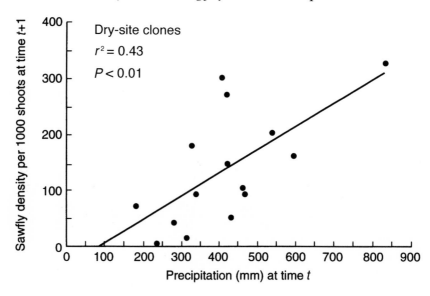

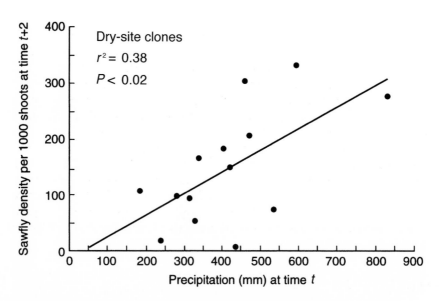

Fig. 14.5 Effects of winter precipitation (October–May) at time t on sawfly populations at $t+1$ (above) and $t+2$ (below) on 12 clones in the relatively dry sites. (Reproduced with permission from Hunter and Price (1998), courtesy of the Royal Entomological Society and Blackwell Science.)

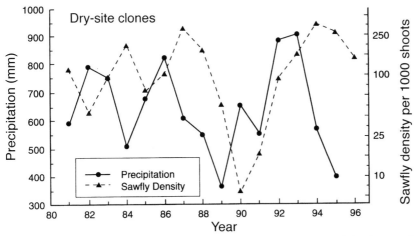

Fig. 14.6 Periodic fluctuations in precipitation are followed with a delay by sawfly populations on the 12 clones in the relatively dry sites. (Reproduced with permission from Hunter and Price (1998), courtesy of the Royal Entomological Society and Blackwell Science.)

to be resolved in terms of population dynamics. We can offer the following three points. (i) The net result of competition and facilitation appears to be density-dependent. (ii) We observe local extinction in individual clones after a series of dry winters, which could not be driven by contest competition. In dry sites in 1990, when populations reached their nadir (Fig. 14.4), three clones supported no sawfly individuals, whereas 3 years before and after they supported more than 100 sawflies in two cases and 10 or more in the third clone. (iii) If competition plays a role at all in the population dynamics of this sawfly, it must be to limit populations in the rare site and year when they are very high. But the effect is inevitably minor relative to the strong impact of resource supply. The evidence yields a view that lateral effects in the trophic system through competition are clearly a weak force in population dynamics, but they cannot be discounted completely.

14.6. RELATED SPECIES

Once we understood the basic pattern of preference–performance linkage in relation to module quality for *Euura lasiolepis*, we started comparative studies on phylogenetically close species in the sawfly gall-inducing group. The genera included *Euura*, with species that form galls in stems, buds, petioles, and midribs of willows; *Phyllocolpa*, with species that form leaf-edge galls by folding the leaf; and *Pontania*, with species initiating galls on the leaf lamina usually associated with a major or minor leaf vein. The adaptive radiation of this group has been discussed by Price and Roininen (1993). We have studied species inducing each type of gall in the genus *Euura*: stem gallers; and a bud, petiole, and midrib galler. We have also studied one *Phyllocolpa* species and two *Pontania* species. In each case we have measured the preference–performance linkage in relation to module length and/or ramet age, and we have evaluated the role of natural enemies. Studies on population dynamics of species have been limited to *E. lasiolepis*, as reported in this chapter, and *Euura amerinae* (Roininen *et al.*, 1993, 1996).

Including *E. lasiolepis*, we have studied 10 species in detail (Table 14.3). In nine cases we found a strong preference–performance linkage in relation to module quality, and a weak impact of natural enemies. In the tenth case, the performance linkage was significant but weak. Carnivore attack does not help to explain the patterns of oviposition and survivorship in relation to shoot length. The patterns are remarkably similar among these species. However, the preference–performance linkage remains strong at all densities in species of *Euura* and *Phyllocolpa*, but in some species of *Pontania*, such as *P.* nr *pacifica*, the relationship may be lost at high densities (c.f. Clancy *et al.*, 1993; Price *et al.*, 1994; Stein and Price, 1995). Five additional species in the genera *Euura* and *Phyllocolpa* show a positive relationship between shoot length and probability of attack (Price and Roininen, 1993), and *Euura mucronata* shows the same pattern on each of the four host species studied. The comparative ecology of this distinctive phylogenetic group of gall-forming sawflies shows a very strong and consistent general pattern.

Remarkably, related free-feeding sawflies in the genus *Nematus* show the same preference–performance linkage in two of the three species studied by Carr (1995), and all three species show a strong preference for longer, rarer shoots. This was an unexpected discovery, suggesting that the phylogenetic constraint of a saw-like ovipositor may predate the evolution of the gall-forming group. However, patterns in oviposition and larval performance in relation to plant module quality need more scrutiny involving many more species.

14.7 A COMPARATIVE MACROEVOLUTIONARY PERSPECTIVE ON POPULATION DYNAMICS

The phylogenetic constraints hypothesis, which invokes a flow of influences from ancestral evolved phylogenetic constraints to the resulting adaptive syndrome and then to the emergent properties of the population dynamics, argues in essence for an evolutionary theatre and an eco-

Table 14.3 Ten studies showing plant-quality effects and preference–performance linkage in gall-inducing sawflies

Species	Gall Type	Location	Reference
Euura sp. 1	Midrib	Lees Ferry, Arizona	Woods *et al.*, 1996
Euura sp. 2	Petiole	Flagstaff, Arizona	Stein and Price, 1995
Euura amerinae	Stem	Joensuu, Finland	Roininen *et al.*, 1993
Euura atra	Stem	Joensuu, Finland	Price *et al.*, 1997a
Euura exiguae	Stem	Weber River, Utah	Price, 1989
Euura lasiolepis	Stem	Flagstaff, Arizona	Craig *et al.*, 1989
Euura mucronata	Bud	Joensuu, Finland	Price *et al.*,1987a,b
Phyllocolpa sp.	Leaf edge	Sapporo, Japan	Price and Ohgushi, 1995
Pontania nr *pacifica*	Leaf lamina	Flagstaff, Arizona	Stein and Price, 1995
Pontania sp.	Leaf lamina	Flagstaff, Arizona	Price, unpublished data

logical play. This is quite the reverse of Hutchinson's (1965) influential view of "the ecological theater and the evolutionary play". But Hutchinson emphasized microevolutionary adjustments related to the co-existence of species within communities. Our perspective is macroevolutionary in scope; a comparative approach across different species and genera.

It may seem axiomatic to argue that the evolutionary history of species will play a large role in current ecological events, and yet a strong macroevolutionary and comparative approach has developed relatively recently in the study of population dynamics in general. Perhaps we must credit Southwood (1975, 1977; Southwood and Comins, 1976) for his pioneering comparative work using a gradient of *r–K*-selected species. Wallner (1987) used a similar gradient for comparing characteristics of rare and outbreak species. We have emphasized a comparative evolutionary approach since 1990 (e.g. Price *et al.*, 1990, 1995; Price, 1994), and Hunter (1991, 1995a, b) has moved in the direction of using strict phylogenetic analysis for studying population dynamics and the evolution of flightlessness in the Macrolepidoptera.

Certainly, broad comparative studies provide a much stronger basis for the development of theory on insect population dynamics than do studies of species in isolation. Empirically based comparative studies have yielded at least well-supported hypotheses, or even factually based theory, on such topics as the evolution of flightlessness in forest Lepidoptera (Barbosa *et al.*, 1989; Hunter 1995a), and the population dynamics of bark beetles (Berryman, 1982), and an attempt has been made to compare latent and eruptive species over a broad range of taxa based on evolutionary traits involving life history, ovipositional behaviour, and the presence or absence of a preference–performance linkage (Price, 1997). Comparative studies on related taxa have certainly provided greater scope for generalization than the use of certain ecological traits such as shared host-plant species (Hunter *et al.*, 1997; Hunter, 1998).

A phylogenetic approach to comparative population dynamics not only aids in the detection of pattern in the absence of confounding variables (Harvey and Pagel, 1991), but it recognizes that every adaptation has evolved within the context of the evolutionary history of the lineage (Gould and Lewontin, 1979). Ultimately, then, a comparative macroevolutionary approach to insect population dynamics could encompass all species on a gradient from rare and/or latent, to common and/or eruptive. We believe that this should be our long-term goal.

In the meantime, we have restricted our comparative studies in this chapter to one group of sawflies, but they may be representative of several hundred species in the Holarctic realm. A strict phylogenetic analysis using free-feeding nematine sawflies as the ancestors of the gall-forming species would provide a sound basis for a theory on the population dynamics of the group. More studies will be required on the population dynamics of individual species, the preference–performance linkage, and the patch dynamics and modular display of host-plant resources through time and space (e.g. Price *et al.*, 1997b). Nevertheless, we argue that we have developed a sound basis for an empirically grounded mechanistic explanation for patterns in nature relating to sawfly and host-plant interactions, and population dynamics.

ACKNOWLEDGEMENTS

We thank the editors of this book for their invitation to make a contribution. This research has been supported since 1980 by the US National Science Foundation; the current grants are DEB-9318188 to P.W.P. and DEB-9527522 to M.D.H.

15

The population ecology of
Trichochermes walkeri

Ian F. G. McLean

15.1 INTRODUCTION

The choice of insect species for intensive, long-term population studies has often been determined by their ecological, economic or conservation significance rather than their tractability for study. Practical considerations include ease of counting individuals at different life stages, of determining mortality factors, and of assessing what influences reproductive success, and the importance of local or large-scale movement on their abundance in time and space. Gall-forming insects have a number of advantages in terms of their suitability for population studies, and it was primarily ease of study which resulted in the choice of *Trichochermes walkeri* (Psylloidea: Triozidae) at the inception of this project in 1982. Prior to this choice being made, it was the serendipitous discovery of the then undescribed *Leucopis* (Diptera, Chamaemyiidae) predator of *T. walkeri* which first revealed the potential for unravelling the population ecology of its psyllid prey.

T. walkeri is a gall-forming psyllid confined to *Rhamnus catharticus* (Rhamnaceae) in Britain (McLean, 1993), with a widespread distribution within the range of its host plant. It has an annual life cycle (Fig. 15.1). The larva of *Leucopis psyllidiphaga* is a specialist predator which attacks only *T. walkeri*, and it has a number of adaptations which enable it to feed on the nymphs concealed inside *T. walkeri* leaf-roll galls. These adaptations include a remarkably extended time for larval growth compared with most *Leucopis* species, which typically kill (and at the same time consume) their homopteran prey during a larval development time of about

Insect Populations, Edited by J. P. Dempster and I. F. G. McLean. Published in 1998 by Kluwer Academic Publishers, Dordrecht. ISBN 0 412 83260 7.

one-third the length of that for *L. psyllidiphaga*, 8–10 weeks (Tanasijtshuk, 1986; I.F.G. McLean, unpublished data). The larvae of *L. psyllidiphaga* behave as external parasites while feeding on the first four instars of *T. walkeri* nymphs, rather than killing and consuming their prey immediately, as is typical for other predatory *Leucopis*. *L. psyllidiphaga* kill and consume *T. walkeri* nymphs only when they reach the final (fifth) nymphal instar immediately before the galls open.

R. *catharticus* is a deciduous bush, or small tree, which occurs in a number of habitats including fen peat, calcareous grassland and scrub, hedges and broadleaved woodlands. It is a frequent species in East Anglia in the UK, where this study has been undertaken, and has a distinctive growth form illustrated in Fig. 15.2, where the nomenclature is given for identifying the alternative egg-laying sites for *T. walkeri*. Because *R. catharticus* is a relatively long-lived species (with a lifespan of several decades), it represents a stable and predictable food resource for *T. walkeri*, thereby contributing to the temporal stability of the populations reported here. *R. catharticus* supports a range of insect herbivores, including a number of Lepidoptera which chew or mine the leaves, and there are various generalist predators (including Heteroptera, Coleoptera and Araneae) which forage over the bark and foliage. As well as the vertical interactions between the host plant, herbivore and predator (which are the main theme of this chapter), there are significant lateral interactions between *T. walkeri* galls and other sap-sucking Homoptera (McLean, 1994b). The galls represent a high-quality food source for sap suckers to exploit, which is available for most of the time when leaves are present. Those free-living sap suckers which feed preferentially on galled leaves or on the galls themselves can develop more rapidly than on ungalled leaves (McLean, 1994b). Whether there is any metabolic or other cost to the *T. walkeri* nymphs from this feeding has yet to be investigated.

The aim of this chapter is to test alternative population theories using findings from population surveillance and field experiments. From an evolutionary perspective, the close relationships which have developed between *T. walkeri* and its sole host plant *R. catharticus*, and *T. walkeri* and its specific predator *L. psyllidiphaga*, suggest that this is a tri-trophic system where complex interrelationships between these species may have developed over long periods of time. The emergence and nature of these relationships is shaped by the spatial stability of *R. catharticus*, combined with the effects of unpredictable weather factors (from month to month and between years) which affect timing and duration of life stages, as well as survival and reproduction of these insects.

The questions addressed in this chapter are:

how important are density-dependent processes in the population dynamics of *T. walkeri*?

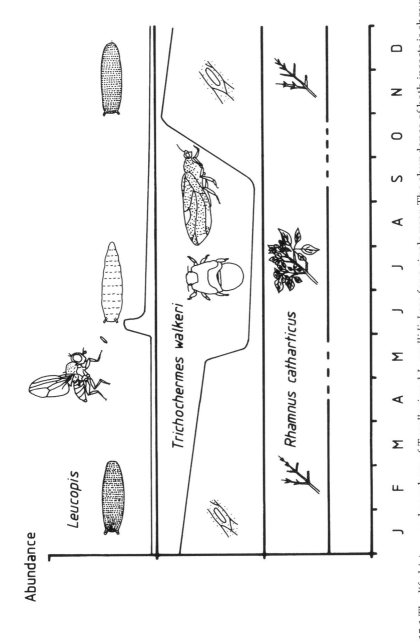

Fig. 15.1 The life history and numbers of *T. walkeri* and *L. psyllidiphaga* for a single year. The abundance of both insects is shown schematically as mean density on the same arithmetic scale.

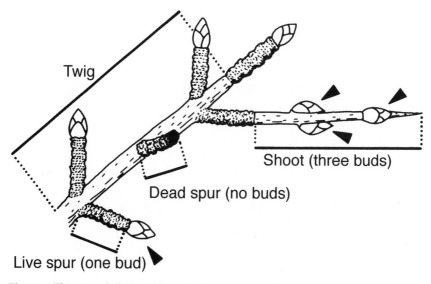

Fig. 15.2 The growth form of *R. catharticus* showing the nomenclature and position of the different locations where *T. walkeri* females lay their eggs.

are *T. walkeri* populations regulated about an equilibrium density, or is
   their increase limited by competition or other factors?
what is the relative importance of top-down and bottom-up processes for
   determining population density for the herbivore and its predator, and
   its changes from generation to generation?

The first and third questions can be tackled with some confidence on the
basis of the available evidence; the second question is more problematic
and can be answered only partially at present.

The regulation of populations by density-dependent processes, resulting in the return of numbers towards an equilibrium density level after
perturbation events, is a theory which many ecologists accept as accounting for the observed behaviour of most populations. The role of density-
independent factors in causing substantial population changes within
and between generations for many species is also widely accepted.
Whether populations can be stabilized by a combination of density-independent factors, as postulated by Den Boer (1968) in the 'spreading of
risk' concept, is both disputed and hard to test. The interaction between
density-dependent and density-independent factors, and their relative
importance in determining population density and the size and direction
of population changes, are still debated frequently, and will be considered briefly in the discussion of the population dynamics of this psyllid
herbivore and its dipteran predator.

15.2 METHODS

15.2.1 Site descriptions

The four sites where *T. walkeri* and *L. psyllidiphaga* were studied are in the western part of East Anglia, and are separated by distances ranging from 12–63 km. They have different numbers of *R. catharticus* bushes and hence different population structures for insects confined to this host plant. The sites also have contrasting soils and vegetation communities. Chippenham Fen National Nature Reserve, Cambridgeshire (grid reference TL 647694) has many bushes of *R. catharticus* scattered throughout the site, of which four neighbouring examples in the south-east corner of compartment 5 have been used for counts and/or experiments. Two bushes have had annual counts of *T. walkeri* eggs, followed by counts of galls and their occupants. The bushes are on fen peat, which is not generally waterlogged in winter, next to mixed fen vegetation. Foulden Common, West Norfolk (TF 7600) also has many *R. catharticus* bushes, two of which have had annual counts of *T. walkeri* galls and their occupants from marked branches (one bush was substituted by a replacement in summer 1987 after the original had been cut down during the winter of 1986–87). The bushes are on a mixture of chalky boulder clay next to calcareous grassland and damp pingo hollows. At Chippenham and Foulden, *T. walkeri* and *L. psyllidiphaga* have populations where the adult insects have the potential to move freely between adjacent bushes. A single *R. catharticus* bush studied at Bromholme Lane, Brampton, Huntingdonshire (TL 225709) has two near-by bushes at 3 and 5 m away, and a bush at 30 m and another at 290 m distance. It is situated on alluvial clays in a green lane and is subject to winter flooding. At Cavenham Heath National Nature Reserve, West Suffolk (TL 757728), two *R. catharticus* bushes 40 m apart have had annual counts of *T. walkeri* galls and their occupants, with no other bushes recorded so far within 200 m. The bushes are on dry sandy soil amongst sparse heathy grassland. At the last two sites, the low number of *R. catharticus* bushes reduces the potential for extensive movement of insects between bushes, and hence it is possible that the bushes studied there have populations of *T. walkeri* and *L. psyllidiphaga* which are largely confined to the individual bushes studied, in contrast to the populations at Chippenham Fen and Foulden Common, where extensive movement between bushes is possible. Weather data were abstracted from annual summaries by Clarke (1986) and subsequent annual reports for Swaffham Prior, which is about 9 km from Chippenham Fen National Nature Reserve.

15.2.2 *T. walkeri* egg counts

The eggs of *T. walkeri* are small (0.38 mm long by 0.14 mm wide) and are usually laid in small crevices in the bark of *R. catharticus*, which means that they cannot be counted accurately in the field. For both counts of eggs laid by *T. walkeri* on two study bushes at Chippenham Fen (to measure changes in egg numbers from year to year) and on a bush at Brampton (to investigate the phenology of egg laying), as well as for experiments confining females in sleeves to record their fecundity, branches were cut from the bushes at the end of the egg-laying season and the eggs counted on live and dead spurs, shoots and twigs, using a binocular microscope. For the counts of natural populations, samples were taken in December, and in early Spring of the following year, to assess over-wintering egg loss. A minimum combined length of 3.7 m of spurs, shoots and twigs constituted a sample from each bush.

15.2.3 Measuring the abundance and phenology of *T. walkeri* and *L. psyllidiphaga*

The number of galls has been recorded on marked branches for each of the study bushes, with galled leaves removed for measurement and dissection. *T. walkeri* nymphs and *L. psyllidiphaga* eggs and larvae were counted, and any failed or damaged galls noted.

From 1985 onwards, six water traps (diameter 205 mm) under bush 1 at Chippenham Fen have been used to measure the phenology of leaf fall and premature abscission of galled leaves. They have also been used to record numbers of leaves with and without galls for each sampling period (of 1–3 weeks) and the number of *L. psyllidiphaga* larvae falling to the ground when the galls open.

From Spring 1987, six emergence traps (diameter 205 mm) under bush 1 at Chippenham Fen have recorded the number of *L. psyllidiphaga* adults emerging, thus enabling the over-wintering mortality of *L. psyllidiphaga* puparia to be estimated.

15.2.4 *T. walkeri* emergence from galls removed before opening

To record the emergence of *T. walkeri* from galls removed before opening, two samples of 12 and 14 galls were removed from a bush at Bromholme Lane, Brampton on 8 and 12 August, 1994, respectively, and then placed on grass cuttings overlying soil in a flower pot. The pot was placed under an *R. catharticus* bush in a garden at Brampton, and records made twice daily of any emerging *T. walkeri*.

15.2.5 *T. walkeri* fecundity

To measure the fecundity of individual female *T. walkeri*, or of females crowded to different degrees, adult females were placed in terylene sleeves (1 m long) placed over *R. catharticus* branches at Chippenham

Fen, from which insects and spiders had been previously removed by hand. Individual female *T. walkeri* were confined with two male *T. walkeri*; crowded *T. walkeri* females were confined with approximately equal numbers of males. Female *T. walkeri* were obtained from galls with known numbers of occupants by enclosing individual mature galls in small terylene sleeves (about 100×80 mm) during July, and removing the insects within 2 days of their emergence.

15.2.6 *L. psyllidiphaga* egg laying

To restrict *L. psyllidiphaga* egg laying on a sample of *T. walkeri* galls to a period of a week, adult *L. psyllidiphaga* were excluded from galls on single branches by terylene sleeves. These were removed from different branches for a week at a time to allow *L. psyllidiphaga* to lay eggs for a single week before the exclusion sleeve was replaced.

15.3 RESULTS

Results are presented here in turn for *T. walkeri* and *L. psyllidiphaga*, and are given in a broadly chronological sequence, following the life history of both species.

15.3.1 Population dynamics of *T. walkeri*

The proportion of eggs estimated to give rise to *T. walkeri* nymphs in galls varies from year to year (Table 15.1). It is thought that this variation is mainly the result of the degree of synchronization between *T. walkeri* egg hatch and bud burst of *R. catharticus*, with the optimal timing for egg hatch coinciding with the presence of rapidly expanding leaf tissue for gall initiation. In 1988, when numbers of *T. walkeri* galls per 100 leaves were universally low at the study sites (Fig. 15.3), bud burst of *R. catharticus* at Chippenham Fen was late relative to bud burst for other plants, and this may have increased the mortality of first-instar nymphs, with only 41% of eggs estimated to have given rise to nymphs in galls (Table 15.1). These synchronization effects are analogous to those discussed below for the synchronization of *L. psyllidiphaga* egg laying with *T. walkeri* gall development.

The degree of synchronization between egg hatch and bud burst probably combines with the effects of these weather factors which influence the number of tiny first-instar nymphs surviving their potentially hazardous journey – from where the eggs hatch on bark, to the nearest suitable leaf margin where they initiate their galls. First-instar nymphs are likely to be very vulnerable to being dislodged by wind and rain during their search for a suitable expanding leaf, where they can begin to feed and initiate the development of a protective gall.

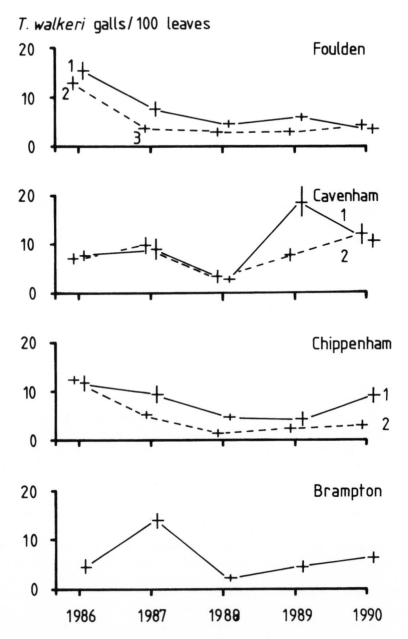

Fig. 15.3 The density of *T. walkeri* as galls per 100 leaves at four sites from 1986–90 (vertical bars±1 s.e.). Foulden: bush 1 (1986–90), solid line; bush 2 (1986) and bush 3 (1987–90), dashed line. Cavenham: bush 1 (1986–90), solid line; bush 2 (1986–90), dashed line. Chippenham: bush 1 (1986–90), solid line; bush 2 (1986–90), dashed line. Brampton: bush B (1986–90), solid line.

Table 15.1 Proportion of *T. walkeri* eggs giving rise to nymphs in galls estimated for Bush 1 at Chippenham Fen, 1985–90

Year	Eggs→nymphs (%)
1985	84
1986	72
1987	61
1988	41
1989	56
1990	67

Number of *T. walkeri* eggs estimated from twig samples, number of *T. walkeri* nymphs estimated from gall counts and dissecting gall samples.

The suggestion that weather factors are important in determining the proportion of nymphs which establish galls is supported by a further line of evidence, in that the numbers of *T. walkeri* galls at the four study sites show similar annual changes (Fig. 15.3), though with some differences within and between some sites in some years. Given the similarity of weather patterns experienced at all the study sites, it is to be expected that, if weather factors are important in determining the number of *T. walkeri* galls, there will be synchronous and similar population fluctuations at the four sites. Such synchronous changes are known to occur for other insects where weather factors are thought to be important in causing annual changes in abundance, a well-studied example being the synchronous year-to-year fluctuations displayed by many adult British butterflies over Great Britain (Pollard and Yates, 1993).

After gall establishment in May, the leaf-margin galls (formed by the edge of the leaf turning upwards and inwards to enclose the sap-sucking nymphs) grow rapidly through June and then more slowly through July and August until the gall opens (McLean, 1993). *T. walkeri* in galls with three nymphs grow more rapidly than in galls with two nymphs, which in turn grow more rapidly than those living singly (I.F.G. McLean, unpublished observations). Galls with multiple occupants open before those with single nymphs, with the first galls opening in late July in early years and continuing to open until early or later in September, according to the prevailing weather. The proportion of galls with multiple occupants increases at higher population densities, so that the average number of nymphs per gall increases as the number of galls per hundred leaves rises. It is likely that the dispersed distribution of eggs (with many spurs having only a single egg), and the limited mobility of first-instar nymphs, combined with variation in date of egg hatch, precludes a greater degree of aggregation of *T. walkeri* nymphs in multiple galls.

There is significant spatial variability in the distribution of galls within and between bushes. The number of galls is usually higher on the south side compared with the north side of a bush (Fig. 15.4). The only consistent exception to this has been bush 1 at Cavenham, which is a slightly smaller bush than others sampled, and which is in a dry and hot, south-facing position. This may mean that the microclimate on the north side of bush 1 at Cavenham is more similar to that on the south sides of other bushes. Furthermore, the south side of bush 1 at Cavenham is both hot and likely to be under stress through lack of water in the summer months of most years. There is a strong tendency for the ranking of bushes within a site to remain the same from year to year with respect to the number of *T. walkeri* galls per 100 leaves, but this is not absolutely consistent (Fig. 15.3).

Once *T. walkeri* galls are established, the nymphs are relatively well protected from the direct effects of wind and rain, though there are a number of mortality agents which kill significant numbers of nymphs in total. After gall initiation, which results from feeding on the leaf margin by nymphs, a variable percentage of galls fail at an early stage from unknown causes. Gall failure is not correlated with gall abundance, nor with the number of *L. psyllidiphaga* eggs laid per gall. Possible causes of gall failure include premature or excessive feeding by small *L. psyllidiphaga* larvae; attack by generalist predators before the gall completely encloses the nymph(s); fungal or other microbial pathogens (some failed galls contain a shrivelled first-instar nymph which looks as though it may have been infected by a pathogen); a defensive plant response against gall formation; and wet or windy weather which may dislodge nymphs before they are completely enclosed by the protective gall. It has not yet been possible to discriminate between these (or other) causes of gall failure.

Premature leaf abscission is another cause of mortality for both *T. walkeri* nymphs and *L. psyllidiphaga* larvae. There are two major peaks in leaf fall for *R. catharticus*: the first during the period of flower fall in late June and early July when many small leaves are dropped, and the second in the autumn from late September through into November (after all adult *T. walkeri* have emerged from their galls). For the purposes of analysing the results obtained for premature leaf abscission, it has been assumed that for all galls retrieved from a water trap which contain *T. walkeri* nymphs (with or without an *L. psyllidiphaga* larva) the occupants would have died. However, when the galls are close to opening it is possible that at least some *T. walkeri* and *L. psyllidiphaga* may emerge successfully. A simple experiment undertaken in August 1994 showed that the *T. walkeri* within 9 out of 26 galls (35%), on leaves removed in mid-August, emerged within 6 days. This indicates that at least some galls may open

T. walkeri galls / 100 leaves

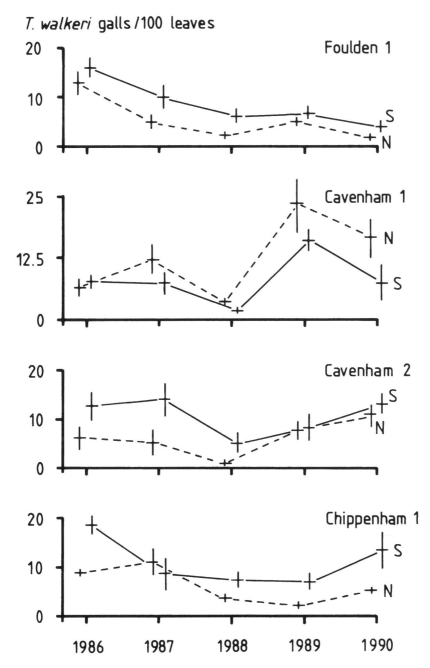

Fig. 15.4 The density of *T. walkeri* galls per 100 leaves on north (N) compared with south (S) sides of bushes at three sites (Foulden bush 1, Cavenham bushes 1 and 2, Chippenham bush 1) from 1986–90 (vertical bars ±1 s.e.).

soon after falling prematurely and thereby allow their occupants to emerge and survive. However, these figures relate to a time when about one-third of the galls had already opened, and most *T. walkeri* nymphs would be in their final instar and hence ready to emerge from their galls. Galls which fall earlier would contain nymphs of younger instars which would not be able to produce adult psyllids. No relationship between the density of *T. walkeri* galls and the proportion of galled leaves falling prematurely has been discovered (Fig. 15.5). This suggests that an induced plant-defence response is not involved, in contrast to the poplar gall aphids investigated by Williams and Whitham (1986), where much higher mortalities due to leaf abscission were reported for the gall-forming insects. It is possible that extreme weather conditions (for instance drought or heavy rain) might lead to an increased rate of premature leaf abscission, but no clear evidence for this is apparent from a preliminary examination of the Chippenham Fen results in relation to rainfall data for nearby Swaffham Prior.

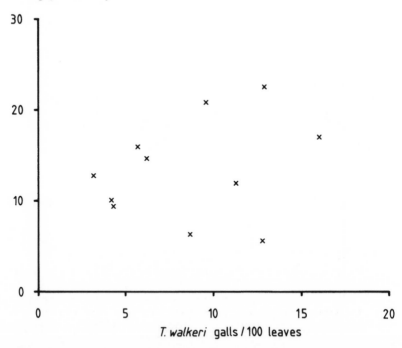

Fig. 15.5 The proportion of *T. walkeri* galls falling prematurely, in relation to the number of galls present, recorded from water traps under bush 1 at Chippenham Fen, 1986–96.

T. walkeri nymphs are subject to predation by *L. psyllidiphaga* larvae, with some evidence for higher rates of predation at greater densities of *T. walkeri* galls (Fig. 15.6). In some years, substantial numbers of *T. walkeri* nymphs are killed. Factors which affect the proportion of *T. walkeri* nymphs killed by *L. psyllidiphaga* are discussed in the section below dealing with the population dynamics of this predator. As discussed by McLean (1994a), the mortality imposed by *L. psyllidiphaga* does not affect the number of *T. walkeri* entering the next generation because of the strong density-dependent response at the next (adult) stage of the life cycle. This is confirmed by results from the Brampton study site where *L. psyllidiphaga* was present in 1986 and 1987 but not in 1988–90, without apparently changing the abundance of the *T. walkeri* population on this bush relative to other sites (Fig. 15.3; McLean, 1994a). Also, the mean number of galls per 100 leaves did not differ between 1986–87 and 1988–90 (*t*-test, *P*=0.3).

Adult *T. walkeri* emerge from galls as fifth-instar nymphs, which typically remain close to the gall while they shed the nymphal skin and then expand and harden the adult integument. Adults have been observed

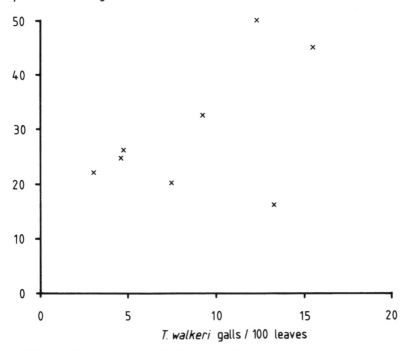

Fig. 15.6 The number of *T. walkeri* killed by *L. psyllidiphaga* at different population densities, from bush 1 at Chippenham Fen, 1983–90.

feeding on the midrib and major veins of buckthorn leaves, and mating has been seen during September. Adults have been found in water traps until November, showing that they can remain alive and active for 3–4 months. When resting on spurs and shoots of *R. catharticus* they closely resemble the leaf buds of their host plant.

Previous analysis has shown that density-dependent processes operate during the period between emergence of *T. walkeri* adults from galls and the conclusion of egg laying in late October (McLean, 1994a). This results in many more eggs per female *T. walkeri* being laid in years of low gall abundance (see McLean, 1994a, pp. 101–2, for results from bush 2 at Chippenham Fen). This variation in estimated fecundity between years, combined with little change in egg density from year to year, suggests that females may prefer to lay their eggs at a limited number of favoured egg-laying sites. Therefore, some experiments and observations have been undertaken to investigate the factors which influence the fecundity and choice of oviposition sites by *T. walkeri* females. Just as emergence from galls is spread over about 6 weeks (McLean, 1993), so egg laying is also a lengthy process, extending over about 8 weeks (Fig. 15.7). During this time *R. catharticus* is steadily shedding leaves, resulting in fewer leaves being available on each spur or shoot for female *T. walkeri* to feed on. As there is likely to be some significant mortality of adult females during the period of oviposition, the relatively steady rate of egg laying until towards the end of October suggests that a few females may be increasing their rate of egg laying during October, these individuals thereby contributing a high proportion of the eggs from which the next generation will be formed. The relatively long duration of oviposition may also contribute towards the fairly stable numbers of eggs observed from year to year (McLean, 1994a, Figure 5) because females do not have a narrow window of opportunity to mature eggs and find suitable egg-laying sites. Instead, they have a longer time to achieve their egg-laying potential, though the causes of adult mortality (whether due to generalist natural enemies including Heteroptera, Coleoptera and Araneae, or due to adverse weather such as strong winds, heavy rain or frosts), which may affect the number of eggs laid per female, have not yet been investigated.

An experiment in summer 1996 investigated whether maternal history during nymphal development could influence the number of eggs laid. The results obtained showed that there was no significant difference in the number of eggs laid per female when comparing females originating from galls with one nymph per gall with females originating from galls with two nymphs per gall (Table 15.2). Clark (1963) also found that crowding nymphs of the test-forming psyllid *Cardiaspina albitextura*, which feeds on *Eucalyptus blakelyi*, had little effect on the fecundity of the resulting females. The 1996 experiment also gives an estimate of the mean fecundity of uncrowded *T. walkeri* (186.6±11.3 eggs per female) as

T. walkeri eggs / 100mm

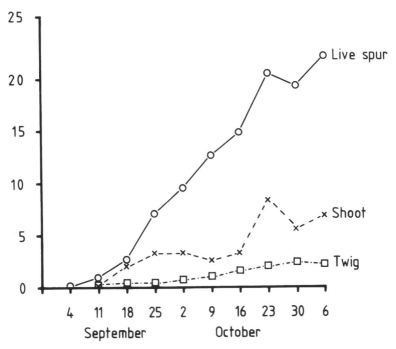

Fig. 15.7 Number of *T. walkeri* eggs from Bromholme Lane, Brampton, September–November, 1990.

well as information about the individual egg-laying preferences of females kept in isolation. Both groups of adult *T. walkeri* were remarkably similar in terms of the distribution of eggs between different egg-laying sites (Table 15.3).

In order to investigate the effects of crowding adults on female fecundity, different numbers of *T. walkeri* were placed in sleeves and the resulting eggs counted and their positions recorded. The results indicate that crowding reduces average female fecundity (Table 15.4). Watmough (1968) also found that increased crowding of adult broom psyllids reduced their fecundity, as did Clark (1963) for *Cardiaspina albitextura*. The reduction in fecundity due to crowding of *T. walkeri* may be the result of competitive interactions between females at favoured egg-laying sites near to buds on live spurs and shoots. It is also possible that when males and females are crowded together, male–female interactions change (for example, males attempting to mate with females more frequently could interfere with oviposition behaviour). The behaviour and reproductive condition of *T. walkeri* under different levels of crowding require investi-

Table 15.2 Total fecundity of female *T. walkeri* originating from galls with one nymph per gall, compared with females originating from galls with two nymphs per gall, recorded from sleeved females on a bush at Chippenham Fen in 1996

Number of eggs laid by females from galls with one nymph	Number of eggs laid by females from galls with two nymphs
110	273
113	119
201	213
194	127
174	279
112	142
262	229
183	158
175	253
174	185
200	230
Mean[a] 172.5±13.9 (±S.E.)	Mean[a] 200.7±17.5 (±S.E.)

[a] Two-sample *t*-tests for means.
$t = -1.263$, not significant at 5% level.

Table 15.3 Distribution of eggs laid by *T. walkeri* females under different conditions at Chippenham Fen in 1996

Conditions	Total eggs	Live spurs (%)	Dead spurs (%)	Shoots (%)	Twigs (%)
'Wild' population (bush 1)	171	69.0	0.0	19.3	11.7
11 Females ex single galls	1898	84.8	1.9	9.8	3.5
11 Females ex double galls	2208	83.0	2.1	10.1	4.8
27 Females per sleeve	781	90.4	0.3	7.8	1.5
138 Females per sleeve	7209	29.8	4.4	21.5	44.3

gation to discover the specific mechanisms responsible for reducing fecundity at high population densities.

In addition to the differences in total fecundity recorded in the crowding experiment, the most crowded females laid their eggs at different sites, compared with less crowded females or with 'wild' females sampled from Chippenham Fen in the same year (Table 15.3). However, sleeved female *T. walkeri* retain a preference for laying on spurs, at adult densities equivalent to those observed in the field (up to 27 females per sleeve in this experi-

Table 15.4 Number of eggs laid by female *T. walkeri* under different degrees of crowding (sleeved on bushes at Chippenham Fen in 1996)

Crowding level	Eggs laid per female (±s.e.)
One per sleeve (*n*=22)	186.6±11.3
27 per sleeve (*n*=1)	28.9
138 per sleeve (*n*=1)	52.2

ment). The most crowded females laid many more eggs on twigs, where they are substantially further from live buds on spurs or shoots, compared with the other experimental females or wild females. After nymphs hatch from eggs laid on twigs, they will have further to crawl to reach leaf buds, where the first-instar nymphs must find expanding leaves for gall initiation. The changed distribution of eggs may have been the result of competition for a limited number of favoured oviposition sites (on live spurs and shoots), where in these crowded conditions (138 females per sleeve) the density of eggs was much higher than has been observed in the field. It is also possible that the similar egg density observed each year, on two bushes at Chippenham Fen, is the result of females ovipositing until a limited number of favoured egg-laying sites (on spurs and shoots) have eggs present. Once favoured sites have been utilized in this way, females may disperse (rather than laying eggs in less-favourable sites, as happened in the high-density sleeves where females were unable to leave).

The pattern of oviposition responses observed in *T. walkeri* is consistent with intraspecific competition reducing average female fecundity. While this competition may be responsible for the density-dependent response observed at this point in the life cycle, the role of generalist natural enemies, which may reduce numbers of adult *T. walkeri* at high population densities, has not yet been investigated. Thus, although there is some evidence for crowding and competition for oviposition sites acting on *T. walkeri* in a density-dependent manner as a bottom-up process, top-down effects cannot be dismissed as unimportant without further work.

15.3.2 Population dynamics of *L. psyllidiphaga*

L. psyllidiphaga females lay their eggs near the ends of *T. walkeri* galls, typically tucked in the fold between the margin of the gall and the upper surface of the leaf. The eggs are distributed more evenly between galls than would be expected by chance, with fewer galls lacking eggs and more galls with a single egg (Table 15.5). The number of *L. psyllidiphaga* eggs per 100 *R. catharticus* leaves depends upon the number of *T. walkeri* galls per 100 leaves (Fig. 15.8), indicating that predator numbers are determined by a bottom-up process at this stage of the life cycle.

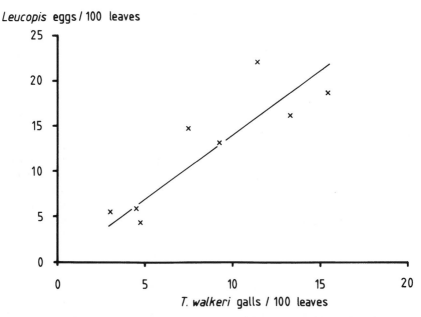

Fig. 15.8 *L. psyllidiphaga* eggs per 100 leaves in relation to number of *T. walkeri* galls per 100 leaves, from bush 1 at Chippenham Fen, 1983–90 (r^2=0.76, P=0.005).

The percentage of *L. psyllidiphaga* larvae successfully entering *T. walkeri* galls differs considerably from year to year (Fig. 15.9). These differences are not correlated with the abundance of *T. walkeri* or with the number of *L. psyllidiphaga* eggs laid. A field experiment undertaken in 1995 to investigate the entry success of *L. psyllidiphaga* larvae into galls showed that larvae hatching from eggs laid early were more likely to enter galls than larvae hatching from eggs laid late (Table 15.6), though there was considerable variation between the samples for each date class. Nevertheless, the results indicate that the degree of synchrony between *L. psyllidiphaga* oviposition and *T. walkeri* gall development is a major factor determining the numbers of *T. walkeri* nymphs killed by this specialist predator. The mechanism underlying the higher rate of entry success of larvae from eggs laid early is the stage of gall development. Small galls appear to be softer and their ends are less tightly rolled than larger, older galls. Therefore, older galls are likely to offer a greater physical barrier to the entry of *L. psyllidiphaga* larvae than young galls. Conversely, it is possible that there are disadvantages to female *L. psyllidiphaga* laying their eggs too early on very small galls. Very small *T. walkeri* nymphs may not be able to withstand feeding by *L. psyllidiphaga* larvae, leading to the premature death of nymphs and consequent gall failure.

Only one *L. psyllidiphaga* larva is found per gall, although a substantial proportion of galls have more than one egg laid on them (Table 15.5).

Table 15.5 Distribution of *L. psyllidiphaga* eggs between galls sampled from two bushes at Chippenham Fen in 1986

Galls with eggs of L. psyllidiphaga	Observed eggs	Poisson expected	Deviation (%)	
Bush 1				
0	16	31	−15	(−6.8)
1	81	61	+20	(+9.1)
2	58	59	−1	(−0.5)
3	38	39	−1	(−0.5)
4	15	19	−4	(−1.8)
5	9	7	+2	(+0.9)
6	2	2	0	(0)
7	0	1	−1	(−0.5)
$X^2_{(218)} = 6757$ ($P<0.001$)				
Bush 2				
0	12	19	−7	(−3.8)
1	56	42	+14	(+7.7)
2	53	48	+5	(+2.7)
3	23	37	−14	(−7.7)
4	17	21	−4	(−2.2)
5	13	10	+3	(+1.6)
6	3	4	−1	(−0.5)
7	4	1	+3	(+1.6)
$X^2_{(180)} = 82.66$ ($P<0.001$)				

Presumably the first larva to enter a gall kills those larvae which gain access subsequently. This cannibalism is a classic example of contest competition, which ensures that an individual *L. psyllidiphaga* larva grows to the maximum size possible rather than sharing the limited resources available with other larvae. As the majority of *T. walkeri* galls contain a single nymph, this is a good strategy for a *L. psyllidiphaga* larva because only a small proportion of galls could potentially support more than one larva, and one *T. walkeri* nymph would be insufficient for more than one larva to reach maturity.

Both *T. walkeri* nymphs and *L. psyllidiphaga* larvae suffer mortality from galls falling prematurely before their occupant(s) have emerged. This mortality is discussed above for *T. walkeri*; all that can be added here is that there is no evidence from the analysis of water trap results that galls occupied by *L. psyllidiphaga* are any more or less likely to fall prematurely than those without the enclosed predator larvae.

The number of *L. psyllidiphaga* larvae per 100 *R. catharticus* leaves (i.e. their absolute abundance rather than their abundance relative to numbers of their prey) is determined to a large extent by the number of *T.*

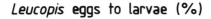

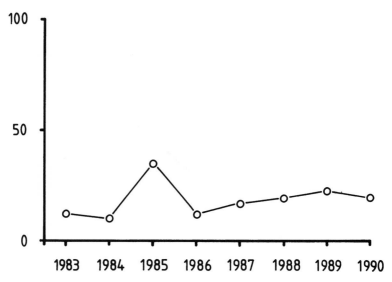

Fig. 15.9 Entry success of *L. psyllidiphaga* from bush 2 at Chippenham Fen, 1983–90.

walkeri galls per 100 leaves (Fig. 15.10). Thus, their abundance is essentially determined by prey numbers (a bottom-up process).

The difference between the number of *L. psyllidiphaga* larvae falling into water traps and the number of adult *L. psyllidiphaga* which emerge the following spring into emergence traps under bush 1 at Chippenham Fen constitutes the total mortality from August through to May in the following year. In most years there is about 13% survival of *L. psyllidiphaga* for this period, but in some years with low numbers of larvae at the start of autumn, a greater proportion survive to become adults in the follow-

Table 15.6 Entry success of *L. psyllidiphaga* larvae during successive periods throughout duration of oviposition by *L. psyllidiphaga* at Chippenham Fen in 1995

Oviposition period	L. psyllidiphaga eggs	L. psyllidiphaga larvae	Percentage eggs to larvae
Before 4 June	73	30	41 (27–73)
4–11 June	33	9	27 (0–40)
11–18 June	49	17	35 (25–50)
After 18 June	171	25	15 (0–22)

Each date class the mean of four replicates, two from bush 1 and two from bush 2, the range of values given in parentheses.

Leucopis larvae / 100 leaves

Fig. 15.10 Number of *L. psyllidiphaga* larvae per 100 leaves in relation to number of *T. walkeri* galls per 100 leaves, from bush 1 at Chippenham Fen, 1983–90 ($r^2=0.68$, $P=0.012$).

ing spring (Fig. 15.11). Thus survival of puparia over winter is more variable (and can be higher) at low population levels, with a greater potential for recovery from low numbers in at least some years. The causes of this pattern of over-winter survival have yet to be investigated, but it is possible that generalist predators (such as beetle larvae and adults) may feed on *L. psyllidiphaga* puparia, and that these predators are better able to locate and feed on puparia when this food source is more abundant.

15.4 DISCUSSION

The effects of *T. walkeri* and *L. psyllidiphaga* on each other's population dynamics are contrasting and may be summarized as follows. Although *L. psyllidiphaga* can kill a substantial proportion of *T. walkeri* nymphs in at least some years, this does not depress the numbers of *T. walkeri* eggs laid in the following autumn because there is such a strong density-dependent response at this time each year, resulting in similar numbers of eggs being laid each autumn irrespective of the number of *T. walkeri* emerging from galls in late summer (McLean, 1994a). There is some evidence pre-

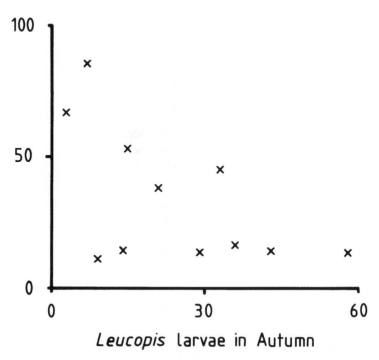

Fig. 15.11 Over-winter mortality for *L. psyllidiphaga*, from bush 1 at Chippenham Fen, 1986–87 to 1996–97.

sented in this chapter that interactions between ovipositing females may be responsible for the relative stability in *T. walkeri* egg numbers from year to year, with competition for favoured oviposition sites. If this is the case then bottom-up processes will be of greater importance than top-down in determining the density of *T. walkeri* eggs entering the next generation. On the other hand, if generalist predators act to reduce the number of *T. walkeri* adults in years when they are more abundant, then top-down processes may prevail. However, the numbers of *L. psyllidiphaga* eggs, larvae and puparia in each generation are greatly dependent upon the number of *T. walkeri* galls in the same summer. The number of adults emerging after winter disappearance is of less significance, so even if different levels of predation of *L. psyllidiphaga* puparia are responsible for the different proportions of puparia which are lost over winter, the effects of these predators do not extend to altering the number of *L. psyllidiphaga* eggs in the next generation. Hence, for determining population size in each generation for this predator, bottom-up processes are more significant than top-down.

Thus, to summarize the responses to the questions posed in 15.1:

density-dependent processes are important in the population dynamics
of *T. walkeri*, particularly via the oviposition responses of adult females
which result in very similar densities of eggs being laid each year;
there is no strong evidence for regulation of *T. walkeri* populations about
an equilibrium density, although further investigation of the possible
effects of generalist natural enemies upon adult numbers is required to
test this possibility more thoroughly;
the balance of available evidence suggests that bottom-up processes are
more important for determining population density in both the herbivore and its predator, though the evidence is stronger for the predator
because, as stated above, the effects of generalist natural enemies upon
T. walkeri have not been quantified.

Which population theory best fits the behaviour of *T. walkeri* and *L. psyllidiphaga* populations? While experimental tests have yet to probe some
aspects of the populations of these two species, the available evidence
indicates that weather factors determine the numbers of *T. walkeri* galls
up to a maximum set by the number of eggs laid the previous autumn.
Hence the number of *L. psyllidiphaga* larvae, which depends mainly upon
the number of *T. walkeri* galls, is also dependent upon those weather factors which influence the success of gall establishment in spring. Thus we
would expect *T. walkeri* mortality, caused by asynchrony between egg
hatch and bud burst, and resulting from adverse weather factors in
spring, to vary unpredictably from year to year.

For both *T. walkeri* and *L. psyllidiphaga*, weather factors are also important because of their influence on the degree of synchronization between
egg hatch and the availability of suitable resources for nymphs and larvae, respectively. In both cases there is strong selection pressure in favour
of those individuals which succeed in timing their appearance with the
availability of high-quality resources during a short window of opportunity. The role of synchronization of an insect's emergence (from a resting
stage to an exploiting stage) in determining both population and evolutionary outcomes, has often been underestimated.

Following the density-independent spring mortality for *T. walkeri*, the
summer is a period when both species show a steady decline in numbers due to interactions between the physiology of *R. catharticus* and
weather factors. Although substantial numbers of *T. walkeri* can be killed
by *L. psyllidiphaga* in some years, this mortality precedes the strong density-dependent response which results in similar egg numbers in most
years irrespective of the number of adults emerging from galls in
August and September.

Previous work has already demonstrated a strong density-dependent
relationship between the number of adult *T. walkeri* emerging from galls

and an index of the number of eggs laid per female (McLean, 1994a). Thus, in answer to the first question posed in section 15.1, density-dependent processes are important in the population dynamics of *T. walkeri*.

T. walkeri has populations which are relatively stable, varying over approximately an order of magnitude during this study. This contrasts with many insect populations, including the two free-living broom psyllids studied by Watmough (1968) and Dempster (1968); the test-forming psyllid *C. albitextura* (Clark and Dallwitz, 1975); and the free-living heather psyllid *Strophingia ericae* (which has a 2-year life cycle) reported by Whittaker (1985), all of which have populations varying more abruptly over a wider range of densities. In part this may be due to the evolution of the gall-forming habit in *T. walkeri*, which may reduce mortality due to adverse weather and generalist natural enemies during the nymphal growth period for this species. However, the complex array of factors which affect numbers of *T. walkeri* at different life stages may also play a part in stabilizing numbers. The interaction between weather and the phenology and physiology of *R. catharticus* sets the range of conditions within which *T. walkeri* and *L. psyllidiphaga* must exploit the narrow range of resources which they have evolved to depend upon. The stable distribution of *R. catharticus* in time and space may also be of significance in stabilizing the numbers of *T. walkeri* and *L. psyllidiphaga*, in comparison with those psyllids associated with shorter-lived shrubs (such as broom and heather) whose distribution, size, and possibly nutritional status can change more quickly.

Even if generalist natural enemies killed substantial numbers of *T. walkeri* adults, it is very probable that female behavioural responses would compensate for this mortality, resulting in more eggs laid per surviving individual female. This is similar to the findings of Dempster (1968) reporting on two broom psyllids, though in this case adult dispersal was also shown to be important at high densities. The interrelationships between numbers of *T. walkeri* killed by natural enemies and the oviposition rate of surviving females are likely to be complex and difficult to unravel, because these processes operate simultaneously and will interact with each other over a range of *T. walkeri* densities. Although more evidence is required for confirmation, it appears that the reduction in female fecundity brought about by crowding of *T. walkeri* adults is mainly responsible for the density-dependent response which results in a similar egg density each year, irrespective of the density of the adult psyllids which emerge from their galls.

T. walkeri populations are apparently most affected by density-independent factors when first-instar nymphs move after egg hatch to establish galls; this leads to changes in numbers of galls from year to year, irrespective of the number of eggs surviving the winter. Thus, gall numbers from year to year vary independently of *T. walkeri* egg density, up to a limit set by the number of eggs laid the previous autumn. The within-gall phase of the life cycle (with variable mortality from gall failure, *L. psyllidiphaga* predation, and premature leaf abscission) is followed by strong density dependence which acts to produce similar numbers of eggs laid in the autumn, whether adults numbers are high or low. This fits a ceiling model if, as seems likely, there is a limited number of favoured egg-laying sites. Adult female behaviour (with respect to both oviposition and dispersal) is likely to interact with mortality agents (generalist natural enemies) in complex ways which may not fit easily with either equilibrium or ceiling models. For *L. psyllidiphaga* the picture may be simpler, in that there is clear evidence for bottom-up processes determining numbers of eggs and larvae, with the synchronization hurdle (timing egg laying with a window in gall development) acting in a density-independent way as the main modifying factor at this stage of the life history. A ceiling model of limited resources fits the observations here, while over-wintering success of puparia may fit an equilibrium model, with better survival, on average, observed at low densities (though with considerable variation at these low densities).

Although this study of *T. walkeri* and *L. psyllidiphaga* has identified a number of factors which are important in determining their numbers at different life stages, two major aspects remain to be investigated.

The first is the movement of adults between bushes, which is difficult to assess directly. The numbers staying on or leaving a bush may be influenced by the population density on this bush, which may result in higher mortality and reduced fecundity for dispersing adults, and which may redistribute adults between bushes. However, there is a little indirect evidence on the role of movement in relation to the population ecology of these species. The four study sites are different with respect to the number of *R. catharticus* bushes (see section 15.2.1), which suggests that individual *T. walkeri* and *L. psyllidiphaga* have more opportunity to move between bushes with different characteristics at Chippenham Fen and Foulden Common where there are many bushes, than at Cavenham Heath and Bromholme Lane, Brampton where there are few bushes. The changes in numbers of *T. walkeri* from year to year have been similar at all four sites, so if there is significant movement between bushes within these sites, the results in terms of influencing the trends and stabilities of these populations do not appear to be very significant. This does not rule out the possibility that longer-distance movement between sites may occur, which might be investigated by recording colonization rates for *R. catharticus* bushes at a range of distances from source populations of *T. walkeri*. The distances which adult *T. walkeri* can travel by flying remain unknown, as only local flight from branch to branch within the same bush has been observed. Alternatively, the duration of flight could be investigated by laboratory trials, though this would still not determine whether longer flights are undertaken under field conditions.

The second major aspect yet to be investigated is the possible effects of generalist predators on adult *T. walkeri* and on puparia and adults of *L. psyllidiphaga*. Predator-exclusion experiments are the most likely means of making an initial assessment of the significance of these predators.

15.5 ACKNOWLEDGEMENTS

The Nature Conservancy Council (since 1991 the Nature Conservancy Council for England) kindly granted permission for my studies of *T. walkeri* at Chippenham Fen and Cavenham Heath National Nature Reserves. I am grateful to Martin Twyman-Musgrave and Malcolm Wright, successive wardens for these reserves, who supported and helped with the project throughout. Comments from Jack Dempster and two referees significantly improved this chapter. My wife Christine, and daughters Ailsa and Iona, assisted with setting up experiments and with collecting data; their support and involvement made this study possible as well as even more enjoyable!

16

Bottom-up population regulation of a herbivorous lady beetle: an evolutionary perspective

Takayuki Ohgushi

16.1 INTRODUCTION

Population regulation is fundamental to most phenomena in ecology (Murdoch, 1994). Since the well-known debate on density-dependent population regulation in the late 1950s (Andrewartha and Birch, 1954; Nicholson, 1954; see Sinclair, 1989 for a historical review), a number of long-term studies on natural insect populations have been carried out to search for the factors that determine population persistence and fluctuations (e.g. Klomp, 1966; Baltensweiler *et al.*, 1977; Dempster, 1982; Southwood *et al.*, 1989). These studies have attempted to quantify the relative importance of density-dependent and density-independent processes. In particular, intensive studies on natural populations of insect herbivores have made a major contribution to our understanding of the causes of fluctuation and regulation of populations, population outbreaks, and host–parasitoid interactions (see papers in Southwood, 1968; Den Boer and Gradwell, 1971; Barbosa and Schultz, 1987; Cappuccino and Price, 1995). While stimulating conclusions and insights have been gained, the underlying mechanisms of insect herbivore population regulation are still poorly understood.

This chapter addresses bottom-up population regulation through resource availability for a lady beetle, *Epilachna niponica*, which is a specialist herbivore of thistle plants. I have chosen this topic because insect population ecologists have paid little attention to variability of host-plant

Insect Populations, Edited by J. P. Dempster and I. F. G. McLean. Published in 1998 by Kluwer Academic Publishers, Dordrecht. ISBN 0 412 83260 7.

characteristics as factors that govern the population dynamics of herbivorous insects. In particular, I will illustrate how changes in the behaviour
and physiology of adult females in response to host-plant conditions are
important in generating population regulation through resource availability. In addition, I will discuss the importance of incorporating evolutionary perspectives into the understanding of the population regulation
of herbivorous insects.

16.2 MECHANISTIC APPROACH TO POPULATION REGULATION

Empirical studies often consider population regulation to be indicated by
a temporal constancy of population size (Connell and Sousa, 1983;
Hanski, 1990). Regulation implies that a negative feedback mechanism
acts upon a population as a result of density-dependent mortality, reproduction or dispersal. In this context, many insect ecologists have long
searched for density-dependent processes as evidence for population
regulation. However, the existence of density-dependent processes does
not, in itself, ensure the existence of regulatory mechanisms.
Demonstrating population regulation requires more than simply providing evidence that a population remains relatively constant through time,
and that temporal or spatial density-dependent processes exist. As population regulation is defined as the tendency for the population to return
towards an equilibrium density following perturbation, perturbation
experiments provide the most crucial test for population regulation
(Murdoch, 1970; Harrison and Cappuccino, 1995).

Although perturbation experiments may demonstrate population regulation, population manipulation will not necessarily reveal the underlying factors generating it. Thus, it is essential to continue to search for
causal mechanisms of population regulation (Gaston and Lawton, 1987;
Murdoch and Walde, 1989; Ohgushi, 1992). The traditional approach to
this has been key-factor analysis (Varley and Gradwell, 1960), which rests
largely on seeking correlations between population densities and factors
that affect survival and reproduction. Correlation studies, however, have
a weakness in that they do not prove causal mechanisms generating the
observed population fluctuations (Royama 1977; Ohgushi, 1992). A more
mechanistic approach to understanding population regulation requires
greater attention to differences between individuals within the population, because important causes of population regulation may be over-
looked when using mean numbers per generation (Hassell, 1986a). To
incorporate differences between individuals into population ecology, we
should recognize how individual attributes of behaviour and physiology
affect demographic parameters through survivorship and reproduction
(Hassell and May, 1985; Lomnicki, 1988; Sutherland, 1996). Having advocated such a mechanistic approach, Schoener (1986) argued that each

demographic parameter at the population level must be translated into behavioural and physiological parameters at the individual level.

16.3 BOTTOM-UP POPULATION REGULATION OF HERBIVOROUS INSECTS

Recent studies of insect populations have emphasized that a more exact understanding of the population dynamics of insect herbivores requires a thorough understanding of the dynamics of their food plants, i.e. bottom-up influences between the two trophic levels (Hawkins, 1992; Hunter *et al.*, 1992; Ohgushi, 1995; Harrison and Cappuccino, 1995). It is increasingly evident that a wide variety of plant characteristics greatly affect survivorship and reproduction of insect herbivores (Ohgushi, 1992). These characteristics are related to plant quality in terms of nitrogen, water and defensive chemicals (Scriber and Slansky, 1981; Haukioja and Neuvonen, 1987; Mattson and Scriber, 1987). In addition, large impacts of their host plants on the survivorship of herbivorous insects are caused by variations in spatial dispersion of their food plants (Root, 1973) and their phenology (Feeny, 1970; Connor *et al.*, 1994). Furthermore, recent discussions of multitrophic interactions have suggested that host-plant quality is of great importance to insect herbivores through both direct and indirect interactions (Price *et al.*, 1980; Faeth, 1987; Ohgushi, 1997).

Population ecologists have paid less attention to the possible effects of plant resources on the dynamics of insect herbivore populations than to the role of natural enemies as the principal agents for population regulation (Hairston *et al.*, 1960; Lawton and Strong, 1981; Hassell, 1985). However, Harrison and Cappuccino (1995) found that evidence for bottom-up regulation by resources appears to be much more common than evidence for top-down regulation by natural enemies. Evaluation of published life-table data on herbivorous insects has shown that host-plant characteristics are often important density-dependent agents for population regulation of insect herbivores, through intraspecific competition, reduced fecundity, and adult dispersal (Dempster, 1983; Stiling, 1988). Several authors have illustrated temporal and/or spatial resource tracking at the population level, thereby highlighting the importance of bottom-up influences of host plants on herbivore population dynamics (Mattson, 1980; Dempster and Pollard, 1981; Ohgushi and Sawada, 1985a, 1997a). Thus it is necessary to recognize the relative contributions of both top-down effects caused by natural enemies and bottom-up effects caused by host-plant dynamics on their survival and reproduction, to understand the population dynamics of herbivorous insects.

Insect tactics in resource use, in response to spatial and temporal variation of plants or plant parts, can play a significant role in determining survival and/or reproductive processes, so we need detailed knowledge of how variability and heterogeneity of resources affect the survivorship and reproduction of individuals (Wiens, 1984). For example, the availability and quality of host plants greatly affects the fecundity of adult herbivorous insects by affecting larval nutrition and also by acting on the physiology and behaviour of the reproductive female (Leather, 1994). Since a wide variety of life-history tactics in resource use have evolved in insect–plant interactions, an individual-based mechanistic approach should provide a better insight into evolutionary perspectives of the population dynamics of herbivorous insects (Price, 1994).

16.4 RESOURCE VARIABILITY AND LIFE-HISTORY TRAITS

Because of variations in the quality and quantity of plant food, insect life-history traits in resource use can play a dominant role in determining survivorship and/or reproduction of herbivorous insects (Hassell and May, 1985; Smith and Sibly, 1985; Leather, 1994). There is growing evidence that resource-use tactics of adult insects are critical in determining the population dynamics of many herbivorous insects (Price *et al.*, 1990; Ohgushi, 1995, see also Dempster, Chapter 4 in this volume). For example, a number of studies testing Root's (1973) 'resource concentration hypothesis' have suggested that the searching behaviour of adult insects for favourable resources is an important determinant of subsequent population densities in different vegetation structures (Kareiva, 1983; Stanton, 1983). In particular, oviposition behaviour has recently been hypothesized to generate the fundamental patterns of population dynamics in insect herbivores, by affecting offspring performance in terms of survivorship and reproduction (Preszler and Price, 1988; Craig *et al.*, 1989; Price *et al.*, 1990; Ohgushi, 1995).

For many herbivorous insects, the searching abilities of larvae are poor compared with those of adults. Adult oviposition behaviour is therefore of paramount importance in selecting suitable host plants or plant parts for their offspring (Renwick and Chew, 1994). Female herbivores whose offspring develop at the oviposition site are strongly favoured by natural selection to optimize their choice of oviposition site. Thus, the relationship between oviposition preference and growth, survival, and reproduction of offspring has been the crux of the evolution of insect–plant associations (Thompson, 1988). The preference–performance linkage has been recently explored in terms of the host-plant selection or site selection on a plant, revealing a positive correlation between oviposition preference and offspring performance (Rausher, 1980; Whitham, 1980; Craig *et al.*, 1989; Roininen and Tahvanainen, 1989). For example, ovipositing females of a willow-galling sawfly *Euura lasiolepis* have a strong oviposition preference for long shoots of young and vigorous willows, associated with a high larval survival (Craig *et al.*, 1989; see also Price *et al.*, Chapter 14 in this volume). Similarly, the aphid *Pemphigus betae* shows a strong oviposition preference for young leaves of *Populus angustifolia* that will grow to be large, and more progeny of larger size are produced on these larger leaves (Whitham, 1980).

However, in some herbivorous insects there is a lack of correlation between preference and offspring performance, with females laying their eggs rapidly, irrespective of quality of oviposition site for their offspring. A poor preference–performance correlation may be brought about by oviposition onto introduced host plants, or by a relative shortage of suitable plants or plant parts (Thompson, 1988). There also may be ecological constraints, such as impacts of natural enemies which are independent of plant quality (Denno *et al.*, 1990), or life-history constraints such as a short time available for oviposition and a poor capacity for directed flight (Larsson and Ekbom, 1995), that may result in a weak preference–performance relationship. A weak correlation may also result from specific oviposition behaviours, so that a female lays her eggs away from the larval feeding site, and frequently long before foliage is available for larval feeding (Price, 1994). Oviposition behaviours of these kinds of insects therefore show poor correlation with larval survival (Karban and Courtney, 1987; Auerbach and Simberloff, 1989; Valladares and Lawton, 1991). It should be noted that evaluation of the relationship between oviposition preference and offspring performance is still useful in understanding the population dynamics in those species without apparent adult preference. If we can remove oviposition preference from our analyses, we can concentrate on resource-use tactics of the immature stage (Schultz, 1983), or on the effects of natural enemies and host-plant characteristics on the survival of immature insects (Auerbach and Simberloff, 1989; Denno *et al.*, 1990).

The traditional life-table approach has long ignored the important consequences of oviposition site selection by adult females on insect population dynamics (Price *et al.*, 1990). To understand the behavioural and physiological mechanisms that determine the selection of host plants or host-plant parts requires an evolutionary approach. Behavioural ecology investigates the evolutionary relationships between fitness and behaviour, and other variables including population density (Krebs and Davies, 1997). Therefore, the study of the behavioural ecology of herbivorous insects should reveal the underlying evolutionary mechanisms that potentially determine their population dynamics (Smith and Sibly, 1985; Ohgushi, 1992; Sutherland, 1996). This approach, which focuses on how life-history tactics and resource availability can result in population regulation, bridges the two disciplines of behavioural ecology and population ecology.

16.5 POPULATION REGULATION OF A HERBIVOROUS LADY BEETLE

Epilachna niponica is a univoltine species and a specialist herbivore of thistle plants. Over a period of 10 years, this species has been censused in several areas in central Japan (Nakamura and Ohgushi, 1981; Ohgushi and Sawada, 1981, 1995; Ohgushi, 1992, 1995). Here I will discuss the behavioural and physiological mechanisms of bottom-up population regulation in *E. niponica*, and its evolutionary implications. The results are from two populations (A and F) in Kutsuki, in the northern part of Shiga prefecture, central Japan, together with those from a population in the Botanical Garden introduced from Asiu (Fig. 16.1). The Botanical Garden of Kyoto University is located in the north-eastern part of Kyoto City, 10 km south of the southern limit of the natural distribution of *E. niponica*.

The study sites A and F are located in different valleys along the River Ado. Site A (60×30 m) is situated at 220 m elevation on an accumulation of sandy deposits resulting from a dam construction in 1968. The surface of the rather flat and open area consists mainly of unhardened, sandy deposits. Floods caused by heavy rainfall often submerge and wash away the ground flora along the watercourse, and most of the surviving vegetation is composed of annual and perennial herbs. Site F (90×15 m) is situated at 350 m elevation, about 10 km upstream from site A. The more hardened soil deposits at this site mean that most grasses and shrubs successfully escape serious flood damage, except during large-scale floods. Vegetation in and around the site includes various deciduous trees such as *Quercus mongolica* and *Quercus salicina*.

In the study area, the lady beetle feeds exclusively on leaves of its host plant, *Cirsium kagamontanum*, which is a perennial herb, patchily distributed along the riverside. It grows rapidly from sprouting in late April to late June, becoming full-sized at 1.5–1.8 m in height by late August, and then flowers over 2 months from mid-August. Old leaves begin to wither after summer. Although the number of thistle leaves gradually increases until late August, leaf quality (in terms of amino acid and water content) consistently declines during the growing season (Ohgushi, 1986).

Over-wintering adult females emerge from hibernacula in the soil in early May and begin to lay eggs in clusters on the under surfaces of thistle leaves. Larvae pass through four instars. New adults emerge from early July to early September, feeding on thistle leaves through the autumn. They enter hibernation in the soil by early November. Seasonal changes in numbers of adults and immature stages are given by Ohgushi and Sawada (1981).

Each population was censused from early May to early November in each year between 1976 and 1980. All thistle plants growing in the study sites were carefully examined; the numbers of eggs, fourth-instar larvae, pupae, pupal exuviae and adult beetles were recorded separately for each plant. Each adult beetle was individually marked with four small dots of lacquer paint on its elytra. Newly marked adults were released immediately on the thistle plant where they had been captured. Sex, body size and subsequent capture history (date and place) were recorded for individual

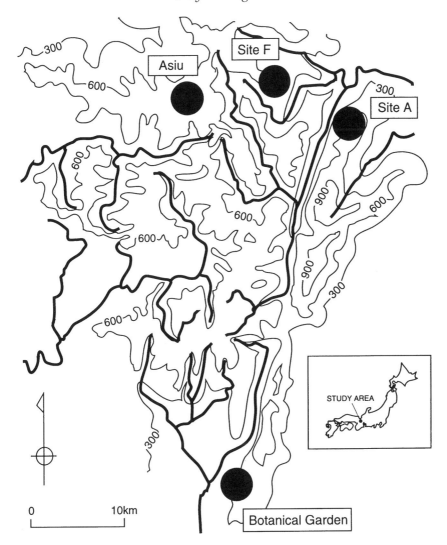

Fig. 16.1 Location of study sites A and F in Kutsuki, and an introduced population in the Botanical Garden, which came from Asiu in 1971. Thin lines show altitude in metres above sea level; thick lines show rivers.

beetles. The survival rate and numbers of adult beetles were estimated by the Jolly–Seber stochastic model based on the mark and recapture data (Jolly, 1965; Seber, 1973). In each spring, >98% of over-wintered adults that emerged from hibernation were marked 1 week after the census commenced. Also, >92% of newly emerged adults were successfully marked. Recapture rate on each census date (the number of marked adults that were recaptured divided by the estimated number of adults) was >85% throughout the census period. Because of the exceptionally high marking and recapture ratios, the estimated survival rates were highly reliable. The same capture–recapture experiments were carried out on the Asiu population between 1974 and 1976 and the Botanical Garden population between 1975 and 1981. Based on these data, detailed life tables were then constructed for every census year (Nakamura and Ohgushi, 1981; Ohgushi, 1986; T. Ohgushi and H. Sawada, unpublished data).

16.5.1 Demonstration of population regulation

(a) High level of stability in population density

Egg density of the two populations at sites A and F remained remarkably stable over a 5-year study period (Fig. 16.2). To examine their stability, the standard deviation of log-transformed egg densities during the study period, which is an appropriate index for temporal variability of a population (Gaston and McArdle, 1994), was calculated. The variability indices of sites A and F, respectively, were 0.135 and 0.051, indicating the highest level of population stability that has been reported for an insect species (Hanski, 1990). The other five populations studied, including Asiu and the Botanical Garden populations, also showed exceptionally high stability, ranging from 0.0002–0.190.

Note that the average egg density per thistle shoot at site A (29.8±3.34 [mean±s.e.]) was almost identical to that at site F (29.4±1.49) over the 5-

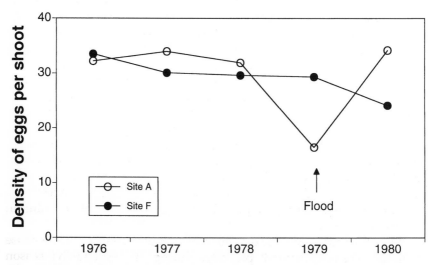

Fig. 16.2 Temporal changes in egg density per shoot at sites A and F. A large flood in late June 1979 considerably reduced reproductive adults at site A. Modified from Ohgushi (1995).

year study period, suggesting that the two different populations are maintained at a certain level in relation to resource availability. This 'equilibrium' density was, however, far below the level where defoliation of host plants occurs. Low leaf herbivory continued through late June when old larvae reached a peak density, averaging 30% of the total leaf area at site A and less than 20% at site F.

(b) Population regulation in relation to resource availability

Population regulation is defined as the return of a population to an equilibrium density, following departure from the density, as a result of density-dependent processes. To determine whether a population is regulated, we need to look at two points: (i) the tendency of the population to return towards the equilibrium density when disturbed from it, and (ii) the existence of a density-dependent process or processes that cause the population to return to equilibrium.

(i) A return toward an equilibrium density: a large-scale flood in June 1979 washed away all of the reproductive females at site A, and considerably reduced egg density. Despite this large population reduction, the egg population quickly returned to the previous density in the next year (Fig. 16.2). This implies that the beetle population has an effective regulatory mechanism to return it to the previous level of density following disturbance.

(ii) Density-dependent reproductive processes: to determine the life stage at which a population is stabilized, the year-to-year variability of densities was compared among different life stages over a 5-year study period. The variability index declined sharply from reproductive adult to egg stage at both sites. This indicates that the populations were highly stabilized during the reproductive process. In contrast, a destabilization of density occurred during the survival process from egg to adult stage. In other words, population stabilization was completed in the reproductive process, and no other regulatory agents operated in the survival from egg to reproductive age in the following spring.

The reproductive process involved two density-dependent relationships. Firstly, lifetime fecundity was negatively correlated with adult density (Fig. 16.3a). Secondly, female survival sharply decreased with adult density (Fig. 16.3b). On the other hand, there were no density-dependent survival processes during the period from egg to over-wintered adult in the following year.

The population meets the definition of a regulated population because it shows density-dependent processes during the reproductive season and because it returns to its previous density following perturbation. Density-dependent population regulation in the reproduction of *E. niponica* has been demonstrated in seven independent populations: the

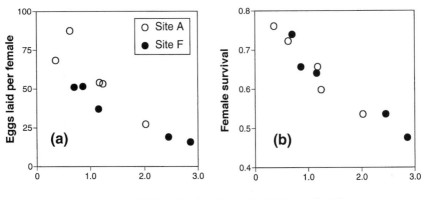

Fig. 16.3 Density dependence in reproductive processes. (a) Relationship between lifetime fecundity and adult density (site A: $r=-0.90$, $F=12.49$, $P=0.038$; site F: $r=-0.98$, $F=61.00$, $P=0.004$). (b) Relationship between female survival per 10 days and adult density (site A: $r=-0.98$, $F=59.27$, $P=0.005$; site F: $r=-0.97$, $F=46.65$, $P=0.006$).

two populations described here, and five other populations (Nakamura and Ohgushi, 1981; Ohgushi and Sawada, 1997a, 1998; T. Ohgushi, unpublished data).

16.5.2 Bottom-up population regulation

(a) The influence of resource limitation on population size

Next, let us consider how resource availability affects the lady beetle's population size (Fig. 16.4). Resource abundance changed independently at the nearby sites, which were separated by 10 km. Shoot numbers of thistles at site A consistently increased over the study period; while at site F they remained fairly constant from 1976–79, and then dropped in 1980 as a result of two large floods in the previous autumn. Shoot numbers showed 3.6- and 2.3-fold variations throughout the study period at sites A and F, respectively. Despite these two very different patterns of change in host abundance, the egg populations closely tracked the variation in resource availability through time. Indeed, the annual variation in egg population was mostly determined by the variation in resource abundance. Host abundance explains 66 and 98% of variation in population size at sites A and F, respectively. Population regulation was completed during the oviposition process when resource limitation determined egg density. This conclusively demonstrates bottom-up regulation of the lady beetle population.

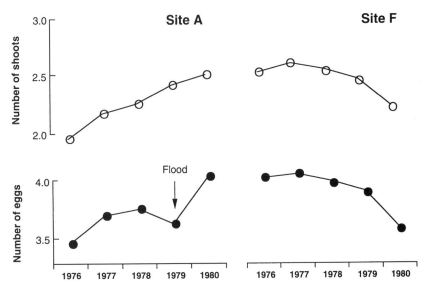

Fig. 16.4 Temporal changes in host-plant abundance (number of shoots) and egg population at sites A and F. Data were log transformed. Note that a large flood in 1979 considerably reduced the egg population at site A. From Ohgushi (1992) by courtesy of Academic Press.

(b) Top-down influences of natural enemies on population regulation

Since the role of natural enemies as regulatory agents of herbivorous insect populations has long been emphasized, I will summarize how top-down influences contribute to regulation of the lady beetle populations. Eggs and larvae were frequently subjected to heavy arthropod predation, mainly by nymphs of an earwig, *Anechura harmandi*. A cage experiment illustrated that arthropod predation was a main cause of mortality during the larval period (Ohgushi and Sawada, 1985b). The higher predation on immature stages at site F resulted in a significantly lower density of new adults, compared with that at site A where there was less predation (Fig. 16.5).

In contrast to their importance in determining new adult density, natural enemies are unlikely to function as a regulatory agent for three reasons. Firstly, arthropod predation operates in neither a spatially nor a temporally density-dependent manner (Ohgushi, 1988). Secondly, despite the significant difference in adult densities at emergence due to differential predation, egg densities in the next generation of the two populations were almost identical (Fig. 16.2). In other words, the large difference in adult density between the two populations caused by natural enemies was mostly compensated for by the regulatory process in the reproductive season. Thirdly, in contrast to egg and larval stages, natural

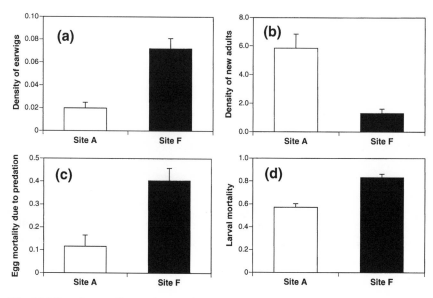

Fig. 16.5 Top-down effects of natural enemies. (a) Density of earwigs (number of earwigs per shoot). (b) Density of new adults of the lady beetle (number of newly emerged adults per shoot). (c) Egg mortality due to arthropod predation. (d) Larval mortality. Each column represents mean and s.e. in 1976–80. There are significant differences in density or mortality between the two sites (Mann–Whitney U test: $U=25$, $P<0.01$ for each case).

enemies had little effect on adult survival during the reproductive period, when population regulation occurred.

16.5.3 Mechanisms of population regulation

Next, let us search for the underlying mechanisms of population regulation. Density-dependent reduction in fecundity and female survival suggest that oviposition tactics play an important role. The density and spatial distribution of eggs among plants by herbivorous insects is determined by the strategy employed by females in searching for and choosing oviposition sites. The evolution of this oviposition strategy is determined by the life-history constraints of the herbivore. In *E. niponica* the trade-off between energy allocation to movement and reproduction, combined with an avoidance of high densities of conspecific eggs, regulates population density and spatial distribution.

(a) Behavioural mechanism: female movement for oviposition

The strategy used by a female in searching for an oviposition site determines the spatial distribution of eggs. Females avoid ovipositing on plants with high egg densities. Therefore as egg densities increase, female movement increases and egg distribution becomes more uniform. Several pieces of evidence support this hypothesis. Firstly, female mobility, expressed as the variance of distances travelled per day, increased with adult density (site A: $r=0.80$, $F=5.38$, $P=0.103$; site F: $r=0.96$, $F=34.98$, $P=0.01$). Secondly, the seasonal change in the movement pattern of ovipositing females was correlated with changing egg density. Female movement increased from early May to mid-June as egg densities increased, but thereafter it consistently decreased (Fig. 16.6a). In early June, female movement was positively correlated with cumulative egg density (site A: $r=0.90$, $F=13.26$, $P=0.036$; site F: $r=0.98$, $F=65.16$, $P=0.004$). This enhanced movement was synchronized with an increasingly uniform distribution of eggs among plants (Fig. 16.6b).

As energy allocation is shifted from reproduction to movement, with increasing egg density, female survival and/or fecundity decreases in a

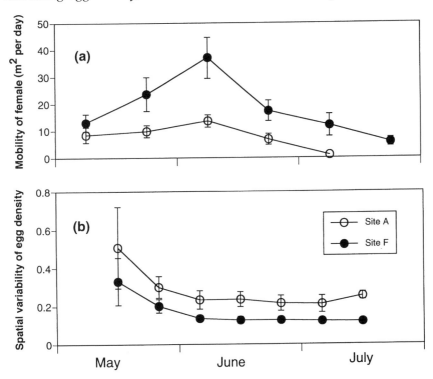

Fig. 16.6 Female mobility and spatial stabilization of egg density. (a) Seasonal changes in mobility of ovipositing females. Each point represents mean and S.E. in 1976–80. Mobility is expressed by the variance of distances travelled per day. From Ohgushi (1992). (b) Seasonal changes in spatial variability of cumulative egg density in terms of individual plants. Each point represents mean and S.E. in 1976–80. Spatial variability is expressed by the standard deviation of the log-transformed egg densities of individual plants. Modified from Ohgushi (1995).

density-dependent manner. Ohgushi and Sawada (1985a) demonstrated that increased female movement results in a reduction in oviposition rate through decreased fecundity and/or increased female loss. This is the result of the trade-off between egg production and female survival (Ohgushi, 1996a).

(b) Physiological mechanism: egg resorption

Egg resorption by females is a physiological mechanism that contributes to the density-dependent reduction in oviposition rate. Field-cage experiments showed that as host plants deteriorate in quality, females resorb eggs in the ovaries (Ohgushi and Sawada, 1985a), and this finding was confirmed by sampling females in the field. This process is reversible, so that when host plant quality improves the ovaries again become productive. Ohgushi and Sawada (1985a) found that egg resorption increased after mid-June as leaf damage increased, and the proportion of females that resorb eggs increases in years with high egg densities because of the high levels of early-season leaf damage in those years. Thus egg resorption directly contributes to the density-dependent changes in oviposition rate.

16.5.4 Evolutionary background of population regulation

We would expect that natural selection would have produced an oviposition strategy that would maximize lifetime fitness. To test this hypothesis, offspring lifetime fitness was calculated to evaluate the contribution of oviposition traits to the lifetime reproductive success of a female. Lifetime fitness of offspring is expressed as the total number of eggs produced in the following generation due to the expected reproductive contribution of one egg at the moment of birth.

(a) Oviposition site selection

In the previous section, I established that female avoidance of plants with high egg densities substantially influences egg distribution. The adaptive nature of this behaviour is demonstrated by the decline of lifetime fitness of offspring as egg density on a plant increases (Fig. 16.7). Fitness decreased sharply when egg density rose beyond the lowest levels. The egg density where fitness declined sharply corresponds well with the equilibrium densities of the two populations.

The decrease in fitness with increasing egg density is caused by both larval and adult mortality up to reproductive age. It is likely that the decrease in fitness is determined by changes in the plant caused by increasing lady beetle densities, as leaf damage results in decreased leaf quality in terms of amino-acid concentration and water content

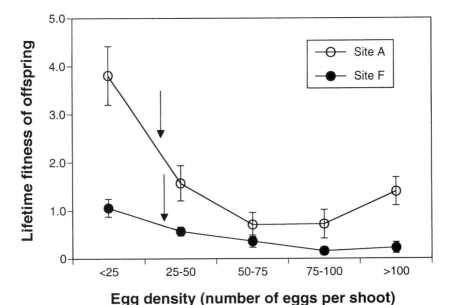

Fig. 16.7 Lifetime fitness of offspring which were born on plants with different egg densities. Each point represents mean and S.E. in 1976–80. Vertical arrows show the average egg densities at sites A and F. The lifetime fitness of offspring grown on the *i*th plant (F_i) is defined as follows: $F_i = E_i \times L_i \times A_i \times R_i$, where E_i=egg survival to hatching, L_i=larval survival from egg hatching to adult emergence, A_i=survival of adult females from emergence to the reproductive season in the following year, and R_i=lifetime fecundity of females estimated by reproductive lifespan. From Ohgushi (1995) by courtesy of Academic Press.

(Ohgushi, 1986). As a consequence, as leaf damage increases with increasing larval density, we see density-dependent mortality on a spatial scale. Also, thistles with high egg densities produced small-sized adults, which suffered higher mortality before reproductive age (Ohgushi, 1987, 1996b). In contrast, arthropod predation does not operate in a density-dependent manner (Ohgushi, 1988), and thus cannot be responsible for the observed spatially related patterns of mortality.

(b) Oviposition time selection

Egg resorption causes females to stop laying eggs and so it will not lead to increased lifetime fitness unless they are able to resume ovipositing later. Future oviposition may be classified into two categories. The first category is within-season oviposition, which is the resumption of oviposition in the same season; the second is inter-season oviposition, which is oviposition by a female that survives long enough to oviposit in the next reproductive season.

Females resorb eggs when host plants are highly exploited, or when the plants are destroyed or damaged by other forces. A large flood at site F in June 1979 washed away some thistle plants and buried others with soil. Following this disturbance females resorbed eggs and ceased oviposition for a 2-week period (Fig. 16.8a). When the damaged plants recovered and reflushed new leaves in mid-July, females resumed oviposition. Females thus avoided reproduction when food was unavailable to larvae, and resumed oviposition when offspring fitness was no longer reduced by food shortage.

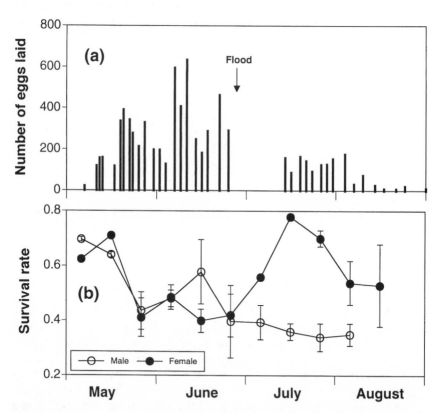

Fig. 16.8 Egg resorption caused by a large flood at site *F* and survival of reproductive adults. (a) Number of eggs laid on each census date. (b) Seasonal changes in survival rate of reproductive adults per 10 days. Each point represents estimated survival rates and 95% confidence limits, which were calculated by the Jolly–Seber stochastic model based on the mark–recapture data. From Ohgushi (1996a) by courtesy of Springer-Verlag.

Egg resorption can also be adaptive if it increases female survival so that reproduction is possible in a second season. There are two lines of evidence that a trade-off exists between egg production and female survival (Ohgushi, 1996a). Firstly, periods of low oviposition due to egg resorption were correlated with periods of higher female survival, but not higher male survival. Secondly, the severe flooding at site F in 1979 and the resulting egg resorption was correlated with changes in female survival (Fig. 16.8b). When females stopped laying eggs, their survival increased immediately up to mid-July. Then, after oviposition resumed, it began to decline. Again, male survival was unaffected. This indicates that reduced investment in reproduction through egg resorption increases investment in survival.

Females which resorbed eggs and then over-wintered oviposited in the following season. At site A, where egg resorption was low, most of the females died by mid-July and none survived until the next reproductive season. In contrast, at site F, where females resorbed eggs, 56 reproductive females were alive until mid-August, and nearly 40% survived until the next spring (Ohgushi, 1996a). Most of these females were observed ovipositing in the second year, thus demonstrating that long-lived females gain fitness by ovipositing in a second reproductive season.

Egg resorption is an adaptive response that increases lifetime fitness. No adults emerged from the cohort that was initiated in August (Fig. 16.9). This indicates that females will have higher reproductive success if they stop ovipositing late in the reproductive season and invest energy in surviving to the next spring, when an egg cohort has a higher probability of producing survivors.

The results of these studies indicate that there is a strong link between oviposition preference and offspring performance on both spatial and temporal scales.

16.5.5 Evolutionary implications of reproductive schedule

Since population regulation is largely due to the timing of reproduction, next I will examine the relationship between the reproductive schedule and lifetime fitness. There were different patterns of lifetime fitness among cohorts that emerged at different times at the two sites (Fig. 16.9). At site A, early cohorts enjoyed consistently higher lifetime fitness than later ones. In contrast, later cohorts at site F had higher lifetime fitness, except for the last cohorts in August. If the schedule of oviposition is a heritable trait, we would hypothesize that natural selection would favour early reproduction in population A, and delayed and/or prolonged reproduction in population F.

A laboratory experiment where environmental variation had been eliminated showed that populations A and F had different reproductive

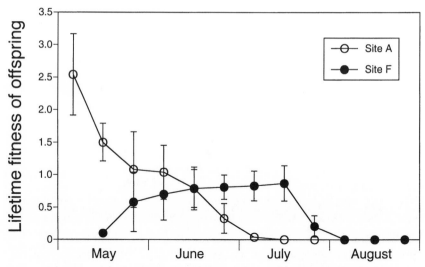

Fig. 16.9 Lifetime fitness of offspring which were born at different times in the reproductive season. Each point represents mean and s.e. for each cohort in 1976–80. The lifetime fitness of the ith cohort (F_i) is defined as follows: $F_i = E_i \times L_i \times A_i \times R_i$, where E_i=egg survival to hatching, L_i=larval survival from egg hatching to adult emergence, A_i=survival of adult females from emergence to the reproductive season in the following year, and R_i=lifetime fecundity of females living to average age of the ith cohort. Data from Ohgushi (1991).

schedules. When females from both populations were reared under constant laboratory conditions, the survivorship curves of females differed significantly (Fig. 16.10). Females from site A had a significantly shorter reproductive lifespan (Mann–Whitney U test: U=332, P=0.002). As a result, the distribution of oviposition differed between sites (Fig. 16.11). At site A, the oviposition rate declined slightly during the first 60 days and then declined sharply during the last 15 days. At site F, the oviposition rate peaked during the first 20 days and then declined until a second peak of oviposition near the end of the reproductive period at day 180. The maintenance of different reproductive schedules under identical environmental conditions indicates that oviposition timing is a heritable trait.

16.5.6 Natural selection alters population regulation

Since oviposition traits are heritable and influence population regulation, we expected that natural selection which alters egg-laying behaviour would indirectly influence population regulation. An introduction experiment strongly supported this hypothesis (Ohgushi and Sawada, 1997a, b, 1998). In May 1971, 15 males and 30 females from the Asiu

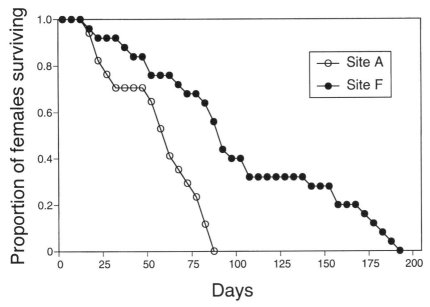

Fig. 16.10 Survivorship curves of adult females collected from sites A and F (Logrank test: $\chi^2=16.022$, d.f.$=1$, $P<0.0001$). Adult beetles were kept under constant conditions of 20 °C and LD 16:8 in an environmental chamber. Data from Ohgushi (1991).

Experimental Forest were released in the Kyoto University Botanical Garden (see Fig. 16.1). This is 30 km south of Asiu and 10 km beyond the southern limits of the lady beetle's natural range. There was no evidence of the occurrence of *E. niponica* in the Botanical Garden before introduction. The lady beetle has an extremely limited dispersal ability; mean distance travelled throughout its lifetime is less than 10 m for both sexes (T. Ohgushi, unpublished data), suggesting that 10 km is a sufficient distance for population isolation. Hence, it is most likely that the introduced population has been genetically isolated from any other *E. niponica* populations since the introduction. The introduced population was successfully established, and because of low arthropod predation the thistles were heavily defoliated. The introduced population had a significantly higher density at adult emergence than the source population that is maintained at a relatively low density (Nakamura and Ohgushi, 1981).

Because Botanical Garden larvae in late cohorts suffered from a severe food shortage resulting in high mortality, we predicted that selection would favour early reproduction in this population, leading to a higher oviposition rate and shorter reproductive lifespan. A laboratory experiment strongly supported this prediction. Firstly, females from the intro-

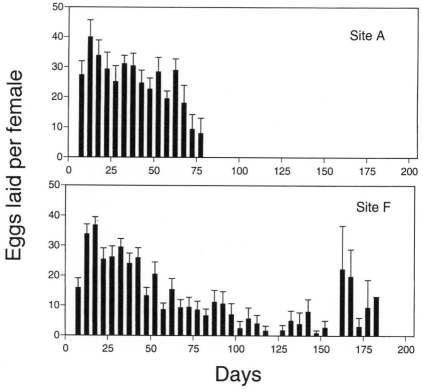

Fig. 16.11 Temporal changes in the number of eggs laid per female in consecutive 5-day intervals for females collected from sites A and F (Kolmogorov–Smirnov test: $\chi^2=14.69$, d.f.$=2$, $P=0.001$). Each point represents mean and s.e. Adult beetles were kept under constant conditions of $20°$ C and LD 16:8 in an environmental chamber. Data from Ohgushi (1991).

duced population had a higher oviposition rate early in their reproductive period in the laboratory (Fig. 16.12). Secondly, the lifespan of reproductive females from the introduced population was significantly shorter than that of the source population (Ohgushi and Sawada, 1997b; see also Fig. 16.13a), suggesting that reduced longevity had selected for earlier reproduction, as we expected.

Female mobility was also decreased in the introduced population (Mann–Whitney U test: $U=19.5$, $P=0.039$) (see Fig. 16.13b). We did not measure egg resorption in the introduced population, but we believe that it does not occur for two reasons (Ohgushi and Sawada, 1997a). Firstly, females resorb eggs in response to leaf damage that occurs late in the season. However, in the introduced population the density-dependent reduction in fecundity occurred early in the season. Moreover, a cage

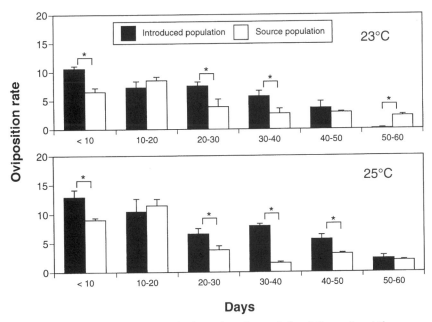

Fig. 16.12 Oviposition rate (eggs laid per female per 2 days) throughout the experimental period. Each point represents mean and s.e. Asterisks show significant differences between the introduced and source populations (Scheffe's test: $P<0.05$). Modified from Ohgushi and Sawada (1997b).

experiment showed that oviposition was greatly reduced even at low levels of leaf damage. Secondly, egg resorption would be disadvantageous in the introduced population. Females in this population had a shorter lifespan and thus would have little chance of using the energy saved by resorption in future oviposition. Indeed, we found no females that survived to the second reproductive season.

We predicted that reduced female movement and the probable lack of egg resorption would prevent efficient population regulation in the introduced population. In support of this prediction, we found that the introduced population had 6.8 times higher temporal variability in egg density than the source population (Fig. 16.13c), suggesting a lack of an efficient mechanism of population regulation in the introduced population. As a result, the introduced population had the lowest temporal stability in egg density among the seven populations that we studied (Nakamura and Ohgushi, 1981; Ohgushi and Sawada, 1981, 1997a; Ohgushi, 1992). However, the introduced population was stabilized by density-dependent reduction in fecundity and female survival. No other regulatory agents operated, as was the case for the source population, and for site A and F populations. This indicates that the population density of the introduced

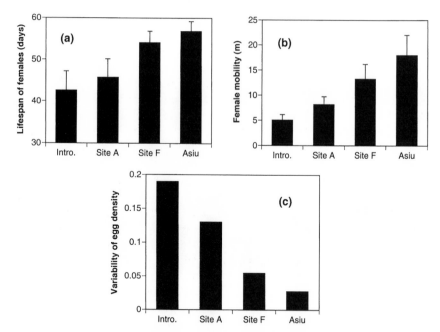

Fig. 16.13 Relationship between reproductive traits and temporal variability in four populations of the lady beetle. (a) Reproductive lifespan of females; (b) female mobility; (c) temporal variability of egg density. Intro. = Botanical Garden population introduced from Asiu. Means and s.e. are presented in (a) and (b). Female mobility was measured by mean distance moved per week by individual females. Reproductive lifespan was measured under constant conditions of 20 °C and LD 16:8 in an environmental chamber

population was also regulated through reproduction, even though survival was greatly modified by a lack of effective predation. This demonstrates again that top-down effects by natural enemies are unlikely to influence population regulation of the lady beetle populations.

Clearly, oviposition traits are the mechanism of population regulation of this lady beetle. We found strong correlations between oviposition traits responsible for population regulation (i.e. female movement and reproductive lifespan) and the temporal variability of four populations (Fig. 16.13). Thus we conclude that the oviposition traits that respond to resource availability regulate the population of the lady beetle *E. niponica*.

16.6 CONCLUSIONS

In this chapter, I have shown how the behavioural and physiological responses to resource availability are important in population regulation.

Our results also suggest that natural selection has altered oviposition traits that regulate population density.

Population studies have long ignored the important consequences of behavioural and physiological characteristics of adult insects in population regulation. However, recent studies have highlighted the role of oviposition tactics of females as regulatory processes generating stability in herbivorous insect populations (Preszler and Price, 1988; Craig *et al.*, 1989; Price *et al.*, 1990; Ohgushi, 1995). Recent analyses of density-dependent factors have also concluded that many insect populations are regulated in the adult stage by such factors as adult dispersal and reduced fecundity (Dempster, 1983; Stiling, 1988; Denno and Peterson, 1995). Individual-based population models have shown that life-history, physiology and behaviour have significant consequences for population dynamics (Murdoch and Nisbet, 1996; Uchmanski and Grimm, 1996). These models provide a framework for bringing together evolutionary and ecological studies to further our understanding of population regulation.

In the past, studies of insect population dynamics have focused exclusively on direct effects of biological agents such as killing power, and this has given us an inadequate understanding of population regulation. In the lady beetle, the lifetime fitness of offspring on different plants or cohorts is largely determined by the specific patterns of seasonal changes in predation and host-plant quality (Ohgushi, 1991). Thus, both these biological processes may operate as selective forces on the evolution of oviposition behaviours, including between-plant movement and egg resorption. This demonstrates that we should pay much more attention to natural enemies and host-plant quality as selective forces, through their influence on resource-use tactics of herbivorous insects, when seeking to understand population dynamics. This approach will undoubtedly provide new insights, giving an evolutionary perspective to insect population dynamics.

ACKNOWLEDGEMENTS

I thank Tim Craig, Peter Price, Hefin Jones, Jack Dempster and Ian McLean for valuable comments on the earlier version of this chapter. Financial support was provided by Japanese Ministry of Education, Science, Sports, and Culture Grant-in-Aid for Creative Basic Research (An integrative study on biodiversity conservation under the global change and bio-inventory management system).

17

The population ecology of *Megaloprepus coerulatus* and its effect on species assemblages in water-filled tree holes

Ola M. Fincke

17.1 INTRODUCTION

The role of competition, either in the form of intraspecific interactions that may lead to density-dependent population regulation, or in the form of interspecific interactions that may lead to niche differentiation, has recently come under renewed scrutiny from both theoreticians and empiricists (e.g. Cappuccino and Price, 1995; Denno *et al.*, 1995). Alternative explanations now abound for patterns of population persistence and species co-existence, and testing between these alternatives is a current challenge for insect population ecologists. Because many processes may lead to the same pattern, knowing the processes that give rise to population or community patterns is critical to understanding the dynamics of ecological assemblages. Too often, community ecologists infer process from pattern, whereas behavioural ecologists usually ignore the consequences of other species on individual fitness. The goal of my long-term research on tropical odonates is to bridge this gap.

Because they are discrete and relatively simple habitats, water-filled plant containers (phytotelmata), such as leaf axils, tank bromeliads, and tree holes, are particularly amenable to investigations into the processes controlling populations and structuring communities (e.g. Bradshaw and Holzapfel, 1983, 1988; Naeem, 1988; Mogi and Young, 1992). For the past

Insect Populations, Edited by J. P. Dempster and I. F. G. McLean. Published in 1998 by Kluwer Academic Publishers, Dordrecht. ISBN 0 412 83260 7.

15 years, I have studied a guild of odonates whose larvae are top predators in tree holes. As a behavioural ecologist, my initial interest was in the
adult reproductive behaviour of the three common pseudostigmatid
damselflies, which were known primarily for their unusual habit of feeding on orb-weaving spiders (e.g. Calvert, 1911; Stout, 1983). I asked why
males of *Megaloprepus coerulatus*, the world's largest damselfly, defend
water-filled tree holes, whereas males of the two *Mecistogaster* species do
not. One might think that all three species should be territorial, because
they all use potentially defendable and limiting tree-hole oviposition sites
(Fincke, 1992a). I then asked how larval dynamics affected adult fitness:
for example, why do *M. coerulatus* lay as many as five times the number
of eggs in a single tree hole as do *Mecistogaster* females, which oviposit a
relatively small number of eggs in any hole, regardless of its size? Also,
given that its larval offspring are cannibalistic, how long should a territorial *Megaloprepus* defend a tree hole, where progeny from later matings
are likely to be consumed by their older half-sibs? Such questions could
not be answered by focusing only on adults, or on a single species.
Because intraguild predation (*sensu* Polis and McCormick, 1987) is a
major source of mortality for tree-hole predators, studying the community dynamics at the larval stage was required to understand how guild
members, especially *M. coerulatus*, are regulated. Thus I have come to
population and community ecology through the back door, so to speak.
My initial goal was not to test theoretical hypotheses about population
regulation or community assemblages, but rather to understand adult
reproductive behaviour. I hope to show that regulation of *M. coerulatus* is
best understood in terms of competition between larvae. In the first part
of this chapter, I summarize population data on *M. coerulatus* collected
over an 11-year period from the tropical moist forest of Barro Colorado
Island, Panama. I identify life-history characteristics of larval and adult
guild members, which enable *Megaloprepus* to dominate tree holes, and
suggest that seasonal tree-hole drying prevents it from excluding less-
competitive species. I then test this hypothesis by comparing the Barro
Colorado guild with the tree-hole assemblage at La Selva Biological
Station in Costa Rica, where tree holes retain water throughout the year.

17.2 NATURAL HISTORY

Tree holes that collect water form in rotting burls, branch break-offs, or
convolutions in the trunk of fallen trees. Those harbouring odonates
range in volume from 0.01 to over 50 litres. This system lends itself to
controlled field experiments, because artificial tree holes are colonized by
a fauna nearly identical to that of natural holes (Fincke *et al.*, 1997). Tree
holes on Barro Colorado Island harbour a more complex array of top
predators than at previously studied tropical sites (e.g. Kitching, 1990).

The major macropredators, from the largest to the smallest (as measured by maximal larval size), are two aeshnid dragonflies, *Gynacantha membranalis* and *Triacanthagyna dentata*, and the pseudostigmatid damselflies *Megaloprepus coerulatus*, *Mecistogaster linearis* and *Mecistogaster ornata*, tadpoles of *Dendrobates auratus*, and the mosquito *Toxorhynchites theobaldi*. Another pseudostigmatid, *Pseudostigma accedens*, is extremely rare on Barro Colorado Island: during my 14-month stay on the island in 1983–84 I found two larvae, each in a small tree hole, and since then I have seen only five adults, so this species is ignored here. In past work, I referred only to the aeshnid *G. membranalis*, but I recently discovered that most of the aeshnids in Barro Colorado Island tree holes are the smaller and more abundant species *T. dentata*.

Size ratios of the final developmental stages among the common macropredators are *Gynacantha:Triacanthagyna*, 1.2; *Triacanthagyna:Megaloprepus*, 1.3; *Megaloprepus:Mecistogaster*, 1.2; *Mecistogaster:Dendrobates*, 1.4; *Dendrobates:Toxorhynchites*, 1.2. Although it has been suggested that such ratios indicate that these predators partition the feeding niche (e.g. Hutchinson, 1959; but see Lawton and Strong, 1981), the two *Mecistogaster* species contradict this pattern, being nearly identical in final instar size and growth rate (Fincke, 1992a).

Mosquito larvae are the most abundant and ubiquitous prey in tree holes (Fincke *et al.*, 1997). Syrphid fly larvae, chironomid midge larvae, tadpoles of *Physalaemus pustulosus*, and smaller individuals of the predator guild are also taken if available. In large holes, newly hatched *Megaloprepus* and *Mecistogaster* can emerge within 3.5 and 4 months, respectively, whereas the aeshnids require at least 5.5 months. In small holes, odonates may take 8 months or more to emerge as adults (Fincke 1992a; Fincke *et al.*, 1997). *Dendrobates auratus* eggs hatch within 11 days, after which the male carries the tadpoles to tree holes where they metamorphose after 1–3 months (Summers, 1990). *Toxorhynchites* develop within a month (Lounibos *et al.*, 1987).

The tropical moist forest on Barro Colorado Island receives an average of 2600 mm of rainfall annually, but experiences a dry season that lasts from January to late April (Rand and Rand, 1982). Most tree holes dry out by mid-March, and larvae that have not emerged by this time usually die because they cannot withstand more than 3 weeks of totally dry conditions (Fincke, 1994). For example, in one year only 5% of the tree holes contained larvae (either *Megaloprepus* or aeshnids) that survived the dry season. Because two to three cohorts of *M. coerulatus* can emerge over a wet season, tree-hole drying affects only the last generation. *Megaloprepus* and aeshnid adults are reproductive throughout the year except late in the dry season (i.e. March–April), when they aestivate. They reappear in May or June to produce the first wet-season generation of the year. In contrast, *Mecistogaster linearis* and *M. ornata* seem to produce only one generation per year. Adults of both species emerge in late wet or early dry season, and forage throughout the dry season. *M. linearis* begins mating in mid-December and apparently lays diapause eggs until the following wet season. This is the only species seen to oviposit regularly into holes with little water that soon thereafter dry out completely. *M. ornata* adults remain in reproductive diapause from the time they emerge until shortly before or after the first wet-season rains. The onset of reproduction in *M. ornata* is indicated by sexually dimorphic changes in wing pigmentation. The ventral side of the yellow wing tips of the male turn black, whereas those of females remain yellow (Fincke, 1984).

Of the odonates, *M. coerulatus* and *T. dentata* (possibly *G. membranalis*) males defend tree holes, usually large ones in light-gap areas where females can be reliably found (Fincke, 1992b; O.M. Fincke, unpublished data). Males of both species typically defend a hole for 2 weeks, although *Megaloprepus* may stay as long as 3 months, mating with any female before permitting her to lay eggs in the defended hole. *Megaloprepus* is one of the few odonates whose males are significantly larger than females. Sexual selection favours large males. Although females also oviposit in undefended holes, they mate only at defended sites. Because body size rather than prior residency best predicts the winner of territorial disputes, mated males are larger than males not seen to mate. Small males defend large holes until displaced by a larger male; they also play a satellite role at very large territories. Body size of males, but not females, is correlated with the volume of their larval habitat (Fincke, 1992b).

17.3 METHODS

During five wet seasons, the distribution of tree-hole organisms was quantified by repeatedly checking the contents of natural holes. Detritus and standing water were first removed, and emptied into white pans for inspection. The inside of the hole was then searched with a torch to detect odonate larvae, which typically are not sucked out with the water. A total of 331 unique tree holes were censused. Of the tree holes sampled in 1983, 52% were included in the 1984 sample. Similarly, 65% of holes sampled in 1992 were also sampled in 1993. However, of the holes sampled in 1992 and 1993, only 21 and 27%, respectively, were sampled in 1984. These yearly censuses can be considered independent for the present purpose because they are cleared of odonates in the dry season, and prior colonization of a hole by a predator was not predictive of its colonization by that predator in subsequent years (Fincke, 1992a; in press).

The distribution of tree holes from which odonates successfully emerged was determined by collecting final instars and allowing them to emerge in an outdoor insectary. To determine whether tree-hole predators used alternative habitats on Barro Colorado Island, the contents of freshly fallen, water-filled fruit husks of the liana *Tontelea ovalifolia* (*richardii*) were sampled repeatedly between 2 May and 10 July 1990. Fallen palm fronds were also censused irregularly during the wet seasons of 1983, 1984 and 1997.

Larval *Mecistogaster linearis* and *M. ornata* are not easily distinguished; data refer to pooled samples of the two species. Similarly, *G. membranalis* and *T. dentata* larvae were pooled. The term 'aeshnid' is used when no distinction was made between the two genera. Newly hatched odonates range from 2–3 mm in size, and no effort was made to detect very tiny odonate larvae or first-instar larvae of *Toxorhynchites*. Thus, species' occupancies may be underestimated, especially for the mosquito. Throughout, mean numbers are reported ±s.e. 'Small pots' refers to 400-ml plastic pots; 'large pots' were 7-litre tubs that usually contained 4–5 litres of water. All pots also contained leaves and a stick perch. Detailed methods for studies summarized here are described elsewhere (Fincke, 1984, 1992a, b, 1994; Fincke *et al.*, 1997).

17.4 PATTERNS OF EMERGENCE AND OVIPOSITION ON BARRO COLORADO ISLAND

The largest odonate species consistently dominate large tree holes within 2 months of the first rains of wet season (Table 17.1). The aeshnids and *Megaloprepus* begin to emerge about September, and continue to do so until tree holes dry up in March. The two smaller *Mecistogaster* species, which emerge from relatively small holes, have a single, peaked emergence from late wet season to early dry season (Fig. 17.1, Fincke, 1992a). These patterns suggest that *Mecistogaster* preferentially oviposits in smaller tree holes than either *Megaloprepus* or the aeshnids. However, a field experiment designed to detect very early wet-season colonization demonstrated this was not the case. Loosely covering tree holes with netting a month after the first rains of the wet season, prevented any subsequent oviposition by odonates, but not by their mosquito prey. When the larvae had grown large enough to identify, it was found that *Mecistogaster* had occupied about half of both large and small holes, significantly more than had been colonized by *Megaloprepus* or the aeshnids (Fincke, 1992a). Thus ovipositing *Mecistogaster* females do not preferentially use small tree holes. I have seen both *M. linearis* and *M. ornata* oviposit in large, defended holes. *Megaloprepus* males inspect the females but do not chase them away. Moreover, any female can oviposit when territorial males are absent, as they often are in late afternoon.

Indeed, the only niche partitioning by odonate females that was detected, was an absence of aeshnid larvae from holes with slit openings. Because of their short abdomens and inability to hold their wings together vertically, the aeshnids are prevented access to these holes (tree-hole

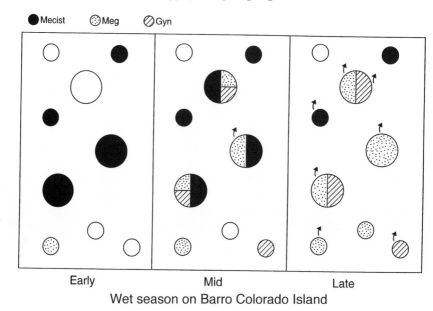

Early   Mid   Late
Wet season on Barro Colorado Island

Fig. 17.1 Schema of seasonal changes in odonate occupancy of large and small tree holes on Barro Colorado Island. 'Early dry season' refers to less than 2 months from the first wet-season rains. Arrows indicate emergence of adults, which first occurs from large holes in mid-wet season, and continues until the holes dry out in March. Mecist: *Mecistogaster*; Meg: *Megaloprepus*; Gyn: *Gynacantha* and *Triacanthagyna*.

Table 17.1 Mean volume of tree holes by site and year that were occupied by predator genera. On Barro Colorado Island, sampling was at least 2 months after the start of rainy season

Predator	*Barro Colorado Island*					*La Selva*
	1982 (78)	*1983* (110)	*1984* (129)	*1992* (92)	*1993* (97)	*1991* (64)
Tree-hole aeshnids	6.3±2.6	7.4±2.1	5.8±2.1	3.6±1.9	3.5±1.7	20.0*
Megaloprepus	7.4±2.1	3.2±0.4	2.4±0.7	1.5±0.4	3.0±1.4	2.3±0.8
Mecistogaster	0.2±0.0	0.8±0.2	0.6±0.1	0.7±0.2	0.6±0.1	0.2*
Dendrobates	–	5.4*	6.7±4.7	5.4±4.9	0.7±0.3	3.5±2.8
Toxorhynchites	–	0.3±0.1	1.8±1.3	0.5±0.2	0.7±0.2	0.8±0.2

Number of holes sampled is in parentheses.
*, Larvae found in only a single hole.
Volume of tree holes occupied by odonate genera did not differ across years on Barro Colorado Island ($F_{3,68}=1.1$, $P>0.3$; $F_{3,194}=0.8$; $P>0.4$, $F_{4,133}=0.4$, $P>0.7$, for aeshnids, *Megaloprepus* and *Mecistogaster*, respectively).

odonates on Barro Colorado Island insert their eggs into bark just above the water line). Because all the sampled tree holes were below about 4 m above ground, niche partitioning by height is a possibility. Sampling of tree holes and pots in the forest canopy on Barro Colorado Island failed to find any *Megaloprepus* larvae, although *Mecistogaster* and aeshnids were found there (S.P. Yanoviak, unpublished data). The absence of *Megaloprepus* larvae in canopy holes does not seem to result from an inability of adults to reach such holes, as on several occasions I watched a territorial *Megaloprepus* male fly from the hole it was defending in the lower trunk of a *Ceiba pentandra* tree, to the top of the tree, where there was a very large tree hole occupied by *D. auratus* tadpoles (R. Wirth, personal communication). Moreover, when foraging for spiders, *Megaloprepus* flies at least as high as *Mecistogaster* (Fincke, 1992c).

I argue below that differential survivorship of larvae with respect to tree-hole volume, rather than niche partitioning by adults, is the most likely explanation for the observed changes in tree-hole occupancy by mid wet season (Fig. 17.1). I systematically eliminate alternative explanations for differential survivorship and conclude that observed patterns of emergence are most likely to be the result of competition among odonate larvae.

17.5 CAUSES OF LARVAL MORTALITY DURING THE WET SEASON ON BARRO COLORADO ISLAND

17.5.1 Tree-hole chemistry

Differential survivorship of species with respect to tree-hole volume might result from physiological differences in larval tolerance of abiotic factors characteristic of large and small holes. However, although oxygen content, pH, and temperature varied significantly among tree holes on Barro Colorado Island, these factors did not differ consistently between large and small tree holes, nor were they predictive of species occupancy. Moreover, switching an odonate from its original hole to one previously occupied by another odonate genus did not affect its survivorship, contrary to what would be expected if tree-hole odonates were differentially adapted to the physical conditions of tree holes (Fincke, in press).

17.5.2 Priority effects and intraguild predation

If intraguild predators are equal in their competitive ability, then the species that is the first to colonize a tree hole should have an advantage over latecomers. This is typically the case for small tree holes, where the first colonist can patrol the entire hole and nutrient input is typically low. Experiments using pairs of predators in small pots provided with alternative prey demonstrated a strong advantage to the larger individual (i.e. the first colonist). When *D. auratus* was paired with an odonate, or any two odonate genera were paired, the largest individual typically killed the smaller individual within 2 weeks. Occasionally however, small odonates co-existed with large tadpoles until the latter metamorphosed. *Toxorhynchites theobaldi* was the least competitive guild member. These mosquito larvae were killed in about half of the trials involving smaller odonates or *Dendrobates* (Fincke, in press).

Competitive asymmetries among odonates were revealed when prey abundance and/or hole size (and detritus) was increased. In small pots with superabundant prey, smaller *Megaloprepus* eventually eliminated larger *Mecistogaster* in about 50% of the trials (Fincke, 1992a). Similarly, in an experiment using large pots, smaller *Megaloprepus* often overtook larger but slower-growing individuals of *M. linearis* and *M. ornata* in size, and eventually killed them (Fincke, 1992a). In large, natural holes, *Mecistogaster* is found occasionally with smaller *Megaloprepus* or aeshnid individuals, but in these cases the *Mecistogaster* typically disappear before they have time to emerge. Based on their maximal growth rates, in large holes both the aeshnids and *Dendrobates* should also be able to overtake a larger *Mecistogaster*, as *Megaloprepus* was shown to do. However, *Dendrobates* tadpoles occupied less than 10% of available tree holes sampled well after the frogs began to breed, whereas by mid-wet season, odonates collectively occupied from 55–75% of the tree holes sampled (Fig. 17.2). Thus odonates pose the greatest threat to developing larvae. Asymmetrical competition among odonate guild members in large tree holes best explains the premature disappearance of *Mecistogaster* from these microhabitats.

17.5.3 Obligate siblicide

On Barro Colorado Island, where *Megaloprepus* is the dominant tree-hole predator (Fig. 17.2), its larvae are more likely to encounter a conspecific than another predator species. In the absence of other predators, possible factors limiting the abundance of *Megaloprepus* include (i) interference competition in the form of cannibalism; (ii) feeding inhibition; and (iii) exploitative competition for food. Field experiments demonstrated that all three phenomena occur, but that cannibalism is the most important proximal mechanism limiting the number of larvae in a tree hole (Fincke, 1994). In small pots supplied with *ad libitum* alternative prey (i.e. newly hatched *P. pustulosus* tadpoles) the larger individual of a pair of *Megaloprepus* typically killed the smaller one. Intermediate-sized *Megaloprepus* have the fastest growth rates, and were the quickest to kill conspecifics. In contrast, larvae less than *ca* 7 mm in size often did not kill until they grew to ≥10 mm. Final instars, which stop feeding about 10–14

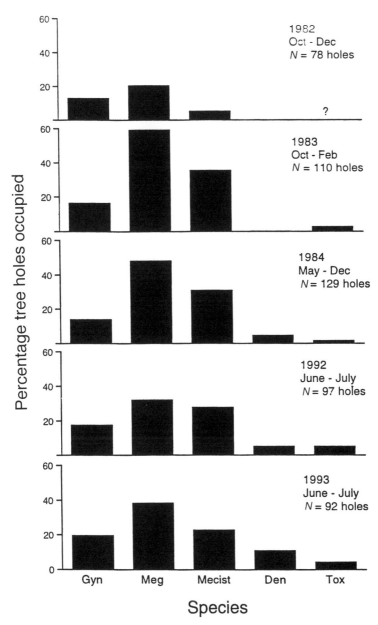

Fig. 17.2 Predator occupancy in Barro Colorado Island tree holes in five wet seasons. The presence or absence of *Toxorhynchites* was not noted consistently in 1982. *Dendrobates* was in only one hole in the 1983 census, and was absent in 1982; both censuses were at the end of the frog's breeding season. Mecist: *Mecistogaster*; Meg: *Megaloprepus*; Gyn: *Gynacantha* and *Triacanthagyna*; Den: *Dendrobates auratus*; Tox: *Toxorhynchites theobaldi*.

days before they emerge, often co-existed with smaller larvae. In 40% of the replicates in which killing occurred, the carcasses were not eaten, suggesting that large larvae kill potential competitors before they become a threat. Indeed, if cannibalism is prevented experimentally, exploitative competition for prey limits larval growth (Fincke, 1994), and under semi-natural conditions a single odonate significantly depresses the abundance of its mosquito prey (Fincke *et al.*, 1997).

Several lines of evidence suggest that cannibalism is the cause of the low larval density in tree holes, which rarely exceeds about one to two individuals per litre (Fig. 17.3). Territoriality by adult males results in an aggregation of *Megaloprepus* eggs in defended holes, with as many as 13 females known to oviposit in the same hole. Oviposition duration is positively correlated with tree-hole volume. A female may lay as many as 250 eggs in a large hole from which no more than 15 adults emerge in a season, and as many as 60 eggs in a small hole that subsequently produces only a single adult per season (Fincke, 1992b). Moreover, in tree holes that contained only *Megaloprepus* predators, small and intermediate larval size classes were often absent, despite recent known ovipositions by females. Finally, at such sites one also finds individuals missing caudal lamellae, legs, or most of their abdomen, damage which is caused by conspecifics in laboratory experiments (e.g. Fincke, 1994, 1995). Density-dependent regulation does not result from larval territoriality. Within holes, larvae exhibit size-dependent dominance. Larger individuals displace smaller ones as they move around in search of prey (Fincke, 1995). Small larvae often 'freeze' in the presence of a larger one, resulting in feeding inhibition.

Even though cannibalism occurs at relatively low densities, it is density dependent. When an excess of *Megaloprepus* larvae were placed in 3-litre artificial holes, individuals killed conspecifics until the density of larvae was reduced to 2–3 larvae (Fincke, 1994). Such densities permit maximum growth of at least the largest larvae in large holes, perhaps even when nutrient input is low. The fastest developmental rates in the field are similar to those with *ad lib* prey (Fincke, 1992a), and mean body size of adult *Megaloprepus* did not differ significantly between years (Table 17.2), despite a massive fruiting failure in 1994 (J. Wright, personal communication).

Because a male *Megaloprepus* defends a given hole for as little as a day or for as long as 3 months, and will mate with several females that then lay in the same hole, full sibs, half sibs and unrelated larvae may all occur together. Although kin recognition should be adaptive in this situation, it has not evolved in *Megaloprepus*. In experimental holes seeded with one larger sib and an equal number of smaller sibs and unrelated individuals, the larger sib displayed to, and killed, conspecifics closest to themselves in size, regardless of kinship (O.M. Fincke, unpublished data). Thus can-

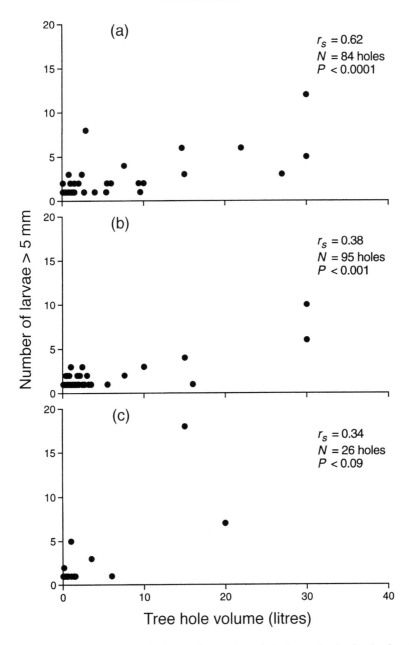

Fig. 17.3 Maximum number of odonate larvae found during a single check of natural tree holes as a function of tree-hole volume. (a) Barro Colorado Island, 1983 data; (b) Barro Colorado Island, 1984 data; (c) La Selva, 1991 data. Many holes were <1 litre, and single points for each hole sampled are not always distinguishable at the scale shown. r_s, Spearman rank coefficient.

Table 17.2 Mean size of adult males and females marked, and the relative abundance of adults on Barro Colorado Island and La Selva. Other than 1994, when sampling was biased towards the end of the year, sampling was throughout the rainy season

Year		*Males*			*Females*		*Encounter rate (Individuals/ observer day)*
	N	*Wing*	*Abdomen*	N	*Wing*	*Abdomen*	
Barro Colorado Island							
1983–84	115	67.4±0.4	80.6±0.5	78	61.2±0.6	72.0±0.6	2.0
1994	83	66.9±0.6	79.5±0.6	49	61.7±0.5	72.6±0.6	
1995	140	68.4±0.5	81.3±0.5	77	61.8±0.5	72.3±0.6	
1996	238	66.0±0.4	80.5±0.4	82	61.7±0.5	73.2±0.5	1.3
La Selva							
1991	38	75.9±1.0	87.9±1.0	5	67.2±2.0	77.5±2.6	1.9

On Barro Colorado Island, body size did not differ among years ($F_{3,612}$=1.29, $P>0.27$ for male wing length, $F_{3,281}$-0.67, $P>0.55$ for female wing length). Males and females at La Selva were larger than the pooled sample from Barro Colorado Island ($t=-8.4$, $P<0.001$, $t=-2.77$, $P<0.005$ for male and female wing length, respectively).

nibalism in this species is often siblicide, promoted by females which lay an excess of eggs.

Are *Megaloprepus* mothers promoting cannibalism or simply trying to swamp any predator already in the hole? The latter is unlikely, because the timing of hatching rather than clutch size *per se* is the variable critical to survivorship. When single clutches of *Megaloprepus* (each producing about 80 hatchlings) were added to large pots seeded with five aeshnid larvae, *Megaloprepus* rarely survived unless the initial aeshnids were very small, or very large and soon to emerge (O.M. Fincke, unpublished data). Moreover, under field conditions, some holes known to receive multiple clutches of *Megaloprepus* over an 8-month span produced only aeshnid adults (Fincke, 1992b). Finally, in large pots with a staggered input of *Megaloprepus* clutches, offspring survivorship was not correlated with the mother's clutch size (Fincke and Hadrys, unpublished data).

I hypothesized that laying excess eggs is an adaptive response by *Megaloprepus* to sexual selection for large male size or fast developmental time. Even among larvae hatching on the same day, slight differences in prey capture early in life result in a size hierarchy among sibs. Because tiny sibs feed on cladocerans and other prey too small for larger individuals to catch easily, harvesting smaller sibs may be an efficient way for the largest larva to exploit all of the prey available in a hole. I am currently testing this 'multiple mouths' hypothesis by determining whether cannibals emerge faster, or at a larger size, than individuals reared singly.

17.6 TEMPORAL AND SPATIAL REFUGIA

Although it is possible that over a wet season, all members of the predator guild could emerge from the same large tree hole, more commonly,
one or two eliminate the others (Fincke, 1992a). The behaviour and life-
history traits of these predators appear to be adapted to their asymmetrical competitive ability and the uncertainty of tree-hole occupancy.
Egg-laying odonate females do not discriminate between holes known to
contain odonate larvae and those that do not (Fincke, 1992a), as if they
cannot detect larvae already present. Because small odonates can be
found in holes with large *Dendrobates*, females apparently do not avoid
holes occupied by tadpoles. However, *Dendrobates* adults typically repeatedly swim about in tree holes before transporting their tadpoles to them
(Summers, 1989). On Barro Colorado Island, the infrequent overlap of
Dendrobates and larger odonates suggests that an adult frog rejects a hole
if it encounters an odonate predator (Fincke, in press; see also Caldwell,
1993). Similarly, of 14 tree holes frequented by adult *Dendrobates pumilio* at
La Selva, only three contained *Megaloprepus* larvae, fewer than expected
based on the frequency of odonates in holes at this site ($\chi^2=4.6$, 1 d.f.,
$P<0.05$).

On Barro Colorado Island, *Mecistogaster*, *Dendrobates* and *Toxorhynchites*
begin breeding earlier in the wet season than either *Megaloprepus* or the
aeshnids, resulting in an initial temporal refugium from the dominant
competitors. All tree holes, even those at territorial sites, must be redis-
covered by *Megaloprepus* and the aeshnids the following wet season.
Thus, *Mecistogaster* might emerge from large tree holes before any later-
colonizing heterospecifics get too big (Fincke, 1992a). This may explain
why *Mecistogaster linearis* and *M. ornata* do not totally ignore large oviposition sites. They lay a few eggs in any hole encountered, thereby spreading their risk across a large number of holes.

Megaloprepus has evolved a unique way of coping with the unpredictable presence of predators in large tree holes. *Megaloprepus* eggs take
a minimum of 18 days to hatch. The time span between first- and last-
hatched *Megaloprepus* eggs, all laid within 2 h of each other and kept sub-
merged in water, is as great as 184 days (O.M. Fincke, unpublished data).
Such extreme hatching asynchrony, unique among odonates, increases
the chances of at least some of a female's offspring encountering a preda-
tor-free window of opportunity. In contrast, egg hatch in a few
Triacanthagyna clutches occurred over a span of less than 2 weeks. The
5–6-month developmental time of the aeshnids means they must hatch
quickly to have time to develop before tree holes dry out. Similarly, once
tree holes fill in the wet season, *Mecistogaster* should benefit from fast egg
hatching because the earlier in the wet season they enter a hole, the
greater their chance of surviving. Because of their slower growth rate,

even if they successfully 'invaded' a large hole in mid-season, it is likely that they would be overtaken by later colonists of *Megaloprepus* or aeshnids. *M. ornata* eggs hatch in as little as 12 days and over a span of 78 days. Unfortunately, *Mecistogaster* rarely oviposit in captivity, and I have been unable to collect eggs from *M. linearis*.

Dendrobates and *Toxorhynchites* also have spatial refugia from tree-hole odonates, whose larvae have not been found in other phytotelmata on Barro Colorado Island. Only on one occasion, in July 1990, did I see a *Mecistogaster ornata* oviposit in a water-filled *Tontelea* husk that harboured a 11-mm *Dendrobates* tadpole. However, I assume this was a case of confusion, because the husk was adjacent to a live *Platypodium elegans* tree that contained several holes near ground level, where the female subsequently laid eggs. Furthermore, *Tontelea* husks decomposed rapidly, leaking water after about 2 months, which is barely enough time for a *Dendrobates* to develop, much less a tree-hole odonate. By early June 1990, 44% of the nine *Tontelea* fruit husks sampled contained *D. auratus* tadpoles, and two (22%) harboured *Toxorhynchites*. The largest tadpole found at that time had a snout-vent length of 10 mm; it metamorphosed a month later, in early July. Although some husks lost all of their water to evaporation between rains (Table 17.3), a drying experiment demonstrated that tadpoles could withstand 7–20 days of completely dry conditions (O.M. Fincke, unpublished data).

Larval *Toxorhynchites* were the only tree-hole predator also found in water-filled fallen palm fronds, occupying 6% of those checked at least twice. Palm fronds often have a large, exposed surface area and thus can collect more detritus than fruit husks and even tree holes of comparable volume. The macrofauna of fronds included scirtid (heliodid) beetle larvae and adults, occasionally syrphid fly larvae, and chironomids, in addition to mosquito larvae, particularly *Trichoprosopon*, which were often very abundant. However, between wet-season rains, fronds dried out more frequently than fruit husks, which might be why none was found harbouring *Dendrobates* or odonates.

17.7 STABILITY OF THE TREE-HOLE GUILD ON BARRO COLORADO ISLAND

Cannibalism and priority effects play important roles in stabilizing and organizing the assemblage of tree-hole predators on Barro Colorado Island. This conclusion is supported by consistently low larval densities within tree holes, and the remarkable year-to-year similarity in species occupancy over an 11-year period (Table 17.1; Fig. 17.2). Population stability attributed to cannibalism has also been documented in a lake dragonfly (Crowley and Johnson, 1992; Hopper *et al.*, 1996). The apparent stability of tree-hole odonate populations contrasts with many herbivo-

Table 17.3 Comparison of abiotic characteristics of phytotelmata

Habitat	N	Volume (litre)	Persistence	O_2 (p.p.m.)	pH	Temperature (°C)
Bromeliads	20	0.19±0.1 (0.05–0.6)	0.9 (0.66–1)	–	–	–
Palm fronds						
La Selva	13	0.26±0.5[a] (1.8–0)	0.4±0.2 (0–0.8)	4.9±1.3 (2.7–6.5)	5.6±0.7 (4.4–6)	27.8±0.1 (27.3–28.3)
Barro Colorado Island	17	0.31±0.1[a] (0.05–0.9)	0.2±0.2 (0–1.0)	–	–	–
Fruit husks						
Lecythis						
Fresh	14	0.60±0.1[b] (0.1–0.9)	0.69±0.2 (0–0.9)	0.51±0.2 (0.1–2.5)	6.3±0.3[d] (4.7–7.2)	26.0±0.0 (24–29)
Old	17	0.50±0.0[b] (0.1–1.2)	0.22±0.1 (0–0.4)	1.41±0.2[c] (0.1–3.1)	5.0±0.4[d] (1.4–7.2)	25.7±0.1 (24–29)
Tontelea	9	0.08±0.0 (0.04–0.17)	0.11±0.01	–	–	–
Tree holes						
La Selva	64	1.5±0.4[e] (0.01–20)	0.76±0.03 (0.11–1.0)	1.1±0.1[c] (0.3–2.0)	4.6±0.07 (3.4–6.0)	25.1±0.2[f] (24.3–29.7)
Barro Colorado Island	331	1.9±0.3[e] (0.01–40)	–	3.4±0.4 (0.6–11)	7.7±0.0 (5.1–9.0)	25.0±0.1[f] (20.1–29.6)

For bromeliads, N is the number of leaf axils sampled in five individuals (ranges). Persistence is the minimum volume during wet season/maximum volume. Letters a–f indicate means are not significantly different (Bonferroni tests, $P>0.05$). Phytotelmata differed in volume ($F_{3,124}=2.9$, $P<0.04$), O_2 ($F_{2,70}=112.5$, $P<0.0001$), pH ($F_{2,83}=11.8$, $P<0.001$) and temperature ($F_{1,46}=61.2$, $P<0.001$).

rous insects on Barro Colorado Island whose populations fluctuated over a 14-year period in a seemingly unregulated way (Wolda, 1992; but see Ray and Hastings, 1996).

On Barro Colorado Island, co-existence of tree-hole odonates appears to be mediated by trade-offs between larval competitive ability and colonization efficiency among adult guild members, coupled with a predictable disturbance in the form of a dry season, which clears the holes of predators. Tree-hole drying provides the opportunity for *Mecistogaster* to realize its colonization advantage over *Megaloprepus* and the aeshnids, whose larvae would otherwise be predicted to exclude the two smaller species. Seasonal drying results in the wholesale, but temporary loss of larval habitats, a disturbance that is less than total destruction of the forest, but more serious than intermittent drying. Thus my data support Hutchinson's (1959) multiple-niche hypothesis and Connell's (1978) intermediate-disturbance hypothesis for species co-existence. Another relevant hypothesis is that of Shorrocks (1990), who explained co-existence of competitors that use the same temporally and spatially patchy resources. If the eggs of superior competitors are aggregated, they cannot swamp all sites, providing refugia for the less-competitive species, as occurs in the *Drosophila* community on Barro Colorado Island (Sevenster and Van Alphen, 1996). Tree holes are a temporally and spatially patchy resource used by all odonates, and both the aeshnids and *Megaloprepus* aggregate eggs in large, defended holes in light gaps, which are easier to find than holes in understorey (Fincke, 1992b). However, because *Megaloprepus* and, to a lesser extent, the aeshnids also colonize small holes, it seems only a matter of time before they would exclude the two smaller species. Marked *Megaloprepus* adults have been re-sighted after 8 months, can travel a kilometre in less than a week, and disperse over the entire island. Large pots that were kept watered during the dry season (see Fincke, 1992b) contained only *Megaloprepus* or aeshnid larvae the following wet season, suggesting that continuous occupancy by dominants precludes establishment of *Mecistogaster*. I predict that seasonal drying, rather than egg aggregation, maintains odonate diversity in tree holes on Barro Colorado Island.

17.8 COMPARATIVE STUDY AT LA SELVA: A TEST OF THE ROLE OF COMPETITION

To test the role of tree-hole drying on species co-existence, and to determine if *Megaloprepus* retains its dominant position in a different forest type, I examined the distribution of tree-hole predators at La Selva Biological Station in Costa Rica. This lowland forest receives an average of 3900 mm of rain annually (1.5 times the rainfall on Barro Colorado Island). The driest months are February and March, which commonly have a 12-

day stretch with less than 5 mm of rainfall daily, but no month averages less than 100 mm of rainfall (Sanford *et al.*, 1994). Thus, although tree holes may experience intermittent drying during these months, they should remain sufficiently moist for odonates to survive. I predicted that, at La Selva, *Megaloprepus* and the aeshnids should locally exclude *Mecistogaster* from the tree-hole habitat. If present, *Mecistogaster* was expected to occupy alternative sites such as fallen palm fronds or fruit husks, which support odonates at other tropical sites (Santos, 1981; Caldwell, 1993).

17.8.1 Methods

Field censuses and observations were made between 3 June and 16 July 1991. To determine whether adult males of *M. coerulatus* were as territorial as they are on Barro Colorado Island, the location of marked males was noted. Any male seen at the same site over a span of at least 3 days was considered to be territorial, the same criterion as was used on Barro Colorado Island (Fincke, 1992b). I measured wing and abdomen length of any odonate I captured. I also noted the presence of odonates and *Dendrobates* adults around tree holes and other phytotelmata.

At weekly or bi-monthly periods, I censused 64 tree holes that held at least 10 ml of standing water, located in mature forest and in the arboretum. I used the same protocol as on Barro Colorado Island, with one exception. Many of the tree holes were full of thick, muddy detritus, which in some cases was >2 litres, making it difficult to detect organisms. In 26% of the holes, I removed some of the mud after thoroughly inspecting it for organisms. Although this procedure subsequently resulted in artificially greater water volume, for analysis I used the initial volume of standing water. Dissolved oxygen was measured three times a day with a portable OXAN analyser, and water temperature and pH were recorded using an electronic unit. To determine which species could colonize holes quickly, I also censused the contents of plastic pots (0.4 and 2.0 litres in volume) that were filled with water and leaves, and attached to trees between 1 and 1.5 m above the ground.

To determine if tree-hole odonates also used other phytotelmata at La Selva, I censused the contents of 20 water-filled leaf axils in five large tank bromeliads (*Aechmea* sp.) that were attached low enough on trees in the laboratory clearing to sample easily. I censused water-filled fruit husks within a 10×10 m plot around two *Lecythis costaricensis* or 'monkeypot' trees (Hartshorn, 1983), noting whether husks were freshly fallen or old. Those from the previous season were dull in colour and softer in texture. Finally, I checked the contents of 13 water-filled fallen palm fronds for the presence of odonates and other predators.

To determine whether odonate larvae could survive to emergence in fruit husks, I placed single *Megaloprepus* larvae (7–24.8 mm) in newly fallen

Lecythis husks kept on a table in the laboratory clearing, and measured the larvae bimonthly until they emerged. Mosquito prey colonized the husks naturally. Husks were periodically topped up with water because the eaves of the laboratory building prevented them from receiving rainfall.

17.8.2 Size and behaviour of adult *M. coerulatus*

I saw a total of 75 *Megaloprepus* over a 40-day span, an average of 1.9 individuals per observer day. Although sexual size dimorphism was similar between sites, adult *Megaloprepus* were larger at La Selva than on Barro Colorado Island (Table 17.2). Body size of Barro Colorado adults seen for the first time declines over the wet season, probably due to declining nutrient input, which should reflect the phenology of leaf and fruit drop (e.g. Foster, 1982). Because the sampling period at La Selva was limited to a 6-week period, a more appropriate comparison is between La Selva males and Barro Colorado males marked in June and July. Even when all La Selva males were compared with only territorial males marked during this time on Barro Colorado Island, the size difference between the sites remained highly significant ($t=-3.7$ and -2.9, $P<0.005$, for male wing and abdomen length, respectively).

Behavioural observations did not suggest that sexual selection was stronger on the La Selva males. I found *M. coerulatus* males defending large holes at six natural sites and two sites supplied with plastic tubs; all were in light gaps. As on Barro Colorado Island, size best determined the outcome of male–male encounters. In all but two of the 11 fights for which I knew the size of both contestants, the larger male won the encounter ($\chi^2=4.4$, 1 d.f., $P<0.05$). Although I repeatedly checked sites at three *Lecythis* trees, each surrounded by at least 10 water-filled fruit husks, I only once saw a *Megaloprepus* male in the area; he was absent on subsequent checks. However, one marked male was repeatedly sighted over an 8-day span, at a fallen palm frond that was near to a large fallen branch in the arboretum. I assume this was a case of confusion rather than a geographic difference in behaviour. I never saw any females at this site, and found no larvae in the frond. The male left when the frond started to leak water.

At La Selva, males were present as early as 8 am, whereas on Barro Colorado Island males were rarely at territories before 9:30 am. On Barro Colorado Island, I never witnessed any direct competition between females, but at La Selva, one ovipositing female fought with another female when she approached an undefended tree. These minor differences aside, reproductive behaviour of *Megaloprepus* at La Selva was essentially the same as on Barro Colorado Island (see Fincke, 1992b).

17.8.3 Distribution of predators in tree holes at La Selva

Mecistogaster ornata was absent at La Selva, whereas *Mecistogaster modesta* (which is absent on Barro Colorado Island) was present. I saw 14 adult *M. modesta* over a 40-day period. This species was the smallest pseudostigmatid at this site (mean wing length = 44.3±0.4 mm, mean abdomen length = 70.2±0.6 mm, $N=7$ males; mean wing length = 43.4±0.6 mm, mean abdomen length = 65.3±1.9 tmm, $N=3$ females). *M. modesta* larvae were easy to distinguish from those of other pseudostigmatids. The long-stalked caudal lamellae had a white spot on each tip, considerably smaller than the spot on the short-stalked caudal lamellae of *Megaloprepus*. *Mecistogaster linearis* lacks any spots on its dark brown lamellae. I commonly found *M. modesta* larvae in bromeliads, but found none in any of the tree holes, husks or palm fronds sampled, corroborating Calvert's (1911) conclusion that this species is a bromeliad specialist.

Thus the tree-hole odonates present at La Selva were at least one aeshnid and three pseudostigmatid damselflies, *Megaloprepus coerulatus*, *Mecistogaster linearis* and *Pseudostigma aberrans*. The latter two species were extremely rare; I saw only two adults of each species. Although *G. membranalis* is known from La Selva, I saw no adults during my study. Non-odonate tree-hole predators at La Selva were tadpoles of the poison arrow frog, *D. pumilio*, and larvae of the mosquito *Toxorhynchites* sp.. Although *T. theobaldi*, which occurs in Barro Colorado Island tree holes, is found in Costa Rica (Darsie, 1993), I identified the mosquitoes to genus only.

La Selva received 56 mm of rain in January and 53 mm in March 1991, well below the averages for these months (273 and 152 mm, respectively; Sanford *et al.*, 1994). Nevertheless, in July 18% of the holes harboured *Megaloprepus* larvae that were larger than 25 mm. Even if they grew at maximum rates, these larvae must have been from eggs laid in March at the latest. Because females of neither species lay eggs in dry holes, the tree holes must have held water during the short dry season at La Selva. The 64 holes sampled did not differ from those on Barro Colorado Island in water volume ($t=0.88$, $P>0.35$; Table 17.3) or temperature, but they had lower pH and much less dissolved oxygen than those on Barro Colorado Island (Fig. 17.4). Many of the holes were in *Pentaclethra macroloba*, which accounts for about 35% of total basal area of trees at La Selva (Hartshorn and Hammel, 1994).

Overall, predators occupied 58–69% of the tree holes at La Selva (depending on the accuracy of my estimate for *Dendrobates*), similar to occupancy rates on Barro Colorado Island ($F_{1,2}=0.37$, $P>0.6$). However, the distribution of odonate genera in tree holes at La Selva differed significantly from that on Barro Colorado Island (Fig. 17.5, pooled data from Fig. 17.2, $\chi^2=94.0$, 9 d.f., $P<0.001$). As predicted, the dominant predator in tree holes was *Megaloprepus*, which occupied nearly half of all the holes sampled, sim-

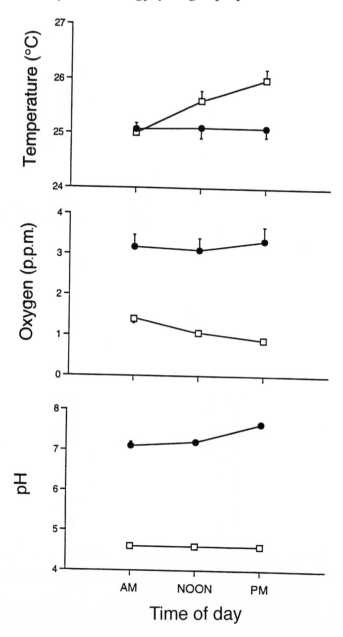

Fig. 17.4 Comparison of tree-hole chemistry over the day and between sites. Tree holes on Barro Colorado Island (solid circles) differed from those at La Selva (open squares) in mean pH ($t=26.6$, $P>0.0001$) and oxygen ($t=5.8$, $P<0.0001$), but not in temperature ($t=-0.38$, $P>0.7$). Bars represent $\pm$s.e., often too small to be displayed at the scale shown.

ilar to its frequency on Barro Colorado Island ($F_{1,3}=0.22$, $P>0.6$). However, *Mecistogaster linearis* and aeshnids were rare. I found only a single *M. linearis* (19.3 mm) in a 0.24-litre tree hole. The only hole with aeshnids was a 20-litre tree hole in a fallen tree, which contained at least five larvae of *Gynacantha membranalis* (largest = 35 mm) and several *Megaloprepus*, in addition to *Agalychnis calcarifera* tadpoles. Although I never found a larva of *Pseudostigma aberrans* in any of the phytotelmata sampled, I saw a female lay eggs in a 1.5-litre hole. Within holes, the mean density of odonates was one per 1.1±0.2 litres, no different from larval densities on Barro Colorado Island ($x=1.2\pm 0.1$, $t=0.2$, $P>0.8$; Fig. 17.3).

By mid-July, tadpoles of *D. pumilio* occupied 6.3% of the natural and 21% of the artificial tree holes at La Selva. *Dendrobates pumilio* typically begins breeding in July (Waclawski, personal communication; see also Donnelly, 1989a), shortly before my study ended. I repeatedly saw adult frogs in or immediately around 14 of the 64 natural tree holes. Within 2 weeks, two of these holes harboured one to two tadpoles (3–5 mm). Similarly, I found an adult swimming in one of the 400-ml plastic pots

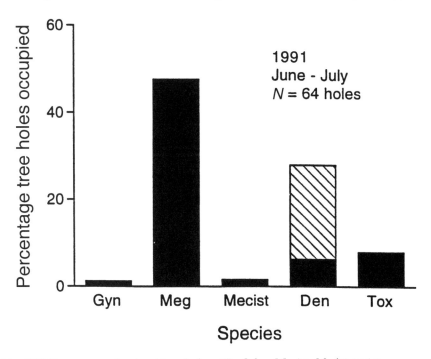

Fig. 17.5 Occupancy of natural tree holes at La Selva. Mecist: *Mecistogaster linearis*; Meg: *Megaloprepus*; Gyn: *Gynacantha membranalis*; Den: *Dendrobates pumilio*; Tox: *Toxorhynchites* sp. Striped bar indicates proportion of holes repeatedly frequented by adult *D. pumilio* but in which no tadpoles were found before the study ended (see text).

and later found two tadpoles (3 and 4.2 mm) there. Another 2-litre pot contained three tropic eggs in addition to a 3-mm tadpole. Because adult *Dendrobates* are territorial, I assume that all of the 14 individuals would have eventually deposited tadpoles in the tree holes they frequented. In that case, my estimate of tree holes eventually occupied by this genus increases to 28%, significantly higher than *Dendrobates* occupancy of Barro Colorado Island tree holes ($\chi^2=17.2$, 2 d.f., $P<0.005$). *Toxorhynchites* occupied 8% of the tree holes sampled at La Selva, marginally higher than on Barro Colorado Island ($\chi^2=3.2$, 1 d.f., $0.1<P>0.05$). This mosquito was also found in two (15%) of the 13 pots, which were easier to sample than tree holes.

17.8.4 Alternative habitats for tree hole fauna

The phytotelmata sampled at La Selva differed in volume, O_2, pH and temperature (Table 17.3). All of the possible alternative habitats were smaller, on average, than tree holes. Palm fronds, old *Lecythis* husks, and probably bromeliads (e.g. Laessle, 1961) were richer in dissolved oxygen than either fresh husks or tree holes. At La Selva, as on Barro Colorado Island, fruit husks and palm fronds were the least persistent phytotelmata, often losing water between rains. Old fruit husks were the most barren of the microhabitats, rarely supporting even a few mosquitoes. The only tree-hole predator that used any of these alternative habitats was *Toxorhynchites*, which I found in two (15%) of the 15 husks censused (1–2 individuals per occupied husk). This genus is also known to occupy bromeliads (Lounibos *et al.*, 1987), though I did not find any during my 6-week census.

Although I found no tree-hole odonates in any other phytotelmata, it was probably not because they could not tolerate the chemical conditions in these microhabitats. All of the six *Megaloprepus* larvae maintained in *Lecythis* husks survived to emergence, and grew very quickly. For example, the four 9–11 mm larvae all emerged in about 2 months, slightly faster than similarly sized Barro Colorado *Megaloprepus* provided with *ad libitum* prey (Fincke, 1992a). The fresh husks were nutrient-rich, and typically had an abundance of mosquito prey, particularly the husk specialist *Trichoprosopon* sp. (Lounibos and Machado-Allison, 1983). However, even freshly fallen *Lecythis* husks began to leak water after 4 months, slightly less than the minimum span required for a *Megaloprepus* or *M. linearis* to develop from egg to adult. Three *Gynacantha* larvae (29–32 mm), individually placed in old husks that leaked water, disappeared within 3 days, probably due to predation.

Seven of the 20 bromeliad leaf axils sampled (35%) harboured a *Mecistogaster modesta* larva, ranging in size from 4–17.6 mm. Two of these were final instars, both over 17 mm in length. The behaviour of *M. modes-*

ta appeared to be specialized for bromeliads. Individuals maintained singly in 400-ml pots in the laboratory clearing were repeatedly missing from the original pot, only to reappear in an adjacent one. All eventually disappeared. In a bromeliad, *M. modesta* are likely to climb from one leaf axil to another, searching for food or escaping a cannibalistic conspecific (see Calvert, 1911). In contrast, tree-hole odonates crawl out of pots only when they emerge. The only other odonate larvae found in bromeliads were three libellulid larvae (3–5 mm) in a single axil. These were likely to be the same species as a small female with black wing bands that I once saw oviposit in a leaf axil, but could not catch.

17.9 DISCUSSION

Current results and those published previously (Fincke, 1992a, b, 1994) support the view that the tree-hole predator fauna on Barro Colorado Island contains two dominant competitors; others survive by rapid colonization early in the wet season. This led to the prediction that the superior competitor(s) would dominate tree holes in wetter sites, where tree holes do not dry out completely. Inferior competitors were expected to be rare or absent, or to use alternative phytotelmata. This hypothesis was tested at La Selva. There, as predicted, *Megaloprepus* dominated the tree-hole habitat, apparently even out-competing the aeshnids. *Mecistogaster linearis* was rare and *M. ornata* was absent altogether. The latter species co-occurs with *Megaloprepus* in the drier transitional forests on the Pacific slopes of Costa Rica (De La Rosa and Ramirez, 1995), which is consistent with the hypothesis that seasonal drying provides it with a colonization advantage. *Mecistogaster linearis* did not use other phytotelmata. This is not surprising given that bromeliads, the only alternative habitat that persisted long enough for the development of a tree-hole odonate, were frequented by at least one odonate specialist, *M. modesta*.

Mecistogaster linearis might be rare simply because its larvae are only marginally tolerant of the abiotic conditions in La Selva tree holes, which differed significantly from those on Barro Colorado Island. This scenario seems unlikely given that most tree-hole organisms tolerate a broad range of conditions (Fincke, in press), as *M. coerulatus* did in the current study. Alternatively, conditions may be only marginal for adults, and *M. linearis* may persist as a sink population, renewed by immigration from some source population in dryer areas. Physiological intolerance of adults to the wetter conditions at La Selva is a more plausible explanation for the absence of *M. ornata*. On Barro Colorado Island, *M. ornata* often forages for long periods, fully exposed in hot light gaps, whereas the other pseudostigmatids retreat to shade after short foraging bouts in full sun. *Mecistogaster ornata* characteristically inhabits very dry sites, such as the lowland dry forest of Guanacaste, Costa Rica. Similarly, I suspect the low abundance of *Pseudostigma* at La Selva and Barro Colorado Island reflects some geographic range limitation, because *P. accedens* and *P. aberrans* are common in Mexico (E. Gonzalez-Soriano, unpublished data).

Assuming that *M. linearis* can tolerate the conditions at La Selva, its rarity there fails to support Shorrocks's (1990) hypothesis for species coexistence. If competitor aggregation were the mechanism mediating coexistence among tree-hole odonates, then there is no apparent reason why *M. linearis* should be less abundant at La Selva. At the wetter site, *Megaloprepus* were as territorial as on Barro Colorado Island, and consequently, eggs are likely to be as aggregated in large holes at La Selva as on Barro Colorado Island. I thus conclude that the rarity of *Mecistogaster* and tree-hole aeshnids at La Selva result from *Megaloprepus*'s competitive advantage at that site. Failure of *Megaloprepus* to exclude *Mecistogaster* competitively may be due to infrequent, but periodic dry spells, which could temporarily restore *Mecistogaster*'s colonization advantage, as seasonal drying does on Barro Colorado Island. Twice within a 22-year period, La Selva experienced less than 2900 cm of annual rainfall (i.e. levels similar to Barro Colorado Island; Sanford *et al.*, 1994). The 10-year maximum dry spell was 30 consecutive days with less than 5 mm of rain in 1983, perhaps sufficient to kill any odonate larvae present in tree holes during that time. Canopy tree holes might also provide *M. linearis* with refugia from *Megaloprepus* (but not from the aeshnids). Exposed canopy holes should also be more likely to dry out.

Given the low abundance of tree-hole aeshnids at La Selva, it is surprising that *Megaloprepus* did not occupy more than about 50% of the available tree holes, if adults were breeding throughout the year. Although more extensive sampling is needed to confirm that *Megaloprepus* does not increase its occupancy later in the year, similar encounter rates of adults at the two sites suggested similar population densities. Within tree holes, characteristically low larval densities suggested that cannibalism regulates the *M. coerulatus* population at La Selva, as it does on Barro Colorado Island. Tree-hole occupancy rates at La Selva might be limited by the rate at which females can locate small tree holes in forest understorey (e.g. Fincke, 1992a).

Alternatively, competition from *D. pumilio* may prevent *Megaloprepus* from occupying more than 50% of the holes during the frog's breeding season. Unlike *D. auratus*, *D. pumilio* were never found in fruit husks, which are sub-optimal habitats because periodic drying decreases prey and exposes tadpoles to terrestrial predators. Previous work on *D. pumilio* at La Selva assumed that adult *D. pumilio* deposits tadpoles in water-filled bromeliads in the canopy (Donnelly, 1989a, b). My results indicate that this species regularly uses tree hole sites below 2 m. Although I found no *Dendrobates* in or around the *Aechmea* axils which I sampled in the laboratory clearing, these sites were very hot and exposed, and so were unlikely to be attractive to the adult frogs. Donnelly also seemed unaware that predatory *Mecistogaster modestus* are common in bromeliads at La Selva. Their presence might explain why she failed to find an increased recruitment of *D. pumilio* juveniles after experimentally augmenting the number of bromeliad breeding sites (Donnelly, 1989b).

17.10 CONCLUSIONS

Although its larvae are restricted to tree holes, the influence of *M. coerulatus* extends beyond that microhabitat via its effects on intraguild predators, such as *Dendrobates* and *Toxorhynchites*, that also breed in other phytotelmata. The evidence to date suggests that the abundance of *M. coerulatus* is primarily affected by biotic factors during the larval stage, specifically obligate siblicide and cannibalism followed by intraguild predation (Fig. 17.6). Within this competitive framework, at a local level, population size should be affected by abiotic factors. The number of generations a tree hole can support annually reflects rainfall patterns as well as nutrient input (e.g. leaf and fruit detritus), which influences growth rate and adult body size via increased prey productivity. Among forests, body size probably reflects evolutionary responses to differences in tree-hole nutrient levels. For example, *Megaloprepus* were larger at La Selva, where flowering and fruiting phenologies are relatively more even across the year than on Barro Colorado Island (Newstrom *et al.*, 1994). Finally, changes in forest composition would affect the abundance of

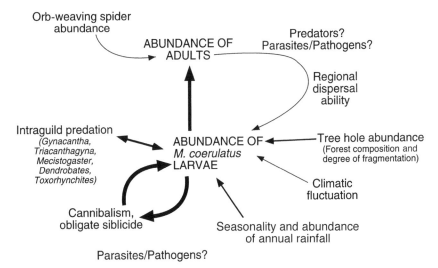

Fig. 17.6 Biotic and abiotic factors affecting abundance of *Megaloprepus coerulatus*. Thicknesses of arrows indicate the probable relative importance of each factor.

Megaloprepus because tree holes are non-randomly distributed with respect to tree species (Fincke, 1992a).

I have concentrated on causes of mortality during the larval stage because I have no evidence that adult *Megaloprepus* are limited by the availability of prey (i.e. spiders) or by predation. Their ability to find tree-hole oviposition sites may be more limiting than their capacity to produce excess eggs. However, in seasonal forests, persistence of *Megaloprepus* depends on adults surviving the dry season. Adults are also the dispersal stage. Because *Megaloprepus* avoids large, man-made clearings, it may be particularly vulnerable to habitat fragmentation. Future work will focus on the metapopulation dynamics of this keystone predator.

ACKNOWLEDGEMENTS

This chapter is dedicated to the memory of Leslie K. Johnson, who introduced me to the richness of the Neotropics. The work was supported by the University of Oklahoma, and NSF grant IBN-9408143. I thank M. Albrecht for field assistance, and D. Graham for measuring odonates after I left La Selva. M. May kindly confirmed the identity of the two tree-hole aeshnid species on Barro Colorado Island. C. Thomas and an anonymous reviewer provided helpful comments that greatly improved the text.

Concluding remarks

Although the contributors to this Symposium were selected to provide a variety of views on insect population dynamics, they are certainly not a random sample of population ecologists – if such a thing is possible. This does mean that any conclusions which we draw from the preceding chapters of this book will be somewhat biased by the opinions of the authors that we have chosen.

In view of the long history of controversy in this subject, it would be unbelievable if we now claimed that unanimous agreement had been reached on the processes determining insect abundance. Indeed this is not so: nor was it anticipated when we put the Symposium together. However, some points of agreement can be identified, and the differences of opinion, aired here, are perhaps less divisive than might appear at first sight.

A number of questions were posed in the Introduction to this volume, which were aimed at highlighting those aspects of insect population ecology that have led to controversy in the past. Having read the intervening chapters of this book, which demonstrate the complexity of interacting factors that determine the abundance of insects, these questions may appear somewhat simplistic. Nevertheless, we think that they provide a useful framework for discussing where general agreement has been reached and where uncertainty still persists.

Firstly, we asked how important are density-dependent processes in governing the extent of fluctuations in population size and the persistence of populations? This question strikes at the heart of the long-standing debate polarized in the writings of Andrewartha and Birch and of Nicholson, and summarized in the first four chapters of this book.

As shown by Den Boer (Chapter 3), density-independent factors, such as weather, have enormous direct and indirect impacts on insect popula-

Insect Populations, Edited by J. P. Dempster and I. F. G. McLean. Published in 1998 by Kluwer Academic Publishers, Dordrecht. ISBN 0 412 83260 7.

tions. This topic is also taken up by Leather and Awmack (Chapter 8), who show how the 'quality' of individuals making up a population can be affected by variations in their environment, including their own density. There can be no doubt that there is enormous heterogeneity in the impact of all factors affecting population size, resulting from spatial variations in the environment and from the different susceptibilities of individuals making up a population. Den Boer argues that this heterogeneity can buffer the population against environmental change and so contribute to both population stability and persistence, simply by the spreading of risk. The potentially stabilizing effect of this heterogeneity has been demonstrated in Den Boer's study of carabid beetles, but at present it is very difficult to estimate its contribution generally. No doubt the time will come when ecologists can quantify the amount of heterogeneity in populations and in their habitats and so test whether populations of species with greater heterogeneity actually do fluctuate less, and by how much, than those with less heterogeneity, but at present we cannot do this. Nevertheless, the potential of heterogeneity to buffer the impacts of environmental variations is acknowledged by all of our early contributors, Hanski (Chapter 1), Hassell (Chapter 2) and Dempster (Chapter 4), although they argue that density-dependent processes are more important in determining the stability and persistence of populations. In their view, any process that is linked to changes in density has a greater potential for stabilizing a population than one which is independent of density.

As discussed by Rothery in Chapter 5, there are analytical problems in identifying density dependence in population data, which need to be borne in mind when assessing the frequency of the occurrence, and the likely role, of density dependence. One such problem concerns the scale at which individual studies have been carried out. As Hanski makes clear in Chapter 1, populations of different species can differ greatly in their population structures, and in the extent to which their local populations depend upon immigration for their persistence. Movement is very difficult to quantify, and many field ecologists still ignore the effects of movement into, and out of, their study populations. Sometimes, density dependence will not be present in population data, simply because the latter have been collected at an inappropriate scale. Most insects are very mobile, so that density dependence may not be present if the area sampled is small compared to the mobility of the insect. In such circumstances, movement may result in numbers in one generation being totally independent of those in the previous generation (see for example, *Tephritis bardanae*; Chapter 4). One would expect movement to have a larger impact, the smaller the scale of the study, thus reducing the likelihood of density dependence at that scale. Hassell's data on the viburnum whitefly (Chapter 2) go against this view, but we have to be very careful in interpreting such results without knowledge of the mobility of the insect. As Rothery points out, movement could add a random component to observed density, which would cause problems akin to sampling error, and so lead to spurious detection of density dependence. Thus we need to know more about the structure of populations, and mobility within them, before we can interpret the presence or absence of density dependence in population change.

Nevertheless, despite these problems we think that there really is little doubt that density-dependent processes are actually present in the population dynamics of most, if not all, species, including all of those considered in our case studies. However, as pointed out by several of our authors, identification of density dependence does not, in itself, provide evidence that a process is capable of stabilizing a population. For this, the density-dependent process must have a sufficiently large impact to counter the effects of other factors affecting changes in population size from one generation to the next. The timing of a density-dependent process may be critical in determining its impact, since its effect will be greater if it follows a large density-independent change in population size than if it precedes one (see Straw, Chapter 10; McLean, Chapter 15). The impact of a density-dependent process can be determined satisfactorily only by modelling the processes concerned.

If density-dependent processes are important for population stability, how are they likely to act? Do they provide a mechanism for regulating populations about a theoretical equilibrium, or do they simply limit the upward growth of populations? In attempting to answer these questions we need to distinguish between the likely effects of natural enemies, and of intraspecific competition.

As pointed out in the Introduction, there is a commonly held belief that herbivores are kept below the limits of their food resources by natural enemies (i.e. top-down control), whilst other organisms (plants, decomposers, carnivores, etc.) are normally limited by their food resources (i.e. bottom-up control). We can now be certain that this generalization is untrue, as can be seen from the high frequency of bottom-up control recorded in the case studies of phytophagous insects (e.g. aphids, Chapter 9; Tephritidae, Chapter 10; *Tyria*, Chapter 12; *Euura*, Chapter 14; *Epilachna*, Chapter 16). There is some uncertainty in the case of *Trichochermes* (Chapter 15) and, in contrast, the control of *Operophtera* is thought to be top-down (Chapter 13). Added to these, Dempster (Chapter 4) and Price (Chapter 14) give many other examples of bottom-up control of phytophagous insect populations.

More population studies have been carried out on phytophagous species than on those at any other trophic level. However, studies of carnivores suggest that many of their populations are also controlled by bottom-up processes. This is seen in the specific predator of *Trichochermes*, *Leucopis psyllidiphaga* (McLean, Chapter 15); in *Maculinea arion*, a specific ant predator (Thomas *et al.*, Chapter 11); and in the generalist predator *Megaloprepus coerulatus* (Fincke, Chapter 17). It is also described for a number of parasitoids (Dempster, Chapter 4), and in *Maculinea rebeli*, a specific parasite of *Myrmica* ants (Thomas *et al.*, Chapter 11).

Twenty years ago, top-down control by natural enemies was thought to be the rule for phytophagous insects, and this has undoubtedly contributed to the view that they commonly regulate populations of their prey. Although natural enemies have been central in the theoretical study of population regulation (Chapters 2 and 6), field evidence that they can act in this way is provided in only one of our case studies. We would hasten to add that this may not be an unbiased sample of field studies. The many cases of successful biological control (Chapter 8) are sometimes quoted as evidence of population regulation by natural enemies, and there is no doubt that enemies can cause very high mortalities to their prey, which may lead to greater population stability at lower mean densities. However, there is little evidence to suggest that this mortality is consistently density-dependent. Several of our contributors would claim that more subtle interactions between enemy and prey can lead to regulation of prey populations, but in our opinion these interactions must be translated into direct temporal density dependence for regulation, as defined here, and this should be detectable in between-generation growth rates. However, there is no agreement on this amongst our contributors.

Intraspecific competition for resources is always density-dependent and so has the potential to regulate or limit population size. The distinction between these two theoretical processes depends on the shape of the density-dependent response curve in mortality, reproduction or dispersal to increasing density. Competition for resources is recorded in many of the case studies, but there is disagreement as to how this is likely to affect the species concerned. Dixon (Chapter 9), Thomas *et al.* (for *Maculinea rebeli*, Chapter 11), Ohgushi (Chapter 16) and Fincke (Chapter 17) consider that competition can regulate their populations; whilst Straw (Chapter 10), Thomas *et al.* (for *Maculinea arion*, Chapter 11), Van der Meijden *et al.* (Chapter 12), and Price *et al.* (Chapter 14) think that it simply limits the upward growth of populations. Clearly, we need more information on how precisely intraspecific competition acts. Subtle mechanisms of resource use and resource partitioning have become apparent in some species which, if widespread amongst insects, need to be taken into account when assessing the role of competition in population regulation or limitation. There is also growing evidence that the supply of resources is often so inadequate that population growth of many insects is reduced even at low densities.

Past reviews of the population dynamics of closely related insect taxa [e.g. grasshoppers and locusts (Acrididae), Dempster, 1963; butterflies and moths (Lepidoptera), Dempster, 1983] suggest that closely related species have very similar processes operating on their populations, due to the similarity of their evolved behaviour and their ecological requirements. Chapters by Dixon *et al.* on aphids, by Straw on flower-inhabiting gall flies (Tephritidae), by Price *et al.* on the stem-galling sawfly, *Euura*, and its relatives, and to a lesser extent by Thomas *et al.* on butterflies of the genus *Maculinea*, support this view. In all cases, their population dynamics are broadly determined by their phylogenetic backgrounds.

Added to this, studies of the same species in different locations appear to have very similar, but not identical population ecologies. Van der Meijden *et al.* (Chapter 12) show how a very simple population model for *Tyria jacobaeae*, with only three variables, adequately describes its population dynamics at two sites, but not at a third. Roland (Chapter 13) shows remarkably synchronized population cycles of *Operophtera brumata* in Britain, Scandinavia and Germany, but this is less marked in the more limited data from North America, where the moth has been introduced. We need far more comparative studies of this sort to determine the extent to which we can generalize between species and locations.

Unfortunately, different population theories are not mutually exclusive. Thus, for example, it is perfectly possible for population stability to be the combined result of both density-independent spreading of risk and density-dependent processes limiting or regulating population growth, acting at the same time. The inherent complexity of interacting processes that determine insect numbers makes it difficult to design experimental or observational tests which will reveal which theory offers the best explanation of how the world works. Many insects have life stages with different requirements, that often exploit different resources in different places, so that different population processes may be important to different stages and/or generations. Whilst it is desirable to seek the simplest explanation which fits the available evidence, such explanations inevitably oversimplify and give an impression of lack of realism. Added to this, even very simple models can display very complex properties and behaviours. These problems place special requirements on both theoretical and field ecologists. Theorists need to place far more emphasis than in the past on ensuring that their hypotheses are testable, and that the basic assumptions underlying their ideas are spelt out carefully, so that they, too, can be tested. For their part, field ecologists need to become far more 'hypothesis-testing' in their approach. In the past, too many field studies have been little more than long-term monitoring exercises, with little regard for unravelling the processes underlying population changes. There are too few experimental studies in which hypotheses are tested by population manipulation. We cannot stress too strongly the need for testing different theoretical ideas, and the assumptions on which they are based. Without this, we can expect another 50 years of argument, in which our science is based upon opinion rather than proven facts.

Probably the most important single line of research required in population ecology is to find better ways of quantifying the effects of dispersal.
Dispersal from and to a population may be density-dependent, and so
may provide a very sensitive mechanism for population stability, but we
know little about the quantitative effects of emigration and immigration
on insect populations, and even less about the rates of survival of individuals that move. In the past, both field and theoretical ecologists have
ignored the effects of movement, by assuming that they are dealing with
closed populations. This has largely been forced on them by the difficulty
in assessing rates of immigration and emigration. Those studies that have
been made have used mark and recapture techniques to assess dispersal
from a release site. This can give highly biased impressions of lack of
movement, because larger and larger areas have to be sampled as the
insects move further and further away from the point of release, and
these have not been searched at the same intensity as the release site.
Dempster *et al.* (1995b) tried to overcome this problem by measuring rates
of immigration of unmarked individuals to their study site, but this was
only partly successful because of loss of marks. The quantification of
movement continues to be a challenge to our ingenuity. The next
Symposium of the Royal Entomological Society will be on the subject of
insect dispersal, and so perhaps we can look to this event to increase
research interest in this topic.

Problems of funding long-term research are currently facing ecologists
throughout the world. Indeed, we had some difficulty in finding current
long-term studies when putting this Symposium together, and even then
we finished up with very large gaps in our coverage. Few new, long-term
population studies are likely to be started, so we would advocate building on past studies of those species for which good life-table data are
already available. We would like to see studies on dispersal, and experimental population manipulation made on these species to test earlier
conclusions. Short-term studies of this sort could be extremely valuable in
making previous field research more comparable with current theory.

There are still many gaps in our knowledge, often as a result of the
inadequate techniques at our disposal for measuring parameters in the
field. We close with the general point that we need to be constantly on
the look-out for techniques developed in other areas of science, that
might help us to test some of the ideas that have been presented in this
volume.

Jack Dempster and Ian McLean

References

Abrams, P.A. (1994) The fallacies of "ratio-dependent" predation. *Ecology*, **75**, 1842–50. [Ch. 6]

Adler, F.R. (1993) Migration alone can produce persistence of host–parasitoid models. *American Naturalist*, **141**, 642–50. [Ch. 6]

Akçakaya, H.R., Arditi, R. and Ginzburg, L.R. (1995) Ratio-dependent predation: an abstraction that works. *Ecology*, **76**, 995–1004. [Ch. 6]

Altenkirch, V.W. (1991) Zyklische Fluktuation beim Kleinen Frostspanner (*Operophtera brumata* L.) [Cyclic fluctuation of winter moth (*Operophtera brumata* L.)]. *Allgemeine Forst und Jagdzeitung*, **162**, 2–7. [Ch. 13]

Aluja, M. (1994) Bionomics and management of *Anastrepha*. *Annual Review of Entomology*, **39**, 155–78. [Ch. 10]

Alverson, D.R. and English, W.R. (1990) Dynamics of pecan aphids, *Monelliopsis pecanis* and *Monellia caryella*, on field-isolated single leaves of pecan. *Journal of Agricultural Entomology*, **7**, 29–38. [Ch. 9]

Anderson, R.M. and May, R.M. (1978) Regulation and stability of host–parasite population interactions: I. Regulatory processes. *Journal of Animal Ecology*, **47**, 219–47. [Ch. 2]

Anderson, R.M. and May, R.M. (1991) *Infectious Diseases of Humans: Dynamics and Control*, Oxford University Press, Oxford. [Ch. 1]

Andow, D. A. (1990) Vegetational diversity and arthopod population response. *Annual Review of Entomology*, **36**, 561–86. [Ch. 7]

Andrewartha, H.G. and Birch, L.C. (1954) *The Distribution and Abundance of Animals*, University of Chicago Press, Chicago. [Intro., Ch. 1, 2, 3, 16]

Andrewartha, H.G. and Birch, L.C. (1984). *The Ecological Web. More on the Distribution and Abundance of Animals*, University of Chicago Press, Chicago. [Ch. 3]

Andrewartha, H.G. and Browning, T.O. (1961) An analysis of the idea of "resources" in animal ecology. *Journal of Theoretical Biology*, **1**, 83–97. [Ch. 4]

Anscombe, F.J. (1959) Sampling theory of the negative binomial and logarithmic series distributions. *Biometrika*, **37**, 358–82. [Ch. 2]

Argov, Y. and Rossler, Y. (1993) Biological control of the Mediterranean black scale *Saisssetia oleae* Hom. Coccidae in Israel. *Entomophaga*, **38**, 98–100. [Ch. 7]

Insect Populations, Edited by J. P. Dempster and I. F. G. McLean. Published in 1998 by Kluwer Academic Publishers, Dordrecht. ISBN 0 412 83260 7.

Armstrong, R.A. and McGehee, R. (1980) Competitive exclusion. *American Naturalist*, **115**, 151–70. [Ch. 6]

Atkinson, W.D. and Shorrocks, B. (1981) Competition on a divided and ephemeral resource: a simulation model. *Journal of Animal Ecology*, **50**, 461–71. [Ch. 1]

Atkinson, W.D. and Shorrocks, B. (1984) Aggregation of larval Diptera over discrete and ephemeral breeding sites: the implications for coexistence. *American Naturalist*, **124**, 336–51. [Ch. 2]

Auerbach, M. and Simberloff, D. (1989) Oviposition site preference and larval mortality in a leaf-mining moth. *Ecological Entomology*, **14**, 131–40. [Ch. 16]

Auger, M.A., Geri, C., Jay-Allemand, C. and Bastien, C. (1990) Comestibilité de différénts clones de pin sylvestre pour *Diprion pini* L. (Hym., Diprionidae). 1. Incidence de la consommation des aiguilles de différénts clones de pin sylvestre sur le developpement de *Diprion pini* L. *Journal of Applied Entomology*, **110**, 489–500. [Ch. 8]

Averill, A.L. and Prokopy, R.J. (1989) Host-marking pheromones, in *Fruit Flies: their Biology, Natural Enemies and Control* (eds A.S. Robinson and G. Hooper), Elsevier, Amsterdam, pp. 207–17. [Ch. 10]

Awmack, C.S., Harrington, R., Leather, S.R. and Lawton, J.H. (1996) The impact of elevated CO_2 on aphid–plant interactions. *Aspects of Applied Biology*, **45**, 317–22. [Ch. 8]

Awmack, C.S., Woodcock, C.M. and Harrington, R. (1997a) Climate change may increase vulnerability of aphids to natural enemies. *Ecological Entomology*, **22**, 367–9. [Ch. 8]

Awmack, C.S., Harrington, R. and Leather, S.R. (1997b) Host plant effects on the performance of the aphid *Aulacorthum solani* (Homoptera:Aphididae) at ambient and elevated CO_2. *Global Change Biology* (in press). [Ch. 8]

Axelsen, J.A. (1994) Host–parasitoid interactions in an agricultural ecosystem – a computer simulation. *Ecological Modelling*, **73**, 189–203. [Ch. 6]

Ayal, Y. and Green, R.F. (1993) Optimal egg distribution among host patches for parasitoids subject to attack by hyperparasitoids. *American Naturalist*, **141**,120–38. [Ch. 9]

Baars, M.A. (1979a). Patterns of movement of radioactive carabid beetles. *Oecologia*, **44**, 125–40. [Ch. 3]

Baars, M.A. (1979b). Catches in pitfall traps in relation to mean densities of carabid beetles. *Oecologia*, **41**, 25–46. [Ch. 3]

Baars, M.A. and van Dijk, Th.S. (1984) Population dynamics of two carabid beetles at a Dutch heathland. II. Egg production and survival in relation to density. *Journal of Animal Ecology*, **53**, 389–400. [Ch. 3]

Baguette, M. and Nève, G. (1994) Adult movements between populations in the specialist butterfly *Proclossiana eunomia* (Lepidoptera, Nymphalidae). *Ecological Entomology*, **19**, 1–5. [Ch. 1]

Bailey, V.A., Nicholson, A.J. and Williams, E. (1962) Interaction between hosts and parasites when some host individuals are more difficult to find than others. *Journal of Theoretical Biology*, **3**, 1–18. [Ch. 7]

Bakker, K. (1964) Backgrounds of controversies about population theories and terminologies. *Zeitschrift für angewandte Entomologie*, **53**, 187–206. [Ch. 3]

Bakker, K. (1971) Some general remarks on the concepts 'population' and 'regulation', in *Dynamics of Populations* (eds P.J. den Boer and G.R. Gradwell), Centre for Agricultural Publishing and Documentation, Wageningen, pp. 565–7. [Ch. 3]

Baltensweiler, W. (1964) *Zeiraphera griseana* Hübner (Lepidoptera: Tortricidae) in the European Alps. A contribution to the problem of cycles. *Canadian Entomologist*, **96**, 792–800. [Ch. 1]

Baltensweiler, W. (1993) A contribution to the explanation of the larch bud moth cycles, the polymorphic fitness hypothesis. *Oecologia*, **93**, 251–5. [Ch. 8]

Baltensweiler, W., Benz, G., Bovey, P. and Delucchi, V. (1977) Dynamics of larch bud moth populations. *Annual Review of Entomology*, **22**, 79–100. [Ch. 16]

Barbosa, P. and Schultz, J.C. (eds) (1987) *Insect Outbreaks*, Academic Press, San Diego. [Ch. 16]

Barbosa, P., Krischik, V. and Lance, D. (1989) Life history traits of forest-inhabiting flightless Lepidoptera. *American Midland Naturalist*, **122**, 262–74. [Ch. 14]

Barlow, N.D. and Dixon, A.F.G. (1980) *Simulation of lime aphid population dynamics*, Centre for Agricultural Publishing and Documentation, Wageningen. [Ch. 6, 9]

Barlow, N.D., Moller, H. and Beggs, J.R. (1996) A model for the effect of *Sphecophaga vesparum vesparum* as a biological control agent of the common wasp in New Zealand. *Journal of Applied Ecology*, **33**, 31–44. [Ch. 6]

Barron, J.R. (1989) Status of the parasite *Agrypon flaveolatum* (Gravenhorst) (Hymenoptera: Ichneumonidae), introduced to control the winter moth in Nova Scotia and British Columbia. *Canadian Entomologist*, **121**, 11–26. [Ch. 13]

Bartlett, B.R. (1953) Natural control of the Citricola scale in California. *Journal of Economic Entomology*, **46**, 25–8. [Ch. 7]

Bauer, G. (1986) Life-history strategy of *Rhagoletis alternata* (Diptera: Trypetidae), a fruit fly operating in a 'non–interactive' system. *Journal of Animal Ecology*, **55**, 785–94. [Ch. 10]

Beddington, J.R. and Hammond, P.S. (1977) On the dynamics of host–parasite–hyperparasite interactions. *Journal of Animal Ecology*, **46**, 811–21. [Ch. 6]

Beddington, J.R., Free, C.A. and Lawton, J.H. (1978) Modelling biological control: on the characteristics of successful natural enemies. *Nature*, **273**, 513–9. [Ch. 6]

Begon, M., Sait, S.M. and Thompson, D.J. (1995) Persistence of a parasitoid–host system: refuges and generation cycles? *Proceedings of the Royal Society, London, Series B*, **260**, 131–7. [Ch. 6]

Begon, M., Bowers, R.G., Sait, S.M. and Thompson, D.J. (1996a) Population dynamics beyond two species: hosts, parasitoids and pathogens, in *Frontiers of Population Ecology* (eds R.B. Floyd, A.W. Sheppard and P.J. De Barro), pp. 115–26, CSIRO, Melbourne. [Ch. 6]

Begon, M., Sait, S.M. and Thompson, D.J. (1996b) Predator–prey cycles with period shifts between two- and three-species systems. *Nature*, **381**, 311–5. [Ch. 6, 11]

Bellows, T.S. (1981) The descriptive properties of some models for density dependence. *Journal of Animal Ecology*, **50**, 139–56. [Ch. 5]

Bellows, T.S. (1982a) Analytical models for laboratory populations of *Callosobruchus chinensis* and *C. maculatus* (Coleoptera, Bruchidae). *Journal of Animal Ecology*, **51**, 263–87. [Ch. 2]

Bellows, T.S. (1982b) Simulation models for laboratory populations of *Callosobruchus chinensis* and *C. maculatus. Journal of Animal Ecology*, **51**, 597–623. [Ch. 2]

Bellows, T.S. and Hassell, M.P. (1988) The dynamics of age-structured host–parasitoid interactions. *Journal of Animal Ecology*, **57**, 259–68. [Ch. 6]

Bennettova, B. and Fraenkel, G. (1981) What determines the number of ovarioles in a fly ovary? *Journal of Insect Physiology*, **27**, 403–10. [Ch. 8]

Bernays, E.A. and Minkenberg, O.J.M. (1997) Insect herbivores: different reasons for being a generalist. *Ecology*, **78**, 1157–69. [Ch. 8]

Bernstein, C. (1986) Density dependence and the stability of host–parasitoid systems. *Oikos*, **47**, 176–80. [Ch. 6]

Bernstein, C., Kacelnik, A. and Krebs, J.R. (1988) Individual decisions and the distribution of predators in a patchy environment. *Journal of Animal Ecology*, **57**, 1007–26. [Ch. 6]

Bernstein, C., Kacelnik, A. and Krebs, J.R. (1991) Individual decisions and the distribution of predators in a patchy environment. II. The influence of travel costs and the structure of the environment. *Journal of Animal Ecology*, **60**, 205–26. [Ch. 6]

Berryman, A.A. (1982) Population dynamics of bark beetles, in *Bark Beetles in North American Conifers: A System for the Study of Evolutionary Biology* (eds. J.B. Mitton and K.B. Sturgeon), University of Texas Press, Austin, pp. 264–314. [Ch. 14]

Berryman, A.A. (1991) Vague notions of density-dependence. *Oikos*, **62**, 252–4. [Ch. 2]

Berryman, A.A. (1992) On choosing models for describing and analysing ecological time series. *Ecology*, **73**, 694–8. [Ch. 5]

Berryman, A.A. (1994) Population dynamics: forecasting and diagnosis from time-series, in *Individuals, Populations and Patterns in Ecology* (eds S.R. Leather, A.D. Watt, N.J. Mills and K.F.A. Walters), Intercept, Andover, pp. 119–28. [Ch. 14]

Berryman, A.A. (1997) What causes population cycles of forest Lepidoptera? *Trends in Ecology and Evolution*, **11**, 28–32. [Ch. 6]

Berube, D. (1980) Interspecific competition between *Urophora affinis* and *U. quadrifasciata* (Diptera: Tephritidae) for ovipositional sites on diffuse knapweed (*Centaurea diffusa*: Compositae). *Zeitschrift für angewandte Entomologie*, **90**, 299–306. [Ch. 10]

Birch, L.C. (1971) The role of environmental heterogeneity and genetical heterogeneity in determining distribution and abundance, in *Dynamics of Populations* (eds P.J. den Boer and G.R. Gradwell), Centre for Agricultural Publishing and Documentation, Wageningen, pp. 109–28. [Ch. 2]

Blank, T.H., Southwood, T.R.E. and Cross, D.J. (1967) The ecology of the partridge. 1. Outline population processes with particular reference to chick mortality and nest density. *Journal of Animal Ecology*, **37**, 549–56. [Ch. 2]

Boerlijst, M.C., Lamers, M.E. and Hogeweg, P. (1993) Evolutionary consequences of spiral waves in a host–parasitoid system. *Proceedings of the Royal Society, London, Series B*, **253**, 15–18. [Ch. 6]

Boggs, C.L. (1981) Nutritional and life-history determinants of resource allocation in holometabolous insects. *American Naturalist*, **117**, 692–709. [Ch. 8]

Boggs, C.L. (1995) Male nuptial gifts: phenotypic consequences and evolutionary implications, in *Insect Reproduction* (eds S.R. Leather and J. Hardie), CRC Press, Boca Raton, Florida, pp. 215–42. [Ch. 8]

Boggs, C.L. and Gilbert, L.E. (1979) Male contribution to egg production in butterflies: evidence for transfer of nutrients at mating. *Science,* **206**, 83–4. [Ch. 8]

Boller, E.F. and Prokopy, R.J. (1976) Bionomics and management of *Rhagoletis*. *Annual Review of Entomology*, **21**, 223–46. [Ch. 10]

Bonsall, M.B. and Hassell, M. P. (1995) Identifying density-dependent processes: a comment on the regulation of winter moth. *Journal of Animal Ecology*, **64**, 781–4. [Ch. 7]

Bonsall, M.B. and Hassell, M.P. (1997) Apparent competition structures ecological assemblages. *Nature*, **388**, 371–3. [Ch. 6]

Boswell, M.T. and Patil, G.P. (1970) Chance mechanisms generating the negative binomial distribution, in *Random Counts in Models and Structures* (ed. G.P. Patil), Pennsylvania University Press, Philadelphia, pp. 1–22. [Ch. 2]

Box, G.E.P. and Draper, N.R. (1987) *Empirical Model-Building and Response Surfaces*, Wiley, New York. [Ch. 5]

Bradshaw, W.E. and Holzapfel, C.M. (1983) Predator-mediated, non-equilibrium coexistence of tree hole mosquitoes in southeastern North America. *Oecologia*, **57**, 239–56. [Ch. 17]

Bradshaw, W.E. and Holzapfel, C.M. (1988) Drought and the organization of tree hole mosquito communities. *Oecologia*, **74**, 507–14. [Ch. 17]

Briggs, C.J. (1993) Competition among parasitoids on a stage-structured host, and its effect on host suppression. *American Naturalist*, **141**, 372–97. [Ch. 6, 7]

Briggs, C.J. and Godfray, H.C.J. (1995) The dynamics of insect–pathogen interactions in stage-structured populations. *American Naturalist*, **145**, 855–87. [Ch. 6]

Briggs, C.J. and Godfray, H.C.J. (1996) The dynamics of insect–pathogen interactions in seasonal environments. *Journal of Animal Ecology*, **65**, 149–77. [Ch. 2]

Briggs, C.J., Nisbet, R.M. and Murdoch, W.W. (1993) Coexistence of competing parasitoid species on a host with a variable life cycle. *Theoretical Population Biology*, **44**, 341–73. [Ch. 6]

Briggs, C.J., Nisbet, R.M., Murdoch, W.W., Collier, T.R. and Metz, J.A.J. (1995) Dynamical effects of host-feeding in parasitoids. *Journal of Animal Ecology*, **64**, 403–16. [Ch. 6, 7]

Briggs, C.J., Murdoch, W.W. and Nisbet. R.M. (1998) Recent developments in theory for biological control of insect pests by parasitoids, in *Theoretical Approaches to Biological Control* (eds B.A. Hawkins and H.V. Cornell), Cambridge University Press, Cambridge (in press). [Ch. 7]

Brodmann, P.A., Wilcox, C.V. and Harrison, S. (1997) Mobile parasitoids may restrict the spatial spread of an insect outbreak. *Journal of Animal Ecology*, **66**, 65–72. [Ch. 6, 13]

Brown, C.E. (1962) The life history and dispersal of the Bruce spanworm, *Operophtera bruceata* (Hulst) (Lepidoptera: Geometridae). *Canadian Entomologist*, **94**, 1103–7. [Ch. 13]

Brown, J.H. (1995) *Macroecology*, University of Chicago Press, Chicago. [Ch. 1]

Brown, V.C. (1995) Insect herbivores and gaseous air pollutants – current knowledge and predictions, in *Insects in a Changing Environment* (eds R. Harrington and N. E. Stork), Academic Press, London, pp. 220–51. [Ch. 8]

Bulmer, M.G. (1975) The statistical analysis of density dependence. *Biometrics*, **31**, 901–11. [Ch. 2, 3, 5, 13]

Bumroongsook, S. and Harris, M.K. (1991) Nature of the conditioning effect on pecan by the blackmargined aphid. *Southwestern Entomologist*, **16**, 267–75. [Ch. 9]

Burnett, T. (1958) A model of host–parasite interaction. *Proceedings, 15th International Congress of Entomology*, **2**, pp. 679–86. [Ch. 6]

Burpee, D.M. and Sakaluk, S.K. (1993) Repeated matings offset costs of reproduction in female crickets. *Evolutionary Ecology*, **7**, 240–50. [Ch. 8]

Buse, A. and Good, J.E.G. (1996) Synchronization of larval emergence in winter moth (*Operophtera brumata* L.) and budburst in pedunculate oak (*Quercus robur* L.) under simulated climate change. *Ecological Entomology*, **21**, 335–43. [Ch. 13]

Bylund, H., Tenow, O. and Strong, D.R. (1995) Multiple mechanisms in cycles of the autumnal moth, cited in Bylund, H. (1995) *Long–term interactions between the autumnal moth and mountain birch: the roles of resources, competitors, natural enemies, and weather*, PhD thesis, Swedish University of Agricultural Sciences, Uppsala. [Ch. 13]

Byrne, D.N., Bellows, T.S. and Parella, M.P. (1990) Whiteflies in agricultural systems, in *Whiteflies: their Bionomics, Pest Status and Management*, (ed. Gerling, D.), Intercept, Andover, pp. 227–61. [Ch. 7]

Caldwell, J.P. (1993) Brazil nut fruit capsules as phytotelmata: interactions among anuran and insect larvae. *Canadian Journal of Zoology*, **71**, 1193-201. [Ch. 17]

Calvert, P.P. (1911) Studies on Costa Rican Odonata. II. The habits of the plant-dwelling larva of *Mecistogaster modestus*. *Entomological News*, **22**, 402–11. [Ch. 17]

Cameron, E. (1935) A study of the natural control of ragwort (*Senecio jacobaea* L.). *Journal of Ecology*, **23**, 265–322. [Ch. 12]

Cammell, M.E. and Knight, J.D. (1992) Effects of climatic change on the population dynamics of crop pests. *Advances in Ecological Research*, **22**, 117–62. [Ch. 8]

Canning, E.U. (1960) Two new microsporidian parasites of the winter moth, *Operophtera brumata* (L.). *Journal of Parasitology*, **46**, 755–63. [Ch. 13]

Canning, E.U. and Barker R.J. (1982) Transmission of microsporidia between generations of winter moth, *Operophtera brumata*. *Proceedings of the British Society of Parasitology*, (1981) **84**, xiv (abstracts). [Ch. 13]

Cappuccino, N. (1992) The nature of population stability in *Eurosta solidaginis*, a non-outbreaking herbivore of goldenrod. *Ecology*, **73**, 1792–801. [Ch. 14]

Cappuccino, N. and Price, P.W. (eds) (1995) *Population Dynamics: New Approaches and Synthesis*, Academic Press, San Diego. [Ch. 16, 17]

Carpenter, S.R., Cottingham, K.L. and Stow, C.A. (1994) Fitting predator–prey models to time series with observation error. *Ecology*, **75**, 1254–64. [Ch. 5]

Carr, T.G. (1995) *Oviposition preference–larval performance relationships in three free-feeding sawflies*, MSc thesis, Northern Arizona University, Flagstaff, Arizona. [Ch. 14]

Carroll, R.J., Ruppert, D. and Stefanksi, L.A. (1995) *Measurement Error in Nonlinear Models*, Chapman & Hall, London. [Ch. 5]

Carton, Y., Frey, F. and Nappi, A. (1992) Genetic determinism of the cellular immune reaction in *Drosophila melanogaster*. *Heredity*, **69**, 393–9. [Ch. 6]

Casas, J. (1989) Foraging behaviour of a leafminer parasitoid. *Ecological Entomology*, **14**, 257–65. [Ch. 6]

Caughley, G. and Lawton, J.H. (1981) Plant–herbivore systems, in *Theoretical Ecology* (ed. R.M. May), Blackwell, Oxford, pp. 132–66. [Ch. 10]

Chambers, R.J., Wellings, P.W. and Dixon, A.F.G. (1985) Sycamore aphid numbers and population density. II. Some processes. *Journal of Animal Ecology*, **54**, 425–42. [Ch. 9]

Charnov, E.L. (1982) *The Theory of Sex Allocation*, Princeton University Press, Princeton, New Jersey. [Ch. 8]

Charnov, E.L., Los-den Hartogh, R.L., Jones, W.T. and van den Assem, J. (1981) Sex ratio evolution in a variable environment. *Nature*, **289**, 27–33. [Ch. 6, 8]

Chesson, P. (1994) Multispecies competition in variable environments. *Theoretical Population Biology*, **45**, 227–76. [Ch. 2]

Chesson, P.L. and Huntly, N. (1989) Short-term instabilities and long-term community dynamics. *Trends in Ecology and Evolution*, **4**, 293–8. [Ch. 6]

Chesson, P.L. and Murdoch, W.W. (1986) Aggregation of risk: relationships among host–parasitoid models. *American Naturalist*, **127**, 696–715. [Ch. 6, 7]

Chitty, D. (1955) Adverse effects of population density upon the viability of later generations, in *The Numbers of Man and Animals* (eds J.B. Cragg and N.W. Pirie), Oliver & Boyd, Edinburgh, pp. 57–67. [Ch. 8]

Clancy, K.M., Price, P.W. and Sacchi, C.F. (1993) Is leaf size important for a leaf-galling sawfly (Hymenoptera: Tenthredinidae)? *Environmental Entomology*, **22**, 116–26. [Ch. 14]

Clark, L.R. (1963) The influence of population density on the number of eggs laid by females of *Cardiaspina albitextura* (Psyllidae). *Australian Journal of Zoology*, **11**, 190–201. [Ch. 15]

Clark, L.R. (1964) The population dynamics of *Cardiaspina albitextura* (Psyllidae). *Australian Journal of Zoology*, **12**, 362–80. [Ch. 1]

Clark, L.R. and Dallwitz, M.J. (1975) The life system of *Cardiaspina albitextura* (Psyllidae), 1950–74. *Australian Journal of Zoology*, **23**, 523–61. [Ch. 15]

Clark, L.R., Geier, P.W., Hughes, R.D. and Morris, R.F. (1967) *The Ecology of Insect Populations in Theory and Practice*, Methuen, London. [Ch. 1]

Clarke, J.W. (1986) Weather notes for Cambridgeshire 1985. *Nature in Cambridgeshire*, No. 28, 63–4. [Ch. 15]

Clarke, R.T., Thomas, J.A., Elmes, G.W. and Hochberg, M.E. (1997) The effects of spatial patterns in habitat quality on community dynamics within a site. *Proceedings of the Royal Society, London, Series B*, **264**, 247–54. [Ch. 11]

Clarke, R.T., Thomas, J.A., Elmes, G.W., Wardlaw, J.C., Munguira, M.L. and Hochberg, M.E. (1998) Population modelling of the spatial interactions between *Maculinea*, their initial food plant and *Myrmica* ants within a site. *Journal of Insect Conservation*, **2**, 29–37. [Ch. 11]

Clausen, C.P. (1978) Introduced Parasites and Predators of Arthropod Pests and Weeds: A World Review, Handbook 480, US Department of Agriculture, Washington, D.C. [Ch. 7]

Clutton-Brock, T.H., Guinness, F.E. and Albon, S.D. (1982) *Red Deer: Behaviour and Ecology of Two Sexes*, Edinburgh University Press, Edinburgh. [Ch. 2]

Cohen, J.E. (1990) Food webs and community structure, in *Community Food Webs* (eds J.E. Cohen, F. Briand and C.M. Newman), Springer-Verlag, Berlin, pp. 1–20. [Ch. 6]

Cole, L.C. (1951) Population cycles and random oscillations. *Journal of Wildlife Management*, **15**, 233–52. [Ch. 3]

Cole, L.C. (1954) Some features of random population cycles. *Journal of Wildlife Management*, **18**, 2–24. [Ch. 3]

Comins, H.N. and Hassell, M.P. (1996) Persistence of multispecies host–parasitoid interactions in spatially distributed models with local dispersal. *Journal of Theoretical Biology*, **183**, 19–28. [Ch. 6]

Comins, H.N. and Wellings, P.W. (1985) Density-related parasitoid sex-ratios: influence of host–parasitoid dynamics. *Journal of Animal Ecology*, **54**, 583–94. [Ch. 6]

Comins, H.N., Hassell, M.P. and May, R.M. (1992) The spatial dynamics of host–parasitoid systems. *Journal of Animal Ecology*, **61**, 735–48. [Ch. 2, 6, 7]

Connell, J.H. (1978) Diversity in tropical rain forests and coral reefs. *Science*, **199**, 1302–10. [Ch. 17]

Connell, J.H. and Sousa, W.P. (1983) On the evidence needed to judge ecological stability or persistence. *American Naturalist*, **121**, 789–824. [Ch. 16]

Connor, E.F., Adams-Manson, R.H., Carr, T.G. and Beck, M.W. (1994) The effects of host plant phenology on the demography and population dynamics of the leaf-mining moth, *Cameraria hamadryadella* (Lepidoptera: Gracillariidae). *Ecological Entomology*, **19**, 111–20. [Ch. 16]

Cook, J. and Stefanski, L.A. (1995) Simulation extrapolation method for parametric measurement error models. *Journal of the American Statistical Association*, **89**, 1314–28. [Ch. 5]

Corbett, A. and Rosenheim, J.A. (1996) Impact of a natural enemy overwintering refuge and its interaction with the surrounding landscape. *Ecological Entomology*, **21**, 155–64. [Ch. 7]

Coudriet, D.L., Meyerdirk, D.E., Prabhaker, N. and Kishaba, A.N. (1986) Bionomics of sweet potato whitefly (Homoptera; Aleyrodidae) on weed hosts in the Imperial Valley, California. *Environmental Entomology*, **15**, 1179–83. [Ch. 7]

Courtney, S.P. and Kibota, T.T. (1990) Mother doesn't know best: selection of hosts by ovipositing insects, in *Insect–Plant Interactions* (ed. E.A. Bernays), Vol. 2, CRC Press, Boca Raton, Florida, pp. 161–88. [Ch. 14]

Cox, D.R. and Oakes, D. (1984) *Analysis of Survival Data*, Chapman & Hall, London. [Ch. 5]

Cox, D.R. and Snell, E.J. (1981) *Applied Statistics: Principles and Examples*, Chapman & Hall, London. [Ch. 5]

Craig, T.P., Price, P.W. and Itami, J.K. (1986) Resource regulation by a stem-galling sawfly on the arroyo willow. *Ecology*, **67**, 419–25. [Ch. 14]

Craig, T.P., Itami, J.K. and Price, P.W. (1988) Plant wound compounds from oviposition scars used as oviposition deterrents by a stem-galling sawfly. *Journal of Insect Behavior*, **1**, 343–56. [Ch. 14]

Craig, T.P., Itami, J.K. and Price, P.W. (1989) A strong relationship between oviposition preference and larval performance in a shoot-galling sawfly. *Ecology*, **70**, 1691–9. [Ch. 14, 16]

Craig, T.P., Itami, J.K. and Price, P.W. (1990a) Intraspecific competition and facilitation by a shoot-galling sawfly. *Journal of Animal Ecology*, **59**, 147–59. [Ch. 14]

Craig, T.P., Itami, J.K. and Price, P.W. (1990b) The window of vulnerability of a shoot-galling sawfly to attack by a parasitoid. *Ecology*, **71**, 1471–82. [Ch. 14]

Craig, T.P., Itami, J.K. and Price, P.W. (1992) Facultative sex ratio shifts by a herbivorous insect in response to variation in host plant quality. *Oecologia*, **92**, 153–61. [Ch. 14]

Crawley, M.J. (1997) Plant–herbivore dynamics, in *Plant Ecology*, 2nd edn (ed. M.J. Crawley), Blackwell Science, Oxford, pp. 401–74. [Ch. 12]

Crawley, M.J. and Gillman, M.P. (1989) Population dynamics of cinnabar moth and ragwort in grassland. *Journal of Animal Ecology*, **58**, 1035–50. [Ch. 12]

Crawley, M.J. and Nachapong, M. (1985) The establishment of seedlings from primary and regrowth seeds of ragwort (*Senecio jacobaea*). *Journal of Ecology*, **73**, 255–61. [Ch. 12]

Crawley, M.J. and Pattrasudhi, R. (1988) Interspecific competition between insect herbivores: asymmetric competition between cinnabar moth and the ragwort seed-head fly. *Ecological Entomology*, **13**, 243–9. [Ch. 12]

Cronin, H. (1991) *The Ant and the Peacock*, Cambridge University Press, Cambridge. [Ch. 8]

Crowley, P.H. (1992) Density dependence, boundedness, and attraction: detecting stability in stochastic systems. *Oecologia*, **90**, 246–54. [Ch. 5]

Crowley, P.H. and Johnson, D.M. (1992) Variability and stability of a dragonfly assemblage. *Oecologica*, **90**, 260–9. [Ch. 17]

Cunningham, J.C., Tonks, N.V. and Kaupp, W.J. (1981) Viruses to control winter moth *Operophtera brumata* (Lepidoptera: Geometridae). *Journal of the Entomological Society of British Columbia*, **78**, 17–24. [Ch. 13]

Dall, S.R.X. and Cuthill, I.C. (1997) The information costs of generalism. *Oikos*, **80**, 197–202. [Ch. 8]

Darsie, R.F. (1993) *Keys to the mosquitoes of Costa Rica (Diptera, Culicidae)*. International Center for Disease Control, University of South Carolina. [Ch. 17]

Darwin, C. (1859) *The Origin of Species by Natural Selection*, John Murray, London. [Ch. 2]

Davidson, J. and Andrewartha, H.G. (1948a) Annual trends in a natural population of *Thrips imaginis* (Thysanoptera). *Journal of Animal Ecology*, **17**, 193–9. [Ch. 2]

Davidson, J. and Andrewartha, H.G. (1948b) The influence of rainfall, evaporation and atmospheric temperature on fluctations in the size of a natural population of *Thrips imaginis* (Thysanoptera). *Journal of Animal Ecology*, **17**, 200–22. [Ch. 2]

Day, K. and Crute, S. (1990) The abundance of spruce aphids under the influence of an oceanic climate, in *Population Dynamics of Forest Insects* (eds A.D. Watt, S.R. Leather, M.D. Hunter and N.A.C. Kidd), Intercept, Andover, pp. 25–33. [Ch. 9]

De Jong, G. (1979) The influence of the distribution of juveniles over patches of food on the dynamics of a population. *Netherlands Journal of Zoology*, **29**, 33–51. [Ch. 1, 2, 5]

De Jong, G. (1982) The influence of dispersal pattern on the evolution of fecundity. *Netherlands Journal of Zoology*, **32**, 1–30. [Ch. 2]

De La Rosa, C. and Ramirez, A. (1995) A note on phototactic behavior and on phoretic associations in larvae of *Mecistogaster ornata* Rambur from northern Costa Rica (Zygoptera: Pseudostigmatidae). *Odonatologica*, **24**, 219–24. [Ch. 17]

De Vries, H. (1996) *Viability of ground beetle populations in fragmented heathlands*, PhD thesis, Wageningen Agricultural University, The Netherlands. [Ch. 1]

Dempster, J.P. (1957) The population dynamics of the Moroccan locust (*Dociostaurus maroccanus* Thunberg) in Cyprus. *Antilocust Bulletin*, **27**, 1–60. [Ch. 14]

Dempster, J.P. (1963) The population dynamics of grasshoppers and locusts. *Biological Reviews*, **38**, 490–529 [Intro. and Conclud.]

Dempster, J.P. (1968) Intra-specific competition and dispersal: as exemplified by a psyllid and its anthocorid predator, in *Insect Abundance* (ed. T.R.E. Southwood), Blackwell Scientific Publications, Oxford, pp. 8–17. [Ch. 4, 15]

Dempster, J.P. (1971a) Some effects of grazing on the population ecology of the cinnabar moth, in *The Scientific Management of Animal and Plant Communities for Conservation*, 11th Symposium of the British Ecological Society (eds E. Duffey and A.S. Watt), Blackwell Scientific Publications, Oxford, pp. 517–26. [Ch. 12]

Dempster, J.P. (1971b) The population ecology of the cinnabar moth *Tyria jacobaeae* L. (Lepidoptera, Arctiidae). *Oecologia*, **7**, 26–67. [Ch. 4, 12]

Dempster, J.P. (1975) *Animal Population Ecology*, Academic Press, London. [Ch. 10, 12]

Dempster, J.P. (1982) The ecology of the cinnabar moth *Tyria jacobaeae* (Lepidoptera, Arctiidae). *Advances in Ecological Research*, **12**, 1–36. [Ch. 2, 4, 10, 12, 16]

Dempster, J.P. (1983) The natural control of populations of butterflies and moths. *Biological Reviews*, **58**, 461–81. [Intro. Ch. 1, 2, 4, 5, 6, 16, Conclud.]

Dempster, J.P. (1992) Evidence of an oviposition-deterring pheromone in the orange-tip butterfly, *Anthocharis cardamines* (L). *Ecological Entomology*, **17**, 83–5. [Ch. 4]

Dempster, J.P. (1997) The role of larval food resources and adult movement in the population dynamics of the orange-tip butterfly (*Anthocharis cardamines*). *Oecologia*, **111**, 549–56. [Ch. 4]

Dempster, J.P. and Lakhani, K.H. (1979) A population model for the cinnabar moth and its food plant, ragwort. *Journal of Animal Ecology*, **48**, 143–64. [Ch. 12]

Dempster, J.P. and Pollard, E. (1981) Fluctuations in resource availability and insect populations. *Oecologia*, **50**, 412–6. [Ch. 4, 10, 16]

Dempster, J.P. and Pollard, E. (1986) Spatial heterogeneity, stochasticity and the detection of density dependence in animal populations. *Oikos*, **46**, 413–16. [Ch. 2, 4]

Dempster, J.P., Atkinson, D.A. and Cheesman, O.D. (1995a) The spatial population dynamics of insects exploiting a patchy food resource. I. Population extinctions and regulation. *Oecologia*, **104**, 340–53. [Ch. 1, 4, 5, 10, 14]

Dempster, J.P., Atkinson, D.A. and French, M.C. (1995b) The spatial population dynamics of insects exploiting a patchy food resource. II. Movements between patches. *Oecologia*, **104**, 354–62. [Ch. 4, 10, Conclud.]

Den Boer, P.J. (1968) Spreading of risk and the stabilization of animal numbers. *Acta Biotheoretica (Leiden)*, **18**, 165–94. [Ch. 1, 2, 3, 4, 15]

Den Boer, P.J. (1971) Stabilization of animal numbers and the heterogeneity of the environment: the problem of the persistence of sparse populations, in *Dynamics of Populations* (eds P.J. den Boer and G.R. Gradwell), Centre for Agricultural Publishing and Documentation, Wageningen, pp. 77–97. [Ch. 2]

Den Boer, P.J. (1979) The individual behaviour and population dynamics of some carabid beetles of forests, in *On the Evolution and Behaviour of Carabid Beetles*, Miscellaneous Paper No. 18 (eds P.J. den Boer, H.U. Thiele and F. Weber), Landbouwhogeschool, Wageningen, pp. 151–66. [Ch. 3]

Den Boer, P.J. (1981) On the survival of populations in a heterogeneous and variable environment. *Oecologia*, **50**, 39–53. [Ch. 1, 3]

Den Boer, P.J. (1985). Fluctuations of density and survival of carabid populations. *Oecologia*, **67**, 322–30. [Ch. 3]

Den Boer, P.J. (1986a). Population dynamics of two carabid beetles at a Dutch heathland. The significance of density-related egg production, in *Carabid Beetles, their Adaptation and Dynamics* (eds P.J. den Boer, M.L. Luff, D. Mossakowski and F. Weber), Gustav Fischer, Stuttgart, pp. 361–70. [Ch. 3]

Den Boer, P.J. (1986b) Facts, hypotheses and models on the part played by food in the dynamics of carabid populations, in *Feeding Behaviour and Accessibility of Food for Carabid Beetles* (eds P.J. den Boer, L. Grüm and J. Szysszko), Warsaw Agricultural University Press, Warsaw, pp. 81–96. [Ch. 3]

Den Boer, P.J. (1986c) Density dependence and the stabilization of animal numbers. 1. The winter moth. *Oecologia*, **69**, 507–12. [Ch. 5, 13]

Den Boer, P.J. (1988) Density dependence and the stabilization of animal numbers. 3. The winter moth reconsidered. *Oecologia*, **75**, 161–8. [Ch. 13]

Den Boer, P.J. (1990a). On the stabilization of animal numbers. Problems of testing. 3. What do we conclude from significant test results? *Oecologia*, **86**, 38–46. [Ch. 3]

Den Boer, P.J. (1990b). Density limits and survival of local populations in 64 carabid species with different powers of dispersal. *Journal of Evolutionary Biology*, **3**, 19–40. [Ch. 3]

Den Boer, P.J. (1990c) Reaction to J. Latto and C. Berstein. Regulation in natural insect populations: reality or illusion? *Acta Oecologica*, **11**, 131–2. [Ch. 5]

Den Boer, P.J. (1991) Seeing the trees for the wood: random walks or bounded fluctuations of population size? *Oecologia* **86**, 484–91. [Ch. 2, 3]

Den Boer, P.J. and Gradwell, G.R. (eds) (1971) *Dynamics of Populations*, Centre for Agricultural Publishing and Documentation, Wageningen. [Ch. 16]

Den Boer, P.J. and Reddingius, J. (1989) On the stabilization of animal numbers. Problems of testing. 2. Confrontation with data from the field. *Oecologia*, **79**, 143–9. [Ch. 1, 2, 3, 5]

Den Boer, P.J. and Reddingius, J. (1996) *Regulation and Stabilization Paradigms in Population Ecology*, Chapman & Hall, London. [Ch. 3, 5]

Den Boer, P.J. and Sanders, G. (1970) Structure of vegetation and microweather. *Miscellaneous Papers Landbouwhogeschool, Wageningen*, **6**, 33–53. [Ch. 3]

Den Boer, P.J., Szysszko, J. and Vermeulen, R. (1993) Spreading the risk of extinction by genetic diversity in populations of the carabid beetle *Pterostichus oblongopunctatus* F. (Coleoptera, Carabidae). *Netherlands Journal of Zoology*, **43**, 242–59. [Ch. 3]

Dennis, B. and Taper, M.L. (1994) Density dependence in time series observations of natural populations: estimation and testing. *Ecological Monographs*, **64**, 205–24. [Ch. 2, 3, 5]

Denno, R.F. and Benrey, B. (1997). Aggregation facilitates larval growth in the neotropical nymphalid butterfly *Chlosyne janais*. *Ecological Entomology*, **22**, 133–41. [Ch. 8]

Denno, R.F. and Peterson, M.A. (1995) Density-dependent dispersal and its consequences for population dynamics, in *Population Dynamics: New Approaches and Synthesis* (eds N. Cappuccino and P.W. Price), Academic Press, San Diego, pp. 113–30. [Ch. 16]

Denno, R.F., Larsson, S. and Olmstead, K.L. (1990) Role of enemy-free space and plant quality in host-plant selection by willow beetles. *Ecology*, **71**, 124–37. [Ch. 16]

Denno, R.F., McClure, M.S. and Ott, J.R. (1995) Interspecific interactions in phytophagous insects: competition re-examined and resurrected. *Annual Review of Entomology*, **40**, 297–331. [Ch. 17]

Deseo, K.V. (1971) Study of factors influencing the fecundity and fertility of codling moth, (*Laspeyresia pomonella* L.: Lepidoptera, Tortricidae). *Acta Phytopathologica Academiae Scientiarum Hungaricae*, **6**, 243–52. [Ch. 8]

Dewar, R.C. and Watt, A.D. (1992) Predicted changes in the synchrony of larval emergence and budburst under climatic warming. *Oecologia*, **89**, 557–9. [Ch. 13]

Diamond, J.M. (1984) "Normal" extinctions of isolated populations, in *Extinctions* (ed. M.H. Nitecki), University of Chicago Press, Chicago, pp. 191–246. [Ch. 1]

Dixon, A.F.G. (1963) Reproductive activity of the sycamore aphid, *Drepanosiphum platanoides* (Schr.) (Hemiptera: Aphididae). *Journal of Animal Ecology*, **32**, 33–48. [Ch. 9]

Dixon, A.F.G. (1966) The effect of population density and nutritive status of the host on the summer reproductive activity of the sycamore aphid, *Drepanosiphum platanoides* (Schr.). *Journal of Animal Ecology*, **35**, 105–12. [Ch. 9]

Dixon, A.F.G. (1969) Population dynamics of the sycamore aphid *Drepanosiphum platanoides* (Schr.)(Hemiptera: Aphididae): migratory and trivial flight activity. *Journal of Animal Ecology*, **38**, 585–606. [Ch. 9]

Dixon, A.F.G. (1970) Stabilisation of aphid populations by an aphid induced plant factor. *Nature*, **227**, 1368–9. [Ch. 9]

Dixon, A.F.G. (1971) The role of intra-specific mechanisms and predation in regulating the numbers of the lime aphid, *Eucallipterus tiliae* L. *Oecologia*, **8**, 179–93. [Ch. 9]

Dixon, A.F.G. (1972) Control and significance of the seasonal development of colour forms in the sycamore aphid, *Drepanosiphum platanoides* (Schr.). *Journal of Animal Ecology*, **41**, 689–97. [Ch. 8]

Dixon, A.F.G. (1975) Effect of population density and food quality on autumnal reproductive activity in the sycamore aphid, *Drepanosiphum platanoides* (Schr.). *Journal of Animal Ecology*, **44**, 297–304 [Ch. 9]

Dixon, A.F.G. (1979) Sycamore aphid numbers: the role of weather, host and aphid, in *Population Dynamics* (eds R.M. Anderson, B.D. Turner and L.R. Taylor), Blackwell Scientific Publications, Oxford, pp. 105–21. [Ch. 9]

Dixon, A.F.G. (1985) *Aphid Ecology*, Blackie, Glasgow. [Ch. 8]

Dixon, A.F.G. (1987a) Aphid reproductive tactics, in *Population Structure, Genetics and Taxonomy of Aphids and Thysanoptera* (eds J. Holman, J. Pelikan, A.F.G. Dixon and L. Weismann), SPB Academic Publishing, Branisovska, pp. 3–18. [Ch. 8]

Dixon, A.F.G. (1987b) Evolution and adaptive significance of cyclical parthenogenesis in aphids, in *Aphids, Their Biology, Natural Enemies and Control* (eds A.K. Minks and P. Harrewijn), Elsevier, Amsterdam, pp. 289–97. [Ch. 9]

Dixon, A.F.G. (1990) Population dynamics and abundance of deciduous tree-dwelling aphids, in *Population Dynamics of Forest Insects* (eds M. Hunter, N. Kidd, S.R. Leather and A.D. Watt), Intercept, Andover, pp. 11–23. [Ch. 9]

Dixon, A.F.G. (1992) Constraints on the rate of parthenogenetic reproduction and pest status of aphids. *Invertebrate Reproduction and Development*, **22**, 159–63. [Ch. 9]

Dixon, A.F.G. (1997) *Aphid Ecology*, 2nd edn, Chapman & Hall, London. [Ch. 9]

Dixon, A.F.G. and Barlow, N.D. (1979) Population regulation in the lime aphid. *Zoological Journal of the Linnean Society*, **67**, 225–37. [Ch. 9]

Dixon, A.F.G. and Dharma, T.R. (1980) 'Spreading of the risk' in developmental mortality: size, fecundity and reproductive rate in the black bean aphid. *Entomologia Experimentalis et Applicata,* **28**, 301–12. [Ch. 8]

Dixon, A.F.G. and Glen, D.M. (1971) Morph determination in the bird cherry-oat aphid, *Rhopalosiphum padi* (L.). *Annals of Applied Biology,* **68**, 11–21. [Ch. 8]

Dixon, A.F.G. and Kindlmann, P. (1990) Role of plant abundance in determining the abundance of herbivorous insects. *Oecologia*, **83**, 281–3. [Ch. 9]

Dixon, A.F.G. and Mercer, D.R. (1983) Flight behaviour in the sycamore aphid: factors affecting take-off. *Entomologia Experimentalis et Applicata*, **33**, 43–9. [Ch. 9]

Dixon, A.F.G. and Russel, R.J. (1972) The effectiveness of *Anthocoris nemorum* and *A. confusus* (Hemiptera: Anthocoridae) as predators of the sycamore aphid, *Drepanosiphum platanoides*. II. Searching behaviour and the incidence of predation in the field. *Entomologia Experimentalis et Applicata*, **15**, 35–50. [Ch. 9]

Dixon, A.F.G. and Wratten, S.D. (1971) Laboratory studies on aggregation, size and fecundity in the black bean aphid, *Aphis fabae* Scop. *Bulletin of Entomological Research,* **61**, 97–111. [Ch. 8]

Dixon, A.F.G., Chambers, R.J. and Dharma, T.R. (1982) Factors affecting size in aphids with particular reference to the black bean aphid, *Aphis fabae*. *Entomologia Experimentalis et Applicata,* **32**, 123–8. [Ch. 8]

Dixon, A.F.G., Wellings, P.W., Carter, C. and Nichols, J.F.A. (1993) The role of food quality and competition in shaping the seasonal cycle in the reproductive activity of the sycamore aphid. *Oecologia*, **95**, 89–92. [Ch. 9]

Dixon, A.F.G., Hemptinne, J.-L. and Kindlmann, P. (1995) The ladybird fantasy – prospects and limits to their use in the biological control of aphids. (*Proceedings of an International Symposium on 75 years of Phytopathological and Resistance Research at Aschersleben*) *Züchtungsforschung*, **1**, 395–7. [Ch. 9]

Dixon, A.F.G., Kindlmann, P. and Sequeira, R. (1996) Population regulation in aphids, in *Frontiers of Population Ecology* (eds R.B. Floyd, A.W. Sheppard and P.J. De Barro), CSIRO, Melbourne, pp. 103–14. [Ch. 9]

Dixon, A.F.G., Hemptinne, J.-L. and Kindlmann, P. (1997) Effectiveness of ladybirds as biological control agents: patterns and processes. *Entomophaga*, **42**, 71–83. [Ch. 9]

Dohmen, G.P., McNeill, S. and Bell, J.N.B. (1984) Air pollution increases *Aphis fabae* pest potential. *Nature*, **307**, 52–3. [Ch. 8]

Donnelly, M. (1989a) Reproductive phenology and age structure of *Dendrobates* in Northeastern Costa Rica. *Journal of Herpetology*, **23**, 362–7. [Ch. 17]

Donnelly, M. (1989b) Demographic effects of reproductive resource supplementation in a territorial frog, *Dendrobates pumilio. Ecological Monographs*, **59**, 207–21. [Ch. 17]

Doumbia, M., Hemptinne, J.-L. and Dixon, A.F.G. (1997) Assessment of patch quality by ladybirds: role of larval tracks. *Oecologia*, **113**, 197–202. [Ch. 9]

Doutt, R.L. (1964) The historical development of biological control, in *Biological Control of Insect Pests and Weeds* (ed. P. DeBach), Reinhold, New York, pp. 21–42. [Ch. 9]

Driessen, G. and Hemerik, L. (1992) The time and egg budget of *Leptopilina clavipes*, a parasitoid of larval *Drosophila*. *Ecological Entomology*, **17**, 17–27. [Ch. 6]

Driessen, G. and Visser, M.E. (1997) Components of parasitoid interference. *Oikos*, **79**, 179–82. [Ch. 6]

Driessen, G., Bernstein, C., van Alphen, J.J.M. and Kacelnik, A. (1995) A count-down mechanism for host search in the parasitoid *Venturia canescens*. *Journal of Animal Ecology*, **64**, 117–25. [Ch. 6]

East, R. (1974) Predation on the soil-dwelling stages of the winter moth at Wytham Wood, Berkshire. *Journal of Animal Ecology*, **43**, 611–26. [Ch. 3, 13]

Eber, S. and Brandl, R. (1994) Ecological and genetic spatial patterns of *Urophora cardui* (Diptera: Tephritidae) as evidence for population structure and biogeographical processes. *Journal of Animal Ecology*, **63**, 187–99. [Ch. 1, 10]

Eber, S. and Brandl, R. (1996) Metapopulation dynamics of the tephritid fly *Urophora cardui*: an evaluation of incidence–function model assumptions with field data. *Journal of Animal Ecology*, **65**, 621–30. [Ch. 1, 10]

Eckhardt, R.D. (1979) The adaptive syndromes of two guilds of insectivorous birds in the Colorado Rocky Mountains. *Ecological Monographs*, **49**, 129–49. [Ch. 14]

Efron, B. and Tibshirani, R.J. (1993) *An Introduction to the Bootstrap*, Chapman & Hall, New York. [Ch. 5]

Ehler, L.E. (1977) Natural enemies of cabbage looper in cotton in the San Joaquin Valley. *Hilgardia*, **45**, 73–106. [Ch. 7]

Ehrlich, P.R. and Murphy, D.D. (1987) Conservation lessons from long-term studies of checkerspot butterflies. *Conservation Biology*, **1**, 122–31. [Ch. 1]

Elkington, J.S., Healy, W.M., Buonaccorsi, J.P. *et al.* (1996) Interactions among Gypsy moths, white-footed mice, and acorns. *Ecology*, **77**, 2332–42. [Ch. 5]

Ellington, J.J. and El-Sokkari, A. (1986) A measure of the fecundity, ovipositional behavior and mortality of the bollworm, *Heliothis zea* (Boddie) in the laboratory (Lepidoptera: Noctuidae). *South Western Entomologist*, **11**, 177–94. [Ch. 8]

Elliott, J.M. (1984) Numerical changes and population regulation in young migratory trout *Salmo trutta* in a Lake District stream. *Journal of Animal Ecology*, **53**, 327–50. [Ch. 2]

Ellner, S. and Turchin, P. (1995) Chaos in a noisy world: new methods and evidence from time-series analysis. *American Naturalist*, **145**, 343–60. [Ch. 5]

Elmes, G.W. and Thomas, J.A. (1987a). Die Gattung *Maculinea*, in *Tagfalter und ihr Lebensräum* (ed. W. Geiger), Schweizerisches Bund für Naturschutz, Basel, pp. 354–68. [Ch. 11]

Elmes, G.W. and Thomas, J.A. (1987b). Die Biologie und Ökologie der Ameisen der Gattung *Myrmica*, in *Tagfalter und ihr Lebensräum* (ed. W. Geiger), Schweizerisches Bund für Naturschutz, Basel, pp. 404–9. [Ch. 11]

Elmes, G.W. and Wardlaw, J.C. (1981) The quantity and quality of over-wintered larvae in five species of *Myrmica* (Hymenoptera: Formicidae). *Journal of Zoology, London*, **193**, 429–46. [Ch. 11]

Elmes, G.W. and Wardlaw, J.C. (1982) Variations in populations of *Myrmica sabuleti* and *M. scabrinodis* (Formicidae: Hymenoptera) in southern England. *Pedobiologia*, **23**, 90–7. [Ch. 11]

Elmes, G.W., Thomas J.A. and Wardlaw, J.C. (1991a) Larvae of *Maculinea rebeli*, a large-blue butterfly and their *Myrmica* host ants: wild adoption and behaviour in ant-nests. *Journal of Zoology, London*, **223**, 447–60. [Ch. 11]

Elmes, G.W., Wardlaw, J.C. and Thomas, J.A. (1991b) Larvae of *Maculinea rebeli*, a large-blue butterfly and their *Myrmica* host ants: patterns of caterpillar growth and survival. *Journal of Zoology, London*, **224**, 79–92. [Ch. 11]

Elmes, G.W., Thomas, J.A., Hammarstedt, O., *et al.* (1994) Differences in host-ant specificity between Spanish, Dutch and Swedish populations of the endangered butterfly, *Maculinea alcon* (Denis et Schiff.) (Lepidoptera). *Memorabilia Zoologica*, **48**, 55–68. [Ch. 11]

Elmes, G.W., Clarke, R.T., Thomas, J.A. and Hochberg, M.E. (1996) Empirical tests of specific predictions made from a spatial model of the population dynamics of *Maculinea rebeli*, a parasitic butterfly of red ant colonies. *Acta Oecologica*, **17**, 61–80. [Ch. 11]

Elmes, G.W., Thomas, J.A., Wardlaw, J.C. *et al.* (1998) The ecology of *Myrmica* ants in relation to the conservation of *Maculinea* butterflies. *Journal of Insect Conservation* **2**, 67–78. [Ch. 11]

Elton, C. (1949) Population interspersion: an essay on animal community patterns. *Journal of Ecology*, **37**, 1–23. [Ch. 1, 2]

Embree, D.G. (1965) The population dynamics of the winter moth in Nova Scotia, 1954–1962. *Memoirs of the Entomological Society of Canada*, **46**, 1–57. [Ch. 13]

Embree, D.G. (1966) The role of introduced parasites in the control of the winter moth in Nova Scotia. *Canadian Entomologist*, **98**, 1159–68. [Ch. 13]

Embree, D.G. (1991) The winter moth *Operophtera brumata* in eastern Canada, 1962–1988. *Forest Ecology and Management*, **39**, 47–54. [Ch. 13]

Embree, D.G. and Otvos, I.S. (1984) *Operophtera brumata* (L.), winter moth (Lepidoptera: Geometridae), in *Biological Control Programmes Against Insects and Weeds in Canada 1969–1980* (eds J.S. Kelleher and M.A. Hulme), CAB International, Slough, pp. 353–7. [Ch. 13]

Erwin, T.L. (1982) Tropical forests: their richness in Coleoptera and other arthropod species. *Coleopterists Bulletin*, **36**, 74–5. [Ch. 1]

Evans, E.W. and England, S. (1996) Indirect interactions in biological control of insects: pests and natural enemies in alfalfa. *Ecological Applications*, **6**, 920–30. [Ch. 6]

Faeth, S.H. (1987) Community structure and folivorous insect outbreaks: the roles of vertical and horizontal interactions, in *Insect Outbreaks* (eds P. Barbosa and J.C. Schultz), Academic Press, San Diego, pp. 135–71. [Ch. 16]

Feeny, P. (1970) Seasonal changes in oak leaf tannins and nutrients as a cause of spring feeding by winter moth caterpillars. *Ecology*, **51**, 565–81. [Ch. 13, 16]

Ferguson, C.S., Elkinton, J.S., Gould, J.R. and Wallner, W.E. (1994) Population regulation of gypsy moth (Lepidoptera: Lymantriidae) by parasitoids: does spatial density dependence lead to temporal density dependence. *Environmental Entomology*, **23**, 1155–64. [Ch. 6]

Field, S., Keller, M.A. and Calbert, G. (1997) The pay-off from superparasitism in the egg parasitoid *Trissolcus basalis*, in relation to patch defence. *Ecological Entomology*, **22**, 142–9. [Ch. 8]

Fincke, O.M. (1984) Giant damselflies in a tropical forest: reproductive behavior of *Megaloprepus coerulatus* with notes on *Mecistogaster*. *Advances in Odonatology*, **2**, 13–27. [Ch. 17]

Fincke, O.M. (1992a) Interspecific competition for tree holes: consequences for mating systems and coexistence in neotropical damselflies. *American Naturalist*, **139**, 80–101. [Ch. 17]

Fincke, O.M. (1992b) Consequences of larval ecology for territoriality and reproductive success of a neotropical damselfly. *Ecology*, **73**, 449–62. [Ch. 17]

Fincke, O.M. (1992c) Behavioural ecology of the giant damselflies of Barro Colorado Island, Panama (Odonata: Zygoptera: Pseudostigmatidae), in *Insects of Panama and Mesoamerica: Selected Studies* (eds D. Quintero Arias and A. Aiello), Oxford University Press, Oxford, pp. 102–13. [Ch. 17]

Fincke, O.M. (1994) Population regulation of a tropical damselfly in the larval stage by food limitation, cannibalism, intraguild predation and habitat drying. *Oecologia*, **100**, 118–27. [Ch. 17]

Fincke, O.M. (1995) Larval behaviour of a giant damselfly: territoriality or size-dependent dominance? *Animal Behaviour*, **51**, 77–87. [Ch. 17]

Fincke, O.M. (in press) Organisation of predator assemblages in neotropical tree holes: effects of abiotic factors and priority. *Ecological Entomology*. [Ch. 17]

Fincke, O.M., Yanoviak, S.P. and Hanschu, R.D. (1997) Predation by odonates depresses mosquito abundance in water-filled tree holes in Panama. *Oecologia*, **112**, 244–53. [Ch. 17]

Flanders, S.E. (1942) Biological observations on the Citricola scale and its parasites. *Journal of Economic Entomology*, **35**, 830–3. [Ch. 7]

Fletcher, B.S. (1987) The biology of Dacine fruit flies. *Annual Review of Entomology*, **32**, 115–44. [Ch. 10]

Flinn, P.W. and Hagstrum, D.W. (1995) Simulation model of *Cephalonomia waterstoni* (Hymenoptera: Bethylidae) parasitizing the rusty grain beetle (Coleoptera: Cucujidae). *Environmental Entomology*, **24**, 1608–15. [Ch. 6]

Foley, P. (1994) Predicting extinction times from environmental stochasticity and carrying capacity. *Conservation Biology*, **8**, 124–37. [Ch. 1, 2]

Foley, P. (1997) Extinction models for local populations, in *Metapopulation Biology: Ecology, Genetics and Evolution* (eds I.A. Hanski and M.E. Gilpin), Academic Press, San Diego, pp. 215–46. [Ch. 1]

Foote, R.H. (1984) Family Tephritidae, in *Catalogue of Palaearctic Diptera*, Part 9 (eds A. Soos and L. Papp), Akademiai Kiado, Budapest, pp. 66–149. [Ch. 10]

Foster, R.B. (1982) Seasonal rhythms of fruitfall on Barro Colorado Island in *The Ecology of a Tropical Forest: seasonal rhythms and long-term changes* (eds E.G. Leigh, A.S. Rand and D.M. Windsor), 2nd edn, Smithsonian Institution, Washington, D.C., pp. 151–72. [Ch. 17]

Foster, S.P., Harrington, R., Devonshire, A.L. *et al.* (1996) Comparative survival of insecticide-susceptible and resistant peach-potato aphids, *Myzus persicae* (Sulzer) (Hemiptera: Aphididae), in low temperature field trials. *Bulletin of Entomological Research*, **86**, 17–28. [Ch. 8]

Fox, D.R. and Ridsdill-Smith, J. (1995) Tests for density dependence revisited. *Oecologia*, **103**, 435–43. [Ch. 5]

Frank, J.H. (1967) The effect of pupal predators on a population of winter moth, *Operophtera brumata* (L.) (Hydriomenidae). *Journal of Animal Ecology*, **36**, 611–21. [Ch. 13]

Freese, G. (1995) Structural refuges in two stem-boring weevils on *Rumex crispus*. *Ecological Entomology*, **20**, 351–8. [Ch. 6]

Freidberg, A. (1984) Gall Tephritidae, in *Biology of Gall Insects* (ed. T.N. Ananthakrishnan), Oxford and IBH, New Delhi, pp. 129–67. [Ch. 10]

Fritz, R.S. and Price, P.W. (1990) A field test of interspecific competition on oviposition of gall-forming sawflies on willow. *Ecology*, **71**, 99–106. [Ch. 14]

Fritz, R.S., Sacchi, C.F. and Price, P.W. (1986) Competition versus host plant phenotype in species composition: willow sawflies. *Ecology*, **67**, 1608–18. [Ch. 14]

Fujii, K. (1968) Studies on interspecies competition between the azuki bean weevil and the southern cowpea weevil. III. Some characteristics of strains of two species. *Researches on Population Ecology (Kyoto)*, **10**, 87–98. [Ch. 2]

Fujii, K. (1975) A general simulation model for laboratory insect populations. I. From cohort of eggs to adult emergences. *Researches on Population Ecology (Kyoto)*, **17**, 85–133. [Ch. 2]

Fuller, W.A. (1987) *Measurement Error Models*, Wiley, New York. [Ch. 5]

Furuta, K. (1988) Annual alternating population size of the thuja aphid, *Cinara tujafilina* (Del Guercio), and the impacts of syrphids and disease. *Journal of Applied Entomology*, **105**, 344–54. [Ch. 9]

Gaston, K.J. (1994) *Rarity*, Chapman & Hall, London. [Ch. 1]

Gaston, K.J. and Lawton, J.H. (1987) A test of statistical techniques for detecting density dependence in sequential censuses of animal populations. *Oecologia*, **74**, 404–10. [Ch. 1, 16]

Gaston, K.J. and McArdle, B.H. (1994) The temporal variability of animal abundances: measures, methods and patterns. *Philosophical Transactions of the Royal Society of London, Series B*, **345**, 335–58. [Ch. 16]

Geier, P.W. (1964) Population dynamics of codling moth, *Cydia pomonella* (L.) (Tortricidae), in the Australian Capital Territory. *Australian Journal of Zoology*, **12**, 381–416. [Ch. 1]

Getz, W.M. and Mills, N.J. (1996) Host–parasitoid coexistence and egg-limited encounter rates. *American Naturalist*, **148**, 333–47. [Ch. 6]

Ghent, A.W. (1960) A study of the group-feeding behavior of larvae of the jack pine sawfly, *Neodiprion pratti banksiannae*. *Behaviour*, **16**, 110–48. [Ch. 8]

Gilbert, N. and Raworth, D.A. (1996) Insects and temperature. *Canadian Entomologist*, **128**, 1–13. [Ch. 8]

Gillman, M.P. and Crawley, M.J. (1990) The cost of sexual reproduction in ragwort. *Functional Ecology*, **4**, 585–9. [Ch. 12]

Godfray, H.C.J. (1994) *Parasitoids, Behavioral and Evolutionary Ecology*, Princeton University Press, Princeton, New Jersey. [Ch. 6]

Godfray, H.C.J. and Chan, M.S. (1990) How insecticides trigger single-stage outbreaks in tropical pests. *Functional Ecology*, **4**, 329–37. [Ch. 2, 6]

Godfray, H.C.J. and Hassell, M.P. (1987) Natural enemies can cause discrete generations in tropical insects. *Nature*, **327**, 144–7. [Ch. 6]

Godfray, H.C.J. and Hassell, M.P. (1989) Discrete and continuous insect populations in tropical environments. *Journal of Animal Ecology*, **58**, 153–74. [Ch. 2, 6, 7]

Godfray, H.C.J. and Hassell, M.P. (1990) Encapsulation and host–parasitoid population dynamics, in *Parasite–Host Associations, Coexistence or Conflict* (eds C.A. Toft, A. Aeschlimann and L. Bolis), Oxford University Press, Oxford, pp. 131–47. [Ch. 6]

Godfray, H.C.J. and Hassell, M.P. (1992) Long time series reveal density dependence. *Nature*, **359**, 673–4. [Ch. 2]

Godfray, H.C.J. and Pacala, S.W. (1992) Aggregation and the population dynamics of parasitoids and predators. *American Naturalist*, **140**, 30–40. [Ch. 6, 7]

Godfray, H.C.J. and Waage, J.K. (1991) Predictive modelling in biological control: the mango mealy bug (*Rastrococcus invadens*) and its parasitoids. *Journal of Applied Ecology*, **28**, 434–53. [Ch. 6, 7]

Godfray, H.C.J., Hassell, M.P. and Holt, R.D. (1994) The population dynamic consequences of phenological asynchrony between parasitoids and their hosts. *Journal of Animal Ecology*, **63**, 1–10. [Ch. 6, 7]

Goeden, R.D. (1987) Host–plant relations of native *Urophora* spp. (Diptera: Tephritidae) in southern California. *Proceedings of the Entomological Society of Washington*, **89**, 269–74. [Ch. 10]

Goeden, R.D. and Ricker, D.W. (1987) Phytophagous insect faunas of native *Cirsium* thistles, *C. mohavense*, *C. neomexicanum* and *C. nidulum*, in the Mohave Desert of southern California (Diptera: Tephritidae). *Annals of the Entomological Society of America*, **80**, 161–75. [Ch. 10]

Goeden, R.D. and Teerink, J.A. (1993) Phytophagous insect faunas of *Dicoria canescens* and *Iva axilaris*, native relatives of ragweeds, *Ambrosia* spp., in southern California, with analyses of insect associates of Ambrosiinae. *Annals of the Entomological Society of America*, **86**, 37–50. [Ch. 10]

Gohari, H. and Hawlitzky, N. (1986) Activité reproductrice de la pyrale du mais, *Ostrinia nubilalis*, Hbn. (Lep., Pyralidae) à basses températures constantes. *Agronomie*, **6**, 911–8. [Ch. 8]

Gordon, D.M., Nisbet, R.M., De Roos, A. *et al.* (1991) Discrete generations in host–parasitoid models with contrasting life cycles. *Journal of Animal Ecology*, **60**, 295–308. [Ch. 2, 6]

Gould, J.R., Elkinton, J.S. and Wallner, W.E. (1990) Density-dependent suppression of experimentally created gypsy moth, *Lymantria dispar* (Lepidoptera: Lymantriidae) populations by natural enemies. *Journal of Animal Ecology*, **59**, 213–33. [Ch. 6, 7]

Gould, S.J. and Lewontin, R.C. (1979) The spandrels of San Marco and the Panglossian paradigm: a critique of the adaptationist programme. *Proceedings of the Royal Society of London, Series B*, **205**, 581–98. [Ch. 14]

Gradwell, G.R. (1974) The effect of defoliators on tree growth, in *The British Oak* (eds M.G. Morris and F.H. Perring), Classey, Faringdon, pp. 182–93. [Ch. 13]

Green, R.F. (1986) Does aggregation prevent competitive exclusion? A response to Atkinson and Shorrocks. *American Naturalist*, **128**, 301–4. [Ch. 2]

Griffiths, K.J. and Holling, C.S. (1969) A competition submodel for parasites and predators. *Canadian Entomologist*, **101**, 785–818. [Ch. 7]

Griffiths, K.J., Cunningham, J.C. and Otvos, I.S. (1984) *Neodiprion sertifer* (Geoffroy), European pine sawfly (Hymenoptera: Diprionidae), in *Biological Control Programmes against Insects and Weeds in Canada 1969–1980* (eds J.S. Kelleher and M.A. Hulme), CAB International, Slough. [Ch. 2]

Gross, R.S. and Werner, P.A. (1983) Probabilities of survival and reproduction relative to rosette size in the common burdock (*Arctium minus*: Compositae). *American Midland Naturalist*, **109**, 184–93. [Ch. 10]

Gross, R.S., Werner, P.A. and Hawthorn, W.R. (1980) The biology of Canadian weeds. 38. *Arctium minus* (Hill) Bernh. and *A. lappa* L. *Canadian Journal of Plant Science*, **60**, 621–34. [Ch. 10]

Gurney, W.S.C. and Nisbet, R.M. (1985) Fluctuation periodicity, generation separation, and the expression of larval competition. *Theoretical Population Biology*, **28**, 150–80. [Ch. 2]

Gurney, W.S.C. and Nisbet, R.M. (1998) *Ecological Dynamics*, Oxford University Press, New York (in press). [Ch. 12]

Gurney, W.S.C., Nisbet, R.M. and Lawton, J.H. (1983) The systematic formulation of tractable single-species population models incorporating age structure. *Journal of Animal Ecology*, **52**, 479–96. [Ch. 2, 6]

Gutierrez, A.P. and Baumgärtner, J.U. (1984) Multitrophic level models of predator–prey energetics. II. A realistic model of plant–herbivore–parasitoid–predator interactions. *Canadian Entomologist*, **116**, 933–49. [Ch. 6]

Gutierrez, A.P., Mills, N.J., Schreiber, S.J. and Ellis, C.K. (1994) A physiologically based tritrophic perspective on bottom-up–top-down regulation of populations. *Ecology*, **75**, 2227–42. [Ch. 6]

Gwynne, D.T. (1988) Courtship feeding and the fitness of female katydids (Orthoptera: Tettigoniidae). *Evolution*, **42**, 545–55. [Ch. 8]

Gyllenberg, M. and Hanski, I. (1992) Single-species metapopulation dynamics: a structured model. *Theoretical Population Biology*, **72**, 35–61. [Ch. 1]

Gyllenberg, M., Hanski, I. and Hastings, A. (1997) Structured metapopulation models, in *Metapopulation Biology: Ecology, Genetics and Evolution* (eds I.A. Hanski and M.E. Gilpin), Academic Press, San Diego, pp. 93–122. [Ch. 1]

Hails, R. and Crawley, M.J. (1992) Spatial density dependence in populations of a cynipid gall-former *Andricus quercuscalicis*. *Journal of Animal Ecology*, **61**, 567–83. [Ch. 2]

Hairston, N.G., Smith, F.E. and Slobodkin, L.B. (1960) Community structure, population control and competition. *American Naturalist*, **94**, 421–5. [Intro, Ch. 4, 16]

Halley, J.M. and Dempster, J.P. (1996) The spatial population dynamics of insects exploiting a patchy food resource: a model study of local persistence. *Journal of Applied Ecology*, **33**, 439–54. [Ch. 10]

Hamilton, P.A. (1973) The biology of *Aphelinus flavus* (Hym. Aphelinidae), a parasite of the sycamore aphid *Drepanosiphum platanoidis* (Hemipt. Aphididae). *Entomophaga*, **18**, 449–62. [Ch. 9]

Hamilton, P.A. (1974) The biology of *Monoctonus pseudoplatani*, *Trioxys cirsii* and *Dyscritulus planiceps*, with notes on their effectiveness as parasites of the sycamore aphid, *Drepanosiphum platanoides*. *Annales de la Société Entomologique de France*, **10**, 821–40. [Ch. 9]

Hammond, P. (1992) Species inventory, in *Global Biodiversity: Status of the Earth's Living Resources* (ed. B. Groombridge), Chapman & Hall, London, pp. 17–39. [Ch. 1]

Hanski, I. (1979) *The community of coprophagous beetles*. DPhil thesis, University of Oxford. [Ch. 1]

Hanski, I. (1981) Coexistence of competitors in patchy environment with and without predation. *Oikos*, **37**, 306–12. [Ch. 1]

Hanski, I. (1985) Single-species spatial dynamics may contribute to long-term rarity and commonness. *Ecology*, **66**, 335–43. [Ch. 1]

Hanski, I. (1990) Density dependence, regulation and variability in animal populations. *Philosophical Transactions of the Royal Society, London, Series B*, **330**, 141–50. [Ch. 1, 16]

Hanski, I. (1991) Single-species metapopulation dynamics: concepts, models and observations. *Biological Journal of the Linnean Society*, **42**, 17–38. [Ch. 1, 3]

Hanski, I. (1994a) A practical model of metapopulation dynamics. *Journal of Animal Ecology*, **63**, 151–62. [Ch. 1]

Hanski, I. (1994b) Patch-occupancy dynamics in fragmented landscapes. *Trends in Ecology and Evolution*, **9**, 131–5. [Ch. 1]

Hanski, I. (1994c) Spatial scale, patchiness and population dynamics on land. *Philosophical Transactions of the Royal Society, London, Series B*, **343**, 19–25. [Ch. 2]

Hanski, I. (1997) Metapopulation biology: from concepts and observations to predictive models, in *Metapopulation Biology: Ecology, Genetics, and Evolution* (eds I.A. Hanski and M.E. Gilpin), Academic Press, London, pp. 69–92. [Ch. 1]

Hanski, I. and Cambefort, Y. (eds) (1991) *Ecology of Dung Beetles*, Princeton University Press, Princeton, New Jersey. [Ch. 1]

Hanski, I. and Gilpin, M. (1991) Metapopulation dynamics: brief history and conceptual domain. *Biological Journal of the Linnean Society*, **42**, 3–16. [Ch. 10]

Hanski, I. and Gilpin, M.E. (eds) (1997) *Metapopulation Biology: Ecology, Genetics, and Evolution*, Academic Press, San Diego. [Ch. 1, 2, 7]

Hanski, I. and Gyllenberg, M. (1993) Two general metapopulation models and the core-satellite species hypothesis. *American Naturalist*, **142**, 17–41. [Ch. 1]

Hanski, I. and Hammond, P. (1995) Biodiversity in boreal forests. *Trends in Ecology and Evolution*, **10**, 5–6. [Ch. 1]

Hanski, I. and Kuussaari, M. (1995) Butterfly metapopulation dynamics, in *Population Dynamics: New Approaches and Synthesis* (eds N. Cappuccino and P. Price), Academic Press, San Diego, pp. 149–71. [Ch. 1]

Hanski, I. and Simberloff, D. (1997) The metapopulation approach, its history, conceptual domain and application to conservation, in *Metapopulation Biology: Ecology, Genetics, and Evolution* (eds I.A. Hanski and M.E. Gilpin), Academic Press, San Diego, pp. 5–26. [Ch. 1]

Hanski, I. and Woiwod, I.P. (1993) Spatial synchrony in the dynamics of moth and aphid populations. *Journal of Animal Ecology*, **62**, 656–68. [Ch. 5]

Hanski, I., Kouki, J. and Halkka, A. (1993a) Three explanations of the positive relationship between distribution and abundance of species, in *Community Diversity: Historical and Geographical Perspectives* (eds R.E. Ricklefs and D. Schluter), University of Chicago Press, Chicago, pp. 108–16. [Ch. 1]

Hanski, I., Woiwod, I.P. and Perry, J.N. (1993b) Density dependence, population persistence, and largely futile arguments. *Oecologia*, **95**, 595–8. [Ch. 1, 2, 5]

Hanski, I., Kuussaari, M. and Nieminen, M. (1994) Metapopulation structure and migration in the butterfly *Melitaea cinxia*. *Ecology*, **75**, 747–62. [Ch. 1]

Hanski, I., Pakkala, T., Kuussaari, M. and Lei, G. (1995a) Metapopulation persistence of an endangered butterfly in a fragmented landscape. *Oikos*, **72**, 21–8. [Ch. 1, 6]

Hanski, I., Pöyry, J., Kuussaari, M. and Pakkala, T. (1995b) Multiple equilibria in metapopulation dynamics. *Nature*, **377**, 618–21. [Ch. 1]

Hanski, I., Foley, P. and Hassell, M. P. (1996a) Random walks in a metapopulation: how much density dependence is necessary for long-term persistence? *Journal of Animal Ecology*, **65**, 274–82. [Ch. 1, 2]

Hanski, I., Moilanen, A., Pakkala, T. and Kuussaari, M. (1996b) Metapopulation persistence of an endangered butterfly: a test of the quantitative incidence function model. *Conservation Biology*, **10**, 578–90. [Ch. 1]

Hanski, I., Moilanen, A. and Gyllenberg, M. (1996c) Minimum viable metapopulation size. *American Naturalist*, **147**, 527–41. [Ch. 1]

Harcourt, D.G. (1965) Spatial pattern in the cabbage looper, *Trichoplusia ni*, on crucifers. *Annals of the Entomological Society of America*, **58**, 89–94. [Ch. 2]

Harcourt, D.G. (1971) Population dynamics of *Leptinotarsa decemlineata* (Say) in eastern Ontario. III. Major population processes. *Canadian Entomologist*, **103**, 1049–61. [Ch. 2]

Harcourt, D.G. and Cass, L.M. (1966) Photoperiodism and fecundity in *Plutella maculipennis* (Curt.). *Nature*, **210**, 217–18. [Ch. 8]

Harper, J.L. (1977) *Population Biology of Plants*, Academic Press, London. [Ch. 10]

Harris, P. (1980) Effects of *Urophora affinis* Frfld. and *U. quadrifasciata* (Meig.) (Diptera: Tephritidae) on *Centaurea diffusa* Lam. and *C. maculosa* Lam. (Compositae). *Zeitschrift für angewandte Entomologie*, **90**, 190–201. [Ch. 10]

Harris, P. (1989) The use of Tephritidae for the biological control of weeds. *Biocontrol News and Information*, **10**, 9–16. [Ch. 10]

Harrison, S. (1991) Local extinction in a metapopulation context: an empirical evaluation. *Biological Journal of the Linnean Society*, **42**, 73–88. [Ch. 1]

Harrison, S. (1994) Metapopulations and conservation, in *Large-scale Ecology and Conservation Biology* (eds P.J. Edwards, R.M. May and N.R. Webb), Blackwell Scientific Publications, Oxford, pp. 111–28. [Ch. 1]

Harrison, S. (1997) Persistent, localized outbreaks in the western tussock moth *Orgyia vetusta*: the roles of resource quality, predation and poor dispersal. *Ecological Entomology*, **22**, 158–66. [Ch. 6]

Harrison, S. and Cappuccino, N. (1995) Using density-manipulation experiments to study population regulation, in *Population Dynamics: New Approaches and Synthesis* (eds N. Cappuccino and P.W. Price), Academic Press, San Diego, pp. 131–47. [Ch. 14, 16]

Harrison, S. and Taylor, A.D. (1997) Empirical evidence for metapopulation dynamics, in *Metapopulation Dynamics: Ecology, Genetics and Evolution* (eds I.A. Hanski and M.E. Gilpin), Academic Press, San Diego, pp. 27–42. [Ch. 7]

Harrison, S. and Thomas, C.D. (1991) Patchiness and spatial pattern in the insect community on ragwort (*Senecio jacobaea*). *Oikos*, **62**, 5–12. [Ch. 12]

Harrison, S., Murphy, D.D. and Ehrlich, P.R. (1988) Distribution of the bay checkerspot butterfly, *Euphydryas editha bayensis*: evidence for a metapopulation model. *American Naturalist*, **132**, 360–82. [Ch. 1]

Hartley, S.E. and Gardner, S.M. (1995) The response of *Philaenus spumarius* (Homoptera: Cercopidae) to fertilizing and shading its moorland host-plant (*Calluna vulgaris*). *Ecological Entomology*, **20**, 396–9. [Ch. 8]

Hartmann, T. (1995) Pyrrolizidine alkaloids between plants and insects: a new chapter of an old story. *Chemoecology*, **5/6**, 3/4, 139–46. [Ch. 12]

Hartshorn, G.S. (1983) *Lecythis costaricensis* (Jicaro, Olla de Mono, Monkey Pot) in *A Costa Rican Natural History* (ed. D.H. Janzen), University of Chicago Press, Chicago, pp. 268–9. [Ch. 17]

Hartshorn, G.S. and Hammel, B.E. (1994) Vegetation types and floristic patterns, in *La Selva: ecology and natural history of a neotropical rain forest* (eds L.A. McDade, K.S. Bawa, H.A. Hespendeide and G.S. Hartshorn), University of Chicago Press, Chicago, pp. 73–89. [Ch. 17]

Harvey, P.H. and Pagel, M.D. (1991) *The Comparative Method in Evolutionary Biology*, Oxford University Press, Oxford. [Ch. 14]

Hassell, M.P. (1968) The behavioural response of a tachinid fly (*Cyzenis albicans* (Fall.)) to its host, the winter moth (*Operophtera brumata* (L.)). *Journal of Animal Ecology*, **37**, 627–39. [Ch. 13]

Hassell, M.P. (1975) Density dependence in single-species populations. *Journal of Animal Ecology*, **44**, 283–95. [Ch. 2]

Hassell, M.P. (1976) *The Dynamics of Competition and Predation*, Edward Arnold, London. [Ch. 11]

Hassell, M.P. (1978) *The Dynamics of Arthropod Predator–Prey Systems*, Princeton University Press, Princeton, New Jersey. [Ch. 1, 6, 7]

Hassell, M.P. (1984) Parasitism in patchy environments: inverse density dependence can be stabilizing. IMA Journal of Mathematics Applied in Medicine & Biology, **1**, 123–33. [Ch. 6]

Hassell, M.P. (1985) Insect natural enemies as regulating factors. *Journal of Animal Ecology*, **54**, 323–34. [Ch. 4, 16]

Hassell, M.P. (1986a) Detecting density dependence. *Trends in Ecology and Evolution*, **1**, 90–3. [Ch. 5, 16]

Hassell, M.P. (1986b) Parasitoids and population regulation, in *Insect Parasitoids* (eds J.K. Waage and D.J. Greathead), Blackwell Scientific Publications, Oxford, pp. 201–24. [Ch. 6]

Hassell, M.P. (1987) Detecting regulation in patchily distributed animal populations. *Journal of Animal Ecology*, **56**, 705–13. [Ch. 2, 3, 5]

Hassell, M.P. (1998) *Insect Parasitoid Population Dynamics*, Oxford University Press, Oxford. [Ch. 6]

Hassell, M.P. and Anderson, R.M. (1984) Host susceptibility as a component in host–parasitoid systems. *Journal of Animal Ecology*, **53**, 611–21. [Ch. 6]

Hassell, M.P. and Comins, H.N. (1978) Sigmoid functional responses and population stability. *Theoretical Population Biology*, **14**, 62–6. [Ch. 6]

Hassell, M.P. and Godfray, H.C.J. (1992) Insect parasitoids, in *Natural Enemies* (ed. M.J. Crawley), Blackwell Scientific Publications, Oxford, pp. 265–92. [Ch. 6]

Hassell, M.P. and May, R.M. (1973) Stability in insect host–parasite models. *Journal of Animal Ecology*, **42**, 693–726. [Ch. 6, 7]

Hassell, M.P. and May, R.M. (1974) Aggregation of predators and insect parasites and its effect on stability. *Journal of Animal Ecology*, **43**, 567–94. [Ch. 6]

Hassell, M.P. and May, R.M. (1985) From individual behaviour to population dynamics, in *Behavioural Ecology* (eds R.M. Sibly and R.H. Smith), Blackwell Scientific Publications, Oxford, pp. 3–32. [Ch. 1, 2, 6, 16]

Hassell, M.P. and May, R.M. (1986) Generalist and specialist natural enemies in insect predator–prey interactions. *Journal of Animal Ecology*, **55**, 923–40. [Ch. 6]

Hassell, M.P. and Sabelis, M.W. (1987) Density-vague ecology: a reply. *Trends in Ecology and Evolution*, **2**, 78. [Ch. 2]

Hassell, M.P. and Varley, G.C. (1969) New inductive population model for insect parasites and its bearing on biological control. *Nature*, **223**, 1133–7. [Ch. 6]

Hassell, M.P. and Waage, J.K. (1984) Host–parasitoid population interactions. *Annual Review of Entomology*, **29**, 89–101. [Ch. 6]

Hassell, M.P., Lawton, J.H. and May, R.M. (1976) Patterns of dynamical behaviour in single-species populations. *Journal of Animal Ecology*, **45**, 471–86. [Ch. 2, 5]

Hassell, M.P., Waage, J.K. and May, R.M. (1983) Variable parasitoid sex ratios and their effect on host parasitoid dynamics. *Journal of Animal Ecology*, **52**, 889–904. [Ch. 6]

Hassell, M.P., Southwood, T.R.E. and Reader, P.M. (1987) The dynamics of the viburnum whitefly (*Aleurotrachelus jelinkii*): a case study of population regulation. *Journal of Animal Ecology*, **56**, 283–300. [Ch. 2, 5]

Hassell, M.P., Latto, J. and May, R.M. (1989a) Seeing the wood for the trees: detecting density dependence from existing life-table studies. *Journal of Animal Ecology*, **58**, 883–92. [Ch. 1, 2, 6]

Hassell, M.P., Taylor, V.A. and Reader, P.M. (1989b) The dynamics of laboratory populations of *Callosobruchus chinensis* and *C. maculatus* (Coleoptera: Bruchidae) in patchy environments. *Researches on Population Ecology (Kyoto)*, **31**, 35–52. [Ch. 2]

Hassell, M.P., Comins, H.N. and May, R.M. (1991a) Spatial structure and chaos in insect population dynamics. *Nature*, **353**, 255–8. [Ch. 2, 6]

Hassell, M.P., Pacala, S.W., May, R.M. and Chesson, P.L. (1991b) The persistence of host–parasitoid associations in patchy environments. I. A general criterion. *American Naturalist*, **138**, 568–83. [Ch. 6]

Hassell, M.P., Comins, H.N. and May, R.M. (1994) Species coexistence and self-organizing spatial dynamics. *Nature*, **370**, 290–2. [Ch. 2, 6]

Hastings, A. (1991) Structured models of metapopulation dynamics, in *Metapopulation Dynamics* (eds M. Gilpin and I. Hanski), Academic Press, London, pp. 57–71. [Ch. 1]

Haukioja, E. and Hakala, T. (1975) Herbivore cycles and periodic outbreaks. Formulation of a general hypothesis. *Reports from the Kevo Subarctic Research Station*, **12**, 1–9. [Ch. 13]

Haukioja, E. and Neuvonen, S. (1987) Insect population dynamics and induction of plant resistance: the testing of hypotheses, in *Insect Outbreaks* (eds P. Barbosa and J.C. Schultz), Academic Press, San Diego, pp. 411–32. [Ch. 16]

Hawkins, B.A. (1992) Parasitoid–host food webs and donor control. *Oikos*, **65**, 159–62. [Ch. 16]

Hawkins, B.A. (1994) *Pattern and Process in Host–parasitoid Interactions*. Cambridge University Press, Cambridge. [Ch. 6]

Hawkins, B.A. and Lawton, J.H. (1987) Species richness for parasitoids of British phytophagous insects. *Nature*, **326**, 788–90. [Ch. 6]

Hawkins, B.A., Thomas, M.B. and Hochberg, M.E. (1993) Refuge theory and biological control. *Science*, **262**, 1429–32. [Ch. 6, 7]

Heads, P.A. and Lawton, J.H. (1983) Studies on the natural enemy complex of the holly leaf miner: the effects of scale on the detection of aggregative responses and the implications for biological control. *Oikos*, **40**, 267–76. [Ch. 5, 13]

Heimpel, G.E. and Collier, T.R. (1996) The evolution of host-feeding behaviour in insect parasitoids. *Biological Reviews*, **71**, 373–400. [Ch. 7]

Hemptinne, J.-L., Dixon, A.F.G. and Coffin, J. (1992) Attack strategy of ladybird beetles (Coccinellidae): factors shaping their numerical response. *Oecologia*, **90**, 238–45. [Ch. 9]

Hendel, F. (1927) Trypetidae, in *Die Fliegen der Palaearktischen Region* (ed E. Lindner), Part 49, pp. 1–221 Schweizerbart, Stuttgart. [Ch. 10]

Hengeveld, R. (1980). Polyphagy, oligophagy and food specialization in ground beetles (Coleoptera, Carabidae). *Netherlands Journal of Zoology*, **30**, 564–84. [Ch. 3]

Henter, H.J. and Via, S. (1995) The potential for coevolution in a host–parasitoid system. 1. Genetic variation within an aphid population in susceptibility to a parasitic wasp. *Evolution*, **49**, 427–38. [Ch. 6]

Herms, D.A. and Mattson, W.J. (1992) The dilemma of plants: to grow or defend. *Quarterly Review of Biology*, **67**, 283–335. [Ch. 8]

Hjältén, J. and Price, P.W. (1996) The effect of pruning on willow growth and sawfly population density. *Oikos*, **77**, 549–55. [Ch. 14]

Hjältén, J. and Price, P.W. (1997) Can plants gain protection from herbivory by association with unpalatable neighbours? A field experiment in a willow–sawfly system. *Oikos*, **78**, 317–22. [Ch. 14]

Hochberg, M.E. (1996a) An integrative paradigm for the dynamics of monophagous parasitoid–host interactions. *Oikos*, **77**, 556–60. [Ch. 7]

Hochberg, M.E. (1996b) Consequences for host population levels of increasing natural enemy species richness in classical biological control. *American Naturalist*, **147**, 307–18. [Ch. 1, 6, 11]

Hochberg, M.E. and Hawkins, B.A. (1992) Refuges as a predictor of parasitoid diversity. *Science*, **255**, 973–6. [Ch. 6]

Hochberg, M.E. and Hawkins, B.A. (1993) Predicting parasitoid species richness. *American Naturalist*, **142**, 671–93. [Ch. 6]

Hochberg, M.E. and Holt, R.D. (1995) Refuge evolution and the population dynamics of coupled host–parasitoid associations. *Evolutionary Ecology*, **9**, 633–61. [Ch. 6]

Hochberg, M.E. and Lawton, J.H. (1990) Spatial heterogeneities in parasitism and population dynamics. *Oikos*, **59**, 9–14. [Ch. 6]

Hochberg, M.E., Hassell, M.P. and May, R.M. (1990) The dynamics of host–parasitoid–pathogen interactions. *American Naturalist*, **135**, 74–94. [Ch. 6]

Hochberg, M.E., Thomas, J.A. and Elmes, G.W. (1992) A modelling study of the population dynamics of a large blue butterfly, *M. rebeli*, a parasite of red ant nests. *Journal of Animal Ecology*, **61**, 397–409. [Ch. 11]

Hochberg, M.E., Clarke, R.T., Elmes, G.W. and Thomas, J.A. (1994) Population dynamic consequences of direct and indirect interactions involving a large blue butterfly and its plant and red ant hosts. *Journal of Animal Ecology*, **63**, 375–91. [Ch. 11]

Hochberg, M.E., Elmes, G.W., Thomas, J.A. and Clarke, R.T. (1996) Mechanisms of local persistence in coupled host–parasitoid associations: the case model of *Maculinea rebeli* and *Ichneumon eumerus*. *Philosophical Transactions of the Royal Society of London, Series B*, **351**, 1713–24. [Ch. 11]

Hochberg, M.E., Elmes, G.W., Thomas, J.A. and Clarke, R.T. (1998) Effects of habitat reduction on the persistence of *Ichneumon eumerus*, a specialist parasitoid of *Maculinea rebeli*. *Journal of Insect Conservation*, **2**, 59–66. [Ch. 11]

Hogarth, W.L. and Diamond, P. (1984) Interspecific competition in larvae between entomophagous parasitoids. *American Naturalist*, **124**, 552–60. [Ch. 6, 7]

Hogstad, O. (1997) Population fluctuations of *Epirrita autumnata* Bkh. and *Operophtera brumata* (L.) (Lep., Geometridae) during 25 years and habitat distribution of their larvae during a mass outbreak in a subalpine birch forest in Central Norway. *Fauna Norvegica Series B*, **44**, 1–10. [Ch. 13]

Höller, C., Borgemeister, C., Haardt, H. and Powell, W. (1993) The relationship between primary parasitoids and hyperparasitoids of cereal aphids: an analysis of field data. *Journal of Animal Ecology*, **62**, 12–21. [Ch. 9]

Holliday, N.J. (1975) *The ecology and behaviour of the winter moth* (Operophtera brumata (L.)) *in a cider apple orchard*. PhD thesis, University of Bristol. [Ch. 13]

Holliday, N.J. (1977) Population ecology of winter moth (*Operophtera brumata*) on apple in relation to larval dispersal and time of bud burst. *Journal of Applied Ecology*, **14**, 803–13. [Ch. 13]

Holling, C.S. (1965) The functional response of predators to prey density and its role in mimicry and population regulation. *Memoirs of the Entomological Society of Canada*, **45**, 3–60. [Ch. 6]

Holt, R.D. (1977) Predation, apparent competition and the structure of prey communities. *Theoretical Population Biology*, **12**, 197–229. [Ch. 6, 7]

Holt, R.D. (1996) Demographic constraints in evolution: towards unifying the evolutionary theories of senescence and the niche conservatism. *Evolutionary Ecology*, **10**, 1–11 [Ch. 1]

Holt, R.D. (1997) Community modules, in *Multi–trophic Interactions in Terrestrial Systems* (eds A.C. Gange and V.K. Brown), Blackwell Science, Oxford, pp. 333–50 [Ch. 11]

Holt, R.D. and Gaines, M.S. (1992) Analysis of adaptation in heterogeneous landscapes: implications for the evolution of fundamental niches. *Evolutionary Ecology*, **6**, 433–47. [Ch. 1]

Holt, R.D. and Hassell, M.P. (1993) Environmental heterogeneity and the stability of host–parasitoid interactions. *Journal of Animal Ecology*, **62**, 89–100. [Ch. 6]

Holt, R.D. and Lawton, J.H. (1993) Apparent competition and enemy-free space in insect host–parasitoid communities. *American Naturalist*, **142**, 623–45. [Ch. 6, 7]

Holt, R.D. and Lawton, J.H. (1994) The ecological consequences of shared natural enemies. *Annual Review of Ecology and Systematics*, **25**, 495–520. [Ch. 6, 11]

Holt, R.D. and Polis, J.H. (1997) A theoretical framework for intraguild predation. *American Naturalist*, **149**, 745–64. [Ch. 7]

Holyoak, M. (1993a) New insights into testing for density dependence. *Oecologia*, **93**, 435–44. [Ch. 5]

Holyoak, M. (1993b) The frequency of detection of density dependence in insect orders. *Ecological Entomology*, **18**, 339–47. [Ch. 5]

Holyoak, M. (1994a) Identifying delayed density dependence in time-series data. *Oikos*, **70**, 296–304. [Ch. 5]

Holyoak, M. (1994b) Appropriate time scales for identifying lags in density-dependent processes. *Journal of Animal Ecology*, **63**, 479–83. [Ch. 2, 5]

Holyoak, M. (1996) Factors affecting detection of density dependence in British birds. I. Population trends. *Oecologia*, **108**, 47–53. [Ch. 5]

Holyoak, M. and Lawton, J.H. (1993) Comment arising from a paper by Wolda and Dennis: using and interpreting the results of tests for density dependence. *Oecologia*, **95**, 592–4. [Ch. 2, 5]

Honek, A. (1986) Body size and fecundity in natural populations of *Pyrrhocoris apterus* L. (Heteroptera, Pyrrhocoridae). *Zoologische Jahrbücher Abteilung für Systematik Oekologie und Morphologie*, **113**, 125–40. [Ch. 8]

Hopper, K.R., Crowley, P.H. and Kielman, D. (1996) Density dependence, hatching synchrony, and within-cohort cannibalism in young dragonfly larvae. *Ecology*, **77**, 191–200. [Ch. 17]

Horgan, F.G. (1993) *Factors affecting the mortality of winter moth in the Lower Mainland of British Columbia*. MSc thesis, University of British Columbia, Vancouver. [Ch. 13]

Horton, D.R. (1989) Performance of a willow-feeding beetle, *Chrysomela knabi* Brown as affected by host species and dietary moisture. *Canadian Entomologist,* **121**, 777–80. [Ch. 8]

Howard, L.O. and Fiske, W.F. (1911) The Importation into the United States of the Parasites of the gipsy moth and the brown-tail moth. *Bulletin of the Bureau of Entomology, US Department of Agriculture,* **91**, 1–312. [Ch. 2, 7]

Huffaker, C.B. and Kennett, C.E. (1966) Studies of two parasites of olive scale, *Parlatoria oleae* (Colvee). IV. Biological control of *Parlatoria oleae* (Colvee) through the compensatory action of two introduced parasites. *Hilgardia,* **37**, 283–335. [Ch. 7]

Hughes, R.D. (1963) Population dynamics of the cabbage aphid, *Brevicoryne brassicae* (L.). *Journal of Animal Ecology,* **32**, 393–424. [Ch. 1]

Humble, L.M. (1985) Final-instar larvae of native pupal parasites and hyperparasites of *Operophtera* spp. (Lepidoptera: Geometridae) on southern Vancouver Island. *Canadian Entomologist,* **117**, 525–34. [Ch. 13]

Hunter, A.F. (1991) Traits that distinguish outbreaking and non-outbreaking Macrolepidoptera feeding on northern hardwood trees. *Oikos,* **60**, 275–82. [Ch. 14]

Hunter, A.F. (1995a) The ecology and evolution of reduced wings in forest Macrolepidoptera. *Evolutionary Ecology,* **9**, 275–87. [Ch. 14]

Hunter, A.F. (1995b) Ecology, life history, and phylogeny of outbreak and non-outbreak species, in *Population Dynamics: New Approaches and Synthesis* (eds N. Cappuccino and P.W. Price), Academic Press, San Diego, pp. 41–64. [Ch. 14]

Hunter, M.D. and Price, P.W. (1992) Playing chutes and ladders: heterogeneity and the relative roles of bottom-up and top-down forces in natural communities. *Ecology,* **73**, 724–32. [Ch. 6, 14]

Hunter, M.D. and Price, P.W. (1998) Cycles in insect populations: delayed density dependence or exogenous driving variables? *Ecological Entomology* **23**, 216–222. [Ch. 14]

Hunter, M.D., Watt A.D. and Docherty M. (1991) Outbreaks of the winter moth on Sitka Spruce in Scotland are not influenced by nutrient deficiencies of trees, tree budburst, or pupal predation. *Oecologia,* **86**, 62–9. [Ch. 13]

Hunter, M.D., Ohgushi, T. and Price, P.W. (eds) (1992*) Effects of Resource Distribution on Animal–Plant Interactions,* Academic Press, San Diego. [Ch. 16]

Hunter, M.D., Varley, G.C. and Gradwell, G.R. (1997) Estimating the relative roles of top-down and bottom-up forces on insect herbivore populations: a classic study re-visited. *Proceedings of the National Academy of Sciences,* **94**, 9176–81. [Ch. 13, 14]

Hunter, M.D. (1998) Interactions between *Operophtera brumata* and *Tortrix viridana* on oak: new evidence from time-series analysis. *Ecological Entomology,* **23**, 168–73. [Ch. 14]

Hussey, N.W. (1952) A contribution to the bionomics of the green spruce aphid (*Neomyzaphis abietina* Walker). *Scottish Forestry,* **6**, 121–30. [Ch. 9]

Hutchinson, G.E. (1951) Copepodology for the ornithologist. *Ecology,* **32**, 571–7. [Ch. 6]

Hutchinson, G.E. (1959) Homage to Santa Rosalia; or, why are there so many kinds of animals? *American Naturalist,* **9**, 145–59. [Ch. 17]

Hutchinson, G.E. (1965) *The Ecological Theater and the Evolutionary Play,* Yale University Press, New Haven, Connecticut. [Ch. 14]

Imai, T., Kodama, H., Chuman, T. and Kohno, M. (1990) Female-produced oviposition deterrents of the cigarette beetle, *Lasioderma serricornis* F. (Coleoptera, Anobiidae). *Journal of Chemical Ecology,* **16**, 1237–48. [Ch. 4]

Ims, R.A. and Yoccoz, N.G. (1997) Studying transfer processes in metapopulations: emigration, migration and colonization, in *Metapopulation Biology: Ecology, Genetics and Evolution* (eds Hanski, I.A. and Gilpin, M.E.), Academic Press, San Diego, pp. 247–66. [Ch. 1]

Ingvarsson, P. K. (1997) *Non-equilibrium population dynamics and the genetic structure of natural populations of the mycophagous beetle* Phalacrus substriatus. PhD thesis, Umeå University, Sweden. [Ch. 1]

Ingvarsson, P.K., Olsson, K. and Ericson, L. (1997) Extinction–recolonization dynamics in the mycophagous beetle *Phalacrus substriatus*. *Evolution*, **51**, 187–95. [Ch. 1]

Islam, Z. and Crawley, M.J. (1983) Compensation and regrowth in ragwort (*Senecio jacobaea*) attacked by the cinnabar moth (*Tyria jacobaeae*). *Journal of Ecology*, **71**, 829–43. [Ch. 12]

Ito, Y. (1972) On the methods of determining density dependence by means of regression. *Oecologia*, **10**, 347–72. [Ch. 5]

Ives, A.R. (1992a) Continuous-time models of host–parasitoid interactions. *American Naturalist*, **140**, 1–29. [Ch. 7]

Ives, A.R. (1992b) Density-dependent and density-independent parasitoid aggregation in model host–parasitoid systems. *American Naturalist*, **140**, 912–37. [Ch. 6]

Ives, A.R. (1995) Spatial heterogeneity and host–parasitoid population dynamics: do we need to study behavior? *Oikos*, **74**, 366–76. [Ch. 6]

Ives, A.R. and Settle, W.H. (1996) The failure of a parasitoid to persist with a superabundant host: the importance of the numerical response. *Oikos*, **75**, 269–78. [Ch. 6]

Ives, A.R. and Settle, W.H. (1997) Metapopulation dynamics and pest control in agricultural systems. *American Naturalist*, **149**, 220–46. [Ch. 7]

Iwao, S. (1971) Dynamics of numbers of a phytophagous lady-beetle, *Epilachna vigintioctomaculata*, living in patchily distributed habitats, in *Dynamics of Populations* (eds P.J. den Boer and G.R. Gradwell), Centre for Agricultural Publishing and Documentation, Wageningen, pp. 129–47. [Ch. 2]

Jardon, Y., Filion, L. and Cloutier, C. (1994) Long-term impact of insect defoliation on growth and mortality of eastern larch in boreal Quebec. *EcoScience*, **1**, 231–8. [Ch. 14]

Jeffries, M.J. and Lawton, J.H. (1984) Enemy-free space and the structure of ecological communities. *Biological Journal of the Linnean Society*, **23**, 269–86. [Ch. 6]

Jervis, M.A. and Kidd, N.A.C. (1986) Host-feeding strategies in hymenopteran parasitoids. *Biological Reviews*, **61**, 395–434. [Ch. 6]

Johnston, M.A. (1992) *Rabbit grazing and the dynamics of plant communities*. PhD thesis, University of London. [Ch. 12]

Jolly, G.M. (1965) Explicit estimates from capture–recapture data with both death and immigration–stochastic model. *Biometrika*, **52**, 225–47. [Ch. 16]

Jones, T.H. and Hassell, M.P. (1988) Patterns of parasitism by *Trybliographa rapae*, a cynipid parasitoid of the cabbage root fly, under laboratory and field conditions. *Ecological Entomology*, **13**, 309–17. [Ch. 2]

Jones, T.H., Hassell, M.P. and Pacala, S.W. (1993) Spatial heterogeneity and the population dynamics of a host–parasitoid system. *Journal of Animal Ecology*, **62**, 251–62. [Ch. 4, 6]

Jones, T.H., Hassell, M.P. and Godfray, H.C.J. (1994) Population dynamics of host–parasitoid interactions, in *Parasitoid Community Ecology* (eds B.A. Hawkins and W. Sheehan), Oxford University Press, Oxford, pp. 371–94. [Ch. 7]

Jones, T.H., Godfray, H.C.J. and Hassell, M.P. (1996) Relative movement patterns of a tephritid fly and its parasitoid wasps. *Oecologia*, **106**, 317–24. [Ch. 6]

Julien, M.H. (ed) (1992) *Biological Control of Weeds: a world catalogue of agents and their target weeds*, CAB International, Wallingford, UK. [Ch. 10]

Kakehashi, M., Suzuki, Y. and Iwasa, Y. (1984) Niche overlap of parasitoids in host–parasitoid systems: its consequence to single versus multiple introduction controversy in biological control. *Journal of Applied Ecology* **21**, 115–31. [Ch. 6, 7]

Karban, R. and Courtney, S. (1987) Intraspecific host plant choice: lack of consequences for *Streptanthus tortuosus* (Cruciferae) and *Euchloe hyantis* (Lepidoptera: Pieridae). *Oikos*, **48**, 243–8. [Ch. 16]

Kareiva, P. (1983) Influence of vegetation texture on herbivore populations: resource concentration and herbivore movement, in *Variable Plants and Herbivores in Natural and Managed Systems* (eds R.F. Denno and M.S. McClure), Academic Press, New York, pp. 259–89. [Ch. 16]

Kareiva, P. (1987). Habitat fragmentation and the stability of predator–prey interactions. *Nature*, **326**, 388–91. [Ch. 7]

Kawecki, T. and Stearn, S.C. (1993) The evolution of life histories in spatially heterogeneous environments: optimal reaction norms revisited. *Evolutionary Ecology*, **7**, 155–74. [Ch. 1]

Kendall, M.G. (1955) *Rank Correlation Methods*, 2nd edn, Charles Griffin, London. [Ch. 3]

Kennett, C.E. (1986) A survey of the parasitoid complex attacking black scale, *Saissetia oleae* (Olivier), in central and northern California (Hymenoptera: Chalcidoidea; Homoptera: Coccidae). *Pan-Pacific Entomologist*, **62**, 363–9. [Ch. 7]

Kenten, J. (1955) The effect of photoperiod and temperature on reproduction in *Acyrthosiphon pisum* (Harris) and on the forms produced. *Bulletin of Entomological Research*, **46**, 599–624. [Ch. 8]

Kerslake, J.E., Kruuk, L.E.B., Hartley, S.E. and Woodin, S.J. (1996) Winter moth *Operophtera brumata* (Lepidoptera: Geometridae) outbreaks on Scottish heather moorlands: effects of host plant and parasitoids on larval survival and development. *Bulletin of Entomological Research*, **86**, 155–64. [Ch. 13]

Kidd, N.A.C. (1990a) Population dynamics of the large pine aphid, *Cinara pinea* (Mordv.). I. Simulation of laboratory populations. *Researches on Population Ecology*, **32**, 189–208. [Ch. 9]

Kidd, N.A.C. (1990b) Population dynamics of the large pine aphid, *Cinara pinea* (Mordv.). II. Simulation of field populations. *Researches on Population Ecology*, **32**, 209–26. [Ch. 9]

Kidd, N.A.C. (1990c) A synoptic model to explain long-term population changes in the large pine aphid, in *Population Dynamics of Forest Insects* (eds A.D. Watt, S.R. Leather, M.D. Hunter and N.A.C. Kidd), Intercept, Andover, pp. 317–27. [Ch. 9]

Kidd, N.A.C. and Jervis, M.A. (1991) Host-feeding and oviposition by parasitoids in relation to host stage: consequences for parasitoid–host population dynamics. *Researches on Population Ecology*, **33**, 87–99. [Ch. 6]

Kindlmann, P. and Dixon, A.F.G. (1989) Developmental constraints in the evolution of reproductive strategies: telescoping of generations in parthenogenetic aphids. *Functional Ecology*, **3**, 531–7. [Ch. 9]

Kindlmann, P. and Dixon, A.F.G. (1993) Optimal foraging in ladybird beetles (Coleoptera: Coccinellidae) and its consequences for their use in biological control. *European Journal of Entomology*, **90**, 443–50. [Ch. 9]

Kindlmann, P. and Dixon, A.F.G. (1998) Patterns in the population dynamics of the Turkey-oak aphid, in *Aphids in Natural and Managed Ecosystems* (eds J.M. Nieto Nafria and A.F.G. Dixon), University of Leon, Leon. [Ch. 9]

Kindvall, O. (1995) *Ecology of the bush cricket* Metrioptera bicolor *with implications for metapopulation theory and conservation.* PhD thesis, University of Uppsala, Sweden. [Ch. 1]

Kindvall, O. and Ahlén, I. (1992) Geometrical factors and metapopulation dynamics of the bush cricket, *Metrioptera bicolor* Philippi (Orthoptera: Tettigoniidae). *Conservation Biology*, **6**, 520–9. [Ch. 1]

Kirsten, K. and Topp, W. (1991) Acceptance of willow-species for the development of the winter moth, *Operophtera brumata* (Lep., Geometridae). *Journal of Applied Entomology*, **111**, 457–68. [Ch. 13]

Kitching, R.L. (1990) Foodwebs from phytotelmata in Madang, Papua New Guinea. *Entomologist*, **109**, 153–64. [Ch. 17]

Klijnstra, J.W. and Schoonhoven, L.M. (1987) Effectiveness and persistence of the oviposition deterring pheromone of *Pieris brassicae* in the field. *Entomologia Experimentalis et Applicata*, **45**, 227–36. [Ch. 4]

Klingauf, F. A. (1987). Feeding, adaptation and excretion, in *Aphids, their Biology, Natural Enemies and Control*, Vol. A (eds A. Minks and P. Harrewijn), Elsevier, Amsterdam, pp. 225–53. [Ch. 8]

Klingenberg, C.P. and Spence, J.R. (1997) On the role of body size for life-history evolution. *Ecological Entomology*, **22**, 55–68. [Ch. 8]

Klomp, H. (1966) The dynamics of a field population of the pine looper, *Bupalus piniarius* L. (Lep., Geom.). *Advances in Ecological Research*, **3**, 207–305. [Ch. 2, 3, 5, 16]

Kluyver, H.N. (1951) The population ecology of the great tit, *Parus m. major* L. *Ardea*, **38**, 99–135. [Ch. 3]

Kluyver, H.N. (1971) Regulation of numbers in populations of great tits (*Parus m. major*), in *Dynamics of Populations* (eds P.J. den Boer and G.R. Gradwell), Centre for Agricultural Publishing and Documentation, Wageningen, pp. 507–23. [Ch. 3]

Köhler, G. (1996) The ecological background of population vulnerability in central European grasshoppers and bush crickets: a brief review, in *Species Survival in Fragmented Landscapes*, (eds J. Settele, C. Margules, P. Poschlod and K. Heule), Kluwer, Dordrecht, pp. 290–8. [Ch. 1]

Kowalski, R. (1976) Biology of *Philonthus decorus* (Coleoptera, Staphylinidae) in relation to its role as a predator of winter moth pupae [*Operophtera brumata* (Lepidoptera, Geometridae)]. *Pedobiologia*, **16**, 233–42. [Ch. 13]

Kowalski, R. (1977) Further elaboration of the winter moth population models. *Journal of Animal Ecology*, **46**, 471–82. [Ch. 13]

Kozlowski, M.W. (1989) Oviposition and host marking by the females of *Ceutorynchus floralis* (Coleoptera, Curculionidae). *Entomologia Generalis*, **14**, 197–201. [Ch. 4]

Kraaijeveld, A.R. and Godfray, H.C.J. (1997) Trade off between parasitoid resistance and larval competitive ability in *Drosophila melanogaster*. *Nature*, **389**, 278–80. [Ch. 6]

Kraaijeveld, A.R., van Alphen, J.J.M. and Godfray, H.C.J. (1998) Coevolution of host resistance and parasitoid virulence. *Parasitology* (in press). [Ch. 6]

Krebs, J.R. (1970) Regulation of numbers of the great tit (Aves: Passeriformes). *Journal of Zoology*, **162**, 317–33. [Ch. 2]

Krebs, J.R. and Davies, N.B. (1997) *Behavioural Ecology*, 4th edn, Blackwell Scientific Publications, Oxford. [Ch. 16]

Kuenen, D.J. (1958) Some sources of misunderstanding in the theories of regulation of animal numbers. *Archives Nerlandaises de Zoologie*, **13** (Supplement 1), 335–41. [Ch. 2]

Kuno, E. (1971) Sampling error as a misleading artifact in key factor analysis. *Researches in Population Ecology*, **13**, 28–45. [Ch. 5]

Kuno, E. (1981) Dispersal and the persistence of populations in unstable habitats: a theoretical note. *Oecologia*, **49**, 123–6. [Ch. 1]

Kuussaari, M., Nieminen, M. and Hanski, I. (1996) An experimental study of migration in the butterfly *Melitaea cinxia*. *Journal of Animal Ecology*, **65**, 791–801. [Ch. 1]

Kuussaari, M., Saccheri, I., Camara, M. and Hanski, I. (1998) Demonstration of the Allee effect in small populations of an endangered butterfly. Unpublished manuscript.

Lack, D. (1954) *The Natural Regulation of Animal Numbers*, Clarendon Press, Oxford. [Ch. 1]

Laessle, A.M. (1961) A micro-limnological study of Jamaican bromeliads. *Ecology*, **42**, 499–517. [Ch. 17]

Lalonde, R.G. and Shorthouse, J.D. (1985) Growth and development of larvae and galls of *Urophora cardui* (Diptera: Tephritidae) on *Cirsium arvense* (Compositae). *Oecologia*, **65**, 161–5. [Ch. 10]

Lampo, M. (1994) The importance of refuges in the interaction between *Contarinia sorghicola* and its parasitic wasp *Aprostocetus diplosidis*. *Journal of Animal Ecology*, **63**, 176–86. [Ch. 6]

Lande, R. (1988) Demographic models of the northern spotted owl (*Strix occidental caurina*). *Oecologia*, **75**, 601–7. [Ch. 1]

Lande, R. (1993) Risks of population extinction from demographic and environmental stochasticity and random catastrophes. *American Naturalist*, **142**, 911–27. [Ch. 1]

Larsson, S. and Ekbom, B. (1995) Oviposition mistakes in herbivorous insects: confusion or a step towards a new host plant? *Oikos*, **72**, 155–60. [Ch. 16]

Latto, J. and Berstein, C. (1990a) Regulation of natural insect populations: reality or illusion? *Acta Oecologica*, **11**, 121–30. [Ch. 5]

Latto, J. and Berstein, C. (1990b) Reply to Den Boer. (1990c) *Acta Oecologica*, **11**, 135–6. [Ch. 5]

Latto, J. and Hassell, M.P. (1987) Do pupal predators regulate the winter moth? *Oecologia*, **74**, 153–5. [Ch. 2, 5, 13]

Lawton, J.H. (1993) Range, population abundance and conservation. *Trends in Ecology and Evolution*, **8**, 409–13. [Ch. 1]

Lawton, J.H. and Strong, D.R. (1981) Community patterns and competition in folivorous insects. *American Naturalist*, **118**, 317–38. [Ch. 16, 17]

Leather, S.R. (1985) Oviposition preferences in relation to larval growth rates and survival in the pine beauty moth, *Panolis flammea*. *Ecological Entomology*, **10**, 213–7. [Ch. 8]

Leather, S.R. (1987) Lodgepole pine provenance and the pine beauty moth, in *Population Biology and Control of the Pine Beauty Moth*, Forestry Commission Bulletin 67 (eds S.R. Leather, J.T. Stoakley and H.F. Evans), HMSO, London, pp. 27–30. [Ch. 8]

Leather, S.R. (1988) Size, reproductive potential and fecundity in insects: things aren't as simple as they seem. *Oikos*, **51**, 386–9. [Ch. 8]

Leather, S.R. (1989) Do alate aphids produce fitter offspring? The influence of maternal rearing history and morph on life-history parameters of *Rhopalosiphum padi* (L.). *Functional Ecology*, **3**, 237–44. [Ch. 8]

Leather, S.R. (1994) Life history traits of insect herbivores in relation to host quality, in *Insect–Plant Interactions*, Vol. 5 (ed. E.A. Bernays), CRC Press, Boca Raton, Florida, pp. 175–207. [Ch. 8, 16]

Leather, S.R. (1995) Factors affecting fecundity, fertility, oviposition, and larviposition in insects, in *Insect Reproduction* (eds S.R. Leather and J. Hardie), CRC Press, Boca Raton, Florida, pp. 143–74. [Ch. 8]

Leather, S.R. and Burnand, A.C. (1987) Factors affecting life-history parameters of the pine beauty moth, *Panolis flammea* (D&S): the hidden costs of reproduction. *Functional Ecology,* **1**, 331–8. [Ch. 8]

Leather, S.R. and Dixon, A.F.G. (1982) Secondary host preferences and reproductive activity of the bird cherry-oat aphid, *Rhopalosiphum padi. Annals of Applied Biology,* **101**, 219–28. [Ch. 8]

Leather, S.R. and Knight, J.D. (1997) Pines, pheromones and parasites: a modelling approach to the integrated control of the pine beauty moth. *Scottish Forestry,* **51**, 76–83. [Ch. 8]

Leather, S.R. and Walsh, P.J. (1993) Sub-lethal plant defences: the paradox remains. *Oecologia,* **93**, 153–5. [Ch. 8]

Leather, S.R., Ward, S.A. and Dixon, A.F.G. (1983) The effect of nutrient stress on life history parameters of the black bean aphid, *Aphis fabae* Scop. *Oecologia,* **57**, 156–7. [Ch. 8]

Leather, S.R., Watt, A.D. and Barbour, D.A. (1985) The effect of host plant and delayed mating on the fecundity and life span of the pine beauty moth, *Panolis flammea* (Denis and Schiffermüller) (Lepidoptera: Noctuidae): their influence on population dynamics and relevance to pest management. *Bulletin of Entomological Research,* **75**, 641–51. [Ch. 8]

Leather, S.R., Wellings, P.W. and Walters, K.F.A. (1988) Variation in ovariole number within the Aphidoidea. *Journal of Natural History,* **22**, 381–93. [Ch. 8]

Leather, S.R., Walters, K.F.A. and Bale, J.S. (1993) *The Ecology of Insect Overwintering,* Cambridge University Press, Cambridge. [Ch. 8]

Lebreton, J.-D., Burnham, K.P., Clobert, J. *et al.* (1992) Modelling and testing biological hypotheses using marked animals: a unified approach with case studies. *Ecological Monographs,* **62**, 67–118. [Ch. 5]

Lei, G. (1997) *Metapopulation dynamics of host–parasitoid interactions.* PhD thesis, University of Helsinki, Finland. [Ch. 1]

Lei, G. and Hanski, I. (1997) Metapopulation structure of *Cotesia melitaearum,* a specialist parasitoid of the butterfly *Melitaea cinxia. Oikos,* **78**, 91–100. [Ch. 1, 6]

Lei, G. and Hanski, I. (1998) Spatial dynamics of two competing specialist parasitoids in a host metapopulation. *Journal of Animal Ecology* (in press). [Ch. 1]

Levins, R. (1968) *Evolution in Changing Environments,* Princeton University Press, Princeton, New Jersey. [Ch. 2]

Levins, R. (1969) Some demographic and genetic consequences of environmental heterogeneity for biological control. *Bulletin of the Entomological Society of America,* **15**, 237–40. [Ch. 1, 7]

Levins, R. (1970) Extinction. *Lecture Notes in Mathematics,* **2**, 75–107. [Ch. 1]

Liao, H.T. and Harris, M.K. (1985) Population growth of the black-margined aphid on pecan in the field. *Agriculture, Ecosystems and Environment,* **12**, 253–61. [Ch. 9]

Liebhold, A.M. and Elkington, J.S. (1989) Use of multidimensional life tables for studying insect population dynamics, in *Estimating and Analysis of Insect Populations* (eds L.L. McDonald, B.F.J. Manly, J.A. Lockwood and J.A. Logan), Lecture Notes in Statistics No. 55, Springer-Verlag, Berlin, pp. 360–9. [Ch. 5]

Lincoln, D.E., Fajer, E.D. and Johnson, R.H. (1993) Plant insect interactions in elevated CO_2 environments. *Trends in Ecology and Evolution,* **8**, 64–7. [Ch. 8]

Łomnicki, A. (1988) *Population Ecology of Individuals,* Princeton University Press, Princeton, New Jersey. [Ch. 8, 16]

Losey, J.E., Ives, A.R., Harmon, J., Ballantyne, F. and Brown, C. (1997) A polymorphism maintained by opposite patterns of parasitism and predation. *Nature*, **388**, 269–72. [Ch. 8]

Lounibos, L.P. and Machado-Allison, C.E. (1983) Oviposition and egg brooding by the mosquito *Trichoprosopon digitatum*, in cacao husks. *Ecological Entomology*, **8**, 475–8. [Ch. 17]

Lounibos, L.P., Frank, J.H., Machado-Allison, C.E. *et al.* (1987) Survival, development and predatory effects of mosquito larvae in Venezuelan phytotelmata. *Journal of Tropical Ecology*, **3**, 221–42. [Ch. 17]

Luck, R.F. and Podoler, H. (1985) Competitive exclusion of *Aphytis lingnanensis* by *A. melinus*: potential role of host size. *Ecology*, **66**, 904–13. [Ch. 6, 7]

Ludwig, D., Jones, D.D. and Holling, C.S. (1978) Qualitative analysis of insect outbreak systems: the spruce budworm and the forest. *Journal of Animal Ecology*, **47**, 315–32. [Ch. 7]

Lukefahr, M.J. and Martin, D.F. (1964) The effect of various larval and adult diets on the fecundity and longevity of the bollworm, tobacco budworm and the cotton leafworm. *Journal of Economic Entomology*, **57**, 233–5. [Ch. 8]

MacArthur, R.H. and Levins, R. (1967) The limiting similarity, convergence, and divergence of coexisting species. *American Naturalist*, **101**, 377–85. [Ch. 7]

Mackauer, M. and Völkl, W. (1993) Regulation of aphid populations by aphidiid wasps: does parasitoid foraging behaviour or hyperparasitism limit impact? *Oecologia*, **94**, 339–50. [Ch. 6, 9]

MacPhee, A.W., Newton A. and McRae, K.B. (1988) Population studies on the winter moth *Operophtera brumata* (L.) (Lepidoptera: Geometridae) in apple orchards in Nova Scotia. *Canadian Entomologist*, **120**, 73–83. [Ch. 13]

Maelzer, D.A. (1970) The regression of log N_{n+1} on log N_n as a test of density dependence: an exercise with computer-constructed, density-independent populations. *Ecology*, **51**, 810–22. [Ch. 2, 5]

Malthus, T.R. (1798) *An Essay on the Principle of Population as it affects the Future Improvements of Society*, reprinted by Macmillan, New York. [Ch. 2]

Manly, B.F.J. (1977) The determination of key factors from life table data. *Oecologia*, **31**, 111–7. [Ch. 5]

Manly, B.F.J. (1990) *Stage-Structured Populations: Sampling, Analysis and Simulation*, Chapman and Hall, London. [Ch. 5]

Maquelin, C. (1974) *Observations sur la biologie et l'ecologie d'un puceron utile a l'apiculture: Buchneria pectinatae (Nördl.) (Homoptera, Lachnidae)*. PhD thesis, L'Ecole Polytechnique Federale De Zurich. [Ch. 9]

Markkula, M. and Tiittanen, K. (1969) Effect of fertilizers on the reproduction of *Tetranychus telarius* (L.), *Myzus persicae* (Sulz.) and *Acyrthosiphon pisum* Harris. *Annales Agricultuare Fenniae*, **8**, 9–14. [Ch. 8]

Mattson, W.J. (1980) Cone resources and the ecology of the red pine cone beetle, *Conophthorus resinosae* (Coleoptera: Scolytidae). *Annals of the Entomological Society of America*, **73**, 390–6. [Ch. 16]

Mattson, W.J. and Scriber, J.M. (1987) Nutritional ecology of insect folivores of woody plants: nitrogen, water, fiber, and mineral considerations, in *Nutritional Ecology of Insects, Mites, Spiders, and Related Invertebrates* (eds F.J. Slansky and J.G. Rodriguez), Wiley, New York, pp. 105–45. [Ch. 16]

May, R.M. (1973) On the relationship between various types of population models. *American Naturalist*, **107**, 46–57. [Ch. 2]

May, R.M. (1974) Biological populations with non-overlapping generations: stable points, stable cycles and chaos. *Science*, **186**, 645–7. [Ch. 2]

May, R.M. (1975) Patterns of species abundance and diversity, in *Ecology and Evolution of Communities* (eds M.L. Cody and J.M. Diamond), Belknap, Cambridge, Massachussetts, pp. 81–120. [Ch. 1]

May, R.M. (1976) Models for single populations, in *Theoretical Ecology: Principles and Applications* (ed. R.M. May), Blackwell Scientific Publications, Oxford, pp. 4–25. [Ch. 5]

May, R.M. (1978) Host–parasitoid systems in patchy environments: a phenomenological model. *Journal of Animal Ecology*, **47**, 833–43. [Ch. 1, 6, 7]

May, R.M. (ed.) (1981) *Theoretical Ecology*, Blackwell Scientific Publications, Oxford. [Ch. 10]

May, R.M. (1989) Detecting density dependence in imaginary worlds. *Nature*, **338**, 16–7. [Ch. 5]

May, R.M. and Hassell, M.P. (1981) The dynamics of multiparasitoid–host interactions. *American Naturalist*, **117**, 234–61. [Ch. 6, 7]

May, R.M. and Hassell, M.P. (1988) Population dynamics and biological control. *Philosophical Transactions of the Royal Society, London, Series B*, **318**, 129–69. [Ch. 6]

May, R.M. and Hassell, M.P. (1989) Parasitoid theory: against manichaeism. *Trends in Ecology and Evolution*, **4**, 20–1. [Ch. 2]

May, R.M. and Oster, G.F. (1976) Bifurcations and dynamic complexity in simple ecological models. *American Naturalist*, **110**, 573–600. [Ch. 2]

May, R.M., Hassell, M.P., Anderson, R.M. and Tonkyn, D.W. (1981) Density dependence in host–parasitoid models. *Journal of Animal Ecology*, **50**, 855–65. [Ch. 6]

Maynard Smith, J. (1974) *Models in Ecology*, Cambridge University Press, Cambridge. [Ch. 7]

Maynard Smith, J. and Slatkin, M. (1973) The stability of predator–prey systems. *Ecology*, **54**, 384–91. [Ch. 5]

Mayr, E. (1961) Cause and effect in biology. *Science*, **134**, 1501–6. [Ch. 14]

McCullagh, P. and Nelder, J.A. (1983) *Generalized Linear Models*, 2nd edn, Chapman & Hall, London. [Ch. 5]

McLean, I.F.G. (1993) The host plant association and life history of *Trichochermes walkeri* Förster (Psylloidea: Triozidae). *British Journal of Entomology and Natural History*, **6**, 13–6. [Ch. 15]

McLean, I.F.G. (1994a) The population dynamics of a gall-forming psyllid, in *Individuals, Populations and Patterns in Ecology* (eds S.R. Leather, A.D. Watt, N.J. Mills and K.F.A. Walters), Intercept, Andover, pp. 97–107. [Ch. 15]

McLean, I.F.G. (1994b) Interactions between *Trichochermes walkeri* (Homoptera: Psylloidea) and other Homoptera on *Rhamnus catharticus*, in *Plant Galls* (ed. M.A.J. Williams), Systematics Association Special Volume No. 49, Clarendon Press, Oxford, pp. 151–60. [Ch. 15]

McNair, J.N. (1986) The effects of refuges on predator–prey interactions: a reconsideration. *Theoretical Population Biology*, **29**, 38–63. [Ch. 7]

McNeil, J.N. and Quiring, D.T. (1983) Evidence of an oviposition deterring pheromone in the alfalfa blotch leaf miner *Agromyza frontella* (Diptera, Agromyzidae). *Environmental Entomology*, **12**, 990–2. [Ch. 4]

McPheron, B.A. and Steck, G.J. (eds) (1995) *Economic Fruit-flies: a world assessment of their biology and management*, St Lucie Press, Delray Beach, Florida. [Ch. 10]

Memmott, J. and Godfray, H.C.J. (1992) Parasitoid webs, in *Hymenoptera and Biodiversity* (eds J. Lasalle and I.D. Gauld), CAB International, Wallingford, UK, pp. 217–34. [Ch. 6]

Memmott, J. and Godfray, H.C.J. (1994) The use and construction of parasitoid webs, in *Parasitoid Community Ecology* (eds B.A. Hawkins and W. Sheehan), Oxford University Press, Oxford, pp. 300–18. [Ch. 6]

Memmott, J., Godfray, H.C.J. and Gauld, I.D. (1994) The structure of a tropical host–parasitoid community. *Journal of Animal Ecology*, **63**, 521–40. [Ch. 6]

Middleton, D.A, Veitch, A.R and Nisbet, R.M. (1995) The effect of an upper limit to population size on persistence time. *Theoretical Population Biology*, **48**, 277–305. [Ch. 1]

Mills, N.J. and Getz, W.M. (1996) Modelling the biological control of insect pests: a review of host–parasitoid models. *Ecological Modelling*, **92**, 121–43. [Ch. 6]

Mills, N.J. and Gutierrez, A.P. (1996) Prospective modelling in biological control: an analysis of the dynamics of heteronomous hyperparasitism in a cotton–whitefly–parasitoid system. *Journal of Applied Ecology*, **33**, 1379–94. [Ch. 6]

Milne, A. (1957a) The natural control of insect populations. *Canadian Entomologist*, **89**, 193–213. [Intro, Ch. 1, 4]

Milne, A. (1957b). Theories of natural control of insect populations. *Cold Spring Harbor Symposia on Quantitative Biology*, **22**, 253–71. [Ch. 3]

Milne, A. (1961) Definition of competition among animals. *Symposium Society of experimental Biology*, **15**, 40–71. [Ch. 1, 4]

Milne, A. (1962). On a theory of natural control of insect populations. *Journal of Theoretical Biology*, **3**, 19–50. [Ch. 3, 4]

Milne, A. (1984) Fluctuations and natural control of animal populations as exemplified in the garden chafer *Phyllopertha horticula* (L.). *Proceedings of the Royal Society of Edinburgh*, **82B**, 145–99. [Ch. 5]

Mogi, M. and Yong, H.S. (1992) Aquatic arthropod communities in *Nepenthes* pitchers: the role of niche differentiation, aggregation, predation and competition in community organization. *Oecologia*, **90**, 172–84. [Ch. 17]

Moilanen, A., Smith, A. and Hanski, I. (1998) Long-term dynamics in a metapopulation of the American pika. *American Naturalist* (in press). [Ch. 1]

Moran, P.A.P. (1952) The statistical analysis of game-bird records. *Journal of Animal Ecology*, **21**, 154–8. [Ch. 5]

Moran, P.A.P. (1953a) The statistical analysis of the Canadian lynx cycle. I. Structure and prediction. *Australian Journal of Zoology*, **1**, 163–73. [Ch. 5]

Moran, P.A.P. (1953b) The statistical analysis of the Canadian lynx cycle. II. Synchronization and meteorology. *Australian Journal of Zoology*, **1**, 291–8. [Ch. 13]

Moran, V.C. and Southwood, T.R.E. (1982) The guild composition of arthropod communities in trees. *Journal of Animal Ecology*, **51**, 289–306. [Ch. 1]

Morris, D.W. (1995) Earth's peeling veneer of life. *Nature*, **373**, 25. [Ch. 1]

Morris, R.F. (1959) Single-factor analysis in population dynamics. *Ecology*, **40**, 580–8. [Intro, Ch. 5]

Morris, R.F. (ed.) (1963a) The dynamics of epidemic spruce budworm populations. *Memoirs of the Entomological Society of Canada*, **31**, 1–332. [Ch. 2]

Morris, R.F. (1963b) Predictive population equations based on key factors. *Memoirs of the Entomological Society of Canada*, **32**, 16–21. [Ch. 1]

Morris, R.F. (1969) Approaches to the study of population dynamics, in *Forest Insect Population Dynamics* (ed. W.E. Waters), USDA Forest Service Research Paper NE-125, pp. 9–28. [Ch. 14]

Morris, W.F. (1990) Problems in detecting chaotic behaviour in natural populations by fitting simple discrete models. *Ecology*, **71**, 1849–62. [Ch. 5]

Mountford, M.D. (1988) Population regulation, density dependence and heterogeneity. *Journal of Animal Ecology*, **57**, 845–58. [Ch. 5]

Müller, C.B. and Godfray, H.C.J. (1997) Apparent competition between two aphid species. *Journal of Animal Ecology*, **66**, 57–64. [Ch. 6]

Müller, C.B., Adriaanse, I.C.T., Belshaw, R. and Godfray, H.C.J. (in press) The structure of an aphid-parasitoid community. *Journal of Animal Ecology*. [Ch. 6]

Münster-Svendsen, M. and Nachman, G. (1978) Asynchrony in insect host–parasite interaction and its effect on stability, studied by a simulation model. *Journal of Animal Ecology*, **47**, 159–71. [Ch. 6]

Murdoch, W.W. (1966) Community structure, population control, and competition – a critique. *American Naturalist*, **100**, 219–26. [Intro]

Murdoch, W.W. (1969) Switching in general predators: experiments on predator specificity and stability of prey populations. *Ecological Monographs*, **39**, 335–54. [Ch. 6, 7]

Murdoch, W.W. (1970) Population regulation and population inertia. *Ecology*, **51**, 497–502. [Ch. 16]

Murdoch, W.W. (1990) The relevance of pest-enemy models to biological control, in *Critical Issues in Biological Control* (eds M. Mackauer, L.E. Ehler and J. Roland), Intercept, Andover, pp. 1–24. [Ch. 7]

Murdoch, W.W. (1994) Population regulation in theory and practice. *Ecology*, **75**, 271–87. [Ch. 4, 7, 16]

Murdoch, W.W. and Briggs, C.J. (1996) Theory for biological control: recent developments. *Ecology*, **77**, 2001–13. [Ch. 6, 7]

Murdoch, W.W. and Nisbet, R.M. (1996) Frontiers of population ecology, in *Frontiers of Population Ecology* (eds R.B. Floyd, A.W. Sheppard and P.J. De Barro), CSIRO, Melbourne, pp. 31–43. [Ch. 9, 16]

Murdoch, W.W. and Oaten, A. (1975) Predation and population stability. *Advances in Ecological Research*, **9**, 1–131. [Ch. 7]

Murdoch, W.W. and Reeve, J.D. (1987) Aggregation of parasitoids and the detection of density dependence in field populations. *Oikos*, **50**, 137–41. [Ch. 5]

Murdoch, W.W. and Stewart-Oaten, A. (1989) Aggregation by parasitoids and predators: effects on equilibrium and stability. *American Naturalist*, **134**, 288–310. [Ch. 6, 7]

Murdoch, W.W. and Walde, S.J. (1989) Analysis of insect population dynamics, in *Toward a More Exact Ecology* (eds P.J. Grubb and J.B. Whittaker), Blackwell Scientific Publications, Oxford, pp. 113–40. [Ch. 16]

Murdoch, W.W., Reeve, J.D., Huffaker, C.B. and Kennett, C.E. (1984) Biological control of olive scale and its relevance to ecological theory. *American Naturalist*, **123**, 371–92. [Ch. 6]

Murdoch, W.W., Chesson, J. and Chesson, P.L. (1985) Biological control in theory and practice. *American Naturalist*, **125**, 344–66. [Ch. 6, 7, 13]

Murdoch, W.W., Nisbet, R.M., Blythe, S.P. *et al.* (1987) An invulnerable age class and stability in delay-differential parasitoid–host models. *American Naturalist*, **129**, 263–82. [Ch. 2, 6, 7]

Murdoch, W.W., Briggs, C.J., Nisbet, R.M. *et al.* (1992a) Aggregation and stability in metapopulation models. *American Naturalist*, **140**, 41–58. [Ch. 6, 7]

Murdoch, W.W., Nisbet, R. M., Luck, R. F. *et al.* (1992b) Size-selective sex-allocation and host feeding in a parasitoid–host model. *Journal of Animal Ecology*, **61**, 533–41. [Ch. 6, 7]

Murdoch, W.W., Luck, R.F., Swarbrick, S.L. *et al.* (1995) Regulation of an insect population under biological control. *Ecology*, **76**, 206–17. [Ch. 6, 7]

Murdoch, W.W., Swarbrick, S.L., Luck, R.F. *et al.* (1996a) Refuge dynamics and metapopulation dynamics: an experimental test. *American Naturalist*, **147**, 424–44. [Ch. 6, 7]

Murdoch, W.W., Briggs, C.J. and Nisbet, R.M. (1996b) Competitive displacement and biological control in parasitoids: a model. *American Naturalist*, **148**, 807–26. [Ch. 7]

Murdoch, W.W., Briggs, C.J. and Nisbet, R.M. (1997) Dynamical effects of host size- and parasitoid state-dependent attacks by parasitoids. *Journal of Animal Ecology*, **66**, 542–56. [Ch. 6, 7]

Murdoch, W.W., Briggs, C.J. and Nisbet, R.M. (1998) Individual differences and the dynamics of consumer–resource interactions, in *Herbivores, Plants and*

Predators (eds R.H. Drent, V.K. Brown and H. Olff), Blackwell Science, Oxford (in press). [Ch. 7]

Murphy, B.C., Rosenheim, J.A. and Granett, J. (1996) Habitat diversification for improving biological control: abundance of *Anagrus sepos* (Hymenoptera: Myrmaridae) in grape vineyards. *Environmental Entomology*, **25**, 495–504. [Ch. 7]

Myers, J.H. (1980) Is the insect or the plant the driving force in the cinnabar moth–tansy ragwort system? *Oecologia*, **48**, 151–6. [Ch. 12]

Myers, J.H. and Harris, P. (1980) Distribution of *Urophora* galls in flower heads of diffuse and spotted knapweed in British Columbia. *Journal of Applied Ecology*, **17**, 359–67. [Ch. 10]

Naeem, S. (1988) Resource heterogeneity fosters coexistence of a mite and a midge in pitcher plants. *Ecological Monographs*, **58**, 215–27. [Ch. 17]

Naeem, S. and Fenchel, T. (1994) Population growth on a patchy resource – some insights provided by studies of a histophagous protozoan. *Journal of Animal Ecology*, **63**, 399–409. [Ch. 2]

Nakamura, K. and Ohgushi, T. (1981) Studies on the population dynamics of a thistle-feeding lady beetle, *Henosepilachna pustulosa* (Kono) in a cool temperate climax forest. II. Life tables, key-factor analysis, and detection of regulatory mechanisms. *Researches on Population Ecology*, **23**, 210–31. [Ch. 16]

Nealis, V. (1985) Diapause and the seasonal ecology of the introduced parasite *Cotesia (Apanteles) rubecola* (Hymenoptera: Braconidae). *Canadian Entomologist*, **117**, 333–42. [Ch. 8]

Nealis, V.G. and Lomic, P.V. (1994) Host-plant influence on the population ecology of the jack pine budworm, *Choristoneura pinus* (Lepidoptera: Tortricidae). *Ecological Entomology*, **19**, 367–73. [Ch. 14]

Nee, S., May, R.M and Hassell, M.P. (1997) Two-species metapopulation models, in *Metapopulation Biology: Ecology, Genetics and Evolution* (eds I.A. Hanski and M.E. Gilpin), Academic Press, San Diego, pp. 123–47. [Ch. 1, 2]

Neilson, M.M and Morris, R.F. (1964) The regulation of European spruce sawfly numbers in the maritime provices of Canada from 1937 to 1963. *Canadian Entomologist*, **96**, 773–84. [Ch. 1]

Newstrom, L.E., Frankie, G.W., Baker, H.G. and Caldwell, R.K. (1994) Diversity of long-term flowering patterns, in *La Selva: ecology and natural history of a neotropical rain forest* (eds L.A. McDade, K.S. Bawa, H.A. Hespendeide and G.S. Hartshorn), University of Chicago Press, Chicago, pp. 142–60. [Ch. 17]

Nicholson, A.J. (1933) The balance of animal populations. *Journal of Animal Ecology*, **2** (Supplement 1), 132–78. [Ch. 1, 2, 3, 4, 10]

Nicholson, A.J. (1937) The role of competition in determining animal populations. *Journal of the Council of Scientific and Industrial Research of Australia*, **10**, 101–6. [Ch. 3]

Nicholson, A.J. (1950) Population oscillations caused by competition for food. *Nature*, **165**, 476–7. [Ch. 1]

Nicholson, A.J. (1954) An outline of the dynamics of animal populations. *Australian Journal of Zoology*, **2**, 9–65. [Intro, Ch. 1, 4, 16]

Nicholson, A.J. (1957). The self-adjustment of populations to change. *Cold Spring Harbor Symposia of Quantitative Biology*, **22**, 153–73. [Ch. 3]

Nicholson, A.J. (1958) Dynamics of insect populations. *Annual Review of Entomology*, **3**, 107–36. [Ch. 1, 3]

Nicholson, A.J. and Bailey, V.A. (1935) The balance of animal populations. Part I. *Proceedings of the Zoological Society of London*, **1935**, 551–98. [Ch. 1, 3, 6, 7, 10]

Nieminen, M. (1996) *Metapopulation dynamics of moths*. PhD thesis, University of Helsinki, Finland. [Ch. 1]

Nisbet, R.M. and Gurney, W.S.C. (1982) *Modelling Fluctuating Populations*, Wiley, New York. [Ch. 1, 2]

Nisbet, R.M. and Gurney, W.S.C. (1983) The systematic formulation of population models for insects with dynamically varying instar duration. *Theoretical Population Biology*, **23**, 114–35. [Ch. 2]

Oduor, G.I., Yaninek, J.S., Vandergeest, L.P.S. and Demoraes, G.J. (1996) Germination and viability of capilliconidia of *Neozygites floridana* (Zygomycetes, Entomophthorales) under constant temperature, humidity and light conditions. *Journal of Invertebrate Pathology*, **67**, 267–78. [Ch. 8]

Ohgushi, T. (1986) Population dynamics of an herbivorous lady beetle, *Henosepilachna niponica*, in a seasonal environment. *Journal of Animal Ecology*, **55**, 861–79. [Ch. 16]

Ohgushi, T. (1987) Factors affecting body size variation within a population of an herbivorous lady beetle, *Henosepilachna niponica* (Lewis). *Researches on Population Ecology*, **29**, 147–54. [Ch. 16]

Ohgushi, T. (1988) Temporal and spatial relationships between an herbivorous lady beetle *Epilachna niponica* and its predator, the earwig *Anechura harmandi*. *Researches on Population Ecology*, **30**, 57–68. [Ch. 16]

Ohgushi, T. (1991) Lifetime fitness and evolution of reproductive pattern in the herbivorous lady beetle. *Ecology*, **72**, 2110–22. [Ch. 16]

Ohgushi, T. (1992). Resource limitation on insect herbivore populations, in *Effects of Resource Distribution on Animal–Plant Interactions* (eds M.D. Hunter, T. Ohgushi and P.W. Price), Academic Press, San Diego, pp. 199–241. [Ch. 16]

Ohgushi, T. (1995) Adaptive behavior produces stability in herbivorous lady beetle populations, in *Population Dynamics: New Approaches and Synthesis* (eds N. Cappuccino and P.W. Price), Academic Press, San Diego, pp. 303–19. [Ch. 14, 16]

Ohgushi, T. (1996a) A reproductive tradeoff in an herbivorous lady beetle: egg resorption and female survival. *Oecologia*, **106**, 345–51. [Ch. 16]

Ohgushi, T. (1996b) Consequences of adult size for survival and reproductive performance in a herbivorous ladybird beetle. *Ecological Entomology*, **21**, 47–55. [Ch. 16]

Ohgushi, T. (1997) Plant-mediated interactions between herbivorous insects, in *Biodiversity: An Ecological Perspective* (eds T. Abe, S.A. Levin and M. Higashi), Springer, New York, pp. 115–30. [Ch. 16]

Ohgushi, T. and Sawada, H. (1981) The dynamics of natural populations of a phytophagous lady beetle, *Henosepilachna pustulosa* under different habitat conditions. I. Comparison of adult population parameters among local populations in relation to habitat stability. *Researches on Population Ecology*, **23**, 94–115. [Ch. 16]

Ohgushi, T. and Sawada, H. (1985a) Population equilibrium with respect to available food resource and its behavioural basis in an herbivorous lady beetle, *Henosepilachna niponica*. *Journal of Animal Ecology*, **54**, 781–96. [Ch. 16]

Ohgushi, T. and Sawada, H. (1985b) Arthropod predation limits the population density of an herbivorous lady beetle, *Henosepilachna niponica* (Lewis). *Researches on Population Ecology*, **27**, 351–9. [Ch. 16]

Ohgushi, T. and Sawada, H. (1995) Demographic attributes of an introduced herbivorous lady beetle. *Researches on Population Ecology*, **37**, 29–36. [Ch. 16]

Ohgushi, T. and Sawada, H. (1997a) Population stability in relation to resource availability in an introduced population of an herbivorous lady beetle. *Researches on Population Ecology*, **39**, 37–46. [Ch. 16]

Ohgushi, T. and Sawada, H. (1997b) A shift toward early reproduction in an introduced herbivorous lady beetle. *Ecological Entomology*, **22**, 90–6. [Ch. 16]

Ohgushi, T. and Sawada, H. (1998) What changed the demography of an introduced population of an herbivorous lady beetle? *Journal of Animal Ecology*, **67** (in press). [Ch. 16]

Orians, G.H. (1962) Natural selection and ecological theory. *American Naturalist*, **96**, 257–63. [Ch. 14]

Oyeyele, S.O. and Zalucki, M.P. (1990) Cardiac glycosides and oviposition by *Danaus plexippus* on *Ascelepias fruticosa* in south-east Queensland (Australia), with notes on the effect of plant nitrogen content. *Ecological Entomology*, **15**, 177–85. [Ch. 8]

Pacala, S.W. and Hassell, M.P. (1991) The persistence of host–parasitoid associations in patchy environments. II. Evaluation of field data. *American Naturalist*, **138**, 584–605. [Ch. 4, 6]

Pacala, S.W. and Silander, J.A. (1987) Neighborhood interference among velvet leaf, *Abutilon theophrasti*, and pigweed, *Amaranthus retroflexus*. *Oikos*, **48**, 217–24. [Ch. 2]

Pacala, S.W., Hassell, M.P. and May, R.M. (1990) Host–parasitoid associations in patchy environments. *Nature*, **344**, 150–3. [Ch. 6]

Paine, R.T. (1974) Intertidal community structure. *Oecologia*, **15**, 93–120. [Ch. 2]

Paine, R.T. (1988) Some general problems for ecology illustrated by food web theory. *Ecology*, **69**, 1673–6. [Ch. 6]

Parry, D., Spence, J.R. and Volney, W.J.A. (1997) Responses of natural enemies to experimentally increased populations of the forest tent caterpillar, *Malacosoma disstria*. *Ecological Entomology*, **22**, 97–108. [Ch. 6, 7]

Pearsall, I.A. (1992) *Mortality of winter moth populations in Nova Scotian apple orchards*. MSc thesis, Dalhousie University, Halifax. [Ch. 13]

Pearsall, I.A. and Walde S.J. (1994) Parasitism and predation as agents of mortality of winter moth populations in neglected apple orchards in Nova Scotia. *Ecological Entomology*, **19**, 190–8. [Ch. 13]

Peckarovsky, B.L. and Cowan, C.A. (1991) Consequences of larval intraspecific competition to stonefly growth and fecundity. *Oecologia*, **88**, 277–88. [Ch. 8]

Pedata, P.A., Hunter, M.S., Godfray, H.J.C. and Viggiani, G. (1995) The population dynamics of the white peach scale and its parasitoids in a mulberry orchard in Campania, Italy. *Bulletin of Entomological Research*, **85**, 531–9. [Ch. 7]

Perry, J.H., Woiwod, I.P. and Hanski, I. (1993) Using response-surface methodology to detect chaos in ecological time series. *Oikos*, **68**, 329–39. [Ch. 5]

Pittara, I.S. and Katsoyannos, B.I. (1990) Evidence for a host-marking pheromone in *Chaetorellia australis*. *Entomologia Experimentalis et Applicata*, **54**, 287–96. [Ch. 4, 10]

Podoler, H. and Rogers, D.J. (1975) A new method for the identification of key factors from life-table data. *Journal of Animal Ecology*, **44**, 85–115. [Ch. 2]

Polis, G.A. and McCormick, S. (1987) Intraguild predation and competition among desert scorpions. *Ecology*, **68**, 332–43. [Ch. 17]

Polis, G.A., Myers, C.A. and Holt, R.D. (1989) The ecology and evolution of intraguild predation. *Annual Review of Ecology and Systematics*, **20**, 297–330. [Ch. 7]

Pollard, E. (1991) Synchrony of population fluctuations: the dominant influence of widespread factors on local butterfly populations. *Oikos*, **60**, 7–10. [Ch. 5]

Pollard, E. and Rothery, P. (1994) A simple stochastic model for resource-limited insect populations. *Oikos*, **69**, 287–94. [Ch. 4, 5, 10]

Pollard, E. and Yates, T.J. (1993) *Monitoring Butterflies for Ecology and Conservation*, Chapman & Hall, London. [Ch. 5, 15]

Pollard, E., Lakhani, K.H. and Rothery, P. (1987) The detection of density dependence from a series of annual censuses. *Ecology*, **68**, 2046–55. [Ch. 2, 3, 5]

Preston, F.W. (1948) The commonness, and rarity, of species. *Ecology*, **29**, 254–83. [Ch. 1]

Preszler, R.W. and Price, P.W. (1988) Host quality and sawfly populations: a new approach to life table analysis. *Ecology*, **69**, 2012–20. [Ch. 14, 16]

Price, P.W. (1988) Inversely density-dependent parasitism: the role of plant refuges for hosts. *Journal of Animal Ecology*, **57**, 89–96. [Ch. 14]

Price, P.W. (1989) Clonal development of coyote willow, *Salix exigua* (Salicaceae), and attack by the shoot-galling sawfly, *Euura exiguae* (Hymenoptera: Tenthredinidae). *Environmental Entomology*, **18**, 61–8. [Ch. 14]

Price, P.W. (1990) Evaluating the role of natural enemies in latent and eruptive species: new approaches in life table construction, in *Population Dynamics of Forest Insects* (eds A.D. Watt, S.R. Leather, M.D. Hunter and N.A.C. Kidd), Intercept, Andover, pp. 221–32. [Ch. 14]

Price, P.W. (1991a) Darwinian methodology and the theory of insect herbivore population dynamics. *Annals of the Entomological Society of America*, **84**, 465–73. [Ch. 14]

Price, P.W. (1991b) The plant vigor hypothesis and herbivore attack. *Oikos*, **62**, 244–51. [Ch. 14]

Price, P.W. (1991c) Evolutionary theory of host and parasitoid interactions. *Biological Control*, **1**, 83–93. [Ch. 6]

Price, P.W. (1994) Phylogenetic constraints, adaptive syndromes, and emergent properties: from individuals to population dynamics. *Researches on Population Ecology*, **36**, 3–14. [Ch. 14, 16]

Price, P.W. (1996) Empirical research and factually based theory: what are their roles in entomology? *American Entomologist*, **42**, 209–14. [Ch. 14]

Price, P.W. (1997) *Insect Ecology*, 3rd edn, Wiley, New York. [Ch. 14]

Price, P.W. and Clancy, K.M. (1986a) Multiple effects of precipitation on *Salix lasiolepis* and populations of the stem-galling sawfly, *Euura lasiolepis*. *Ecological Research*, **1**, 1–14. [Ch. 14]

Price, P.W. and Clancy, K.M. (1986b) Interactions among three trophic levels: gall size and parasitoid attack. *Ecology*, **67**, 1593–1600. [Ch. 6, 14]

Price, P.W. and Craig, T.P. (1984) Life history, phenology, and survivorship of a stem-galling sawfly, *Euura lasiolepis* (Hymenoptera: Tenthredinidae), on the arroyo willow, *Salix lasiolepis*, in northern Arizona. *Annals of the Entomological Society of America*, **77**, 712–9. [Ch. 14]

Price, P.W. and Ohgushi, T. (1995) Preference and performance linkage in a *Phyllocolpa* sawfly on the willow, *Salix miyabeana*, on Hokkaido. *Researches on Population Ecology*, **37**, 23–8. [Ch. 14]

Price, P.W. and Roininen, H. (1993) The adaptive radiation in gall induction, in *Sawfly Life History Adaptations to Woody Plants* (eds. M.R. Wagner and K.F. Raffa), Academic Press, San Diego, pp. 229–57. [Ch. 14]

Price, P.W., Bouton, C.E., Gross, P. *et al.* (1980) Interactions among three trophic levels: influence of plants on interactions between insect herbivores and natural enemies. *Annual Review of Ecology and Systematics*, **11**, 41–65. [Ch. 16]

Price, P.W., Roininen, H. and Tahvanainen, J. (1987a) Plant age and attack by the bud galler, *Euura mucronata*. *Oecologia*, **73**, 334–7. [Ch. 14]

Price, P.W., Roininen, H. and Tahvanainen, J. (1987b) Why does the bud-galling sawfly, *Euura mucronata*, attack long shoots? *Oecologia*, **74**, 1–6. [Ch. 14]

Price, P.W., Waring, G.L., Julkunen-Tiitto, R. *et al.* (1989) The carbon-nutrient balance hypothesis in within-species phytochemical variation of *Salix lasiolepis*. *Journal of Chemical Ecology*, **15**, 1117–31. [Ch. 14]

Price, P.W., Cobb, N., Craig, T.P. *et al.* (1990) Insect herbivore population dynamics on trees and shrubs: new approaches relevant to latent and eruptive species and life table development, in *Insect–Plant Interactions*, Vol. 2 (ed. E.A. Bernays), CRC Press, Boca Raton, Florida, pp. 1–38. [Ch. 14, 16]

Price, P.W., Clancy, K.M. and Roininen, H. (1994) Comparative population dynamics of the galling sawflies, in *Ecology and Evolution of Gall–forming Insects* (eds P.W. Price, W.J. Mattson and Y.N. Baranchikov), Northcentral Forest Experiment Station General Technical Report NC-174, USDA Forest Service, pp. 1–11. [Ch. 14]

Price, P.W., Craig, T.P. and Roininen, H. (1995) Working toward theory on galling sawfly population dynamics, in *Population Dynamics: New Approaches and Synthesis* (eds N. Cappuccino and P.W. Price), Academic Press, San Diego, pp. 321–38. [Intro, Ch. 14]

Price, P.W., Roininen, H. and Tahvanainen, J. (1997a) Willow tree shoot module length and the attack and survival pattern of a shoot-galling sawfly, *Euura atra* L. (Hymenoptera: Tenthredinidae). *Entomologica Fennicae* (in press). [Ch. 14]

Price, P.W., Roininen, H. and Carr, T. (1997b) Landscape dynamics, plant architecture and demography, and the response of herbivores, in *Vertical Food Web Interactions: evolutionary patterns and driving forces* (eds K. Dettner, G. Bauer and W. Völkl), Springer Verlag, Berlin (in press). [Ch. 14]

Prins, A.H. and Nell, H.W. (1990) The impact of herbivory on plant numbers in all life stages of *Cynoglossum officinale* L. and *Senecio jacobaea* L. *Acta Botanica Neerlandica*, **39**, 275–84. [Ch. 12]

Prokopy, R.J. (1981) Epideictic pheromones that influence spacing patterns of phytophagous insects, in *Semiochemicals: their role in pest control* (eds D.A. Nordlund, R.L. Jones and W.J. Lewis), Wiley, New York, pp. 181–213. [Ch. 4]

Pschorn-Walcher, H. and Zwölfer, H. (1968) Konkurrenzerscheinungen in Parasitenkomplexen als Problem der biologischen Schädlingsbekampfung. *Anzeiger für Schädlingskunde*, **41**, 71–6. [Ch. 6]

Pulliam, H.R. (1988) Sources, sinks, and population regulation. *American Naturalist*, **132**, 652–61. [Ch. 1]

Quenouille, M.H. (1949) Approximate tests of correlation in time series. *Journal of the Royal Statistical Society, Series B*, **11**, 68–84. [Ch. 5]

Rainey, R.C. (1982) Putting insects on the map: spatial inhomogeneity and the dynamics of insect populations. *Antenna*, **6**, 162–9. [Ch. 14]

Rand, A.S. and Rand, W.M. (1982) Variation in rainfall on Barro Colorado Island, in *The Ecology of a Tropical Forest: seasonal rhythms and long-term changes*, 2nd edn (eds E.G. Leigh, A.S. Rand and D.M. Windsor), Smithsonian Institution, Washington, D.C., pp. 47–60. [Ch. 17]

Rausher, M.D. (1980) Host abundance, juvenile survival, and oviposition preference in *Battus philenor*. *Evolution*, **34**, 342–55. [Ch. 16]

Ravlin, R.W. and Haynes, D.W. (1987) Simulation of interactions and management of parasitoids in a multiple host system. *Environmental Entomology*, **16**, 1255–65. [Ch. 6]

Ray, C. and Hastings, A. (1996) Density dependence: are we searching at the wrong spatial scale? *Journal of Animal Ecology*, **65**, 556–66. [Ch. 5, 17]

Reader, P.M. and Southwood, T.R.E. (1984) Studies on the flight activity of the viburnum whitefly, a reluctant flier. *Entomologia Experimentalis et Applicata*, **36**, 185–91. [Ch. 2]

Reddingius, J. (1971) Gambling for existence. A discussion of some theoretical problems in animal population ecology. *Acta Biotheoretica, (Leiden)*, **20**, (Supplementum 1), 1–208. [Ch. 3, 5]

Reddingius, J. (1990) Models for testing: a secondary note. *Oecologia*, **83**, 50–2. [Ch. 5]

Reddingius, J. and Den Boer, P.J. (1970) Simulation experiments illustrating stabilization of animal numbers by spreading of risk. *Oecologia*, **5**, 240–84. [Ch. 2, 3]

Reddingius, J. and Den Boer, P.J. (1989) On the stabilization of animal numbers. Problems of testing. 1. Power estimates and estimation errors. *Oecologia*, **78**, 1–8. [Ch. 2, 3, 5]

Redfern, M. (1968) The natural history of spear-thistle heads. *Field Studies*, **2**, 669–717. [Ch. 10]

Redfern, M. (1983) *Insects and Thistles*, Cambridge University Press, Cambridge. [Ch. 10]

Redfern, M. and Cameron, R.A.D. (1985) Density and survival of *Urophora stylata* (Diptera: Tephritidae) on *Cirsium vulgare* (Compositae) in relation to flower head and gall size, in *Proceedings of the 6th International Symposium on the Biological Control of Weeds*, Vancouver, Canada, 1994 (ed. E.S. Delfosse), Agriculture Canada, Ottawa, pp. 453–77. [Ch. 10]

Redfern, M., Jones, T.H. and Hassell, M.P. (1992) Heterogeneity and density dependence in a field study of a tephritid–parasitoid interaction. *Ecological Entomology*, **17**, 255–62. [Ch. 10]

Rees, M., Grubb, P.J. and Kelly, D. (1996) Quantifying the impact of competition and spatial heterogeneity on the structure and dynamics of a four-species guild of winter annuals. *American Naturalist*, **147**, 1–32. [Ch. 2]

Reeve, J.D. (1988) Environmental variability, migration, and persistence in host–parasitoid systems. *American Naturalist*, **132**, 810–36. [Ch. 6, 7]

Reeve, J.D. and Murdoch, W.W. (1986) Biological control by the parasitoid *Aphytis melinus*, and population stability of the California red scale. *Journal of Animal Ecology*, **55**, 1069–82. [Ch. 7]

Reeve, J.D., Cronin, J.T. and Strong, D.R. (1994) Parasitism and generation cycles in a salt-marsh planthopper. *Journal of Animal Ecology*, **63**, 912–20. [Ch. 6, 7]

Renwick, J.A.A. and Chew, F.S. (1994) Oviposition behavior in Lepidoptera. *Annual Review of Entomology*, **39**, 377–400. [Ch. 16]

Rhoades, D.F. (1985) Offensive–defensive interactions between herbivores and plants, their relevance in herbivore population dynamics and ecological theory. *American Naturalist*, **125**, 205–38. [Ch. 13]

Richards, O.W. and Waloff, N. (1961) A study of a natural population of *Phytodecta olivacea* (Forster) (Coleoptera: Chrysomeloidea). *Philosophical Transactions of the Royal Society, London, Series B*, **244**, 205–57. [Ch. 2]

Ricker, W.E. (1954) Stock and recruitment. *Journal of the Fisheries Research Board of Canada*, **11**, 559–623. [Ch. 5]

Riley, D., Nava-Camberos, U. and Allen, J. (1996) Population dynamics of *Bemisia* in agricultural systems, in Bemisia: *1995 Taxonomy, Biology, Damage, Control and Management* (eds D. Gerling and R.T. Mayer), Intercept, Andover, pp. 93–109. [Ch. 7]

Rivero-Lynch, A.P. and Jones, T.H. (1993) The choice of oviposition site by *Terellia ruficauda* on *Cirsium palustre*. *Acta Oecologica*, **14**, 643–51. [Ch. 10]

Robinson, A.S. and Hooper, G. (eds) (1989) *Fruit Flies: Their Biology, Natural Enemies and Control*, Elsevier, Amsterdam. [Ch. 10]

Rohani, P. and Miramontes, O. (1995) Host–parasitoid metapopulations – the consequences of parasitoid aggregation on spatial dynamics and searching efficiency. *Proceedings of the Royal Society, London, Series B*, **260**, 335–42. [Ch. 6]

Rohani, P., Godfray, H.C.J. and Hassell, M.P. (1994) Aggregation and the dynamics of host–parasitoid systems: a discrete-generation model with within-generation redistribution. *American Naturalist*, **144**, 491–509. [Ch. 6, 7]

Roininen, H. and Tahvanainen, J. (1989) Host selection and larval performance of two willow-feeding sawflies. *Ecology*, **70**, 129–36. [Ch. 16]

Roininen, H., Price, P.W. and Tahvanainen, J. (1993) Colonization and extinction in a population of the shoot-galling sawfly, *Euura amerinae*. *Oikos*, **68**, 448–54. [Ch. 14]

Roininen, H., Price, P.W. and Tahvanainen, J. (1996) Bottom-up and top-down influences in the trophic system of a willow, a galling sawfly, and parasitoids and inquilines. *Oikos*, **77**, 44–50. [Ch. 14]

Roininen, H., Price, P.W., Julkunen-Tiitto, R. and Tahvanainen, J. (1998) Oviposition stimulant for a galling sawfly, *Euura lasiolepis*, on willow is a phenolic glucoside. *Journal of Chemical Ecology* (in press). [Ch. 14]

Roland, J. (1986a) *Success and failure of* Cyzenis albicans *in controlling its host the winter moth*. PhD thesis, University of British Columbia, Vancouver. [Ch. 13]

Roland, J. (1986b) Parasitism of winter moth in British Columbia during build-up of its parasitoid *Cyzenis albicans*: attack rate on oak *v.* apple. *Journal of Animal Ecology*, **55**, 215–34. [Ch. 13]

Roland, J. (1990a) Parasitoid aggregation: chemical ecology and population dynamics, in *Critical Issues in Biological Control* (eds L. Ehler, M. Mackauer and J. Roland), Intercept, Andover, pp. 185–211. [Ch. 13]

Roland, J. (1990b) Interaction of parasitism and predation in the decline of winter moth in Canada, in *Population Dynamics of Forest Insects* (eds A.D. Watt, S.R. Leather, M.D. Hunter and N.A. Kidd), Intercept, Andover, pp. 289–302. [Ch. 13]

Roland, J. (1994) After the decline: what maintains low winter moth density after successful biological control? *Journal of Animal Ecology*, **63**, 392–8. [Ch. 7, 13]

Roland, J. and Embree, D.G. (1995) Biological control of the winter moth. *Annual Review of Entomology*, **40**, 475–92. [Ch. 7, 13]

Roland, J. and Myers, J.H. (1987) Improved insect performance from host-plant defoliation: winter moth on oak and apple. *Ecological Entomology*, **12**, 409–14. [Ch. 13]

Roland, J. and Taylor, P.D. (1997) Insect parasitoid species respond to forest structure at different spatial scales. *Nature*, **386**, 710–3. [Ch. 6]

Roland, J. and Walde, S.J. (*In preparation*) Predation on winter moth pupae in Nova Scotia oak stands: Embree's sites 30 years later. [Ch. 13]

Romstöck, M. (1984) Zur geographischen Variabilitat des mit *Cirsium heterophyllum* Blutenkopfen assozierten. *Internationales Symposium über Entomofaunistik in Mitteleuropa*, **10**, 123–7. [Ch. 10]

Romstöck, M. and Arnold, H. (1987) Populationsokologie und Wirtswahl bei *Tephritis conura* Loew – Biotypen (Dipt.: Tephritidae). *Zoologischer Anzeiger*, **219**, 83–102. [Ch. 10]

Romstöck-Völkl, M. (1990) Population dynamics of *Tephritis conura* Loew (Diptera: Tephritidae): determinants of density from three trophic levels. *Journal of Animal Ecology*, **59**, 251–68. [Ch. 10]

Romstöck-Völkl, M. and Wissel, C. (1989) Spatial and seasonal patterns in the egg distribution of *Tephritis conura* Loew (Diptera: Tephritidae). *Oikos*, **55**, 165–74. [Ch. 10]

Root, R.B. (1973) Organization of a plant–arthropod association in simple and diverse habitats: the fauna of collards (*Brassica oleracea*). *Ecological Monographs*, **43**, 95–124. [Ch. 16]

Rose, M. and DeBach, P. (1992) Biological control of *Parabemisia myricae* (Kuwana) (Homoptera: Aleyrodidae) in California. *Israel Journal of Entomology*, **25/26**, 73–95. [Ch. 7]

Rosenheim, J.A., Wilhoit, L.R. and Armer, C.A. (1993) Influence of intraguild predation among generalist insect predators on the suppression of an herbivore population. *Oecologia*, **96**, 439–49. [Ch. 7]

Rosenheim, J.A., Kaya, H.K., Ehler, L.E. *et al.* (1995) Intraguild predation among biological-control agents: theory and evidence. *Biological Control*, **5**, 303–35. [Ch. 7]

Ross, M.A. and Harper, J.L. (1972) Occupation of biological space during seedling establishment. *Journal of Ecology*, **60**, 77–88. [Ch. 2]

Rossiter, M.C. (1994) Maternal effects hypothesis of herbivore outbreak. *BioScience*, **44**, 752–63. [Ch. 8]

Rossiter, M.C. (1995) Impact of life-history evolution on population dynamics: predicting the presence of maternal effects, in *Population Dynamics: New Approaches and Synthesis* (eds N. Cappuccino and P.W. Price), Academic Press, San Diego, pp. 251–75. [Ch. 8]

Rossiter, M.C., Cox-Foster, D.L. and Briggs, M.A. (1993) Initiation of maternal effects in *Lymantria dispar:* genetic and ecological components of egg provisioning. *Journal of Evolutionary Biology*, **6**, 577–89. [Ch. 8]

Rothery, P., Newton, I., Dale, L. *et al.* (1997) Testing for density dependence allowing for weather effects. *Oecologia*, **112**, 518–523. [Ch. 5]

Roughgarden, J. and Feldman, M. (1975) Species packing and predation pressure. *Ecology*, **56**, 489–92. [Ch. 7]

Rowell-Rahier, M. and Pasteels, J.M. (1986) Economics of chemical defense in Chrysomelinae. *Journal of Chemical Ecology*, **12**, 1189–203. [Ch. 8]

Rowell-Rahier, M. and Pasteels, J.M. (1990) Phenolglucosides and interactions at three trophic levels: Salicaeae–herbivores–predators, in *Insect–Plant Interactions*, Vol. 2 (ed. E.A. Bernays), CRC Press, Boca Raton, Florida, pp. 75–94. [Ch. 8]

Royama, T. (1977) Population persistence and density dependence. *Ecological Monographs*, **47**, 1–35. [Ch. 16]

Royama, T. (1981) Fundamental concepts and methodology for the analysis of animal population dynamics, with particular reference to univoltine species. *Ecological Monographs*, **51**, 473–93. [Ch. 5]

Royama, T. (1992) *Analytical Population Dynamics*, Chapman & Hall, London. [Ch. 2, 5, 13]

Royama, T. (1996) A fundamental problem in key factor analysis. *Ecology*, **77**, 87–93. [Ch. 13]

Royer, L. and McNeil, J.N. (1993) Male investment in the European corn borer, *Ostrinia nubilalis* (Hübner) (Lepidoptera: Pyralidae): impact on female longevity and reproductive performance. *Functional Ecology*, **7**, 209–215. [Ch. 8]

Saccheri, I., Kuussaari, M., Kankare, M. *et al.* (1998) Inbreeding and extinction in a butterfly metapopulation. Unpublished manuscript. [Ch. 1]

Sait, S.M., Andreev, R.A., Begon, M. *et al.* (1995) *Venturia canescens* parasitizing *Plodia interpunctella*: host vulnerability – a matter of degree. *Ecological Entomology*, **20**, 199–201. [Ch. 6]

Sakaluk, S.K. (1985) Spermatophore size and its role in the reproductive behaviour of the cricket, *Gryllodes supplicans* (Orthoptera: Gryllidae). *Canadian Journal of Zoology*, **63**, 1652–6. [Ch. 8]

Sanford, R.L., Paaby, P., Luvall, J.C. and Phillips E. (1994) Climate, geomorphology, aquatic systems, in *La Selva: ecology and natural history of a neotropical rain forest* (eds L.A. McDade, K.S. Bawa, H.A. Hespendeide and G.S. Hartshorn), University of Chicago Press, Chicago, pp. 19–33. [Ch. 17]

Santos, N.D. (1981) Odonata, in *Aquatic Biota of Tropical South America* (eds S.H. Hurlbert, G. Rodriguez and N.D. Santos), San Diego State University Press, San Diego, pp. 64–85. [Ch. 17]

Scheurer, S. (1964) Untersuchungen zum Massenwechsel einiger Fichten bewohnender Lachnidenarten im Harz. *Biologisches Zentralblatt*, **83**, 427–67. [Ch. 9]

Scheurer, S. (1971) Biologische und ökologische Beobachtungen an auf *Pinus* lebenden Cinarinen im Bereich der Dübener Heide (DDR) während der Jahre 1965–1967. *Hercynia, Leipzig*, **8**, 108–44. [Ch. 9]

Schlumprecht, H. (1989) Dispersal of the thistle gallfly *Urophora cardui* and its endoparasitoid *Eurytoma serratulae* (Hymenoptera: Eurytomidae). *Ecological Entomology*, **14**, 341–8. [Ch. 10]

Schoener, T.W. (1986) Mechanistic approaches to community ecology: a new reductionism? *American Zoologist*, **26**, 81–106. [Ch. 16]

Schoener, T.W. and Spiller, D.A. (1987) High population persistence in a system with high turnover. *Nature*, **330**, 474–7. [Ch. 1]

Schultz, J.C. (1983) Habitat selection and foraging tactics of caterpillars in hetero-geneous trees, in *Variable Plants and Herbivores in Natural and Managed Systems* (eds R.F. Denno and M.S. McClure), Academic Press, New York, pp. 61–90. [Ch. 16]

Scriber, J.M. and Slansky, F.J. (1981) The nutritional ecology of immature insects. *Annual Review of Entomology*, **26**, 183–211. [Ch. 16]

Seber, G.A.F. (1973) *The Estimation of Animal Abundance and Related Parameters*, Griffin, London. [Ch. 16]

Seber, G.A.F. and Wild, C.J. (1989) *Nonlinear Regression*, Wiley, New York. [Ch. 5]

Sequeira, R. and Dixon, A.F.G. (1997) Population dynamics of tree-dwelling aphids: the importance of seasonality and time scale. *Ecology*, **78**, 2603–10. [Ch. 9]

Settle, W.H. and Wilson, L.T. (1990) Invasion by the variegated leafhopper and biotic interactions: parasitism, competition, and apparent competition. *Ecology*, **71**, 1461–70. [Ch. 6, 7]

Settle, W.H., Ariawan, H., Astuti, E.T. *et al.* (1996) Managing tropical rice pests through conservation of generalist natural enemies and alternative prey. *Ecology*, **77**, 1975–88. [Ch. 7]

Sevenster, J.G. and Van Alphen, J.J.M. (1996) Aggregation and coexistence. II. A neotropical *Drosophila* community. *Journal of Animal Ecology*, **65**, 308–24. [Ch. 17]

Shaw, G.G. and Little, C.H.A. (1972) Effect of high urea fertilization of balsam fir trees on spruce budworm development, in *Insect and Mite Nutrition* (ed. J. Rodriguez), North Holland, Amsterdam, pp. 589–97. [Ch. 8]

Shea, K., Nisbet, R.M., Murdoch, W.W. and Yoo, H.J.S. (1996) The effect of egg lim-itation on stability in insect host–parasitoid population models. *Journal of Animal Ecology*, **65**, 743–55. [Ch. 6]

Sheppard, A.W. and Woodburn, T. (1996) Population regulation in insects used to control thistles: can this predict effectiveness? in *Frontiers of Population Ecology* (eds R.B. Floyd, A.W. Sheppard and P.J. De Barro), CSIRO, Melbourne, pp. 277–90. [Ch. 10]

Shimada, M. and Tuda, M. (1996) Delayed density dependence and oscillatory population dynamics in overlapping-generation systems of a seed beetle *Callosobruchus chinensis*: matrix population model. *Oecologia*, **105**, 116–25. [Ch. 2]

Shorrocks, B. (1990) Coexistence in a patchy environment, in *Living in a Patchy Environment* (eds B. Shorrocks and I.R. Swingland), Oxford University Press, Oxford, pp. 91–106. [Ch. 17]

Shorthouse, J.D. (1988) Modification of flowerheads of diffuse knapweed by the gall-inducers *Urophora affinis* and *Urophora quadrifasciata*, in *Proceedings of the 7th International Symposium on the Biological Control of Weeds*, Rome 1988 (ed. E.S. Delfosse), Istituto Sperimentale per la Patologia Vegetale, MAF, Rome, pp. 221–8. [Ch. 10]

Showler, A.T. (1995) Locust (Orthoptera: Acrididae) outbreak in Africa and Asia, 1992–1994: an overview. *American Entomologist*, **41**, 179–85. [Ch. 14]

Siitonen, J. and Martikainen, P. (1994) Occurrence of rare and threatened insects living on decaying *Populus tremula*: a comparison between Finnish and Russian Karelia. *Scandinavian Journal of Forest Research*, **9**, 185–91. [Ch. 1]

Sinclair, A.R.E. (1973) Regulation, and population models for a tropical ruminant. *East African Wildlife*, **11**, 307–16. [Ch. 2]

Sinclair, A.R.E. (1989) Population regulation in animals, in *Ecological Concepts* (ed. J.M. Cherrett), Blackwell Scientific Publications, Oxford, pp. 197–241. [Ch. 1, 16]

Singer, M.C. (1972) Complex components of habitat suitability within a butterfly colony. *Science*, **176**, 75–7. [Ch. 11]

Sjerps, M. and Haccou, P. (1993) A war of attrition between larvae on the same host plant: stay and starve, or leave and be eaten. *Evolutionary Ecology*, **7**, 1–19. [Ch. 12]

Sjögren Gulve, P. and Ray, C. (1996) Large-scale forestry extirpates the pool frog: using logistic regression to model metapopulation dynamics, in *Metapopulations and Wildlife Conservation and Management* (ed. D.R. McCullough), Island Press, Washington, D.C., pp. 111–38. [Ch. 1]

Skellam, J.G. (1951) Random dispersal in theoretical populations. *Biometrika*, **38**, 196–218. [Ch. 6]

Slade, N. (1977) Statistical detection of density dependence from a series of sequential censuses. *Ecology*, **58**, 1094–1102. [Ch. 5]

Smith, F.E. (1961) Density dependence in the Australian Thrips. *Ecology*, **42**, 403–7. [Ch. 2]

Smith, H.S. (1935) The role of biotic factors in the determination of population densities. *Journal of Economic Entomology*, **28**, 873–98. [Ch. 1, 2]

Smith, R.H. and Sibly, R.M. (1985) Behavioural ecology and population dynamics: towards a synthesis, in *Behavioural Ecology*, (eds R.M. Sibly and R.H. Smith), Blackwell Scientific Publications, Oxford, pp. 577–91. [Ch. 16]

Solbreck, C. (1995) Long-term population dynamics of a seed-feeding insect in a landscape perspective, in *Population Dynamics: New Approaches and Synthesis* (eds N. Cappuccino and P.W. Price), Academic Press, San Diego, pp. 279–301. [Ch. 14]

Soldaat, L.L. and Vrieling, K. (1992) The influence of nutritional and genetic factors on larval performance in the cinnabar moth, *Tyria jacobaeae*. *Entomologia Experimentalis et Applicata*, **62**, 29–36. [Ch. 12]

Solé, R.V. and Valls, J. (1992) Spiral waves, chaos and multiple attractors in lattice models of interacting populations. *Physics Letters A*, **166**, 123–8. [Ch. 2]

Solomon, M.E. (1969) *Population Dynamics*, Edward Arnold, London. [Ch. 8]

Solow, A.R. (1990) Testing for density dependence: a cautionary note. *Oecologia*, **83**, 47–9. [Ch. 5]

Solow, A.R. (1991) Response. [To Reddingius, 1990]. *Oecologia*, **86**, 146. [Ch. 5]

Solow, A.R. and Steele, J.H. (1990) On sample size, statistical power, and the detection of density dependence. *Journal of Animal Ecology*, **59**, 1073–6. [Ch. 5]

Southern, H.N. (1970) The natural control of a population of tawny owls (*Strix aluco*). *Journal of Zoology*, **162**, 197–285. [Ch. 2]

Southwood, T.R.E. (1961) The number of species of insect associated with various trees. *Journal of Animal Ecology*, **30**, 1–8. [Ch. 6]

Southwood, T.R.E. (1967) The interpretation of population change. *Journal of Animal Ecology*, **36**, 519–29. [2]

Southwood, T.R.E. (ed.) (1968) *Insect Abundance*, Fourth Symposium of the Royal Entomological Society of London, Blackwell Scientific Publications, Oxford. [Intro, Ch. 13, 14, 16]

Southwood, T.R.E. (1975) The dynamics of insect populations, in *Insects, Science, and Society* (ed. D. Pimentel), Academic Press, San Diego, pp. 151–99. [Ch. 14]

Southwood, T.R.E. (1976) *Ecological Methods*, Chapman & Hall, London. [Ch. 2]

Southwood, T.R.E. (1977) The relevance of population dynamic theory to pest status, in *Origins of Pest, Parasite, Disease and Weed Problems* (eds J.M. Cherrett and G.R. Sagar), Blackwell Scientific Publications, Oxford, pp. 35–54. [Ch. 14]

Southwood, T.R.E. and Comins, H.N. (1976) A synoptic population model. *Journal of Animal Ecology*, **45**, 949–65. [Ch. 14]

Southwood, T.R.E. and Reader, P.M. (1976) Population census data and key factor analysis for the viburnum whitefly, *Aleurotrachelus jelinekii* (Frauenf.) on three bushes. *Journal of Animal Ecology*, **45**, 313–25. [Ch. 2]

Southwood, T.R.E. and Reader, P.M. (1988) The impact of predation of the viburnum whitefly, *Aleurotrachelus jelinekii*. *Oecologia*, **74**, 566–670. [Ch. 2]

Southwood, T.R.E., Hassell, M.P., Reader, P.M. and Rogers, D.J. (1989) Population dynamics of the viburnum whitefly (*Aleurotrachelus jelinekii*). *Journal of Animal Ecology*, **58**, 921–42. [Ch. 2, 16]

St Amant, J.L.S (1970) The detection of regulation in animal populations. *Ecology*, **51**, 823–8. [Ch. 5]

Stacey, P.B., Johnson, V.A. and Taper, M.L. (1997) Migration with metapopulations: the impact upon local population dynamics, in *Metapopulation Biology: Ecology, Genetics and Evolution* (eds I.A. Hanski and M.E. Gilpin), Academic Press, San Diego, pp. 267–92. [Ch. 1]

Stanton, M.L. (1983) Spatial patterns in the plant community and their effects upon insect search, in *Herbivorous Insects* (ed. S. Ahmad), Academic Press, New York, pp. 125–57. [Ch. 16]

Stein, S.J. and Price, P.W. (1995) Relative effects of plant resistance and natural enemies by plant developmental age on sawfly (Hymenoptera: Tenthredinidae) preference and performance. *Environmental Entomology*, **24**, 909–16. [Ch. 14]

Stein, S.J., Price, P.W., Craig, T.P. and Itami, J.K. (1994) Dispersal of a galling sawfly: implications for studies of insect population dynamics. *Journal of Animal Ecology*, **63**, 666–76. [Ch. 14]

Stelter, C., Reich, M., Grimm, V. and Wissel, C. (1997) Modelling persistence in dynamic landscapes: lessons from a metapopulation of the grasshopper *Bryodema tuberculata*. *Journal of Animal Ecology*, **66**, 508–18. [Ch. 1]

Stiling, P. (1987) The frequency of density dependence in insect host–parasitoid systems. *Ecology*, **68**, 844–56. [Ch. 1, 2, 4, 5, 6]

Stiling, P. (1988) Density-dependent processes and key factors in insect populations. *Journal of Animal Ecology*, **57**, 581–94. [Ch. 1, 2, 4, 5, 16]

Stout, J. (1983) *Megaloprepus* and *Mecistogaster* (Gallito Azul, Helicopter Damselfly) in *A Costa Rican Natural History* (ed. D.H. Janzen), University of Chicago Press, Chicago, pp. 734–5. [Ch. 17]

Straw, N.A. (1986) *Resource limitation and competition in tephritid flies*. PhD thesis, University of Cambridge. [Ch. 10]

Straw, N.A. (1989a) Taxonomy, attack strategies and host relations in flowerhead Tephritidae: a review. *Ecological Entomology*, **14**, 455–62. [Ch. 10]

Straw, N.A. (1989b) The timing of oviposition and larval growth by two tephritid fly species in relation to host-plant development. *Ecological Entomology*, **14**, 443–54. [Ch. 10]

Straw, N.A. (1989c) Evidence for an oviposition-deterring pheromone in *Tephritis bardanae* (Schrank) (Diptera: Tephritidae). *Oecologia*, **78**, 121–30. [Ch. 4, 10]

Straw, N.A. (1991) Resource limitation of tephritid flies on lesser burdock, *Arctium minus* (Hill) Bernh. (Compositae). *Oecologia*, **86**, 492–502. [Ch. 10]

Straw, N.A. and Ludlow, A.R. (1994) Small-scale dynamics and insect diversity on plants. *Oikos*, **71**, 188–92. [Ch. 10]

Straw, N.A. and Sheppard, A.W. (1995) The role of plant dispersion pattern in the success and failure of biological control, in *Proceedings of the 8th International Symposium on the Biological Control of Weeds*, Canterbury, New Zealand, 1992 (eds E.S. Delfosse and R.R. Scott), DSIR/CSIRO, Melbourne, pp. 161–8. [Ch. 10]

Strong, D.R. (1984) Density-vague ecology and liberal population regulation in insects, in *A New Ecology: Novel Approaches to Interactive Systems* (eds P.W. Price, C.N. Slobodchikoff and W.S. Gaud), Wiley, New York, pp. 313–27. [Ch. 1, 2]

Strong, D.R. (1986a) Density-vague population change. *Trends in Ecology and Evolution*, **1**, 39–42. [Ch. 2, 5]

Strong, D.R. (1986b) Density vagueness: adding the variance in the demography of real populations, in *Community Ecology* (eds J.M. Diamond and T.J. Case), Harper & Row, New York, pp. 257–68. [Ch. 1, 2]

Strong, D.R., Lawton, J.H. and Southwood, T.R.E. (1984) *Insects on Plants: community patterns and mechanisms*, Blackwell Scientific Publications, Oxford. [Ch. 8]

Sullivan, C.R. and Wellington, W.G. (1953) The light reactions of larvae of the tent caterpillars, *Malacosoma disstria* Hbn., *M. americanum* (Fab.) and *M. pluviale* (Dyar) (Lepidoptera: Lasiocampidae). *Canadian Entomologist*, **85**, 297–310. [Ch. 8]

Summers, K. (1989) Sexual selection and intra-female competition in the green poison-dart frog, *Dendrobates auratus*. *Animal Behaviour*, **37**, 797–805. [Ch. 17]

Summers, K. (1990) Paternal care and the cost of polygyny in the green dart-poison frog. *Behavioral Ecology and Sociobiology*, **27**, 307–13. [Ch. 17]

Sutcliffe, O.L., Thomas, C.D., Yates, T.J. and Greatorex-Davies, J.N. (1997) Correlated extinctions, colonizations and population fluctuations in a highly correlated ringlet butterfly metapopulation. *Oecologia*, **109**, 235–41. [Ch. 11]

Sutherland, W.J. (1996) *From Individual Behaviour to Population Ecology*, Oxford University Press, Oxford. [Ch. 16]

Svärd, L. and Wiklund, C. (1988) Fecundity, egg weight and longevity in relation to multiple matings in females of the monarch butterfly. *Behavioral Ecology and Sociobiology*, **23**, 39–44. [Ch. 8]

Swetnam, T.W. and Lynch, A.M. (1993) Multicentury, regional-scale patterns of western spruce budworm outbreaks. *Ecological Monographs*, **63**, 399–424. [Ch. 14]

Swetnam, T.W., Swetnam, J.R., Lynch, A.M. *et al.* (in press) Western spruce budworm outbreaks are associated with wet periods, not droughts. *Science*. [Ch. 14]

Swinton, J. and Anderson, R.M. (1995) Model frameworks for plant–pathogen interactions, in *Ecology of Infectious Diseases in Natural Populations* (eds B. Grenfell and A. Dobson), Cambridge University Press, Cambridge, pp. 280–294. [Ch. 2]

Tailleux, I. and Cloutier, C. (1993) Defoliation of tamarack by outbreak populations of larch sawfly in subarctic Quebec: measuring the impact on tree growth. *Canadian Journal of Forest Research*, **23**, 1444–52. [Ch. 14]

Takagi, M. and Hirose, Y. (1994) Building parasitoid communities: the complementary role of two introduced parasitoid species in a case of successful biological control, in *Parasitoid Community Ecology* (eds B.A. Hawkins and W. Sheehan), Oxford University Press, Oxford, pp. 437–448. [Ch. 7]

Takashi, S., Stenseth, N.C. and Bjornstad, O. (1997) Density dependence in fluctuating grey-sided vole populations. *Journal of Animal Ecology*, **66**, 14–24. [Ch. 5]

Takens, F. (1981) Detecting strange attractors in turbulence, in *Dynamic Systems and Turbulence* (eds. D.A. Rand and L.S. Young), Springer-Verlag, New York, pp. 366–81. [Ch. 5]

Tanasijtshuk, V.N. (1986) *Family Chamaemyiidae: Fauna USSR*, **14**(7), Academia Nauk, Leningrad. [Ch. 15]

Taylor, A.D. (1988) Large-scale spatial structure and population dynamics in arthropod predator–prey systems. *Annals Zoologici Fennici*, **25**, 63–74. [Ch. 1, 2, 6]

Taylor, A.D. (1990) Metapopulations, dispersal, and predator–prey dynamics: an overview. *Ecology*, **71**, 429–33. [Ch. 2]

Taylor, A.D. (1991) Studying metapopulation effects in predator–prey systems. *Biological Journal of the Linnean Society*, **42**, 305–23. [Ch. 7]

Taylor, A.D. (1993) Heterogeneity in host-parasitoid interactions – 'aggregation of risk' and the '$CV^2 > 1$ rule'. *Trends in Ecology and Evolution*, **8**, 400–5. [Ch. 6]

Taylor, A.D. (1997) Density–dependent parasitoid recruitment per parasitized host: effects on parasitoid–host dynamics. *American Naturalist*, **149**, 989–1000. [Ch. 6]

Taylor, L.R. (1989) Objective and experiment in long-term research, in *Long-term Studies in Ecology* (ed. G.E. Likens), Springer-Verlag, New York, pp. 20–70. [Ch. 9]

Taylor, L.R., Woiwod, I.P. and Perry, J.N. (1979) The negative binomial as an ecological model and the density dependence of k. *Journal of Animal Ecology*, **48**, 289–304. [Ch. 2]

Taylor, T.H.C. (1935) The campaign against *Aspidiotus destructor* Sign. in Fiji. *Bulletin of Entomological Research*, **26**, 1–102. [Ch. 9]

Tenow, O. (1972) The outbreaks of *Oporinia autumnata* Bkh. and *Operophtera* spp. (Lep., Geometridae) in the Scandinavian mountain chain and northern Finland 1862–1968. *Zoologiska Bidrag Från Uppsala*, Supplement 2, 1–107. [Ch. 13]

Ter Braak, C.J.F., Hanski, I. and Verboom, J. (1998) The incidence function approach to modelling of metapopulation dynamics, in *Modelling Spatiotemporal Dynamics in Ecology* (eds J. Bascompte and R.V. Solé), Springer-Verlag, Berlin, pp. 167–88. [Ch. 1]

Thomas, C.D. (1994) Local extinctions, colonizations and distributions: habitat tracking by British butterflies, in *Individuals, Populations and Patterns in Ecology* (eds S.R. Leather, A.D. Watt, N.J. Mills and K.F.A. Walters), Intercept, Andover, pp. 319–36. [Ch. 1]

Thomas, C.D. (1995) The ecology and conservation of butterfly metapopulations in the fragmented British landscape, in *Ecology and Conservation of Butterflies* (ed. A.S. Pullin), Chapman & Hall, London, pp. 46–63. [Ch. 1]

Thomas, C.D. and Hanski, I. (1997) Butterfly metapopulations, in *Metapopulation Biology: Ecology, Genetics and Evolution* (eds I.A. Hanski and M.E. Gilpin), Academic Press, San Diego, pp. 359–86. [Ch. 1]

Thomas, C.D. and Harrison, S. (1992) Spatial dynamics of a patchily distributed butterfly species. *Journal of Animal Ecology*, **61**, 437–46. [Ch. 1]

Thomas, C.D. and Jones, T.M. (1993) Partial recovery of a skipper butterfly (*Hesperia comma*) from population refuges: lessons for conservation in a fragmented landscape. *Journal of Animal Ecology*, **62**, 472–81. [Ch. 1]

Thomas, C.D., Thomas, J.A. and Warren, M.S. (1992) Distributions of occupied and vacant butterfly habitats in fragmented landscapes. *Oecologia*, **92**, 563–7. [Ch. 1]

Thomas, C.D., Singer, M.C. and Boughton, D.A. (1996) Catastrophic extinction of population sources in a complex butterfly metapopulation. *American Naturalist*, **148**, 957–75. [Ch. 1]

Thomas, J.A. (1977) *Second Report on the Ecology and Conservation of the Large Blue Butterfly*. Unpublished report, Institute of Terrestrial Ecology, UK. [Ch. 11]

Thomas, J.A. (1980) Why did the large blue become extinct in Britain? *Oryx*, **15**, 243–7. [Ch. 11]

Thomas, J.A. (1984a) The conservation of butterflies in temperate countries: past efforts and lessons for the future, in *Biology of Butterflies*, Symposia of the Royal Entomological Society No. 11 (eds R. Vane-Wright and P. Ackery), Academic Press, London, pp. 333–53. [Ch. 11]

Thomas, J.A. (1984b) The behaviour and habitat requirements of *Maculinea nausithous* (the dusky large blue) and *M. teleius* (the scarce large blue) in France. *Biological Conservation*, **28**, 325–47. [Ch. 11]

Thomas, J.A. (1991) Rare species conservation: case studies of European butterflies, in *The Scientific Management of Temperate Communities for Conservation* (eds I.F. Spellerberg, F.B. Goldsmith and M.G. Morris), Blackwell Scientific Publications, Oxford, pp. 149–97. [Ch. 11]

Thomas, J.A. (1993) Holocene climate change and warm man-made refugia may explain why a sixth of British butterflies inhabit unnatural early-successional habitats. *Ecography*, **16**, 278–84. [Ch. 11]

Thomas, J.A. (1995) The ecology and conservation of *Maculinea arion* and other European species of large blue butterfly, in *Ecology and Conservation of Butterflies* (ed. A.S. Pullin), Chapman & Hall, London, pp. 180–97. [Ch. 11]

Thomas, J.A. and Elmes, G.W. (1993). Specialised searching and the hostile use of allomones by a parasitoid whose host, the butterfly *Maculinea rebeli*, inhabits ant nests. *Animal Behaviour*, **45**, 593–602. [Ch. 11]

Thomas, J.A. and Elmes, G.W. (in press). Higher productivity at the cost of increased host-specificity when *Maculinea* butterfly larvae exploit ant colonies through trophallaxis rather than by predation. *Ecological Entomology*. [Ch. 11]

Thomas, J.A. and Wardlaw, J.C. (1990) The effect of queen ants on the survival of *Maculinea arion* larvae in *Myrmica* ant nests. *Oecologia*, **85**, 87–91. [Ch. 11]

Thomas, J.A. and Wardlaw, J.C. (1992) The capacity of a *Myrmica* ant nest to support a predacious species of *Maculinea* butterfly. *Oecologia*, **91**, 101–9. [Ch. 11]

Thomas, J.A., Elmes, G.W., Wardlaw, J.C. and Woyciechowski, M. (1989) Host specificity among *Maculinea* butterflies in *Myrmica* ant nests. *Oecologia*, **79**, 452–7. [Ch. 11]

Thomas, J.A., Munguira, M.L., Martin, J. and Elmes, G.W. (1991) Basal hatching by *Maculinea* butterfly eggs: a consequence of advanced myrmecophily? *Biological Journal of the Linnean Society*, **44**, 175–84. [Ch. 11]

Thomas, J.A., Elmes, G.W. and Wardlaw, J.C. (1993) Contest competition among *Maculinea rebeli* butterfly larvae in ant nests. *Ecological Entomology*, **18**, 73–6. [Ch. 11]

Thomas, J.A, Moss, D. and Pollard, E. (1994) Increased fluctuations of butterfly populations towards the northern edges of species' ranges. *Ecography*, **17**, 215–20. [Ch. 11]

Thomas, J.A., Elmes, G.W., Clarke, R.T. *et al.* (1997) Field evidence and model predictions of butterfly-mediated apparent competition between gentian plants and red ants. *Acta Oecologica*. **18**, 671–84. [Ch. 11]

Thomas, J.A., Simcox, D.J., Wardlaw, J.C. *et al.* (1998) Effects of latitude, altitude and climate on the habitat and conservation of the endangered butterfly *Maculinea arion* and its *Myrmica* ant hosts. *Journal of Insect Conservation* **2**, 39–46. [Ch. 11]

Thomas, M.B., Wratten, S.D and Sotherton, N.W. (1991) Creation of "island" habitats in farmland to manipulate populations of beneficial arthropods: predator densities and emigration. *Journal of Applied Ecology*, **28**, 906–17. [Ch. 7]

Thompson, J.N. (1988) Evolutionary ecology of the relationship between oviposition preference and performance of offspring in phytophagous insects. *Entomologia Experimentalis et Applicata*, **47**, 3–14. [Ch. 14, 16]

Thompson, W.R. (1929) On natural control. *Parasitology*, **21**, 269–81. [Ch. 3]

Thompson, W.R. (1939). Biological control and the interactions of populations. *Parasitology*, **31**, 299–388. [Ch. 3]

Tilman, D. (1982) *Resource Competition and Community Structure*, Princeton University Press, Princeton, New Jersey. [Ch. 6]

Tuda, M. and Shimada, M. (1995) Developmental schedules and persistence of experimental host–parasitoid systems at two different temperatures. *Oecologia*, **103**, 283–91. [Ch. 6]

Turchin, P. (1990) Rarity of density dependence or population regulation with lags? *Nature*, **344**, 660–3. [Ch. 5, 9, 13, 14]

Turchin, P. (1995) Population regulation: old arguments and a new synthesis, in *Population Dynamics: New Approaches and Synthesis* (eds N. Cappuccino and P.W. Price), Academic Press, San Diego, pp. 19–40. [Ch. 1, 2]

Turchin, P. and Taylor, A.D. (1992) Complex dynamics in ecological time series. *Ecology*, **73**, 289–305. [Ch. 5, 9]

Uchmanski, J. and Grimm, V. (1996) Individual-based modelling in ecology: what makes the difference? *Trends in Ecology and Evolution*, **11**, 437–41. [Ch. 16]

Utida, S. (1950) On the equilibrium state of the interacting population of an insect and its parasite. *Ecology*, **31**, 165–75. [Ch. 2]

Utida, S. (1953) Interspecific competition between two species of bean weevil. *Ecology*, **34**, 301–7. [Ch. 2]

Utida, S. (1957) Cyclic fluctuations of population density intrinsic to the host parasite system. *Ecology*, **38**, 442–9. [Ch. 6]

Utida, S. (1967) Damped oscillations of population density and equilibrium. *Researches on Population Ecology (Kyoto)*, **9**, 1–9. [Ch. 2]

Uvarov, B.P. (1931). Insects and climate. *Transactions of the Royal Entomological Society of London*, **79**, 1–247. [Ch. 3]

Uvarov, B.P. (1961) Quantity and quality in insect populations. *Proceedings of the Royal Entomological Society of London, Series C*, **25**, 52–9. [Ch. 8]

Valladares, G. and Lawton, J.H. (1991) Host-plant selection in the holly leaf-miner: does mother know best? *Journal of Animal Ecology*, **60**, 227–40. [Ch. 16]

Van Alphen, J.J.M. and Jervis, M.A. (1996) Foraging behavior, in *Natural Enemies* (eds M.A. Jervis and N.A.C. Kidd), Chapman & Hall, London, pp. 1–62. [Ch. 7]

Van Dam, N.M., Van der Meijden, E. and Verpoorte, R. (1993) Induced responses in three alkaloid-containing plant species. *Oecologia*, **95**, 425–30. [Ch. 12]

Van der Drift, J. (1963) The disappearance of litter in mull and mor in connection with weather conditions and the activities of macrofauna, in *Soil Organisms* (eds J. Doeksen and J. van der Drift), North-Holland, Amsterdam, pp. 125–33. [Ch. 3]

Van der Meijden, E. (1979) Herbivore exploitation of a fugitive plant species: local survival and extinction of the cinnabar moth and ragwort in a heterogeneous environment. *Oecologia*, **42**, 307–23. [Ch. 12]

Van der Meijden, E. and van der Waals-Kooi, R.E. (1979) The population ecology of *Senecio jacobaea* in a sand dune system. I. Reproductive strategy and the biennial habit. *Journal of Ecology*, **67**, 131–53. [Ch. 12]

Van der Meijden, E. and Van der Veen-van Wijk, C.A.M. (1997) Tritrophic metapopulation dynamics. A case study of ragwort, the cinnabar moth, and the parasitoid *Cotesia popularis*, in *Metapopulation Biology: Ecology, Genetics and Evolution* (eds I.A. Hanski and M.E. Gilpin), Academic Press, San Diego, pp. 387–405. [Ch. 12]

Van der Meijden, E., de Jong, T.J. and Klinkhamer, P.G.L. (1985) Temporal and spatial dynamics in populations of biennial plants, in *Structure and Functioning of Plant Populations* (eds J. Haeck and J.W. Woldendorp), North Holland, Amsterdam, pp. 91–103. [Ch. 12]

Van der Meijden, E., Wijn, M. and Verkaar, H.J. (1988) Defence and regrowth, alternative plant strategies in the struggle against herbivores. *Oikos*, **51**, 355–63. [Ch. 12]

Van der Meijden, E., van Wijk, C.A.M. and Kooi, R.E. (1991) Population dynamics of the cinnabar moth (*Tyria jacobaea*): oscillations due to food limitation and local extinction risks. *Netherlands Journal of Zoology*, **41**, 158–73. [Ch. 12]

Van der Meijden, E., Klinkhamer, P.G.L., de Jong, T.J. and van Wijk, C.A.M. (1992) Metapopulation dynamics of biennial plants: how to exploit temporary habitats. *Acta Botanica Neerlandica*, **41**, 249–70. [Ch. 12]

Van Dijk, Th.S. (1979) On the relationship between reproduction, age and survival of the carabid beetles, *Calathus melanocephalus* L. and *Pterostichus coerulescens* L. (Coleoptera, Carabidae). *Oecologia*, **40**, 63–80. [Ch. 3]

Van Dijk, Th.S. (1982) Individual variability and its significance for the survival of animal populations, in *Environmental Adaptation and Evolution* (eds D. Mossakowski and G. Roth), Gustav Fischer, Stuttgart, pp. 233–51. [Ch. 3]

Van Dijk, Th.S. (1986) How to estimate the level of food availability in field populations of carabid beetles? in *Carabid Beetles, Their Adaptations and Dynamics* (eds P.J. den Boer, M.L. Luff, D. Mossakowski and F. Weber), Gustav Fischer, Stuttgart, pp. 371–84. [Ch. 3]

Van Dijk, Th.S. (1996) The influence of environmental factors and food on life cycle, ageing and survival of some carabid beetles. *Acta Jutlandica*, **71**, 11–24. [Ch. 3]

Van Dijk, Th.S. and den Boer, P.J. (1992) The life histories and population dynamics of two carabid species on a Dutch heathland. 1. Fecundity and the mortality of immature stages. *Oecologia*, **90**, 340–52. [Ch. 3]

Van Dongen, S., Backeljau T., Matthysen E. and Dhondt, A.A. (1994) Effects of forest fragmentation on the population structure of the winter moth *Operophtera brumata* L. (Lepidoptera: Geometridae). *Acta Oecologica*, **15**, 193–206. [Ch. 13]

Van Dongen, S., Matthysen, E. and Dhondt, A.A. (1996) Restricted male winter moth (*Operophtera brumata* L.) dispersal among host trees. *Acta Oecologica*, **17**, 319–29. [Ch. 13]

Van Lenteren, J.C. and Woets, J. (1988) Biological and integral pest control in greenhouses. *Annual Review of Entomology*, **33**, 239–59. [Ch. 3]

Van Lenteren, J.C., Van Roermund, H.J.W. and Sutterlin, S. (1996) Biological control of greenhouse whitefly (*Trialeurodes vaporariorum*) with the parasitoid *Encarsia formosa*: how does it work? *Biological Control*, **6**, 1–10. [Ch. 6]

Van Roermund, H.J.W., Van Lenteren, J.C. and Rabbinge, R. (1997a) Analysis of foraging behavior of the whitefly parasitoid *Encarsia formosa* on a leaf: a simulation study. *Biological Control*, **8**, 22–36. [Ch. 6]

Van Roermund, H.J.W., Van Lenteren, J.C. and Rabbinge, R. (1997b) Biological control of greenhouse whitefly with the parasitoid *Encarsia formosa* on tomato: an individual-based simulation approach. *Biological Control*, **9**, 25–47. [Ch. 6]

Van San, N. and Sula J. (1993) Allozyme variation in the winter moth, *Operophtera brumata* (Lepidoptera: Geometridae), in isolated populations. *European Journal of Entomology*, **90**, 303–10. [Ch. 13]

Varley, G.C. (1937) The life history of some Trypetid flies, with description of the early stages (Diptera). *Proceedings of the Royal Entomological Society, London*, **12**, 109–22. [Ch. 10]

Varley, G.C. (1947) The natural control of population balance in the knapweed gall-fly (*Urophora jaceana*). *Journal of Animal Ecology*, **16**, 139–87. [Ch. 4, 10]

Varley, G.C. and Gradwell, G.R. (1960) Key factors in population studies. *Journal of Animal Ecology*, **29**, 399–401. [Intro, Ch. 1, 2, 5, 16]

Varley, G.C. and Gradwell G.R. (1963) The interpretation of insect population changes. *Proceedings of the Ceylon Association for the Advancement of Science*, **18**, 142–56. [Ch. 13]

Varley, G.C. and Gradwell G.R. (1968) Population models for the winter moth, in *Insect Abundance*, (ed. T.R.E. Southwood), Blackwell Scientific Publications, Oxford, pp. 132–42. [Ch. 5, 13]

Varley, G.C., Gradwell, G.R. and Hassell, M.P. (1973) *Insect Population Ecology, an analytical approach*, Blackwell Scientific Publications, Oxford. [Ch. 2, 3, 5, 10, 13]

Verboom, J., Schotman, A., Opdam, P. and Metz, J.A.J. (1991) European nuthatch metapopulations in a fragmented agricultural landscape. *Oikos*, **61**, 149–56. [Ch. 1]

Verhulst, P.F. (1838) Notice sur le loi que la population suit dans son accroisse-ment. *Correspondences Mathematiques et Physiques*, **10**, 113–21. [Ch. 2, 4]

Vickery, W.L. (1991) Evaluation of bias in k-factor analysis. *Oecologia*, **85**, 413–8. [Ch. 5]

Vickery, W.L. and Nudds, T.D. (1984) Detection of density-dependent effects in annual duck censuses. *Ecology*, **65**, 96–104. [Ch. 2, 5]

Visser, M.E., van Alphen, J.J.M. and Hemerik, L. (1992) Adaptive superparasitism and patch time allocation in solitary parasitoids: an ESS model. *Journal of Animal Ecology*, **61**, 93–101. [Ch. 6]

Vrieling K, de Vos, H. and van Wijk, C.A.M. (1993) Genetic analysis of concentra-tions of pyrrolizidine alkaloids in *Senecio jacobaea*. *Phytochemistry*, **32**, 1141–4. [Ch. 12]

Waage, J.K. (1983) Aggregation in field parasitoid populations: foraging time allo-cation by a population of *Diadegma* (Hymenoptera: Ichneumonidae). *Ecological Entomology*, **8**, 447–53. [Ch. 6]

Waage, J.K. and Greathead, D.J. (1988) Biological control: challenges and opportu-nities. *Philosophical Transactions of the Royal Society, London (Biology)*, **318**, 111–28. [Ch. 6]

Waage, J.K. and Hassell, M.P. (1982) Parasitoids as biological control agents: a fun-damental approach. *Parasitology*, **84**, 241–68. [Ch. 6]

Wahlberg, N., Moilanen, A. and Hanski, I. (1996) Predicting the occurrence of endangered species in fragmented landscapes. *Science*, **273**, 1536–8. [Ch. 1]

Walde, S.J. (1991) Patch dynamics of a phytophagous mite population: effect of number of subpopulations. *Ecology*, **72**, 1591–8. [Ch. 7]

Walde, S.J. (1994) Immigration and the dynamics of a predator–prey interaction in biological control. *Journal of Animal Ecology*, **63**, 337–46. [Ch. 7]

Walde, S.J. and Murdoch, W.W. (1988) Spatial density-dependence in parasitoids. *Annual Review of Entomology*, **33**, 441–66. [Ch. 6, 7]

Walde, S.J. and Nachman, G. (1998) Dynamics of spatially structured spider mite populations, in *Theoretical Approaches to Biological Control* (eds B.A. Hawkins and H.V. Cornell), Cambridge University Press, Cambridge, (in press). [Ch. 7]

Wallner, W.E. (1987) Factors affecting insect population dynamics: differences between outbreak and non-outbreak species. *Annual Review of Entomology*, **32**, 317–40. [Ch. 14]

Walters, C.J. and Ludwig, D. (1987) The effects of measurement errors on the assessment of stock-recruitment relationships. *Canadian Journal of Fisheries and Aquatic Sciences*, **38**, 704–10. [Ch. 5]

Wang, Y.H. and Gutierrez, A.P. (1980) An assessment of the use of stability analy-ses in population ecology. *Journal of Animal Ecology*, **49**, 435–52. [Ch. 6]

Ward, S.A., Wellings, P.W. and Dixon, A.F.G. (1982) The effect of reproductive investment on pre-reproductive mortality in aphids. *Journal of Animal Ecology*, **52**, 303–13. [Ch. 8]

Ward, S.A., Leather, S.R. and Dixon, A.F.G. (1984) Temperature prediction and the timing of sex in aphids. *Oecologia*, **62**, 230–3. [Ch. 8]

Waring, G.L. and Price, P.W. (1988) Consequences of host plant chemical and physical variability to an associated herbivore. *Ecological Research*, **3**, 205–16. [Ch. 14]

Warren, M.S. (1991) The successful conservation of an endangered species, the heath fritillary butterfly *Mellicta athalia*, in Britain. *Biological Conservation*, **55**, 37–56. [Ch. 1]

Warren, M.S. (1992) The conservation of British butterflies, in *The Ecology of Butterflies in Britain* (ed. R.L.H. Dennis), Oxford Scientific Publications, Oxford, pp. 246–75. [Ch. 11]

Warren, M.S. (1994) The UK status and suspected metapopulation structure of a threatened European butterfly, the marsh fritillary *Eurodryas aurinia*. *Biological Conservation*, **67**, 239–49. [Ch. 1]

Watkinson, A.R. and Sutherland, W.J. (1995) Sources, sinks and pseudo-sinks. *Journal of Animal Ecology*, **64**, 126–30. [Ch. 1]

Watmough, R.H. (1968) Population studies on two species of Psyllidae (Homoptera, Sternorhyncha) on broom (*Sarothamnus scoparius* (L.) Wimmer). *Journal of Animal Ecology*, **37**, 283–314. [Ch. 15]

Watson, A. (1971) Key factor analysis, density dependence and population limitation in red grouse, in *Dynamics of Populations* (eds P.J. den Boer and G.R. Gradwell), Centre for Agricultural Publishing and Documentation, Wageningen, pp. 548–64. [Ch. 2]

Watt, A.D. (1994) The relevance of the stress hypothesis to insects feeding on foliage, in *Individuals, Populations and Patterns in Ecology* (eds S.R. Leather, A.D. Watt, N.J. Mills and K.F.A. Walters), Intercept, Andover, pp. 73–85. [Ch. 8]

Watt, A.D. and Dixon, A.F.G. (1981) The role of cereal growth stages and crowding in the induction of alatae in *Sitobion avenae* and its consequences for population growth. *Ecological Entomology*, **6**, 441–7. [Ch. 8]

Watt, A.D. and McFarlane, A.M. (1991) Winter moth on Sitka spruce: synchrony of egg hatch and budburst, and its effect on larval survival. *Ecological Entomology*, **16**, 387–90. [Ch. 13]

Watt, A.D., Leather, S.R. and Evans, H.F. (1991) Outbreaks of the pine beauty moth on pine in Scotland: the influence of host plant species and site factors. *Forest Ecology and Management*, **39**, 211–21. [Ch. 8]

Watt, A.D., Whittaker, J.B., Docherty, M. *et al.* (1995) The impact of elevated atmospheric CO_2 on insect herbivores, in *Insects in a Changing Environment* (eds R. Harrington and N.E. Stork), Academic Press, London, pp. 198–219. [Ch. 8]

Watt, K.E.F. (1961) Mathematic models for use in insect pest control. *Canadian Entomologist (Supplement 19)*, **93**, 1–62. [Ch. 1]

Watt, T.A. (1987) The biology and toxicity of ragwort (*Senecio jacobaea* L.) and its herbicidal and biological control. *Herbage Abstracts*, **57**, 1–16. [Ch. 12]

Weiss, S.B., Murphy, D.D., Ehrlich, P.R. and Metzler, C.F. (1993) Adult emergence phenology in checkerspot butterflies: the effects of macroclimate, microclimate, and population history. *Oecologia*, **96**, 261–70. [Ch. 1]

Weisser, W.W., Wilson, H.B. and Hassell, M.P. (1997) Interference among parasitoids: a clarifying note. *Oikos*, **79**, 173–8. [Ch. 6]

Wellings, P.W. and Dixon, A.F.G. (1987) Sycamore aphid numbers and population density. III. The role of aphid-induced changes in plant quality. *Journal of Animal Ecology*, **56**, 161–71. [Ch. 9]

Wellings, P.W., Leather, S.R. and Dixon, A.F.G. (1980) Seasonal variation in reproductive potential: a programmed feature of aphid life cycles. *Journal of Animal Ecology*, **49**, 975–85. [Ch. 8]

Wellings, P.W., Chambers, R.J., Dixon, A.F.G. and Aikman, D.P. (1985) Sycamore aphid numbers and population density. I. Some patterns. *Journal of Animal Ecology*, **54**, 411–24. [Ch. 9]

Wellington, W.G. (1957) Individual differences as a factor in population dynamics: the development of a problem. *Canadian Journal of Zoology*, **35**, 293–323. [Ch. 8]

Wells, H., Wells, P.H. and Rogers, S.H. (1993) Is multiple mating an adaptive feature of monarch butterfly winter aggregation? in *Biology and Conservation of the Monarch Butterfly* (eds. S.B. Malcolm and M.P. Zalucki), Natural History Museum of Los Angeles County, Los Angeles, pp. 61–8. [Ch. 8]

White, I.M. (1989) Tephritid flies, in *Handbooks for the Identification of British Insects*, **10** (5a), Royal Entomological Society, London. [Ch. 10]

White, T.C.R. (1969) An index to measure weather-induced stress of trees associated with outbreaks of psyllids in Australia. *Ecology*, **50**, 905–9. [Ch. 14]

White, T.C.R. (1974) A hypothesis to explain outbreaks of looper caterpillars, with special reference to populations of *Selidosema suavis* in a plantation of *Pinus radiata* in New Zealand. *Oecologia*, **16**, 279–301. [Ch. 14]

White, T.C.R. (1984) The abundance of invertebrate herbivores in relation to the availability of nitrogen in stressed food plants. *Oecologia*, **63**, 90–105. [Ch. 8]

White, T.C.R. (1993) *The Inadequate Environment. Nitrogen and the Abundance of Animals*, Springer-Verlag, Berlin. [Ch. 3, 4]

Whitham, T.G. (1980) The theory of habitat selection: examined and extended using *Pemphigus* aphids. *American Naturalist*, **115**, 449–66. [Ch. 16]

Whitlock, M.C. (1992) Nonequilibrium population stucture in forked fungus beetles: extinction, colonization, and genetic variation among populations. *American Naturalist*, **139**, 952–70. [Ch. 1]

Whittaker, J.B. (1985) Population cycles over a 16 year period in an upland race of *Strophingia ericae* (Homoptera: Psylloidea) on *Calluna vulgaris*. *Journal of Animal Ecology*, **54**, 311–21. [Ch. 15]

Wiens, J.A. (1984) Resource systems, populations, and communities, in *A New Ecology: Novel Approaches to Interactive Systems* (eds P.W. Price, C.N. Slobodchikoff and W.S. Gaud), Wiley, New York, pp. 397–436. [Ch. 16]

Wiens, J.A. (1989) *The Ecology of Bird Communities*, Cambridge University Press, Cambridge. [Ch. 2]

Wigley, P.J. (1976) *The epizootiology of a nuclear polyhedrosis virus of the winter moth, Operophtera brumata L. at Wistman's Wood, Dartmoor*. PhD thesis, University of Oxford. [Ch. 13]

Wilbert, H. (1962). Über Festlegung und Einhaltung der mittleren Dichte von Insektenpopulationen. *Zeitschrift für Morphologie und Okologie der Tiere*, **50**, 576–615. [Ch. 3]

Wilbert, H. (1971). Feedback control by competition, in *Dynamics of Populations* (eds P.J. den Boer and G.R. Gradwell), Centre for Agricultural Publishing and Documentation, Wageningen, pp. 174–88. [Ch. 3]

Williams, A.G. and Whitham, T.G. (1986) Premature leaf abscission: an induced plant defense against gall aphids. *Ecology*, **67**, 1619–27. [Ch. 15]

Williams, C.B. (1964) *Patterns in the Balance of Nature*, Academic Press, London and New York. [Ch. 1]

Williams, D.W. and Liebhold, A.M. (1995) Detection of delayed density dependence: effects of autocorrelation in an exogenous factor. *Ecology*, **76**, 1005–8. [Ch. 5, 14]

Williams, D.W. and Liebhold, A.M. (1997) Detection of delayed density dependence: Reply. *Ecology*, **78**, 320–2. [Ch. 14]

Williamson, M. (1981) *Island Populations*, Oxford University Press, Oxford. [Ch. 1]

Wilson, F. (1968) Insect abundance: prospects, in *Insect Abundance* (ed. T.R.E. Southwood), Blackwell Scientific Publications, Oxford, pp. 143–58. [Ch. 14]

Wilson, H.B., Hassell, M.P. and Godfray, H.C.J. (1996) Host–parasitoid food webs: dynamics, persistence, and invasion. *American Naturalist*, **148**, 787–806. [Ch. 6]

Wilson, H.B., Godfray, H.C.J., Hassell, M.P., and Pacala, S.W. (1998) Deterministic and stochastic host–parasitoid dynamics in spatially-extended systems, in *Modeling Spatiotemporal Dynamics in Ecology* (eds J. Bascompte and R.V. Solé), Springer-Verlag, Berlin, pp. 63–82. [Ch. 6]

Wilson, W.G., McCauley, E. and de Roos, A.M. (1995) Effect of dimensionality on Lotka–Volterra predator–prey dynamics: individual based simulation results. *Bulletin of Mathematical Biology*, **57**, 507–26. [Ch. 8]

Wint, W. (1983) The role of alternative host-plant species in the life of a polyphagous moth, *Operophtera brumata* (Lepidoptera: Geometridae). *Journal of Animal Ecology*, **52**, 439–50. [Ch. 13]

Woiwod, I.P. and Hanski, I. (1992) Patterns of density dependence in moths and aphids. *Journal of Animal Ecology*, **61**, 619–29. [Ch. 1, 2, 4, 5]

Woiwod, I.P. and Harrington, R. (1994) *Flying in the Face of Change: the Rothamsted Insect Survey. Long-term Experiments in Agricultural and Ecological Sciences* (eds R.A. Leigh and A.E. Johnston), CAB International, Wallingford, UK, pp. 321–42. [Ch. 5]

Wolda, H. (1989) The equilibrium concept and density dependence tests. What does it all mean? *Oecologia*, **81**, 430–2. [Ch. 2, 3]

Wolda, H. (1991). The usefulness of the equilibrium concept in population dynamics. A reply to Berryman. *Oecologia*, **86**, 144–5. [Ch. 3]

Wolda, H. (1992) Trends in abundance of tropical forest insects. *Oecologia*, **89**, 47–52. [Ch. 17]

Wolda, H. and Dennis, B. (1993) Density dependence tests, are they? *Oecologia*, **95**, 581–91. [Ch. 1, 2, 3, 5]

Wolda, H., Dennis, B. and Taper, M.L. (1994) Density dependence tests, and largely futile comments: answers to Holyoak and Lawton (1993) and Hanski, Woiwod and Perry (1993). *Oecologia*, **98**, 229–34. [Ch. 3, 5]

Wood, B.W., Tedders, W.L. and Thompson, J.M. (1985) Feeding influence of three pecan aphid species on carbon exchange and phloem integrity of seedling pecan foliage. *Journal American Society of Horticultural Science*, **110**, 393–7. [Ch. 9]

Woodman, R.L. (1990) *Enemy impact and herbivore community structure: tests using parasitoid assemblages, predatory ants, and galling sawflies on arroyo willow*. PhD thesis, Northern Arizona University, Flagstaff. [Ch. 14]

Woods, J.O., Carr, T.G., Price, P.W. *et al.* (1996) Growth of coyote willow and the attack and survival of a mid-rib galling sawfly, *Euura* sp. *Oecologia*, **108**, 714–22. [Ch. 14]

Yamaguchi, H. (1976) Biological Studies on the Todo-fir Aphid *Cinara todocola* Inouye, with special reference to its population dynamics and morph determination. *Bulletin of the Government Forest Experiment Station*, (Japan), No. 283, 102 pp. [Ch. 9]

Zalucki, M.P. (1993) Sex around the milkweed patch – the significance of patches of host plants in monarch reproduction, in *Biology and Conservation of the Monarch Butterfly* (eds. S.B. Malcolm and M.P. Zalucki), Natural History Museum of Los Angeles County, Los Angeles, pp. 69–76. [Ch. 8]

Zwölfer, H. (1965) Preliminary list of phytophagous insects attacking wild Cynareae (Compositae) species in Europe. *Technical Publication of the Commonwealth Institute of Biological Control*, **6**, 81–154. [Ch. 10]

Zwölfer, H. (1971) The structure and effect of parasite complexes attacking phytophagous host insects, in *Proceedings of the Advanced Study Institute on Dynamics of Numbers in Populations*, Oosterbeek, 1970, pp. 405–18. [Ch. 6]

Zwölfer, H. (1987) Species richness, species packing and evolution in insect plant systems. *Ecological Studies*, **61**, 301–19. [Ch. 10]

Zwölfer, H. (1988) Evolutionary and ecological relationships of the insect fauna of thistles. *Annual Review of Entomology*, **33**, 103–22. [Ch. 10]

Zwölfer, H. (1994) Structure and biomass transfer in food webs: stability, fluctuations and network control, in *Flux Control in Biological Systems from Enzymes to Populations and Ecosystems* (ed. E.D. Schulze), Academic Press, San Diego, pp. 365–419. [Ch. 10]

Species index

Subject index